CCEA
GCSE
SCIENCE
DOUBLE AWARD

Denmour Boyd
James Napier
Nora Henry
Alyn G McFarland
Frank McCauley
Roy White

HODDER
EDUCATION
AN HACHETTE UK COMPANY

Photo credits are listed on p vi.

Although every effort has been made to ensure that website addresses are correct at time of going to press, Hodder Education cannot be held responsible for the content of any website mentioned in this book. It is sometimes possible to find a relocated web page by typing in the address of the home page for a website in the URL window of your browser.

Hachette UK's policy is to use papers that are natural, renewable and recyclable products and made from wood grown in well-managed forests and other controlled sources. The logging and manufacturing processes are expected to conform to the environmental regulations of the country of origin.

Orders: please contact Hachette UK Distribution, Hely Hutchinson Centre, Milton Road, Didcot, Oxfordshire, OX11 7HH. Telephone: +44 (0)1235 827827. Email education@hachette.co.uk Lines are open from 9 a.m. to 5 p.m., Monday to Friday. You can also order through our website: www.hoddereducation.co.uk

Published in 2017 by
Hodder Education,
An Hachette UK Company
Carmelite House
50 Victoria Embankment
London EC4Y 0DZ

Impression number 10 9
Year 2024

Cover photo © imageBROKER / Alamy Stock Photo

Illustrations by Elektra Media Ltd

Typeset by Elektra Media Ltd

Printed by CPI Group (UK) Ltd, Croydon, CR0 4YY

A catalogue record for this title is available from the British Library.

ISBN 9781471892189

MIX
Paper | Supporting
responsible forestry
FSC™ C104740

CONTENTS

HOW TO GET THE MOST FROM THIS BOOK

Welcome to the CCEA GCSE Double Award Student Book.

This book covers all of the Foundation and Higher-tier content for the 2017 CCEA GCSE Double Award specification.

The following features have been included to help you get the most from this book.

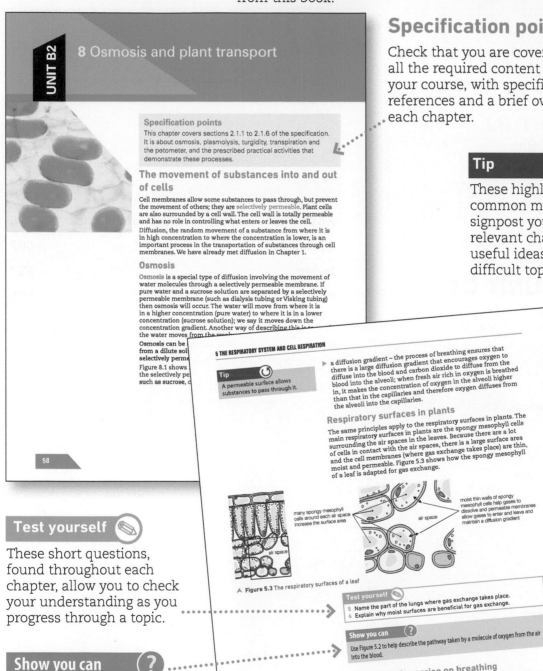

Specification points

Check that you are covering all the required content for your course, with specification references and a brief overview of each chapter.

Tip

These highlight important facts, common misconceptions and signpost you towards other relevant chapters. They also offer useful ideas for remembering difficult topics.

Test yourself

These short questions, found throughout each chapter, allow you to check your understanding as you progress through a topic.

Show you can

Complete the Show you can tasks to prove that you are confident in your understanding of each topic.

Practicals

These practical tasks contain full instructions on apparatus, method and results analysis and will help develop your practical skills.

CCEA's prescribed practicals are clearly highlighted.

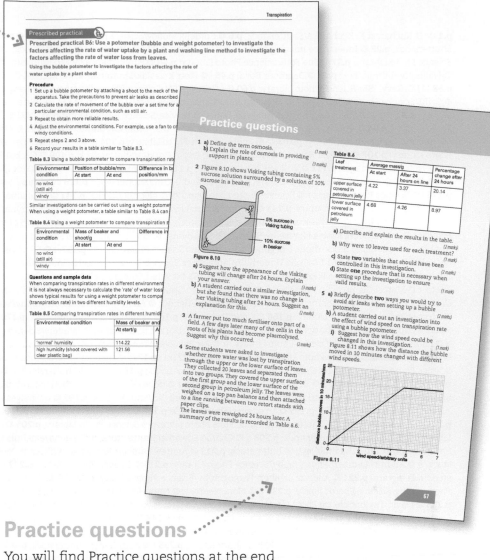

Examples

Examples of questions and calculations that feature full workings and sample answers.

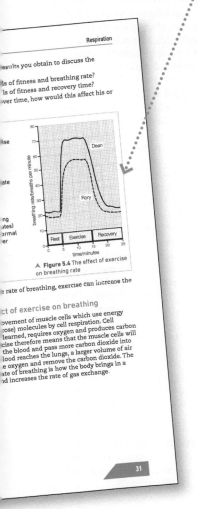

Practice questions

You will find Practice questions at the end of every chapter. These follow the style of the different types of questions you might see in your examination and have marks allocated to each question part.

Level coding

If you are taking GCSE Double Award Foundation-tier you need to study *only* the material with no bars.

If you are taking GCSE Double Award Higher-tier you need to study the material with no bars, plus the material with the purple H bar.

Answers

Answers for all questions in this book can be found online at:

www.hoddereducation.co.uk/cceagcseDoubleAward

1 Cells

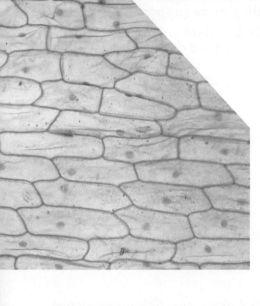

Cells

Living organisms are made up of microscopic units called cells.

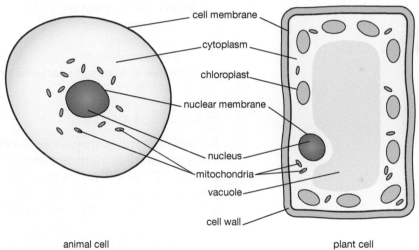

animal cell   plant cell

▲ **Figure 1.1** An animal cell and a plant cell

Animal cells

Animal cells are surrounded by a selectively permeable cell membrane that forms a boundary to the cell and controls what enters or leaves. The main part of the cell is the cytoplasm and this is where chemical reactions take place.

The nucleus, surrounded by a nuclear membrane, contains several threadlike structures: the chromosomes. Each chromosome is made of a molecule of DeoxyriboNucleic Acid (DNA) which has part of the organism's genetic information coded in its structure. For this reason, the nucleus is sometimes referred to as the control centre of the cell.

Mitochondria are structures in the cytoplasm within which the chemical reactions of cell respiration take place.

Ⓗ

Tip

Cells like muscle cells need a lot of energy and so have many more mitochondria in their cytoplasm.

Plant cells

Like animal cells, plant cells have a cell membrane, cytoplasm with mitochondria and a nucleus containing chromosomes. However, in addition they have:

▶ a cellulose cell wall, which is a stiff structure immediately outside the cell membrane that provides support

▶ a large permanent vacuole in the cytoplasm containing cell sap, that when full pushes the cell membrane against the cell wall, making the cell rigid and providing more support

▶ chloroplasts in the cytoplasm that contain chlorophyll, which traps light and helps the plant make food during photosynthesis. Chloroplasts are not found in all plant cells – they are only present in green parts of the plant, particularly the leaves.

> **Test yourself**
>
> 1 Draw a table to compare the structures found in animal, plant and bacterial cells.
> 2 What is the function of:
>   a) a cell membrane
>   b) a chloroplast
>   c) a mitochondrion?
> 3 Give **two** structures in bacterial cells that contain DNA.

Bacterial cells

Bacteria are microscopic single-celled organisms (microorganisms). They are neither plant nor animal, largely because their cell structure is very different. They have a cell membrane surrounding cytoplasm but there is no nucleus. The genetic material (DNA) is in the form of a circular chromosome within the cytoplasm. Small rings of DNA, called plasmids, are also present, as shown in Figure 1.2. A cell wall is present but it is not made of cellulose.

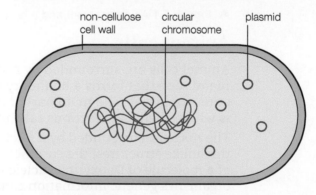

▲ **Figure 1.2** A typical bacterial cell

Observing cells using a microscope

Although a few cells, like birds' eggs, are large and can be seen with the naked eye, the vast majority can only be seen using a microscope. In school, you will use a light microscope in which light shines through a thin layer of cells on a slide.

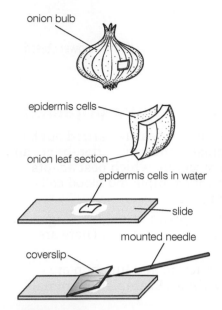

onion bulb

epidermis cells

onion leaf section

epidermis cells in water

slide

mounted needle

coverslip

▲ **Figure 1.3** How to prepare a temporary slide of onion cells

Tip

Sometimes you will see black rings in your slide preparation – these are probably air bubbles trapped under the coverslip.

Tip

When moving the slide on the stage to focus on a different part of the field of view, moving the slide away from you moves what you see towards you and *vice versa*.

Test yourself

4 Why are chemical stains used in temporary slides?
5 What is the field of view?
6 If the microscope you are using has an eyepiece with ×10 magnification, what is the total magnification of the image you see when using an objective lens with a magnification:
   a) ×10?
   b) ×40?

Show you can

Describe how to use a microscope to observe a specimen on a slide.

Making slides

Most slides containing animal cells will already be prepared for you but you should get a chance to prepare temporary slides containing animal and plant cells. Figure 1.3 outlines the procedure for making a temporary slide containing onion skin cells.

Forceps are used to peel the thin, transparent layer of epidermis cells from the inside of a small section of onion leaf. The epidermis cells are then placed on a microscope slide with a few drops of water. Iodine solution or methylene blue can be used instead of water – these chemical stains colour the cells, making certain parts such as the nucleus more obvious.

A coverslip is then lowered onto the onion epidermis using a mounted needle or forceps. It is better to lower the coverslip one end first as this helps to prevent trapping air bubbles. The coverslip will protect the lens of the microscope should it have contact with the slide and will prevent the cells drying out.

Using a microscope

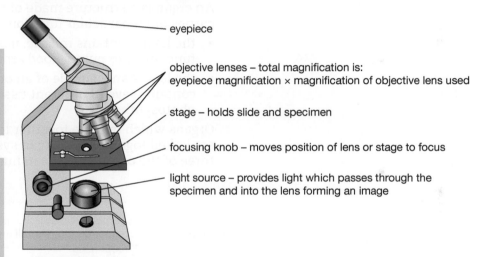

eyepiece

objective lenses – total magnification is: eyepiece magnification × magnification of objective lens used

stage – holds slide and specimen

focusing knob – moves position of lens or stage to focus

light source – provides light which passes through the specimen and into the lens forming an image

▲ **Figure 1.4** A light microscope and how it works

When using a light microscope (Figure 1.4), it is important to start by using the low power objective lens. With the low power lens (×4 or ×10) the circle of light you see in the eyepiece (the field of view) is wider and allows you to see more of the cells on the slide. This makes it easier to find what you are looking for and to focus a sharp image.

After you observe the cells at low power you may want to see some of them in more detail. To do this, first move the slide slightly so that the cell or area of cells you want to look at in greater detail is in the centre of the field of view. One of the larger, high power, objective lenses (×20 or ×40) can then be carefully rotated into place and focused.

Great care is needed while rotating a high-power lens into position above the slide as the end of the high-power lenses are very close to the slide and can be damaged while focusing.

3

Tissues, organs, organ systems and organisms

In single-celled organisms all the life processes are carried out by the one cell but in multi-celled organisms the cells differentiate and become specialised. They develop a structure which best adapts them to the function they carry out, for example red blood cells (page 69) and plant palisade cells (page 12).

Specialised cells are not just randomly spread through an organism but are organised into groups in an effective manner. There are different levels of organisation:

1 Cells with the same specialised structure and function are grouped together and are called a tissue. Examples of animal tissues include blood and skin.

2 An organ is a structure made of several types of tissue that carries out a particular function:
- the heart contains muscle, nerve and blood tissues and its function is to pump blood around the body
- the leaf is an example of an organ in a plant which contains several different tissues and has the function of photosynthesis.

3 Organs which operate together to carry out a particular function are linked together into organ systems. Table 1.1 summarises three of the organ systems in humans.

Table 1.1 The functions and organs in three organ systems

Organ system	Function	Main organs include
nervous	responding to stimuli and making responses	brain, spinal cord, receptors (e.g. eye), effectors (muscles)
reproductive	production of young	testes, ovaries, uterus
excretory	maintaining water balance and removing poisonous wastes	kidney, ureter, bladder, urethra

Practice questions

1 a) Give **two** features of a bacterial cell which are not found in a plant cell.   *(2 marks)*

   b) Give **two** features of a plant cell which are not found in a bacterial cell.   *(2 marks)*

2 The body contains several organ systems. Copy and complete Table 1.2, inserting where appropriate the name of the organ system, its function and two of the main organs involved.   *(6 marks)*

Table 1.2

Organ system	Function	Main organs
	responding to stimuli and making responses	
excretory		
		ovaries, uterus

2 Photosynthesis and plant leaves

Specification points

This chapter covers specification points 1.2.1 to 1.2.6. It covers understanding of the process of photosynthesis, knowledge of the equations involved, investigations into the factors which affect the rate of photosynthesis and how the structure of a plant leaf is adapted for gas exchange and light absorption.

Photosynthesis

Photosynthesis takes place in the green parts of plants, particularly in the leaves, where the raw materials of carbon dioxide and water are made into glucose (sugar). The glucose is usually immediately converted into starch and stored in cells. The green pigment chlorophyll plays an important role in photosynthesis as it traps the light from the Sun that is needed to drive the process. This absorption of energy means photosynthesis is described as endothermic. Oxygen is produced as a waste product.

Photosynthesis can be summarised by the word equation:

$$\text{carbon dioxide + water} \xrightarrow[\text{by chlorophyll}]{\text{light energy trapped}} \text{glucose + oxygen}$$

and the balanced chemical equation:

$$6CO_2 + 6H_2O \rightarrow C_6H_{12}O_6 + 6O_2$$

In photosynthesis, light energy from the Sun is converted into chemical energy (food in the form of sugar and starch).

The glucose that is produced in photosynthesis can be used in several ways or converted into a range of products that the plant requires. Uses include:

▶ respiration – the glucose is used in respiration to provide energy
▶ **storage** – in many plants the glucose is converted into starch and oils for storage
▶ **useful substances** – the glucose can be converted into a range of useful products including cellulose (for cell walls), chlorophyll and protein for growth.

Leaves (and other parts of the plant) that are carrying out photosynthesis in bright light will take carbon dioxide into the leaves and oxygen will pass out. Not surprisingly, the brighter the light, the faster the process will take place.

Tip

An endothermic process or reaction is one which absorbs energy (light energy in the case of photosynthesis) from its surroundings.

Photosynthesis is important for animals as well as plants, as it provides a source of food and releases oxygen back into the atmosphere.

Photosynthesis experiments

It is possible to carry out investigations to show that photosynthesis is taking place or that particular raw materials are needed for the process.

The starch test

This test is used to show the presence of starch in green leaves and allows the conclusion that photosynthesis has taken place. The starch test consists of a series of steps. Remember to wear eye protection for this test.

1 A leaf is removed from a plant that has been in bright light and is placed in boiling water for at least 30 seconds. This kills the leaf, stopping any chemical reactions.
2 The leaf is then placed in boiling ethanol (Figure 2.1). This procedure must take place using a water bath as ethanol is flammable and must not be exposed to a direct flame. The ethanol removes the chlorophyll from the leaf making it a whitish-green colour and brittle.
3 The leaf is then dipped into water again to soften it.
4 The soft leaf can then be spread on a white tile and iodine solution added.

If starch is present, the **yellow-brown** iodine solution will change to a **blue-black** colour. If starch is absent the iodine solution will remain yellow-brown.

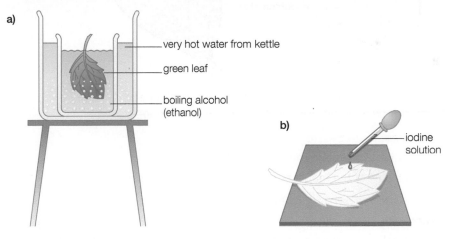

Test yourself

1 Where is chlorophyll found in plants?
2 What are the products of photosynthesis?
3 How does a plant use the products of photosynthesis?
4 Why are the products of photosynthesis also important to animals?
5 How is a plant destarched?

▲ **Figure 2.1** Testing a leaf for starch

Destarching

Show you can (?)

Explain why, after destarching a plant, it is necessary to test one of the plant leaves for the presence of starch before starting the experiment.

To carry out these experiments it is necessary to destarch the leaves of the plant first. Leaves can be destarched by leaving the plant in the dark for at least two days. This will ensure that any starch already in the leaves will either be used by the plant cells or removed and stored elsewhere in the plant. The importance of this is that if the starch test at the end of the experiment is positive, it shows that starch has been produced during the period of the experiment.

Prescribed practical B1: Investigating the need for light and chlorophyll in photosynthesis

Investigate the need for light in photosynthesis

Procedure

1 Destarch a plant by keeping it in a cupboard for two days.

2 Test a leaf from the destarched plant for the presence of starch. If the leaf tests negative continue to **3**.

3 Cover a part of one leaf of the plant with lightproof paper or foil as shown in Figure 2.2a. Make a drawing of the destarched leaf at the start of the experiment.

4 Place the plant in bright light for several hours.

5 Test the leaf from the plant for starch. Make a drawing of the leaf after it is tested for starch.

Sample drawings of the leaf

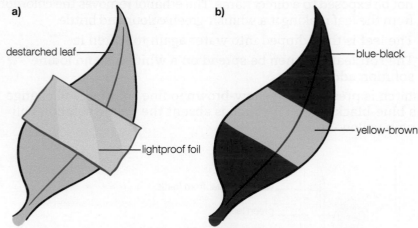

a)
destarched leaf
lightproof foil

b)
blue-black
yellow-brown

▲ **Figure 2.2** Drawings of a leaf investigating the need for light in photosynthesis **a)** before and **b)** after being tested for starch.

Questions

1 Describe and explain the results.

2 Which part of the leaf acts as a control in this experiment?

Investigate the need for chlorophyll in photosynthesis

Procedure

1 Destarch a variegated plant by keeping it in a cupboard for two days. Remember to wear eye protection.

2 Test a leaf from the destarched plant for the presence of starch. If the leaf tests negative, continue to step **3** to proceed with the investigation.

3 Make a drawing of one of the leaves.

4 Place the plant in bright light for several hours.

5 Test the leaf from the plant for starch.

6 Make a drawing of the same leaf after it is tested for starch.

Sample drawings of the leaf

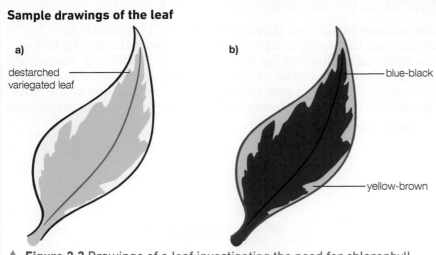

a)

destarched
variegated leaf

b)

blue-black

yellow-brown

▲ **Figure 2.3** Drawings of a leaf investigating the need for chlorophyll in photosynthesis **a)** before and **b)** after being tested for starch

Questions

1 Describe and explain the results.
2 Which part of the leaf acts as the control in this experiment?

Investigating the need for carbon dioxide

To show that carbon dioxide is an essential raw material for photosynthesis it is necessary to compare a leaf that is deprived of carbon dioxide with a leaf that has a good supply of carbon dioxide. This can be achieved by comparing two leaves as shown in Figure 2.4. Sodium hydroxide will absorb the carbon dioxide from the air surrounding the experimental leaf. The control leaf will only have water (or a chemical that increases carbon dioxide levels) in its flask and therefore there will be carbon dioxide present.

Test yourself

6 In Figure 2.4, which flask has the lowest concentration of carbon dioxide? Explain your answer.
7 Why is the cotton wool in the neck of the experimental flask soaked in sodium hydroxide?
8 Explain why the experiment uses leaves from the same plant.
9 Describe and explain the steps involved in testing the leaves for starch.

Show you can

If carbon dioxide is essential for photosynthesis, what results would you expect?

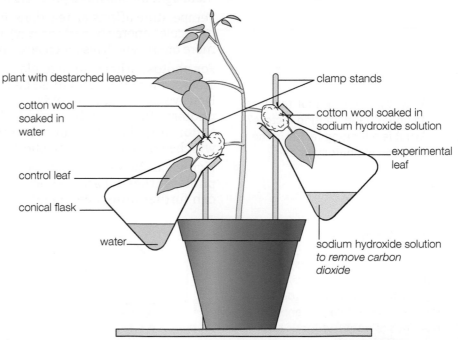

plant with destarched leaves

cotton wool soaked in water

control leaf

conical flask

water

clamp stands

cotton wool soaked in sodium hydroxide solution

experimental leaf

sodium hydroxide solution *to remove carbon dioxide*

▲ **Figure 2.4** Investigating the need for carbon dioxide in photosynthesis

Measuring the rate of photosynthesis

Using apparatus like that shown in Figure 2.5, it is possible to demonstrate that oxygen is produced by photosynthesis. The rate of photosynthesis will affect the rate at which the bubbles of oxygen will be given off and this can be used to compare photosynthesis rates in different conditions. For example, by moving the position of the lamp it is possible to investigate the effect of light intensity on photosynthesis.

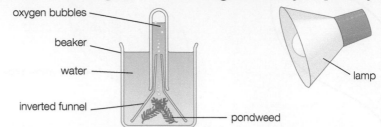

▲ **Figure 2.5** Measuring the rate of photosynthesis

The rate of photosynthesis can be more accurately calculated by measuring the volume of oxygen produced after five minutes. Alternatively, an oxygen electrode connected to a data logger can be used to measure the change of oxygen concentration.

Factors affecting the rate of photosynthesis

The rate at which photosynthesis occurs depends on the availability of the raw materials needed for the process, carbon dioxide, water and light. As we have seen from the equations (page 6), carbon dioxide and water are the substrates which join together chemically to form the product of photosynthesis (sugar). Increases in these substrates will increase the rate at which the products of photosynthesis are formed. Light provides the energy needed to join the substrates together so again more light means more products formed.

Temperature affects all reactions. Increasing temperature also gives molecules energy (kinetic energy), making them move faster and collide more often, which also increases the number of reactions.

For photosynthesis to take place at its maximum rate, all of these environmental factors must be present at peak or optimum levels. However, if one (or more) factor is in short supply, the rate of photosynthesis will be limited. These raw materials become limiting factors and the rate of photosynthesis will be determined by whichever factor is in the shortest supply.

Figure 2.6 shows the effect of light intensity on the rate of photosynthesis. Part a) shows how temperature (cold and hot days) can further influence the rate.

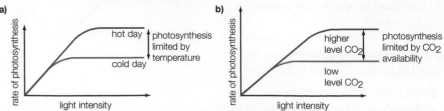

▲ **Figure 2.6** These graphs show how the rate of photosynthesis is affected by increasing light intensity **a)** on a hot and cold day; **b)** at higher and lower CO_2 concentrations

As light intensity increases, irrespective of temperature, the rate of photosynthesis increases up to a point where the graph begins to level off and form a plateau. As an increase in light intensity causes an increase in photosynthesis at the lower light levels, the amount of light must be limiting the rate at which photosynthesis occurs. Within the plateau part of the graph, further increases in light intensity do not lead to an increase in photosynthesis, therefore something else must be limiting the rate.

The effect of temperature can be explained by comparing the rates of photosynthesis on cold and hot days. On a hot day, photosynthesis occurs at a higher rate at higher light intensities when compared to a cooler day. Therefore, we can conclude that temperature is limiting the rate of photosynthesis at the higher light intensities on the cooler day.

It is possible that the rate of photosynthesis may still not be at its maximum where the rate has plateaued on a hot day at the highest light intensities. It is possible that carbon dioxide could be a limiting factor in these conditions. To test this, we would need to increase carbon dioxide levels to see if this has any effect.

Figure 2.6b shows how light intensity affects the rate of photosynthesis at different carbon dioxide levels. Again, light intensity is limiting the rate of photosynthesis at low light levels. The fact that an increased carbon dioxide level leads to a higher rate of photosynthesis at higher light intensities shows that the low carbon dioxide level was a limiting factor once light levels had ceased to be limiting.

It is important for farmers and growers to understand how photosynthesis affects their crops as reducing the effects of limiting factors will increase the growth of a crop.

The leaf: the site of photosynthesis

In most plants the process of photosynthesis takes place in the leaves. Leaves come in many shapes and sizes but, to allow photosynthesis to take place efficiently, they are usually adapted for:

▶ **light absorption**

▶ gas exchange

The way in which leaves are arranged on a plant ensures that each leaf can absorb as much light as possible and that as far as possible each leaf is not in the shade of other leaves. The section through a leaf shown in Figure 2.7 shows many other ways in which a leaf is designed to aid light absorption and encourage gas exchange.

> **Tip**
>
> Leaves are also adapted for defence against disease by the cuticle and cell walls of the epidermis providing a physical barrier, which reduces the entry of disease-causing organisms.

Use a microscope to investigate a cross section of a mesophytic (typical unspecialised) leaf. The leaf you examine may be different from the one described in this book, but you should look for as many adaptations for maximising light absorption and gas exchange as possible.

Show you can ?

Use the information in Figure 2.6 to help you decide which factor might limit the rate of photosynthesis in the following situations:

a) during a bright winter afternoon in a British grassland

b) in a cornfield in mid-summer sunshine in Southern France.

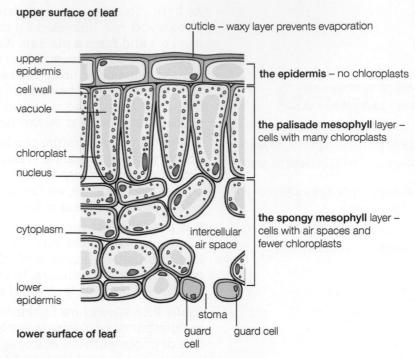

upper surface of leaf

cuticle – waxy layer prevents evaporation

upper epidermis

cell wall

vacuole

chloroplast

nucleus

cytoplasm

the epidermis – no chloroplasts

the palisade mesophyll layer – cells with many chloroplasts

the spongy mesophyll layer – cells with air spaces and fewer chloroplasts

intercellular air space

lower epidermis

lower surface of leaf

stoma

guard cell

guard cell

▲ **Figure 2.7** Cross-section of a mesophytic leaf

Light absorption in a leaf is maximised by:

▶ the short distance between the upper and lower surfaces, which allows all the cells to receive light

▶ the large surface area

▶ the thin transparent cuticle covering the epidermis, which reduces water loss by evaporation, while allowing light to enter the leaf

▶ the epidermis which lacks chloroplasts and so also allows light into the leaf

▶ the presence of chloroplasts rich in the pigment chlorophyll which absorbs the light

▶ the regular structure of the palisade mesophyll (end on to the upper surface), which ensures that many cells rich in chloroplasts are packed together near the upper surface of the leaf.

Gas exchange in a leaf takes place by diffusion and is maximised by:

▶ the intercellular air spaces in the spongy mesophyll, which allow carbon dioxide to enter and oxygen to leave the photosynthesising cells, which are mainly concentrated in the palisade layer

▶ stomata, which allow carbon dioxide and oxygen to enter and leave the leaf. Stomata are small pores that can occur between cells in the epidermis on the lower surfaces of leaves. Each stoma is surrounded by two guard cells that regulate the opening and closing of the stoma. In many plants the stomata are open during the day and closed at night.

In some plants stomata can occur on both the upper and the lower leaf surface. Some plants have all their stomata on their upper leaf surfaces.

Tip

Water also moves through the intercellular air spaces as a gas (water vapour). It is formed by evaporation from the moist surfaces of the spongy mesophyll cells and then diffuses through the air spaces and stomata out of the leaf. This is part of the transpiration stream discussed in Chapter 8.

Test yourself

18 Why are leaves thin?
19 Name the waxy layer on the upper surface of a leaf.
20 Which layer of cells has the highest density of chloroplasts?
21 Which process causes gases to move through the intercellular air spaces and through the stomata?

Show you can ?

Water lilies have leaves which float on the surface of water. Suggest why they only have stomata on their upper leaf surfaces?

The balance between photosynthesis and respiration

All living organisms respire. In plant respiration, the glucose produced in photosynthesis is broken down to release energy. Plants require oxygen to respire and they produce carbon dioxide as a waste product. These gases enter the leaves through the stomata.

During the night when there is no light for photosynthesis, respiration will be the only process involving gas exchange that takes place. Therefore, oxygen will enter the leaf and carbon dioxide will leave. However, during the day when photosynthesis is occurring both processes will take place. When the light intensity is high the rate of photosynthesis will exceed the rate of respiration. When this happens, carbon dioxide enters the leaves and oxygen moves out.

There will be times during the day when the light intensity is low, causing photosynthesis to take place very slowly. At these points, usually at dawn and dusk, the rates of respiration and photosynthesis are equal and there will be no overall, or net, gas exchange. This point is called the compensation point.

The movement of carbon dioxide into and out of plants can be determined using hydrogencarbonate indicator. Hydrogencarbonate indicator is bright red in normal concentrations of atmospheric carbon dioxide. If there is an increase in the carbon dioxide concentration the indicator will change colour to yellow. A decrease in the carbon dioxide concentration will turn the indicator purple. Figure 2.8 shows how the indicator can be used to show gas exchange in living organisms.

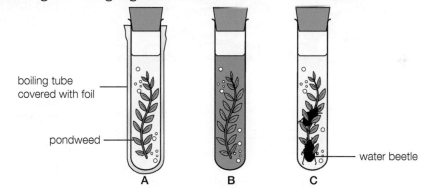

▲ **Figure 2.8** The effect of photosynthesis and respiration on gas exchange. Each boiling tube was filled with hydrogencarbonate indicator and placed in bright light for 1 hour

The results are explained in Table 2.1.

Table 2.1 Results of an experiment on the balance between photosynthesis and respiration using hydrogencarbonate indicator.

Tube	Colour at start	Colour at end	Reason for change
A	red	yellow	The foil strip stops light entering and photosynthesis does not occur. Respiration increases carbon dioxide concentration.
B	red	purple	Both photosynthesis and respiration are taking place in the pondweed. As the rate of photosynthesis is faster than the rate of respiration more carbon dioxide enters the plant than is produced.
C	red	yellow	The water beetles produce more carbon dioxide in respiration than the pondweed takes in for photosynthesis.

Practice questions

1 Photosynthesis occurs in plant leaves.
  a) Name the chemical in leaves that absorbs
     light for photosynthesis. *(1 mark)*
  b) What term is used to describe reactions such
     as photosynthesis that require energy? *(1 mark)*
  c) Copy and complete the boxes in the word
     equation for photosynthesis. *(3 marks)*

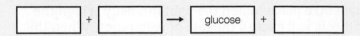

$$\boxed{} + \boxed{} \longrightarrow \boxed{\text{glucose}} + \boxed{}$$

2 Figure 2.9 shows a section of a leaf.

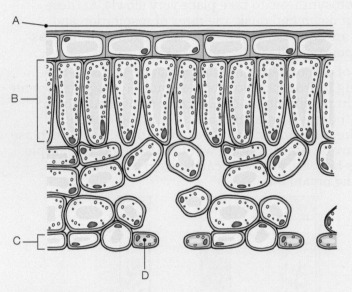

Figure 2.9

  a) Name parts A, B, C and D. *(4 marks)*
  b) Explain how parts A and B adapt the leaf for
     photosynthesis. *(2 marks)*
  c) Describe and explain how carbon dioxide
     moves from the air into the site of
     photosynthesis in B. *(4 marks)*

3 Figure 2.10 shows the setup of an investigation
  into the effect of light on the rate of
  photosynthesis.
  Three sealed tubes containing pondweed and
  equal volumes of red indicator solution were
  placed in different light conditions for 24 hours.

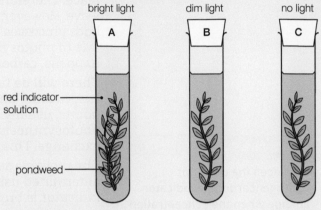

Figure 2.10

  After 24 hours the indicator in A was purple, B
  was red and C was yellow.
  a) Describe the contents of a suitable control for
     this investigation. *(1 mark)*
  b) Suggest **one** reason why it may not be valid to
     compare the results for each tube. *(1 mark)*
  c) Name the indicator solution. *(1 mark)*
  d) Explain the colour change in tube A. *(2 marks)*
  e) Tube B represents the compensation point.
     Explain what the compensation point is in this
     experiment. *(2 marks)*

4 a) What is diffusion? *(2 marks)*
  b) Explain how temperature affects diffusion.
     (2 marks)

5 Figure 2.11 shows a graph of the carbon dioxide
 used by photosynthesis and produced by
 respiration during one day.

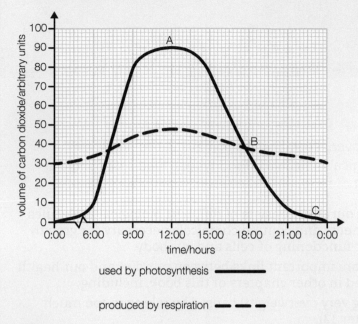

used by photosynthesis ━━━━

produced by respiration ━ ━ ━

Figure 2.11

Use evidence from the graph to help explain
the relationship between respiration and
photosynthesis at:

a) A   *(3 marks)*

b) B   *(3 marks)*

c) C   *(3 marks)*

3 Food and energy

Specification points

This chapter covers specification points 1.3.1 to 1.3.3.
It covers biological molecules (food types), food tests and the energy content of food.

The food we eat, our diet, consists of different biological molecules which provide energy and contain chemicals necessary for growth and the repair or functioning of cells and our body.

There are therefore important links between our diet and our health that are discussed in other chapters of this book, including:

▶ Obesity (being very overweight) caused by taking in too much energy (Chapter 13).

▶ Heart disease and strokes caused by the build-up of fatty substances (cholesterol) in the walls of arteries (Chapter 13).

▶ Diabetes (type 2) caused by poor diet or obesity leading to the body being unable to effectively control blood sugar levels (Chapter 6).

Biological molecules

The foods we eat are made of a range of biological molecules of which there are three important groups: carbohydrates, proteins and fats. The individual molecules of carbohydrates, proteins and fats all contain the elements carbon, hydrogen and oxygen, while all proteins also contain the element nitrogen.

Carbohydrates

These include sugars, such as glucose and lactose (the sugar in milk), and the more complex carbohydrates (cellulose, glycogen and starch). The sugars taste sweet and are soluble in water. The more complex carbohydrates consist of long chains of the sugars joined together (Figure 3.1).

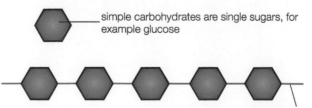

simple carbohydrates are single sugars, for example glucose

complex carbohydrates are long chains of sugars, for example starch

▲ **Figure 3.1** Simple and complex carbohydrates

Starch is a very important part of the human diet and it is broken down to glucose during digestion. In our bodies, we store carbohydrate as glycogen and this can be broken down to glucose when our sugar reserves are low.

Cellulose, found in plant cell walls, cannot be digested by humans but still plays an important role in the diet as fibre, adding bulk to our food and helping the muscles of the intestine wall push the food along.

Carbohydrates provide energy – simple sugars are a fast acting energy source and starch is a slow release source. Examples of foods rich in simple sugars include biscuits, cakes, jam and fizzy drinks. Potatoes, rice, pasta and bread are all rich in starch.

Protein

Protein provides the building blocks for the growth and repair of cells but can be used for energy when reserves of carbohydrate and fat are low. Proteins are complex molecules consisting of long chains of amino acids (Figure 3.2). There are 20 different amino acids which can join in different sequences and produce many thousands of different proteins.

When the proteins we eat are digested, the amino acids are absorbed into the blood, and in our body cells, are reassembled into the different proteins the body needs. Some of the proteins you will learn about are enzymes (Chapter 4), haemoglobin (Chapter 9) and antibodies (Chapter 13).

Good examples of foods rich in protein are lean meat, beans, fish and egg white.

> **Tip**
>
> Carbohydrates and fats provide most of the energy in our diet while the proteins are required for growth and repair.

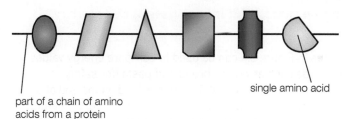

single amino acid

part of a chain of amino acids from a protein

▲ **Figure 3.2** Proteins are made up of amino acids

Fat

Fats are also known as lipids, their structure is shown in Figure 3.3. **(H)**

Fat is an excellent energy store, providing double the energy per gram of carbohydrate and protein. Foods high in fat include streaky bacon, cheese, other dairy products and lard.

When fat is digested, it is broken down into fatty acids and glycerol.

> **Tip**
>
> Each molecule of fat breaks down into one molecule of glycerol and three molecules of fatty acids.

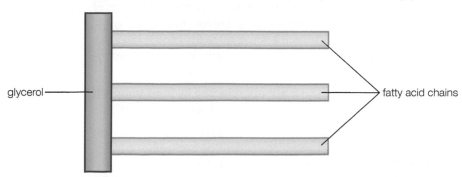

glycerol

fatty acid chains

▲ **Figure 3.3** Fats are made up of fatty acids and glycerol

Test yourself

1 Name the **three** elements found in all biological molecules.
2 Which **two** types of biological molecule provide most of the energy in our food?
3 Name the reagents used in a biuret test.
4 Apart from the reagent used and the colour change, describe one way the Benedict's test differs from the other food tests.

Show you can

Plan an experiment using the Benedict's test to compare how much sugar is present in a potato and an onion. State what you would do to make your results valid (a fair test).

Food tests

You should carry out food tests for sugars, starch, protein and fats on a range of foods. It is often necessary to break the food up using a pestle and mortar and to add a small quantity of water to make it into a solution before carrying out the test. The tests involve adding a chemical known as a reagent to the food solution. The reagent will change colour if a particular biological molecule is present in the food. Table 3.1 outlines the tests you need to know.

Table 3.1 The tests for food molecules

Food	Name of test	Method	Positive result
Starch	starch test	Add iodine solution to the food.	The iodine solution changes from yellow-brown to blue-black.
Reducing (simple) sugar	Benedict's test	Add Benedict's solution to the food and heat in a water bath.	It changes from blue to green then orange to brick red precipitate depending on how much sugar is present.
Protein	biuret test	Add sodium hydroxide to the food solution, then add copper sulphate and shake.	The mixture changes from blue to purple.
Fats	ethanol	Mix the fat with ethanol to dissolve some of it then add to water.	It changes from clear to form a white emulsion.

Prescribed practical

Prescribed practical B2: Investigating the energy content of food by burning food samples

The apparatus in Figure 3.4 can be used to compare energy values in different foods such as crisps, bread and pasta (for safety reasons nuts should not be used). The food is held on the end of a mounted needle, ignited and then placed immediately underneath the test tube. The rise in temperature of the water in the test tube will give an indication of how much energy there is in the food.

The energy released by a food sample is measured in joules (J) and is calculated using the following equation.

Energy released in Joules (J)

= mass of water (g) x rise in water temperature (°C) x 4.2

If the same mass of food is burned each time, then the result of this equation can be used to compare foods.

However, it is not always possible to measure the same mass of food each time. Since the more food is burned, the more energy will be released, it is necessary to calculate the energy released by each gram of food burned. This is done by dividing the energy released (J) by the mass (g) of the food burned.

Energy released (J/g) = $\dfrac{\text{energy released by food sample (J)}}{\text{mass of food sample (g)}}$

Although the energy values of foods can be compared in the laboratory as described above, food labels on packaging can also be used to check the accuracy of your calculations.

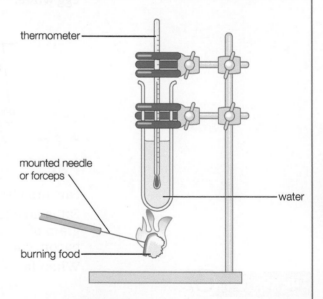
thermometer
water
mounted needle or forceps
burning food

▲ **Figure 3.4** Measuring the energy content of food by burning

Tip

The energy in food is often more than 1000 J. In such cases the energy is expressed in kilojoules (kJ): 1000 J = 1 kJ

Procedure

(SAFETY - wear eye protection)

1 Add 20 cm³ of water to a boiling tube, clamped in a retort stand as shown in Figure 3.4.

2 Record the temperature of the water.

3 Weigh approximately half of a potato crisp.

4 Secure the crisp on a mounted needle.

5 Light the crisp in a Bunsen burner flame and quickly hold it under the boiling tube.

6 Keep the burning crisp under the boiling tube until it is completely burned. If necessary, relight and replace under the boiling tube.

7 Record the maximum temperature of the water after the crisp is burned.

8 Repeat the experiment (steps 1–7) with two more samples of the same type of crisp.

9 Record the results in the table.

10 Copy and complete the table by calculating the energy released by each crisp and the average energy released per gram of crisp.

Table 3.2 The energy content of crisp results

Crisp	Mass of crisp/g	Water temperature/°C			Energy released	
		At start	After burning	Difference	Crisp/J	1 gram of crisp/Jg⁻¹
A						
B						
C						
					Average	

Sample results and questions

Table 3.3 Sample results of energy content of crisps

Crisp	Mass of crisp/g	Water temperature/°C			Energy released	
		At start	After burning	Difference	Crisp/J	1 gram of crisp/Jg⁻¹
A	1.2	12	40	28	2352	1960
B	1.5	13	55	42	3528	2352
C	0.8	11	31	20	1680	2100
					Average	2137

Questions

1 List **three** factors (variables) you would have to keep the same each time you test a food using the apparatus in Figure 3.4, to give valid results (a fair test).

2 If the piece of food stops burning while under the tube, explain why it should be relit and replaced under the tube until it does not light when placed in a Bunsen burner flame.

3 Explain the importance of calculating the average energy released per gram.

4 The typical energy value of crisps given on the packet is 21 kJ/g. What percentage of the typical energy was measured using this method?

5 Suggest why the energy value you get from burning a food is likely to be an underestimation of the energy in the food.

6 Suggest **two** ways the accuracy of this method could be improved.

Practice questions

1 Table 3.4 shows the results of food tests carried out on a biscuit.
   Use the information in the table to draw conclusions about the types of molecule in the biscuit.

Table 3.4 Results of food tests on a biscuit

Test reagent	Reagent colour at start	Result of food test
Benedict's	blue	positive
ethanol	clear	positive
biuret	blue	negative
iodine	yellow-brown	positive

Describe and explain the colour change for each food test.
In this question you will be assessed on your written communication skills, including the use of specialist scientific terms.   *(6 marks)*

4 Digestion

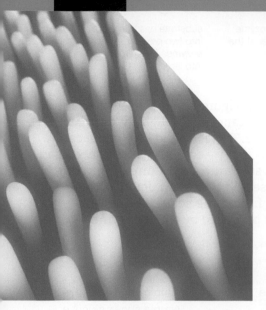

Specification points

This chapter covers specification points 1.4.1 to 1.4.3. It covers enzymes as proteins that are biological catalysts, the effects of temperature, pH, enzyme concentration and inhibitors on the action of enzymes, and their role in digestion and commercial uses.

Enzymes

All **enzymes** are **proteins** that function as biological catalysts, speeding up the rate of reactions such as photosynthesis, respiration, digestion and many others in living organisms.

Some of these reactions, such as respiration and digestion, involve breaking down large molecules, while others, such as photosynthesis, join small molecules together into larger ones. This chapter looks at the enzymes of digestion.

Tip

A catalyst is a substance which increases the rate of a chemical reaction without being changed or consumed during the reaction.

How enzymes work

The molecule on which an enzyme acts is the substrate, while the molecule which is formed by the reaction is the product. Enzymes work because the shape of the substrate matches exactly the shape of a special region in the enzyme molecule, called the active site. The shape of the substrate and the active site are complementary, so when they collide the substrate fits snugly into the active site of the enzyme. This tight fit then enables the enzyme to catalyse the reaction and, in a reaction such as digestion, split the substrate into its products. The enzyme molecule is not changed or used up by the reaction and so is available to repeat the reaction every time it collides with a substrate molecule.

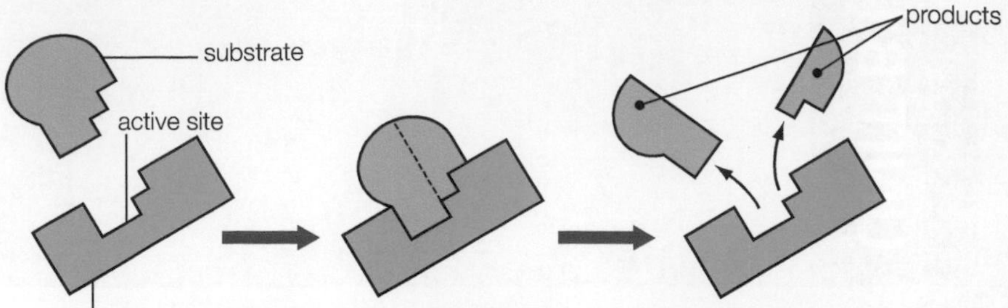

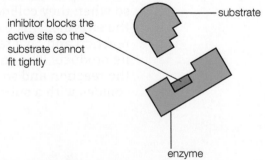

Figure 4.1 How enzymes work

The action of enzymes as described in Figure 4.1 is referred to as the 'lock and key model' due to the importance of the tight fit between the enzyme's active site and the substrate. This lock and key model can explain the principle of enzyme specificity – each enzyme is specific in that it will only work on its normal substrate. For example, only starch will fit into the active site of amylase and be broken down – other molecules such as proteins cannot fit into the active site. Table 4.1 shows the substrates and products produced by some enzymes.

Table 4.1 Enzymes, their substrates and products

Enzyme	Substrate	Product
carbohydrase	carbohydrate	simple sugar, glucose
amylase	starch	simple sugar, glucose
protease	protein	amino acid
lipase	fat (lipid)	glycerol and fatty acids

Inhibitors

However, there are molecules that fit loosely or partially into the active site of some enzymes. Such molecules are called inhibitors because, while they occupy the active site of the enzyme, substrate molecules cannot enter and be broken down, which leads to a reduced (inhibited) rate of reaction.

Figure 4.2 How an inhibitor stops an enzyme working

Tip

The name of most enzymes begins with the name of the substrate and ends with 'ase'.

Test yourself

1 What is an enzyme?
2 What does it mean when the shape of a substrate and the shape of an enzyme's active site are described as complementary?
3 Explain substrate specificity.
4 Name the type of enzyme which digests proteins.

Show you can

Explain the effect of an inhibitor on the action of an enzyme.

The effects of temperature and pH on the action of amylase

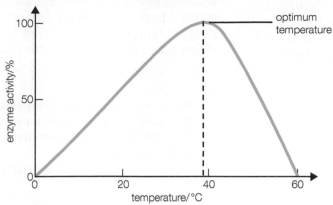

▲ **Figure 4.3** Effect of temperature on enzyme activity

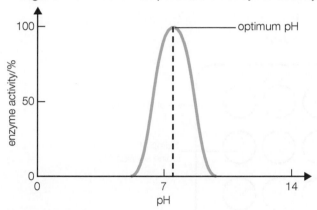

▲ **Figure 4.4** Effect of pH on enzyme activity

Temperature and pH affect the activity of all enzymes. Figures 4.3 and 4.4 show how changing the temperature and pH affect the activity of amylase.

At low temperatures, the enzyme and substrate molecules have reduced kinetic energy and move slowly, resulting in few collisions between them and a low rate of enzyme activity. Increasing the temperature increases the kinetic energy, the number of collisions and the rate of activity. The temperature (or pH) which causes the maximum rate of enzyme activity is the optimum.

Increasing temperatures above the optimum (or changing the pH away from the pH optimum) causes a decrease in the activity of the enzyme. This is due to an irreversible change to the shape of the enzyme's active site, known as denaturation. The further away from the optimum the temperature or pH are, the lower the enzyme activity.

The effect of enzyme concentration on enzyme activity

Figure 4.5 shows that the more enzymes there are, the faster the enzyme reaction (X). This is because there are more active sites for substrates to attach to. This applies up to a limit, when the rate levels off (Y) because there are not enough substrate molecules to react with the extra enzymes.

Show you can ❓

Describe and explain the rate of enzyme action at points X and Y in Figure 4.5.

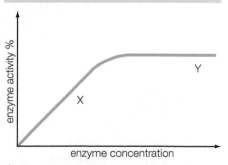

▲ **Figure 4.5** Effect of enzyme concentration on enzyme activity

Tip

Other commercial uses of enzymes include pre-digesting food for babies, extracting juice from fruit, making lactose-free products and softening the centre of chocolates.

Commercial enzymes

Many enzymes have commercial uses. For example, many biological washing powders have enzymes for breaking down difficult-to-remove stains. These enzymes are thermostable – they can work at a wide range of temperatures and they break the complex, large and insoluble stains down into small, soluble molecules that dissolve in the water.

Test yourself

5 Why do enzymes react slowly at low temperatures?
6 What is the optimum pH of an enzyme?
7 Describe how denaturation changes an enzyme and what effect this has on the enzyme action.
8 Why do biological washing powders need thermostable enzymes?

Prescribed practical

Prescribed practical B3: Investigating the effect of temperature on the action of an enzyme

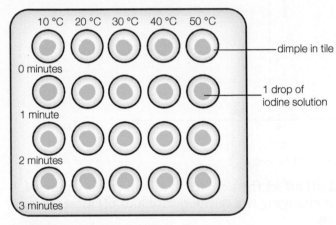

▲ **Figure 4.6** A spotting tile

Procedure

1 Set up five water baths at 10 °C, 20 °C, 30 °C, 40 °C and 50 °C.
2 Label five test tubes with 'starch' and the temperature of one of the water baths.
3 Label five test tubes with 'amylase' and the temperature of one of the water baths.
4 Use a syringe to measure 5 cm³ of 1% starch solution into each of the test tubes labelled starch.
5 Use another syringe to measure 5 cm³ of 1% amylase solution into each of the test tubes labelled amylase.
6 Place one starch and one amylase labelled test tube in each water bath for at least five minutes.
7 Prepare a spotting tile for each of the temperatures by placing one drop of iodine solution in each of the dimples as shown in Figure 4.6.
8 Starting with the 10 °C water bath, pour the amylase solution into the starch solution. Use a clean dropping pipette to sample the mixed solution and start a timer.

Tip

It will be necessary to prepare several spotting tiles as the reactions often last more than three minutes.

9 Add one drop of the sample to the iodine solution in the first dimple of the spotting tile and note any colour change. Return the remainder of the sample into the test tube.

10 Repeat the sampling (steps 8 and 9) every minute until the iodine shows no colour change. Record the time taken for the starch to be digested in minutes.

11 Repeat steps 8, 9 and 10 with each of the water baths in turn.

12 Copy and complete the results table.

Table 4.2 Results table for the effect of temperature on the action of amylase

Temperature/°C	10	20	30	40	50
Time for starch to be digested/minutes					

13 Draw a graph of the time taken for the starch to be digested at different temperatures on a grid, as in Figure 4.7. When drawing this graph, you should use a best fit line (not a point to point graph).

Sample results and questions

1 Explain why a tube of amylase and a tube of starch were placed in the water bath for five minutes before being mixed.

2 What effect will the sample drop have on the iodine solution at zero minutes?

3 Explain why the iodine will show no colour change after a period of time.

Table 4.3 Sample data for the effect of temperature on the action of amylase

Temperature/°C	10	20	30	40	50
Time taken for the starch to be digested/minutes	10	4	3	20	40

4 Explain why it is appropriate to use a line of best fit on a graph of these results.

5 Explain the results between 10 °C and 20 °C.

6 Between which temperatures is the optimum temperature for this amylase enzyme?

7 Suggest how the optimum temperature could be determined more accurately.

8 Explain what happens to the starch after 50 °C.

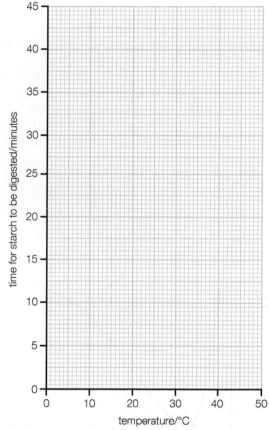

▲ **Figure 4.7** Investigating the effect of temperature on the action of amylase

Digestion

Most of the food we eat is in the form of large, complex, insoluble molecules. We need enzymes to digest and break down these molecules so that they are soluble and small enough to be absorbed into the blood system, where they can be used by the body.

Digestion can be defined as the breakdown of large, complex, insoluble molecules into small, simple, soluble ones.

Practice questions

1 a) The following diagram shows the effect of a protease enzyme on film.

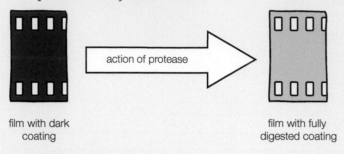

film with dark coating   film with fully digested coating

Figure 4.8

  i) What is a protease enzyme? *(1 mark)*

  ii) Suggest the chemicals produced when the film coating is fully digested. *(1 mark)*

b) The graph in Figure 4.9 plots the effect of pH on the action of the protease enzyme on film.

  i) Explain the shape of the graph. *(3 marks)*

  ii) Describe and explain what would happen to the film if a lipase enzyme was used. *(2 marks)*

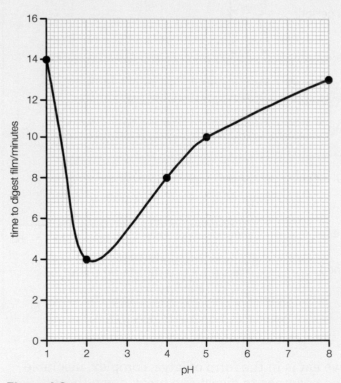

Figure 4.9

2 Soft-centred mint chocolates are made using enzymes. The mint centre is hard when first made so that it will not melt when covered with hot chocolate. After the chocolate has solidified the sweet is kept at 18 °C for 14 days. During this time an enzyme called invertase breaks down the complex sugar in the mint centre making it softer and sweeter. One manufacturer wanted to find a new enzyme to use in this process which would reduce the time needed to soften the mints. Their scientists carried out experiments using four new enzymes: A, B, C and D. Each experiment used the same mass of solid mint centre and the same concentration of enzyme. Table 4.4 shows the time taken by each of the new enzymes to make the mint go soft.

Table 4.4 Time taken by different enzymes to soften mint

Enzyme	Time taken to soften the mint/days
Invertase	14
A	20
B	10
C	18
D	17

a) Suggest which of the new enzymes, A, B, C or D, the manufacturer would use. Give a reason for your answer. *(2 marks)*

Another scientist checked the method and results of the experiment and concluded that some of the factors had not been controlled.

b) Choose **two** factors from the list below which should have been controlled in the manufacturer's experiment. *(2 marks)*

- Temperature
- Humidity
- Light intensity
- Oxygen concentration
- pH
- Softness of the mint at the end

c) What term describes the role of the other scientist checking the work of the manufacturer's scientists? *(1 mark)*

3 Figure 4.10 shows the absorption of glucose into the blood.
Use the information from Figure 4.10 to help explain why starch needs to be digested. *(2 marks)*

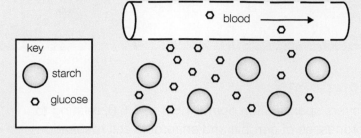

Figure 4.10

5 The respiratory system and cell respiration

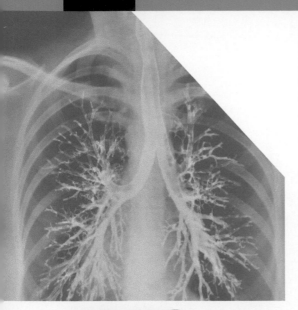

Respiration

Respiration, sometimes called **cell respiration**, is a series of chemical reactions in every cell that continuously releases energy from food molecules. The different cells of the body can use the energy released to produce heat in the body, for movement, growth, reproduction and to carry out active transport.

The reactions of cell respiration are described as exothermic because they release energy. They take place in special structures in the cell cytoplasm, the mitochondria, described in Chapter 1.

The cells in most living organisms use oxygen during aerobic respiration to help release energy. The word equation which summarises aerobic respiration is:

glucose + oxygen → carbon dioxide + water + energy

The balanced chemical equation for aerobic respiration is:

$$C_6H_{12}O_6 + 6O_2 \rightarrow 6CO_2 + 6H_2O + \text{energy}$$

When not enough oxygen is available, some cells, for example muscles and organisms like yeast, can still release a small amount of the energy in food molecules using a series of reactions known as anaerobic respiration.

This can happen during strenuous exercise in human muscles when anaerobic respiration causes lactic acid to build up, leading to muscle soreness. The word equation for anaerobic respiration in muscles is:

glucose → lactic acid + energy

Anaerobic respiration by yeast produces alcohol (ethanol) and is the basis of wine, beer and bread production. The word equation is:

glucose → alcohol + carbon dioxide + energy

Tip

Aerobic respiration is respiration which uses oxygen.

Tip

Anaerobic respiration is respiration without using oxygen.

Tip

The energy released by aerobic respiration is almost 20 times the energy released by anaerobic respiration.

To demonstrate this, a solution of glucose is first boiled to remove any dissolved oxygen and to sterilise it. Yeast cells are added to this solution only after it has cooled to a temperature that will not kill the yeast. The solution of glucose with yeast is then placed in the apparatus shown in Figure 5.1. As the yeast respires anaerobically it produces carbon dioxide and alcohol, and releases energy in the form of heat.

Figure 5.1 shows how anaerobic respiration can be demonstrated using yeast.

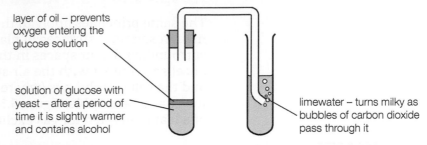

layer of oil – prevents oxygen entering the glucose solution

solution of glucose with yeast – after a period of time it is slightly warmer and contains alcohol

limewater – turns milky as bubbles of carbon dioxide pass through it

▲ **Figure 5.1** Demonstrating anaerobic respiration in yeast

Show you can ?

Compare the products of anaerobic respiration in muscles and yeast.

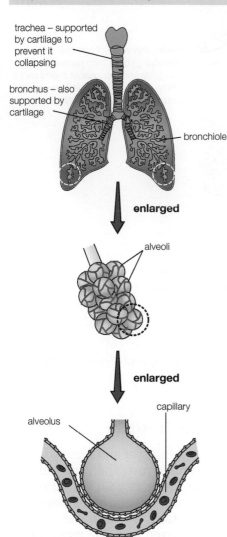

trachea – supported by cartilage to prevent it collapsing

bronchus – also supported by cartilage

bronchiole

enlarged

alveoli

enlarged

capillary

alveolus

site of **gas exchange**

▲ **Figure 5.2** The main parts of the respiratory system

Test yourself ✎

1 List the ways the different cells of the body use the energy released by respiration.
2 Which type of respiration releases less energy from a molecule of glucose?
3 Describe one situation in the human body when anaerobic respiration can take place.
4 Give the balanced chemical equation for aerobic respiration.

H

Respiratory surfaces

In humans **gas exchange** takes place in the alveoli. Oxygen diffuses into the blood and carbon dioxide diffuses from the blood into the alveoli, where it is then breathed out.

Respiratory surfaces are adapted in a number of ways. For example, in humans they have:

▶ a large surface area – there are many alveoli in each lung, and each alveolus has a large surface area; together these give a gas exchange surface (where the alveolar walls are in contact with blood capillaries) in humans of many square metres

▶ thin walls with short diffusion distances – Figure 5.2 shows that there are only two layers of cells separating the oxygen in the alveolus from the red blood cells; this means that there is only a short diffusion distance for the gases involved

▶ moist walls – these help the gases to pass through the respiratory surfaces because the gases dissolve in the moisture

▶ permeable surfaces – the moist, thin walls make the respiratory surfaces permeable

▶ a good blood supply – alveoli are surrounded by capillaries to ensure that any oxygen diffusing through is carried around the body; this also ensures that carbon dioxide is continually taken back to the lungs

▶ a diffusion gradient – the process of breathing ensures that there is a large diffusion gradient that encourages oxygen to diffuse into the blood and carbon dioxide to diffuse from the blood into the alveoli; when fresh air rich in oxygen is breathed in, it makes the concentration of oxygen in the alveoli higher than that in the capillaries and therefore oxygen diffuses from the alveoli into the capillaries.

Respiratory surfaces in plants

The same principles apply to the respiratory surfaces in plants. The main respiratory surfaces in plants are the spongy mesophyll cells surrounding the air spaces in the leaves. Because there are a lot of cells in contact with the air spaces, there is a large surface area and the cell membranes (where gas exchange takes place) are thin, moist and permeable. Figure 5.3 shows how the spongy mesophyll of a leaf is adapted for gas exchange.

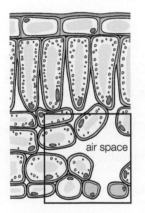
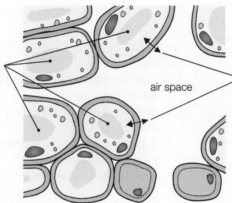

many spongy mesophyll cells around each air space increase the surface area

air space

moist thin walls of spongy mesophyll cells help gases to dissolve and permeable membranes allow gases to enter and leave and maintain a diffusion gradient

air space

▲ **Figure 5.3** The respiratory surfaces of a leaf

Test yourself

5 Name the part of the lungs where gas exchange takes place.
6 Explain why moist surfaces are beneficial for gas exchange.

Show you can ❓

Use Figure 5.2 to help describe the pathway taken by a molecule of oxygen from the air into the blood.

The effect of exercise on breathing

To investigate the effect of exercise on breathing rate and recovery rate you first need to calculate your breathing rate at rest. Then carry out vigorous exercise for a short period of time. Measure your breathing rate immediately after exercising and then at intervals, for example every minute, until it returns to normal. The time taken for the breathing rate to return to normal can be referred to as the recovery time. Compare your results with other members of your class.

You should be able to use the results you obtain to discuss the following points.

▶ Is there a link between levels of fitness and breathing rate?

▶ Is there a link between levels of fitness and recovery time?

▶ If a student became fitter over time, how would this affect his or her recovery time?

Example

Figure 5.4 shows the results of an investigation into how exercise affected the breathing rate of two boys.

The results show Rory is fitter because he shows:
- a lower resting breathing rate (between 0–5 minutes)
- a slower rate of increase during exercise (between 5–10 minutes)
- a lower maximum breathing rate (between 10–15 minutes)
- a faster recovery to his normal resting breathing rate after exercise (between 15–20 minutes).

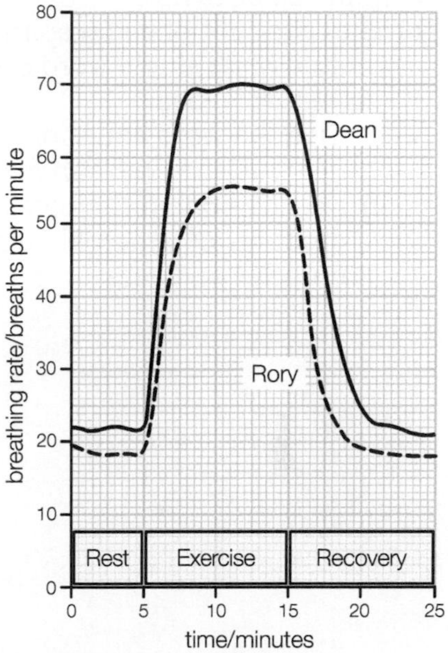

▲ **Figure 5.4** The effect of exercise on breathing rate

Tip

Depth of breathing is the volume of air breathed in during each breath.

As well as increasing the rate of breathing, exercise can increase the depth of breathing.

Explaining the effect of exercise on breathing

Exercise involves the movement of muscle cells which use energy released from food (glucose) molecules by cell respiration. Cell respiration, as we have learned, requires oxygen and produces carbon dioxide. Increased exercise therefore means that the muscle cells will use more oxygen from the blood and pass more carbon dioxide into the blood. When that blood reaches the lungs, a larger volume of air is needed to replace the oxygen and remove the carbon dioxide. The increased depth and rate of breathing is how the body brings in a larger volume of air and increases the rate of gas exchange.

1 a) Copy and complete the balanced chemical equation for aerobic respiration.

_____ + 6O$_2$ → _____ + _____ + Energy   (3 marks)

Aerobic respiration uses oxygen; anaerobic respiration does not.

b) The reactions of cell respiration all produce energy. What single word do biologists use to express this idea?   (1 mark)

c) i) Give the word equation for anaerobic respiration in human muscles.   (2 marks)

ii) Describe **one** difference between the products of anaerobic respiration in human muscles and yeast.   (1 mark)

6 Co-ordination and control

Responding to the environment – the nervous system

We can respond to the environment and anything that we respond to is called a stimulus. In animals, each type of stimulus affects a receptor in the body. There are many types of receptors, each sensitive to a particular type of stimulus or sense (for example sight, sound, touch, taste and smell). If a receptor is stimulated it may cause a different part of the body, an effector (for example **muscles**) to produce a response.

stimulus → receptor → effector → response

The flowchart above is a simplification because it suggests that we will automatically produce a response when we are stimulated. For example, if we hear a sound (the stimulus) we might respond or not, depending on what the sound is.

Coordination

In reality, the receptors and effectors are linked by a coordinator. This coordinator is usually the **brain** but may also be the **spinal cord**. Together the brain and the spinal cord are known as the Central Nervous System (CNS), as shown in Figure 6.1.

Nerve cells or neurones link the receptors and the effectors to the coordinator. A neurone carries information as small electrical charges called nerve impulses. The brain acts as a filter and determines which receptors link up with which effectors and whether or not a particular stimulus brings about a response.

A more complete flowchart is:

stimulus	→	receptor	→	brain	→	effector (muscle)	→	response
Jane texts John		*John's eye reads text*		*John thinks what to do*		*John types a reply*		*John texts Jane*

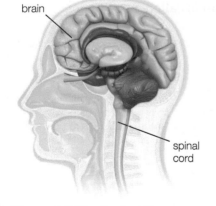

▲ **Figure 6.1** The Central Nervous System

The overall total of our responses to the environment around us is described as our **behaviour**.

Sometimes our receptors are grouped together into complex sense organs. Examples include the nose (smell), the ear (sound) and the eye (sight).

Voluntary and reflex actions

Many of our actions are voluntary. This means we deliberately choose to do them and they involve conscious thought. However, there is another type of action that does not involve conscious thought – these are reflex actions.

Reflex actions

If you accidently touch a very hot object you respond immediately by rapidly withdrawing your hand from the danger area. The advantage of this is that you move your hand away before it can get burned too badly. This type of action does not involve any 'thinking' time, as the time taken to consider the response would cause unnecessary damage to the body. All reflex actions have two main characteristics in common:

▶ they occur very rapidly

▶ they do not involve conscious control (thinking time).

What makes a reflex action so rapid? In a reflex pathway, the total length of the pathway is kept as short as it possibly can be, with the minimum number of neurones involved. In addition, there are relatively few gaps between the neurones (synapses), as they are the places where the nerve impulses travel relatively slowly.

Figure 6.2 shows the nerve pathway when a hand touches a hot object.

There are three types of neurone involved in this response.

The diagram shows that both the association and the motor neurones begin with the cell body. The diagram also shows that only two synapses (short gaps between neurones that slow nervous communication) are involved in this pathway. This system of structures involved in a reflex action is called a reflex arc.

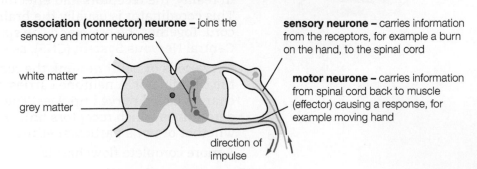

association (connector) neurone – joins the sensory and motor neurones

white matter

grey matter

sensory neurone – carries information from the receptors, for example a burn on the hand, to the spinal cord

motor neurone – carries information from spinal cord back to muscle (effector) causing a response, for example moving hand

direction of impulse

The three neurones link in the grey matter (butterfly-shaped part) of the spinal cord; the outer part is the white matter.

▲ **Figure 6.2** A reflex arc

Hormones

Tip

Each hormone affects different target organs. Some hormones affect many organs while some organs are affected by several hormones.

Another type of communication system is controlled by hormones. Hormones are chemical messages produced by special glands which release them into the blood. Although the hormones travel all around the body in the blood, they can only affect certain organs called target organs.

Hormones usually act more slowly than the nervous system and over a longer period. Good examples to illustrate these points are the sex hormones: oestrogen and testosterone. The changes brought about by testosterone in males and oestrogen in females come about over many years.

Tip

While nervous responses are fast and short term, hormone responses are slow and continue over long periods of time.

Hormones also have an important role in maintaining the internal environment of the body in a relatively constant state in response to changes outside and inside the body. The maintenance of this constant state is referred to as homeostasis. Two examples of the homeostatic role of hormones are controlling the concentration of glucose in the blood by the hormone insulin and controlling the water content of the body, referred to as osmoregulation.

Insulin and blood glucose

Insulin is a hormone that prevents the concentration of glucose (sugar) in the blood becoming too high. Glucose is constantly needed by all cells for respiration and therefore must always be present at a sufficient concentration. However, if there is too much glucose in the blood this can damage body cells due to water loss by osmosis (see Chapter 8).

Tip

Insulin lowers blood glucose concentration – it is not enough to say that it controls blood glucose concentration.

Insulin is produced and released into the blood by special cells in the pancreas in response to increasing or high blood glucose concentrations. This usually occurs after a meal, especially if the meal is rich in carbohydrates.

The main target organ for insulin is the liver where it causes:

Tip

Meals rich in carbohydrates are digested into sugars which can then be absorbed into the blood stream.

▶ **increased absorption** of glucose from the blood, so reducing the blood glucose concentration
▶ the conversion of **excess glucose into** glycogen, which is stored in the liver and to a lesser extent in muscle cells
▶ **increased respiration**.

Figure 6.3 summarises how insulin controls the blood glucose concentration.

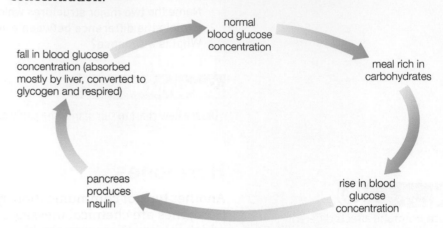

▲ **Figure 6.3** Insulin and blood sugar concentration

Figure 6.4 shows how the concentration of blood glucose typically varies after eating a meal.

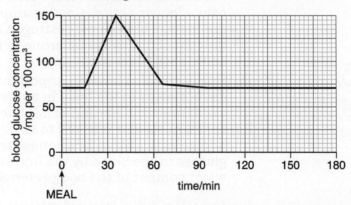

▲ **Figure 6.4** The effect of insulin on blood sugar concentration

When blood glucose concentration is low, less insulin is produced. This means that the above processes do not take place or take place at a slower rate, helping to raise the concentration of glucose in the blood.

Diabetes – when blood glucose regulation fails

Diabetes is a fairly common condition in which the body does not produce enough insulin to keep the blood glucose at the normal concentration. Individuals who develop diabetes are unable to control the concentration of their blood glucose without treatment and the following symptoms are often present:

▶ There is **glucose in the urine**. This happens because their blood glucose concentration is so high that some is filtered out by the kidneys and passed into the urine.

▶ Affected individuals are often thirsty and because they drink so much they need to go to the toilet a lot.

▶ Lethargy may result.

There are two kinds of diabetes, Type 1 and Type 2. Type 1 diabetes normally develops in childhood. Type 2 diabetes usually only develops in older people, but is becoming increasingly common in young people.

▲ **Figure 6.5** Young girl injecting insulin

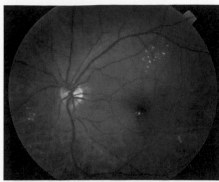

▲ **Figure 6.6** Retina of a person with diabetes. The areas of small yellow dots, caused by leakage from damaged blood vessels, can cause permanent loss of vision

Type 1 diabetes is usually treated by the injection of insulin and by a carefully controlled diet where the intake of carbohydrates is carefully monitored.

Figure 6.5 shows a young girl injecting herself with insulin. Even with the use of insulin injections and a carefully controlled diet, it is difficult for people with diabetes to control their blood sugar concentration very accurately. Problems may arise if too much insulin is injected or if not enough food is eaten at regular intervals. If the blood sugar concentration drops too low, a hypoglycaemic attack (a hypo) may occur and unconsciousness will result. If blood sugar concentrations remain too high for a long period serious medical complications can result.

Type 2 diabetes has a slightly different cause in that insulin is produced but stops working effectively. Type 2 diabetes is often associated with poor diet, obesity and lack of exercise. The treatment of Type 2 diabetes therefore includes changes to diet and exercise to achieve weight loss along with medication in the form of tablets and injections. The increasing number of people with these characteristics largely explains the increase in the number of people with diabetes.

Long-term effects and future trends

People who have had diabetes for a long time (in some cases undiagnosed and unknown) and whose blood sugar concentration is not tightly controlled run the risk of developing long term complications. These include **eye damage** (see Figure 6.6), or even blindness, **heart disease**, **strokes** and **kidney damage**. These complications are usually due to the high blood sugar concentration damaging the capillaries that supply the part of the body involved.

Osmoregulation

Osmoregulation is another homeostatic process in the body. It controls the amount of water in the blood and other body fluids. Water, like the sugars in the blood, has the potential to damage cells due to osmosis causing excessive movements of water.

The body gains water mainly by drinking and from the food we eat. A small amount also comes from cell respiration which produces water as a by-product. At the same time, water is lost from the body through evaporation in the lungs, evaporation while sweating and the production of urine by the kidneys and in faeces. Table 6.1 lists the average volume of water gained and lost by the body each day.

Table 6.1 The average volume of water gained and lost by the body each day

Water gained by the body through	Water lost by the body through
drinking 1500 cm³	urine 1400 cm³
eating 700 cm³	sweat 500 cm³
cell respiration 200 cm³	breath from lungs 400 cm³
	faeces 100 cm³
Total = 2400 cm³	Total = 2400 cm³

In normal conditions the volume of water gained balances the volume lost but if conditions change then osmoregulation brings the volumes back into balance.

Tip

Osmoregulation is the way the body balances the water it gains (drinking, eating and respiration) with the water it loses (urine, sweat, breathing and faeces).

For example, in very warm weather or during vigorous exercise the body will lose more water as sweat. This can be partially balanced by increasing the amount we drink but the kidneys also produce a more concentrated urine containing less water. If we drink larger volumes of liquid than normal the kidneys again act to bring the volumes back into balance by producing large volumes of dilute urine.

Example

Table 6.2 shows the water gained and lost by a boy who has played a game of football on a warm summer day.

Table 6.2

Water gained by the body through	Water lost by the body through
drinking 2500 cm³	urine 2250 cm³
eating 700 cm³	sweat 600 cm³
cell respiration 200 cm³	breath from lungs 450 cm³
	faeces 100 cm³
Total = 3400 cm³	Total = 3400 cm³

In this situation the volumes of water gained and lost have changed because:
- Playing football has increased the water lost as sweat and by breathing more often.
- During the game, he has drunk extra water to compensate for these losses.
- That has increased his total gained to 3400 cm³.
- To maintain the water balance in the body the total lost must also increase to 3400 cm³.
- The kidneys produce a larger volume of dilute urine.

Kidneys – how they affect the volume of urine

The kidneys are part of the excretory system, which functions to remove wastes from the body and to carry out osmoregulation. Figure 6.7 shows the excretory system.

Blood passes into each kidney through the renal artery. When it reaches the cortex of the kidney much of the liquid portion of the blood and many of the substances dissolved in it are filtered out. Then in the medulla region water and other substances are reabsorbed back into the blood in a controlled way until normal concentrations are reached.

The excess water and dissolved substances pass into the urine, which collects in the renal pelvis, and passes down the ureter into the bladder. There it is stored before being passed out via the urethra.

Tip

It is important to learn the correct spelling of ureter and urethra.

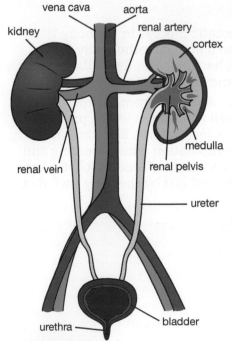

▲ **Figure 6.7** The excretory system

Anti-diuretic hormone (ADH)

The process of reabsorption of water back into the blood by the kidneys is controlled by a hormone, the anti-diuretic hormone (ADH).

ADH is produced by the part of the brain where the amount of water in the blood is monitored. The ADH is released into the blood and travels to its target organ, the kidneys. In the kidney medulla, ADH allows more water to pass from the urine back into the blood (be reabsorbed). As a result, the urine has a lower volume and is more concentrated.

As an example of how ADH works, consider a person who has drunk a lot of water on a hot day. The blood reaching their brain will contain much more water than normal. In response, the brain stops or reduces the production of ADH. In the kidney, low or no ADH means much less water is reabsorbed into the blood leaving large volumes of dilute urine to be produced. Figure 6.8 shows how ADH affects the volume of urine.

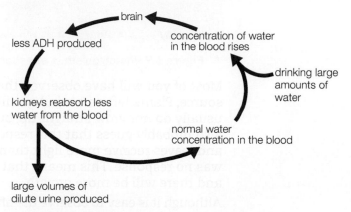

▲ **Figure 6.8** How ADH affects the volume of urine

Test yourself

8 List **three** ways the body gains water.
9 Name the blood vessel that transports blood to the kidney.
10 Name the tube that carries urine from the kidney to the bladder.
11 Where is ADH produced?

Show you can

After drinking a litre of water, explain how ADH would cause a large volume of dilute urine to be produced.

Sensitivity in plants

Plants, like animals, respond to changes in the environment. However, they respond to fewer types of stimuli and in general the response is slower. Plants respond to the environmental stimuli that have the greatest effect on their growth. Roots grow towards water when a moisture gradient exists and shoots tend to grow away from the effects of gravity (they grow upwards). Reasons for these responses are fairly obvious, as they ensure that the plants react in such a way that they receive the best conditions for growth. The response of a plant to light is called phototropism and this response has been investigated in detail to establish how it occurs.

Phototropism – responding to light

▲ **Figure 6.9** Phototropism is a response to unilateral light

Most of you will have observed that plants grow towards a light source. Plants left on a window sill or against the wall of a house usually do not grow straight up, but bend towards the Sun. You can also probably guess that this response ensures that the plant stem and leaves receive more light than they otherwise would do if there was no response. This means that more photosynthesis takes place and there will be more growth.

Although it is easy to observe the effect of phototropism, what causes it to occur? Figure 6.10 shows the growth of young seedlings in unilateral light (light coming from one side or source only) and highlights some of the features of phototropism. Can you use the diagram to identify what part of the plant is sensitive to the light source?

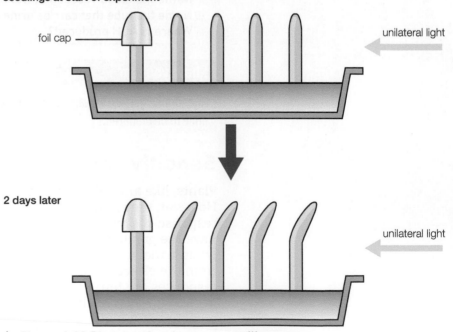

seedlings at start of experiment

foil cap

unilateral light

2 days later

unilateral light

▲ **Figure 6.10** Phototropism in young seedlings

Figure 6.10 shows that it is the shoot tip that is sensitive to light, as when it is covered the phototropic response does not occur. This and many other similar experiments have led to our understanding that the response of phototropism is controlled by a plant hormone called auxin. When the stem is illuminated from one side this hormone tends to accumulate more on the non-illuminated side. The effect of the hormone is to increase the growth of the non-illuminated side of the stem more rapidly than the side receiving most light. This differential growth that occurs when one side of the stem grows more than the other side leads to the stem bending in the direction of the light.

The auxin is actually produced at the tip of the shoot and diffuses downwards. As it does so, light on one side causes the auxin to accumulate on the non-illuminated side. The main effect of this high concentration of auxin is to make each cell grow by elongation (become longer) more than it normally would. The cells in the non-illuminated side of the stem therefore increase in length more than those in the illuminated side, causing the stem to bend towards the light. Figure 6.11 shows how auxin causes a stem to bend towards light.

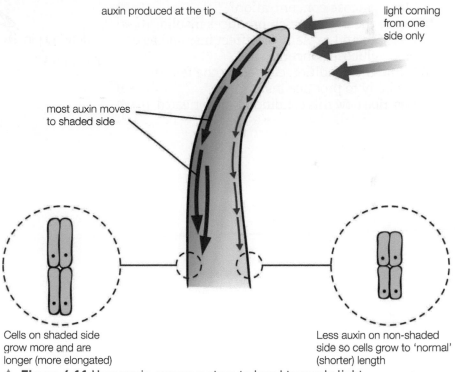

auxin produced at the tip

light coming from one side only

most auxin moves to shaded side

Cells on shaded side grow more and are longer (more elongated)

Less auxin on non-shaded side so cells grow to 'normal' (shorter) length

▲ **Figure 6.11** How auxin causes a stem to bend towards light

Practice questions

1 The word equation summarises a nervous system response.

   stimulus → receptor → Central Nervous System → effector → response

   a) What is a stimulus? *(1 mark)*
   b) i) Name the parts of the Central Nervous System. *(2 marks)*
      ii) What is the role of the Central Nervous System? *(1 mark)*
   c) Give **one** example of an effector. *(1 mark)*
   d) Some responses are voluntary while others are reflex.
      Explain the difference between a voluntary and a reflex response. *(2 marks)*

2 a) What effect would eating chocolate have on blood glucose concentration? *(1 mark)*
   b) Name the organ which produces insulin. *(1 mark)*
   c) Describe and explain the effect insulin has on blood glucose concentration. *(3 marks)*
   d) Name the condition caused by the failure of the body to produce insulin. *(1 mark)*
   e) Describe how this condition can be treated. *(1 mark)*

3 Figure 6.12 shows a reflex arc. Name neurones A, B and C. *(3 marks)* **(H)**

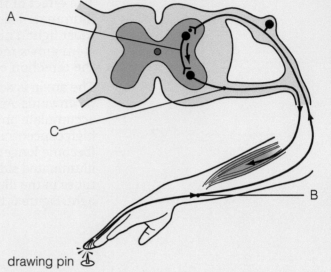

drawing pin

Figure 6.12

7 Ecological relationships and energy flow

Specification points

This chapter covers specification points 1.7.1 to 1.7.12 which include ecological terms, biotic and abiotic factors, sampling with a quadrat, adaptations and competition, the role of the Sun as an energy source, food chains and webs, the carbon cycle, decomposition, the nitrogen cycle, minerals and eutrophication.

Biological terms

Ecology is the study of the relationships between organisms and their surroundings. Biologists often use more precise terms when describing these relationships. Although ecology can involve studying single organisms and their surroundings, more often it involves populations of organisms. A population is the number of organisms of the same species living in the same area. The area where a population lives is its habitat.

Where several populations of different species are found living in habitats close together, the populations are collectively a community and biodiversity is a measure of the number of different species living in the area.

Figure 7.1 shows a tree trunk habitat with a population of orange lichens growing on it. Since there is at least one other species of lichen (light grey and feathery) growing in the same habitat, it is also a community.

The 'surroundings' can also be described in more detail as the environment. This is all the factors which act on an organism or population. They are often further divided into the non-living or abiotic factors such as temperature, humidity, light intensity and living or biotic factors such as the effect of a predator on its prey or where two species compete for the same food. The term ecosystem groups all these ideas together, being an area where a community of organisms live and are affected by a range of environmental factors.

▲ **Figure 7.1** Tree trunk habitat with a community of lichens

Tip

Sometimes abiotic and biotic factors can interact, for example the shade in a woodland (abiotic factor) is created by the trees (biotic factor).

Fieldwork

During your studies, you will be able to investigate a habitat or ecosystem such as a woodland, shoreline or sand dune system, sampling the numbers of different types of organisms, taking measurements of environmental factors and using these to explain the distributions you find.

Sampling

Realistically, when investigating the number of plants and animals in a habitat it is not possible to cover the whole area (usually because of the time required). Only a small fraction, referred to as a sample of the area can be investigated – several subsections within the habitat are sampled to give an overall picture. In most cases, a square frame called a quadrat is used as the subsection. Some quadrats have 1 m sides giving a quadrat area of 1 m²; others have 0.5 m sides giving an area of 0.25 m².

The most important thing about sampling is that the method used should produce reliable results. For this to happen the sample needs to be as large as possible – the number selected will depend on the size of the habitat but for a reasonably large area it is usually necessary to use 20 or more quadrats. The sample should also be representative of the habitat area. In other words, it should be positioned randomly over the whole area and not just in one corner. This can be done by dividing the habitat into a grid and using random numbers to generate coordinates to find the location of each quadrat.

Collecting data

When investigating plant distribution, percentage cover is often used. This is the percentage of the quadrat covered by a particular type of plant. Estimating the exact percentage is a difficult skill so it is normal to round up to the nearest 10%. (An exception is if there are any plants with a percentage cover between one and five – this is recorded as one and not ten.) Therefore, when sampling for percentage cover using a quadrat, the possible values obtained are 0, 1, 10, 20 and so on up to 100%. Figure 7.2 shows the estimated percentage cover of three plant species in a quadrat.

> **Tip**
>
> Random numbers can be created simply. Use the last four digits of phone numbers from a randomly chosen page in a phone book. The first two digits are for the *x*-axis and the last two the *y*-axis.

> **Tip**
>
> In some cases, where some of the plants lie over each other it is possible to record percentage values which give a total greater than 100%.

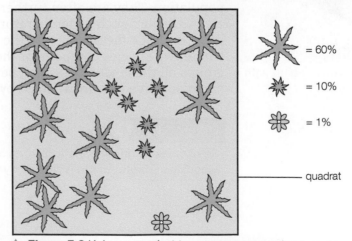

▲ **Figure 7.2** Using a quadrat to measure percentage cover

When the data for all the quadrats is collected, the values for each species can be averaged to give an overall estimate of the percentage cover.

In other cases, where individual plants can be easily identified, percentage cover is not necessary and they can be counted. This data can also be averaged but the area of the quadrat can be used to estimate the average number of plants per square metre, which in turn, if multiplied by the total area of the quadrat, can give an estimate of the total number of that plant species present in the habitat.

Example

Table 7.1 Sample data of plant species

Sample	Number of buttercup plants
1	3
2	2
3	4

Table 7.1 shows the number of buttercup plants recorded in a field measuring 30 m × 20 m.

The square quadrat used had a side of 0.5 m.

The average number of buttercups in each quadrat = (3+2+4) ÷ 3 = 3 plants

The area of each quadrat = 0.5 m × 0.5 m = 0.25 m²

The average number of buttercups in a square metre = 3 × (1 ÷ 0.25) = 12

The field area = 30 m × 20 m = 600 m²

The total number of buttercups = 600 × 12 = 7200

Belt transect sampling

Belt transect sampling is used in habitats where there is a gradual change from one side of the habitat to the other. Randomly placed quadrats would not identify the change in the habit. Figure 7.3 shows a rocky seashore where the types of plants and animals change between the low tide area at the bottom of the photograph and the high tide area at the top of the photograph.

In this habitat, a quadrat can be used to sample the animals as well as the plants because the animals do not move or do so very slowly. In a belt transect (Figure 7.4), quadrats are placed along a line from the bottom of the shore to the top of the shore. Quadrats can be placed continuously or at intervals along the line depending on the distance involved.

▲ **Figure 7.3** Rocky seashore

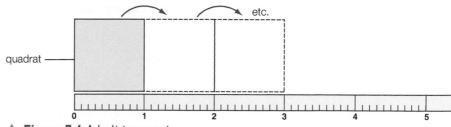
▲ **Figure 7.4** A belt transect

Test yourself

1 What is a population of organisms?
2 Name the apparatus usually used to take a sample of plants.
3 How is percentage cover used when sampling plants?
4 When is a belt transect method of sampling used?

Show you can

Describe what is needed to make a sample reliable.

Prescribed practical B4: Using quadrats to investigate the abundance of plants and animals in a habitat

Estimate the population of weed species in a lawn

Procedure

1 Place two 20 m measuring tapes at right angles to each other to mark out the area of grassland to be sampled.

2 Use random numbers to select five sets of coordinates to sample within the area of grassland.

3 Place a 1 m² quadrat at each coordinate and count the number of dandelion, plantain and daisy plants present.

4 Record the results in a table similar to table 7.2

Table 7.2 Results table for population estimates

Sample	Number of dandelion plants	Number of plantain plants	Number of daisy plants
1			
2			
3			
4			
5			

5 Calculate the average number of each type of plant type per square metre and estimate the total population of each type in the grassland area sampled.

Sample results and questions

1 Explain why a sampling method is used instead of counting all the plants in the area.

2 Explain why random numbers are used to select the position of each quadrat.

3 Table 7.3 shows sample data for this investigation

Table 7.3 Sample data for the number of plants in the area

Sample	Number of dandelion plants	Number of plantain plants	Number of daisy plants
1	3	6	12
2	5	2	8
3	1	5	25
4	4	4	15
5	2	8	10

Use the sample data to calculate the average density of each plant type and the total number of plants in the grassland area. Table 7.4 summarises the results.

Table 7.4 Calculations of average density and population using the sample data

Plant type	Average per m²	Estimate of population
Dandelion	(3+5+1+4+2) ÷ 5 = **3**	3 × (20 × 20) = **1200**
Plantain	(6+2+5+4+8) ÷ 5 = **5**	5 × (20 × 20) = **2000**
Daisy	(12+8+25+15+10) ÷ 5 = **14**	14 × (20 × 20) = **5600**

4 Explain how more accurate results could be obtained in this investigation.

Investigate the abundance of species between high and low tide on a rocky shore

Procedure

1 Extend a measuring tape as a transect, from the high tide mark down to the low tide mark.

2 Place a quadrat (0.5 m × 0.5 m) beside the tape at the 0–0.5 m measurement. Record the position of the quadrat.

3 Identify and list the organisms present in the quadrat.

4 Repeat steps 2 and 3 with quadrats placed at regular intervals down the transect tape until the full length of the transect has been sampled.

5 Record the results in a table similar to Table 7.5.

Table 7.5 Results table for sampling along a transect

Species	Presence of species at position on transect					
	0–0.5 m	3.0–3.5 m	6.0–6.5 m	9.0–9.5 m	12.0–12.5 m	15.0–15.5 m

6 Draw a bar graph of the number of species at each position along the transect.

7 Describe any trends in the results.

Sample results and questions

1 Table 7.6 shows sample data for this investigation.

Table 7.6 Sample data from sampling along a seashore transect

Species	Presence of species at position on transect					
	0–0.5 m	3.0–3.5 m	6.0–6.5 m	9.0–9.5 m	12.0–12.5 m	15.0–15.5 m
A	✓					
B		✓	✓			
C		✓	✓			
D				✓	✓	
E				✓	✓	✓

2 Draw a bar graph of the number of species at each position along the transect.

3 Figure 7.5 shows a bar graph of the sample results. Describe any trend in the results.

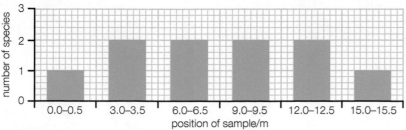

▲ **Figure 7.5**

> **Tip**
>
> Follow instructions given by the teacher. Be aware of the tidal movements during the activity, and exercise care as the rocks may be slippery.

> **Tip**
>
> This method can also be used to investigate any site where one habitat merges into another, for example from a pathway to grassland, conifer to deciduous woodland or stream to field.

The factors acting on organisms

To explain the distribution and abundance of a species we need an understanding of the different factors which affect each species. These factors have been divided into two groups.

Investigating abiotic factors

Most ecological investigations involve the analysis of some of the abiotic (non-living) factors that could affect the distribution of plants and animals. Examples of the abiotic factors that can be investigated include:

▶ Wind – wind speed can be analysed using anemometers. Wind speed can be very important, affecting the numbers and distribution of plants and animals in exposed habitats such as sand dune systems.

▶ Water – soil moisture levels can be calculated by taking soil samples and weighing them to find their mass. The soil samples are then dried in an oven until completely dry and reweighed. The difference in mass as a percentage of the original mass gives a value for the percentage soil moisture. Soil moisture levels can be an important factor in the distribution of many plants and animals.

▶ pH – pH can be measured using soil test kits or probes. The pH of the soil is very important in the distribution of many plants. Some plants will only grow in relatively acidic soils, for example heathers, and some will only grow in relatively alkaline soils, for example some orchids, but most plants prefer soil pH to be around neutral.

▶ Light – can be measured using light meters. Light is particularly important in the distribution of plants. While all plants need light to photosynthesise, some need high light levels to thrive whereas others can survive in very low light levels.

▶ Temperature – temperature can be measured using a thermometer.

Biotic factors

Biotic factors are living features of the environment, or the ways in which the presence of one species interacts with another. One example is competition, where the organisms in a habitat each try to obtain enough of a resource they both need to reproduce and survive. There is usually not enough of the resource to allow both organisms to thrive so the reproduction and ultimately the survival of both organisms is affected.

▲ **Figure 7.6a** A red squirrel

▲ **Figure 7.6b** A grey squirrel

Competition between animals is usually for food, water, territory and mates. The relationship between red and grey squirrels demonstrates how competition can affect populations. Although they both can eat similar types of tree seeds, the red squirrel is not able to digest seeds such as acorns as well as the grey squirrel. This means that in mixed woodland the grey squirrels have an advantage, being able to get more energy from their food, so have a better chance of surviving in this habitat. As a result of grey squirrels out-competing red squirrels in this way, very few red squirrels can survive in mixed woodland

However, grey squirrels are much larger and so need more energy to survive. This becomes a disadvantage in habitats like conifer forests where the seeds are small and so it is more difficult for the grey squirrels to eat enough seeds for the energy they require. In conifer forests red squirrels are therefore at an advantage and are often the larger population.

Another biotic factor affecting the relationship between grey and red squirrels is the disease squirrel pox, which is carried by the grey squirrel with little effect but is lethal if transferred to red squirrels.

Competition in plants is usually for light, carbon dioxide, water, minerals and space. The case study below is an example of how the plants of a sand dune system compete for these resources.

The following case study gives an example of abiotic and biotic data in a habitat.

Case study – plant distribution in a sand dune system

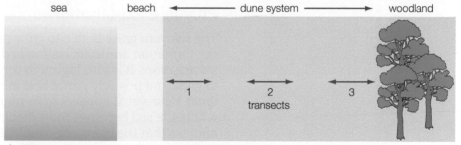

▲ **Figure 7.7** Investigating plant distribution in a sand dune system

The following data was collected from a survey in a sand dune system. As the sand dune system extended for over 1 km the data could not be collected from a continuous belt transect but was collected from three 'interrupted' belt transects. The first belt transect was from the start of the first dune nearest the shore and extended inland. The third belt transect was at the very end of the dune system, just before the typical dune system was replaced by woodland. The second transect was halfway between the other two. Data was collected from 20 quadrats in each transect.

Figure 7.8 shows some of the abiotic information gathered and the distribution of three plants typical of sand dunes. The appropriate abiotic data was collected for each quadrat sampled in each transect.

Marram grass is the grass typical of sand dunes and has an important role in binding the sand within the dunes together and enabling dunes to become stabilised. Unlike most other plants, it can grow in very unstable conditions such as those found near the shore. Heather, a small shrub, grows where the soil becomes more stable and moist. The larger woody shrub, gorse, will thrive when the soil becomes even more stable and contains more nutrients and moisture.

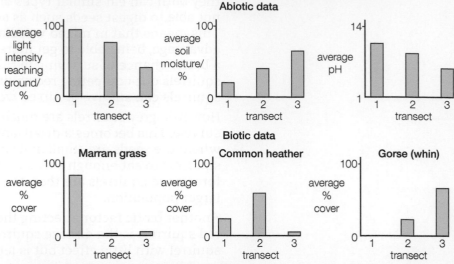

▲ **Figure 7.8** Abiotic data and plant distribution

The graphs show that marram grass is most common in Transect 1, with little coverage in Transect 2 and 3. Heather is the most common plant found in Transect 2. Gorse is the most common species found in Transect 3 but is uncommon or not found in Transects 1 and 2. You should be able to interpret this data and give possible explanations for the plant distributions based on the information given.

Possible explanation

The marram grass is able to grow near the shore and is important in the formation and stabilisation of dunes (you have been given this information). It is also logical to conclude that it requires high light levels to grow and possibly grows best in slightly alkaline pH (from the abiotic data). However, it appears to be much less successful further inland where the soil becomes more stable, more moist and more acidic. It appears that the changing conditions further inland favour other plants such as heather and gorse and these outcompete the marram grass. The low light level in Transect 3 (created by shade from gorse shrubs) probably prevented the marram grass from growing.

The shrub, heather, gains a foothold in the more stable Transect 2 but cannot compete against the larger woody shrubs in Transect 3. Like the marram grass, the heather is probably unable to survive in the shade of the larger gorse.

It is important to note that plants and animals can influence the abiotic data – it is not only the other way around. In this example, the ground light levels in each transect are entirely controlled by the shade produced by plants growing there.

Test yourself

5 What is an abiotic factor?
6 Describe how an oven and a balance can be used to find the water content of soils.
7 Explain why the pH of the soil is important to plants.
8 What is competition?

Show you can

Explain the difference between validity and reliability.

Validity and reliability

During this course, you have come across the term **reliability** on a number of occasions. If data is reliable, someone else could repeat the investigation and get similar results. Reliability can be increased in experiments by doing repeats or taking many measurements.

The validity of information provided is an indication of whether you are actually able to draw conclusions from the information. The concept of a 'fair test' applies as much to ecological experiments as it does to laboratory-based ones. For example, you will only be able to say definitely that a particular abiotic factor is responsible for a change in vegetation if other abiotic factors are controlled.

Transfer of energy and nutrients

Examples of ecosystems include grasslands, woodlands and lakes. If ecosystems can remain stable for long periods of time, there must be some way in which energy continually enters the system to replace the energy that is lost through respiration and the many activities which use energy that occur. Where does this energy come from?

The energy comes from the Sun and is trapped by green plants in the process of photosynthesis. Plants that can photosynthesise are known as producers as they produce their own food and they in turn provide food and energy for other organisms. The **herbivores** (plant-eating animals) that feed on plants are known as **primary** consumers and the **carnivores** (animals that eat other animals) that feed on the primary consumers are known as **secondary consumers**. Animals that feed on secondary consumers are **tertiary consumers** and so on.

The sequence of producers trapping the Sun's energy and this energy then passing into other organisms as they feed is known as energy flow.

The different stages in the feeding sequence can also be referred to as trophic levels. Producers occur in trophic level 1 and primary consumers are trophic level 2.

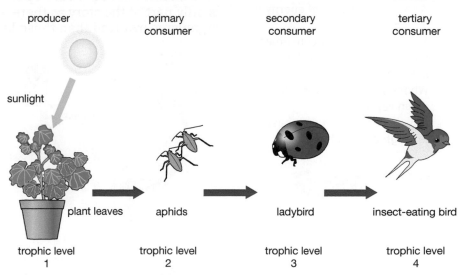

▲ **Figure 7.9** Energy and feeding relationships

Food chains and food webs

Figure 7.9 shows a sequence or chain of living things through which energy passes. It is an example of a food chain. Food chains show the feeding relationships which transfer substances, including carbon and nitrogen, as well as energy between several organisms (represented by the arrows). Examples of some other food chains are shown in Figure 7.10.

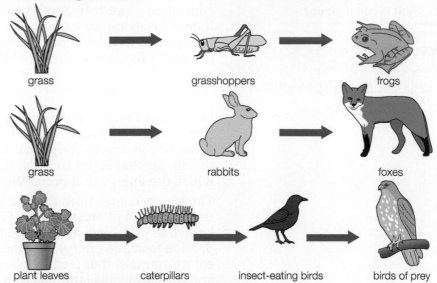

▲ **Figure 7.10** Some common food chains

These examples show that in all food chains the first organism is the producer and it provides food and energy for primary consumers. Of course, food chains are very simplistic in that they do not show the complex interactions that usually exist. In reality very few animals have only one food source.

A food web shows how a number of food chains are interlinked and gives a much more realistic picture. Figure 7.11 shows how the food chains above are built up into a food web. The food web shown is only part of the story as there will still be many more links and organisms involved than those listed.

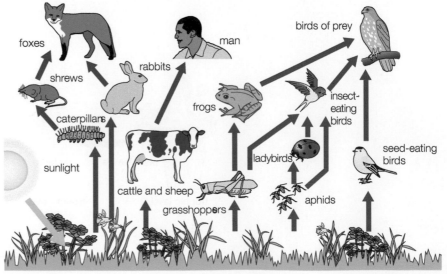

▲ **Figure 7.11** A grassland food web

Test yourself

9 What are trophic levels?
10 What is the source of energy in a food chain?
11 Name **one** primary consumer in the grassland food web shown in Figure 7.11.
12 What do the arrows in a food chain represent?
13 How is energy lost as it passes through a food chain?

Show you can

Explain fully why producers are important to the animals in a food chain.

Show you can

Extract from Figure 7.11 a food chain involving five different organisms.

Nutrient cycles

We have already noticed that energy flows through food chains and webs as part of the feeding process. Figure 7.11 shows that energy must continually enter the system from sunlight as it is lost from all living organisms during the process of respiration. This is why we use the term energy flow.

If we look at the flow of **nutrients** (like carbon and nitrogen) in more detail, we see that it differs from the flow of energy in important ways. In a stable ecosystem, the overall gain or loss of nutrients from the system will be small and, unlike energy, the nutrients can be recycled as part of a nutrient cycle.

Nutrient cycling involves the process of **decay and decomposition**. For recycling to take place, dead organisms must first be broken down during the decay process. Many organisms such as earthworms, woodlice, and various types of insects are involved in the initial stages of breaking down dead organisms into small pieces. **Fungi** and bacteria are the decomposers that break down the organic compounds into their simplest components, which plants can absorb and use again.

These decomposing bacteria and fungi have a special way of breaking down dead organic matter, and so are described as saprophytic. They **secrete enzymes** into the soil or dead organism. The enzymes break down the organic material and it is then absorbed by the bacteria or fungi. When digestion happens like this, outside cells, it is described as extracellular digestion. Humus is the organic content of the soil formed from decomposing plant and animal material. Decomposition takes place more quickly when conditions are optimum.

These include:

▶ a warm temperature
▶ adequate moisture
▶ a large surface area in the decomposing organism.

Figure 7.12 shows one method that can be used to investigate these optimum conditions.

The carbon cycle

The carbon cycle is an example of a very important nutrient cycle. Carbon is an essential element in every living organism. For example, proteins, carbohydrates and fats all contain carbon. The carbon cycle involves the exchange of carbon between living organisms, but also includes transfer between these organisms and carbon dioxide in the atmosphere. The main processes in the carbon cycle are:

▶ **Photosynthesis** – carbon dioxide is taken in by plants and built up into sugar, starch and other organic compounds.
▶ **Feeding** – animals eat the plants (or other animals) and the carbon is built up into other organic compounds that can be transferred further along the food chain.
▶ **Respiration** – when plants, animals and decomposers respire they return carbon compounds to the atmosphere as carbon dioxide (a form of excretion).
▶ **Decomposition** – carbon compounds in dead organisms and from egestion (for example in faeces) are broken down into simpler products. As the decomposers break them down, they respire and release carbon dioxide into the atmosphere.

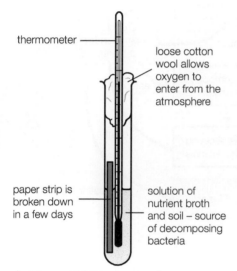

thermometer

loose cotton wool allows oxygen to enter from the atmosphere

paper strip is broken down in a few days

solution of nutrient broth and soil – source of decomposing bacteria

▲ **Figure 7.12** Investigating decomposition

▶ **Fossilisation** – when some plants and animals die, the environmental conditions are not favourable for the process of decomposition (for example the waterlogged, acid conditions of a bog). The dead organisms do not decay or do so very slowly and thus accumulate and are preserved in large quantities (the peat in a bog). Further changes can happen to this fossil material over millions of years producing fossil fuels (peat, lignite, coal, oil and gas).

▶ **Combustion** – when fossil fuels are burned the carbon is returned to the atmosphere as carbon dioxide.

Figure 7.13 shows how the different processes of the carbon cycle are linked.

Tip

Photosynthesis is the only process in the carbon cycle which removes carbon dioxide from the atmosphere.

Test yourself

14 Name **two** types of organism that are decomposers.
15 Explain what is meant by the term fossilisation.

Show you can (?)

Describe saprophytic digestion.

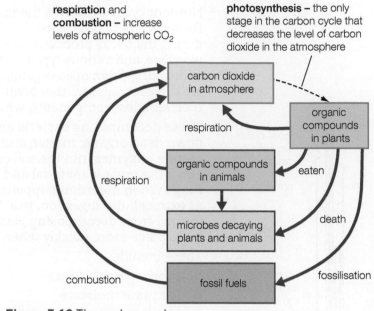

▲ **Figure 7.13** The carbon cycle

The nitrogen cycle

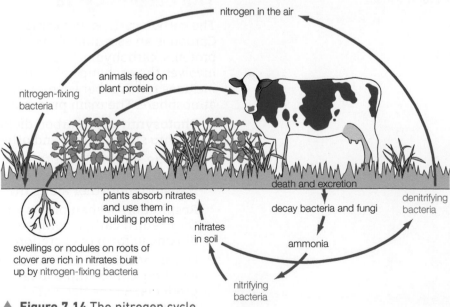

▲ **Figure 7.14** The nitrogen cycle

Most of the nitrogen in plants and animals is in the form of amino acids and protein, while most nitrogen in the environment is in the form of nitrogen gas and soluble nitrates. The ways that nitrogen moves between these different forms are shown in Figure 7.14 and can be split into three phases. These three phases are:

1 Nitrification

The build up of nitrogen into amino acids and protein in plants and animals and the eventual breakdown of these compounds into nitrates. Plants absorb nitrogen as **nitrates** and use them to make protein. As plants (and animals) are eaten, they are digested and then built up into other proteins in sequence. Eventually, the nitrogen is returned to the ground as urine or through the process of death and decay. During decay, bacteria and fungi break down the proteins to release ammonia. A second very important group of bacteria, nitrifying bacteria, convert the ammonia or ammonium compounds into nitrates (**nitrification**) and the cycle can continue.

2 Nitrogen fixing bacteria

Nitrogen fixing bacteria are a special group of bacteria that can convert *nitrogen gas into nitrates.* These bacteria can be found in the soil or frequently in small swellings (root nodules) in the roots of a particular group of plants called legumes. Legumes include peas, beans and clover. The relationship between the legumes and the bacteria is complex, but an important one in which both partners benefit. The bacteria gain carbohydrates from the legumes and they in turn gain a ready source of nitrates for the benefit of the plants. The process of converting nitrogen from the atmosphere into nitrates is called nitrogen fixation and can also be carried out by other bacteria in the soil.

3 Denitrification bacteria

Denitrifying bacteria convert *nitrates into atmospheric nitrogen.* This is a wasteful and undesirable process. Denitrifying bacteria are **anaerobic** and are most commonly found in **waterlogged** soils. Their effect in well drained soils is much reduced. The process of converting nitrates into nitrogen gas is called denitrification.

The processes of nitrification and nitrogen fixation can be accelerated by higher temperatures and aerated soil.

Root hair cells

Root hair cells are specialised cells in the root that are adapted by having a **large surface area** (due to their finger-like shape) for the uptake of nitrates, other minerals and water.

H

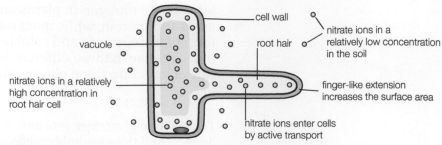

▲ **Figure 7.15** Active uptake of nitrates in a root hair cell

Figure 7.15 shows that there are more nitrate ions inside the root hair cell than outside. The nitrate ions are taken into the root hair cell by the process of active transport (uptake). This process requires **oxygen** for **aerobic respiration** to produce the energy needed to move the nitrates against the concentration gradient.

Water pollution

Water pollution can have a particularly harmful effect on our rivers, lakes and seas as they are relatively fragile environments and are easily damaged. Every so often we hear about substantial fish kills in local rivers or lakes. This may be due to harmful chemicals being accidently released from a factory, but many fish kills are the result of sewage or fertiliser runoff draining from farmland that borders the river or lake. The sewage (or slurry) and fertiliser runoff adds to the nitrate concentration in waterways. But why does the increased nitrate concentration kill the fish?

The high nitrate concentration causes aquatic plants in the water to grow much faster by providing the nitrates needed for growth. The extra nitrates have a particular effect on **algae** (microscopic organisms which live in water). The algae grow so quickly that the water surface becomes green in colour which is described as an algal bloom.

This may block the light from plants lower in the water causing them to die. The excessive growth of the surface plants uses up the nitrates and so they also die. These dead plants are then decomposed by aerobic bacteria which use up the oxygen dissolved in the water and the fish and other animals die due to lack of oxygen. This process is called eutrophication (see Figure 7.16).

This type of pollution can be reduced by increasing the environmental awareness of farmers to encourage better control of fertiliser use and more secure storage of farmyard manure and slurry.

Tip

Slurry is a liquid manure; a mixture of cattle faeces, urine and water.

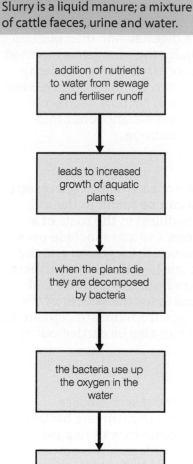

▲ **Figure 7.16** Eutrophication

Practice questions

1 Figure 7.17 shows a decayed leaf found in soil.

Figure 7.17

a) Name **one** type of microorganism which caused the leaf to decay. *(1 mark)*

b) Give **one** way leaf decay improves the soil. *(1 mark)*

2 a) Figure 7.18 shows a food web from a lake.

   i) Name the producer and a secondary consumer in the web. *(2 marks)*

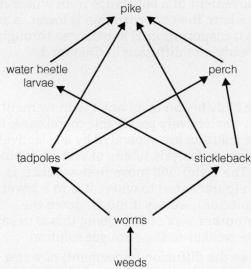

Figure 7.18

   ii) Copy and complete the boxes to show a food chain from the web. *(2 marks)*

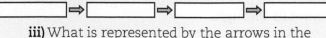

   iii) What is represented by the arrows in the food web? *(1 mark)*

   iv) Explain why producers are important in a food web. *(2 marks)*

   v) Explain what may happen to the stickleback population if pike are removed from the lake. *(2 marks)*

3 Figure 7.19 summarises what happens when large amounts of nitrates pass into a lake.

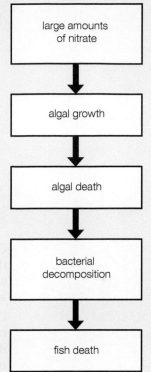

Figure 7.19

a) Suggest **two** sources of the nitrates in the lake. *(2 marks)*

b) Explain why bacterial decomposition in the lake causes the death of fish. *(2 marks)*

c) Name the process summarised in Figure 7.19. *(1 mark)*

8 Osmosis and plant transport

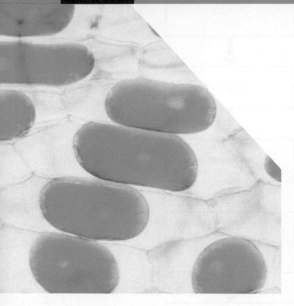

Specification points

This chapter covers sections 2.1.1 to 2.1.6 of the specification. It is about osmosis, plasmolysis, turgidity, transpiration and the potometer, and the prescribed practical activities that demonstrate these processes.

The movement of substances into and out of cells

Cell membranes allow some substances to pass through, but prevent the movement of others; they are selectively permeable. Plant cells are also surrounded by a cell wall. The cell wall is totally permeable and has no role in controlling what enters or leaves the cell.

Diffusion, the random movement of a substance from where it is in high concentration to where the concentration is lower, is an important process in the transportation of substances through cell membranes. We have already met diffusion in Chapter 2.

Osmosis

Osmosis is a special type of diffusion involving the movement of water molecules through a selectively permeable membrane. If pure water and a sucrose solution are separated by a selectively permeable membrane (such as dialysis tubing or Visking tubing) then osmosis will occur. The water will move from where it is in a higher concentration (pure water) to where it is in a lower concentration (sucrose solution); we say it moves down the concentration gradient. Another way of describing this is to say that the water moves from the weaker to the stronger solution.

Osmosis can be defined as the diffusion (movement) of water from a dilute solution to a more concentrated solution through a selectively permeable membrane.

Figure 8.1 shows how, in osmosis, water molecules move through the selectively permeable membrane but other larger molecules, such as sucrose, cannot fit through.

Tip

A selectively permeable membrane allows some substances through but not others. In osmosis, water can pass through the membrane because its molecules are small, but the sucrose cannot pass through as its molecules are large.

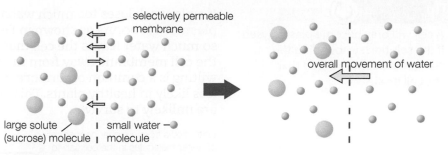

A concentration gradient exists across a selectively permeable membrane. The water molecules can move in any direction across the selectively permeable membrane. There is a higher concentration of water on the right side of the membrane so there will be a net movement of water to the left.

The water molecules have moved through the selectively permeable membrane until there is the same concentration on each side.

▲ **Figure 8.1** Osmosis

Osmosis in cells

Water moves into or out of a cell depending on the concentration of the solution surrounding it.

When water moves into an animal cell, the cell increases in volume, stretching the cell membrane. If this continues, the cell membrane will eventually split and the cell will burst. This is called lysis.

When water enters a plant cell, as shown in Figure 8.2, the vacuole increases in size, pushing the cell membrane against the cell wall. The force of the membrane pushing against the stiff cell wall increases the pressure in the cell, making it firm or turgid. This turgor pressure created by the cell wall prevents too much water from entering and thus stops the cell from bursting, as happens in animal cells. It also gives plant cells support and in non-woody plants this is essential in keeping the plant upright.

Tip

For a plant cell to be turgid the cell must have a more concentrated solution than that surrounding the cell.

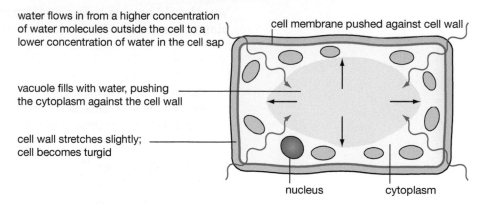

water flows in from a higher concentration of water molecules outside the cell to a lower concentration of water in the cell sap

cell membrane pushed against cell wall

vacuole fills with water, pushing the cytoplasm against the cell wall

cell wall stretches slightly; cell becomes turgid

nucleus   cytoplasm

▲ **Figure 8.2** Turgor in plant cells

The importance of turgor in providing support can be seen when there is a shortage of water. When plant cells do not receive enough water, they cannot remain turgid and **wilting** occurs. Cells that are not turgid are described as being **flaccid**.

Tip

A cell will only become plasmolysed if the solution surrounding the cell is more concentrated than the cell itself.

If a plant cell loses too much water by osmosis, a condition called plasmolysis occurs as shown in Figures 8.3 and 8.4. During plasmolysis, so much water leaves the cell that the cell's contents shrinks, pulling the cell membrane away from the cell wall. Although loss of turgor and wilting is a common occurrence in many plants, plasmolysis is much less likely in healthy plants. This is just as well, as plasmolysed cells are unlikely to survive.

water flows from a higher concentration of water molecules in the cell sap to a lower concentration outside the cell

cell membrane pulls away from the cell wall as the vacuole loses water

cell wall becomes soft; cell becomes plasmolysed

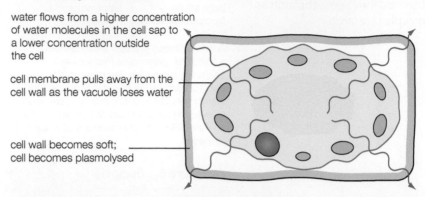

▲ **Figure 8.3** Plasmolysis in plant cells

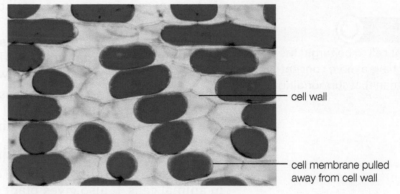

cell wall

cell membrane pulled away from cell wall

▲ **Figure 8.4** Plasmolysis in red onion

You should be able to use a microscope to observe turgid and plasmolysed cells. Revise how to use a microscope in Chapter 1.

Test yourself

1 Define the term osmosis.
2 Explain the difference between osmosis and diffusion.
3 Explain the importance of the plant cell wall in turgor.

Show you can

Explain why it is important that the cells in plant roots have a lower concentration of water molecules than the surrounding soil water.

Prescribed practical

Prescribed practical B5: Investigate the process of osmosis by measuring the change in length or mass of plant tissue or model cells using Visking tubing

Investigating osmosis by measuring the change in mass of plant tissue in solutions of different concentrations

Procedure

1 You will be given a range of sucrose solutions of different concentrations, such as 5%, 10%, 15% and 20%.

2 Set up and label a number of beakers, each with a different sucrose solution. There should also be a beaker containing water only.

3 Using a cork borer, cut five potato cylinders.

4 Weigh each cylinder of potato and add one cylinder to each beaker.

5 Leave the beakers for at least an hour.

6 Pat dry and reweigh the potato cylinders.

7 Copy and complete Table 8.1 with your results.

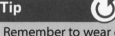

Tip

Remember to wear eye protection when carrying out this practical.

Table 8.1

Percentage concentration of sucrose in beaker/%	Mass of potato cylinder/g			
	Initial	Final (1 hour later)	Change	Percentage change/%
0 (water)				
5				
10				
15				
20				

8 Draw a graph of percentage change in mass against concentration of sucrose using axes similar to the outline drawn in Figure 8.5. When drawing this graph you should use a best-fit line (not a point-to-point graph).

9 Describe and explain your results.

In this investigation you need to be able to calculate percentage change. An example of how to do this is shown below.

Example

The mass of a potato cylinder changed from 11.45 g to 9.04 g. Calculate the percentage change.

Answer

11.45 − 9.04 = 2.41 (decrease)

2.41/11.45 × 100 = −21.05%

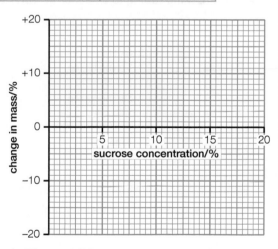

▲ **Figure 8.5** Investigating percentage change in mass of potato cylinders in a range of sucrose solutions

Questions and sample data

1 Explain how you would use your graph to find the concentration of sucrose that has the same concentration as the potato.

2 Suggest why you should use a best-fit line, rather than point to point, in this investigation.

You can complete a similar investigation by measuring the length of the potato cylinders rather than mass.

Table 8.2 shows the results of an investigation measuring length of potato cylinders in different concentrations.

Table 8.2

Concentration of sucrose solution/%	Length of potato cylinder/mm			
	At start	24 hours later	Change	% Change
water	51	57	+6	11.7
5% sucrose	50	53	+3	6
10% sucrose	48	46	−2	−4.2
15% sucrose	55	50	−5	−9.1

3 Using the data in the table, estimate the concentration of sucrose that could have the same concentration as the potato.

4 Explain why it is important in this investigation to use percentage change rather than just change in length.

5 Suggest **two** reasons why it is more accurate to use mass rather than length of potato cylinders in this investigation.

Investigating osmosis using Visking tubing

Visking tubing is selectively permeable and can therefore be used to model the movement of substances through the cell membrane.

Procedure

1 Add 5% sucrose solution to a section of Visking tubing, ensuring the tubing is tied securely at each end.

2 Dry the outside of the tubing if necessary and weigh the tubing and contents.

3 Add the Visking tubing to a beaker of water and leave for at least 1 hour.

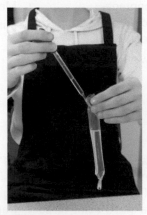

▲ **Figure 8.6** Adding sucrose solution to Visking tubing

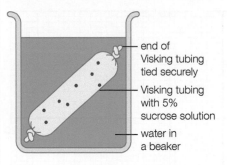

- end of Visking tubing tied securely
- Visking tubing with 5% sucrose solution
- water in a beaker

▲ **Figure 8.7** Using Visking tubing in an osmosis investigation

4 Pat dry the outside of the Visking tubing and reweigh.

5 Describe and explain the results obtained.

Tip

It is very important that the Visking tubing is tied tightly at both ends in this investigation.

Tips

▶ The general principles involving water movement in the three investigations described above are the same: the water will move from a dilute to a more concentrated solution.

▶ In the investigations with potato, the cell membranes of the potato cells are selectively permeable membranes. In the Visking tubing investigation, the Visking tubing itself is selectively permeable.

Transpiration

Much of the water that enters the leaves of a plant evaporates into the atmosphere. This loss of water by evaporation is called transpiration. Transpiration takes place in the **spongy mesophyll** cells in plant leaves, through the air spaces and out of the stomata (small pores).

Transpiration is defined as evaporation from mesophyll cells followed by diffusion through air spaces and stomata.

This continuous movement of water through a plant (the **transpiration stream**) is very important for four reasons:

1 The supply of water to the leaves acts as a raw material for photosynthesis.
2 The movement of water transports minerals through the roots and up the stem to the leaves and other parts of the plant.
3 As water passes through the plant, it enters cells by osmosis to provide support through turgor.
4 Providing water for the process of transpiration itself.

Plants often need to reduce water loss by transpiration. They do this by closing the stomata, which are most common on the underside of leaves. The stomata are necessary to allow gases to enter and leave the leaf, but they can be closed on occasions when it is important to conserve water.

Tip

When covering transpiration it is useful to revise the structure of the leaf; see Chapter 2.

Measuring water uptake using a potometer

The bubble potometer, (Figure 8.8), is a piece of apparatus designed to measure water uptake in a leafy shoot. As water evaporates from the leaves of the cut shoot, the shoot sucks water up through the potometer. The distance an air bubble moves in a period of time can be used to calculate the rate of water uptake.

The reservoir, or syringe, allows the apparatus to be reset so that replicate results can be recorded or the water uptake can be measured in different environmental conditions. Air leaks will hinder the uptake of water into the plant, so it is important that the potometer apparatus is properly sealed, particularly at the junction between the shoot and the neck of the potometer. To prevent the development of unwanted air bubbles in the water column entering the plant, it is necessary to assemble the apparatus under water initially.

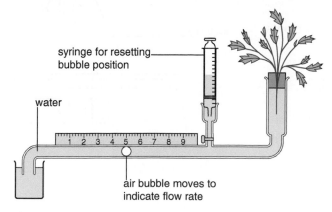

▲ **Figure 8.8** A bubble potometer

A bubble potometer can be used to measure how environmental conditions such as wind speed, temperature or humidity affect the rate of water uptake. Conditions that increase the rate of evaporation and transpiration, such as higher temperatures or higher wind speed (through using a fan), will increase the rate of water uptake. Higher levels of humidity, created through covering the shoot with a polythene bag, will reduce evaporation and transpiration and therefore slow the rate of water uptake by the shoot.

Obviously, when changing a particular environmental factor (such as wind speed) to show how it affects water uptake, it is necessary to keep other environmental factors constant, including temperature, humidity and light. This ensures that the results are **valid**. Replicated results will help ensure that the results are **reliable**.

Leaf surface area, while not an environmental factor, will also affect transpiration rates. Larger leaves will have more stomata, through which water can evaporate and diffuse. You should be able to plan an experiment to investigate the effect of leaf surface area on transpiration rate.

The potometer shown in Figure 8.9, can accurately measure the volume of water taken up by the shoot, but it cannot give an absolute value for transpiration itself. It is impossible to calculate how much of the water taken up by the plant is actually transpired through the leaf surface. Some of the water will be used in photosynthesis and in providing support through turgor, so the volume transpired will inevitably be less than the volume taken into the shoot. However, it is an excellent method for comparing rates in different conditions.

Transpiration rates can also be measured using a weighing method, by weighing how much water a plant loses over a period of time.

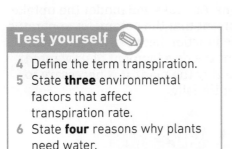

Tip

If you are asked to give three environmental factors that affect transpiration, give wind speed, temperature and humidity. Do not give leaf surface area, as this is not an environmental factor.

Test yourself ✏️

4 Define the term transpiration.
5 State **three** environmental factors that affect transpiration rate.
6 State **four** reasons why plants need water.

Show you can ❓

Explain why the rate of transpiration in a plant and water uptake are closely linked, but will not be exactly the same.

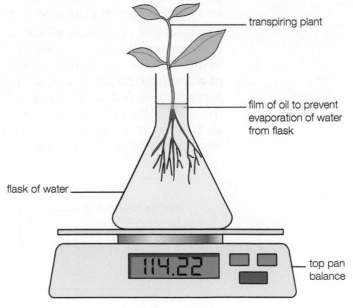

transpiring plant

film of oil to prevent evaporation of water from flask

flask of water

114.22

top pan balance

▲ **Figure 8.9** A weight potometer

Prescribed practical

Prescribed practical B6: Use a potometer (bubble and weight potometer) to investigate the factors affecting the rate of water uptake by a plant and washing line method to investigate the factors affecting the rate of water loss from leaves.

Using the bubble potometer to investigate the factors affecting the rate of water uptake by a plant shoot

Procedure

1 Set up a bubble potometer by attaching a shoot to the neck of the apparatus. Take the precautions to prevent air leaks as described earlier.

2 Calculate the rate of movement of the bubble over a set time for a particular environmental condition, such as still air.

3 Repeat to obtain more reliable results.

4 Adjust the environmental conditions. For example, use a fan to create windy conditions.

5 Repeat steps 2 and 3 above.

6 Record your results in a table similar to Table 8.3.

Table 8.3 Using a bubble potometer to compare transpiration rates

Environmental condition	Position of bubble/mm		Difference in bubble position/mm	Time/minutes	Rate of movement/ mm/minute
	At start	At end			
no wind (still air)					
windy					

Similar investigations can be carried out using a weight potometer.
When using a weight potometer, a table similar to Table 8.4 can be used.

Table 8.4 Using a weight potometer to compare transpiration rates

Environmental condition	Mass of beaker and shoot/g		Difference in mass/g	Time/hours	Rate of change of mass/g/hour
	At start	At end			
no wind (still air)					
windy					

Questions and sample data

When comparing transpiration rates in different environmental conditions, it is not always necessary to calculate the 'rate' of water loss. Table 8.5 shows typical results for using a weight potometer to compare water loss (transpiration rate) in two different humidity levels.

Table 8.5 Comparing transpiration rates in different humidity levels

Environmental condition	Mass of beaker and shoot/g			
	At start/g	After 24 hours/g	Change/g	Percentage change/%
'normal' humidity	114.22	108.47	5.75	5.03
high humidity (shoot covered with clear plastic bag)	121.56	118.02	3.54	2.91

1 Suggest why a top pan balance accurate to two decimal points was used in this investigation.

2 Describe and explain the difference in water loss in the two conditions.

Using the washing line method to investigate the factors affecting the rate of water loss from plant leaves

Procedure

1 Detach six leaves from a tree.

2 Smear petroleum jelly over the cut stalks to make them waterproof.

3 Measure the mass of each leaf and then, using paper clips, hang them on a line of string suspended between two retort stands.

4 Suspend half the leaves on a line of string at a high temperature (such as 30 °C) and the other half at a lower temperature (such as 20 °C), in order to compare the effect of temperature on water loss by a leaf.

5 After 24 hours, reweigh the leaves and calculate the average loss of mass for each environmental condition.

6 Describe and explain the results obtained.

Practice questions

1 a) Define the term osmosis. *(1 mark)*
b) Explain the role of osmosis in providing support in plants. *(3 marks)*

2 Figure 8.10 shows Visking tubing containing 5% sucrose solution surrounded by a solution of 10% sucrose in a beaker.

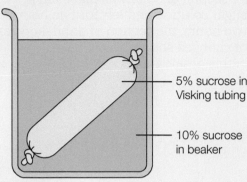

5% sucrose in Visking tubing

10% sucrose in beaker

Figure 8.10

a) Suggest how the appearance of the Visking tubing will change after 24 hours. Explain your answer. *(3 marks)*
b) A student carried out a similar investigation, but she found that there was no change in her Visking tubing after 24 hours. Suggest an explanation for this. *(2 marks)*

3 A farmer put too much fertiliser onto part of a field. A few days later many of the cells in the roots of his plants had become plasmolysed. Suggest why this occurred. *(2 marks)*

4 Some students were asked to investigate whether more water was lost by transpiration through the upper or the lower surface of leaves. They collected 20 leaves and separated them into two groups. They covered the upper surface of the first group and the lower surface of the second group in petroleum jelly. The leaves were weighed on a top pan balance and then attached to a line running between two retort stands with paper clips.
The leaves were reweighed 24 hours later. A summary of the results is recorded in Table 8.6.

Table 8.6

Leaf treatment	Average mass/g		Percentage change after 24 hours
	At start	After 24 hours on line	
upper surface covered in petroleum jelly	4.22	3.37	20.14
lower surface covered in petroleum jelly	4.68	4.26	8.97

a) Describe and explain the results in the table. *(2 marks)*
b) Why were 10 leaves used for each treatment? *(1 mark)*
c) State **two** variables that should have been controlled in this investigation. *(2 marks)*
d) State **one** procedure that is necessary when setting up the investigation to ensure valid results. *(1 mark)*

5 a) Briefly describe **two** ways you would try to avoid air leaks when setting up a bubble potometer. *(2 marks)*
b) A student carried out an investigation into the effect of wind speed on transpiration rate using a bubble potometer.
i) Suggest how the wind speed could be changed in this investigation. *(1 mark)*
Figure 8.11 shows how the distance the bubble moved in 10 minutes changed with different wind speeds.

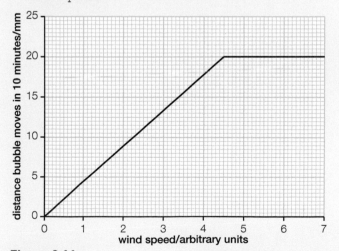

Figure 8.11

ii) Describe and explain these results. *(4 marks)*
iii) State **two** variables that should have been controlled in this investigation. *(2 marks)*
iv) Give **one** advantage in using a weight potometer, compared with a bubble potometer. *(1 mark)*

6 Dandelion stalks have two layers: an outer layer that is largely impermeable to water and an inner layer that is permeable to water. A small section of dandelion stalk was removed, as shown in Figure 8.12, and placed in a beaker of water.

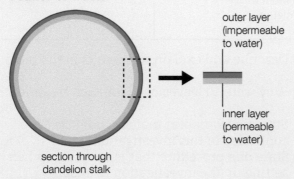

section through
dandelion stalk

Figure 8.12

After two hours in the water, the section of stalk had changed, as shown in Figure 8.13.

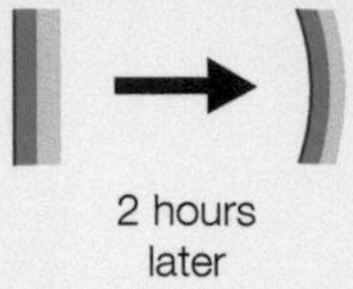

2 hours
later

Figure 8.13

a) Describe and explain the result shown. *(3 marks)*
b) Another section of stalk was placed in a weak sugar solution and left for two hours. After two hours, the section of dandelion stalk had not changed shape. Explain this observation. *(2 marks)*

9 The circulatory system

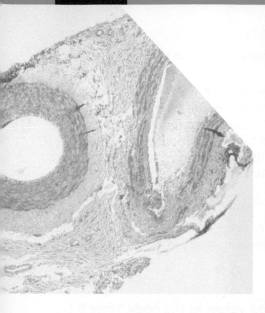

The circulatory system

The circulatory system has three main components: the blood, the blood vessels that carry the blood and the heart that pumps the blood. It has two main functions: transport of blood components and other substances in the blood and protection against disease.

The components of the blood

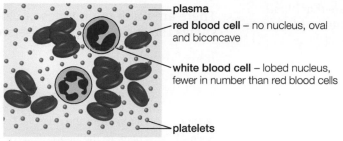

plasma

red blood cell – no nucleus, oval and biconcave

white blood cell – lobed nucleus, fewer in number than red blood cells

platelets

▲ **Figure 9.1** The main components of blood

The main components of blood are as follows.

▶ Red blood cells – the function of these cells is to carry oxygen around the body.
Red blood cells are highly specialised for this function. They:
  1 contain haemoglobin, which is rich in iron, that carries the oxygen
  2 have a biconcave shape that provides a large surface area for diffusion of oxygen
  3 have no nucleus and therefore have more space to pack in the haemoglobin.

▶ White blood cells – the blood contains two types of white blood cell and both are important in defence against disease.
Lymphocytes produce antibodies and phagocytes engulf and digest microorganisms in a process called phagocytosis.

▶ Platelets – these very small structures are important in blood clotting and the formation of scabs. The platelets work by converting the protein fibrinogen to fibrin. The fibrin forms a mesh network that traps other blood components.

▶ Plasma – this is the liquid part of the blood. The plasma is responsible for the transport of the blood cells, absorbed food molecules (for example glucose and amino acids), carbon dioxide, hormones and urea.

> **Tip**
>
> You will learn more about the role of white blood cells in defence in Chapter 13.

You should examine a blood smear using a microscope. Figure 9.2 shows a typical blood smear as viewed using a microscope.

Salts and other chemicals in the plasma keep its concentration stable and at a concentration similar to the blood cells. This is important because if red blood cells are placed in water they will take in water by osmosis and burst in a process called cell lysis.

> **Tip**
>
> Red blood cells (like all animal cells) do not have a cell wall, so there is nothing to stop them swelling and bursting if placed in water or a more dilute solution.

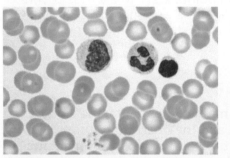

▲ **Figure 9.2** A human blood smear

> **Tip**
>
> You should be aware that blood contains many more red blood cells than white blood cells.

The blood vessels

There are three main types of blood vessel in the body. Table 9.1 shows the main differences between arteries, veins and capillaries.

Table 9.1 Blood vessels

Vessel	Direction of blood flow	Thickness of wall	Blood pressure	Valves
artery	away from the heart	thick	high	none
vein	back to the heart	relatively thin	low	yes
capillary	from arteries to veins	one cell thick	low	none

Figure 9.3 represents the three types of blood vessel in cross section.

> **Tip**
>
> The lumen is the space through which the blood flows.

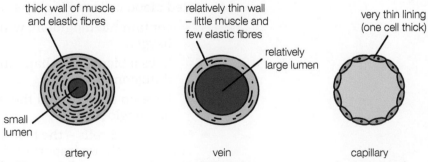

▲ **Figure 9.3** An artery, vein and capillary in cross section (not to scale)

The structure of the blood vessels is closely linked to their function:

▶ Arteries normally carry oxygenated blood (rich in oxygen) to the body organs and veins normally carry deoxygenated blood (blood with little oxygen present) back to the heart. The arteries have relatively thick walls as they carry blood at high pressure; the veins have relatively thin walls as they carry blood at low pressure. The walls of the arteries have muscles and elastic fibres which expand as the blood is pulsed through (the elastic fibres give the strength to prevent bursting). After the high pressure pulse, the elastic fibres recoil and muscles contract returning the wall to its original size while maintaining some pressure in the blood. The overall effect is the smoothing out of the blood flow.

▶ By the time the blood reaches the veins the pressure is low and there is no pulse so the walls need few elastic fibres and muscles. However, valves are necessary to keep the flow in the right direction (unidirectional), preventing backflow. The relatively large lumen of veins reduces friction and further aids the movement of blood.

▶ Diffusion of oxygen, carbon dioxide, dissolved food and urea takes place between the capillaries and the body cells or *vice versa*. The one cell thick walls of the capillaries are thin enough to be permeable and allow diffusion to take place.

artery – note the thick muscular wall and relatively small lumen

vein – note the relatively thin wall and large lumen

▲ **Figure 9.4** A vein and an artery in cross section

Show you can

Explain why veins have much less muscle and fewer elastic fibres than arteries.

Test yourself

1 Describe how platelets work.
2 Describe and explain **two** adaptations of red blood cells.
3 Give **one** advantage of veins having a large lumen.

You should be aware of the functions of the blood vessels shown in Figure 9.5.

Table 9.2 summaries these functions.

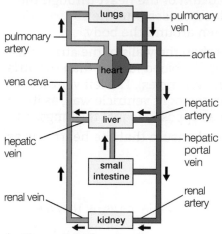

▲ **Figure 9.5** The circulatory system

Table 9.2 The functions of the main blood vessels

Blood vessel	Function
pulmonary vein	carries oxygenated blood from the lungs to the heart
aorta	carries oxygenated blood from the heart around the body
vena cava	carries deoxygenated blood from the body back to the heart
pulmonary artery	carries deoxygenated blood from the heart to the lungs
hepatic artery	carries oxygen and glucose to the liver
hepatic portal vein	carries digested food (for example glucose and amino acids) from the small intestine to the liver
hepatic vein	carries glucose and amino acids from the liver around the body; also carbon dioxide (from respiration) back to the heart and lungs
renal artery	carries blood rich in urea to the kidneys for excretion
renal vein	carries purified (low in urea) blood from the kidney to the vena cava; also carbon dioxide (from respiration) back to the heart and lungs

Tip

Figure 9.5 shows that the pulmonary artery and the pulmonary vein are exceptions to the usual rule in that the pulmonary artery carries deoxygenated blood and the pulmonary vein carries oxygenated blood.

The heart

The **heart** pumps the blood to the lungs and around the body. Figure 9.6 shows why the heart has two sides – the right side pumps the blood to the lungs and the left side pumps the blood that has returned from the lungs around the body.

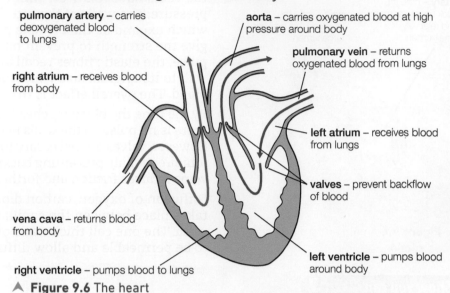

pulmonary artery – carries deoxygenated blood to lungs

aorta – carries oxygenated blood at high pressure around body

pulmonary vein – returns oxygenated blood from lungs

right atrium – receives blood from body

left atrium – receives blood from lungs

valves – prevent backflow of blood

vena cava – returns blood from body

left ventricle – pumps blood around body

right ventricle – pumps blood to lungs

▲ **Figure 9.6** The heart

The right atrium receives deoxygenated blood from the body. This passes into the right ventricle where it is pumped out in the pulmonary artery to the lungs. In the lungs, the blood becomes oxygenated and returns to the left atrium of the heart through the pulmonary vein. The oxygen rich blood passes into the left ventricle and is pumped into the aorta and then around the body.

The walls of the ventricles are thicker than the walls of the atria – this is because the ventricles have to pump the blood further than the atria (the atria only pump the blood into the ventricles). The left ventricle wall is thicker (has more muscle) than the right ventricle wall, as it pumps blood around the whole body as opposed to just the lungs.

The heart valves prevent backflow and ensure that the heart acts as a unidirectional pump.

Tip

The heart valves are similar in structure to the valves in the veins. They are 'flap-like' and will only open one way.

Tip

You need to be able to explain why the walls of the ventricles are thicker than the walls of the atria and also why the wall of the left ventricle is thicker than the right ventricle.

Humans (and other mammals) have a double circulatory system; one circulation system to and from the lungs and one to and from the rest of the body (see Figure 9.7). This means that the blood travels through the heart twice in one circulation of the body.

Figure 9.7 summarises double circulation.

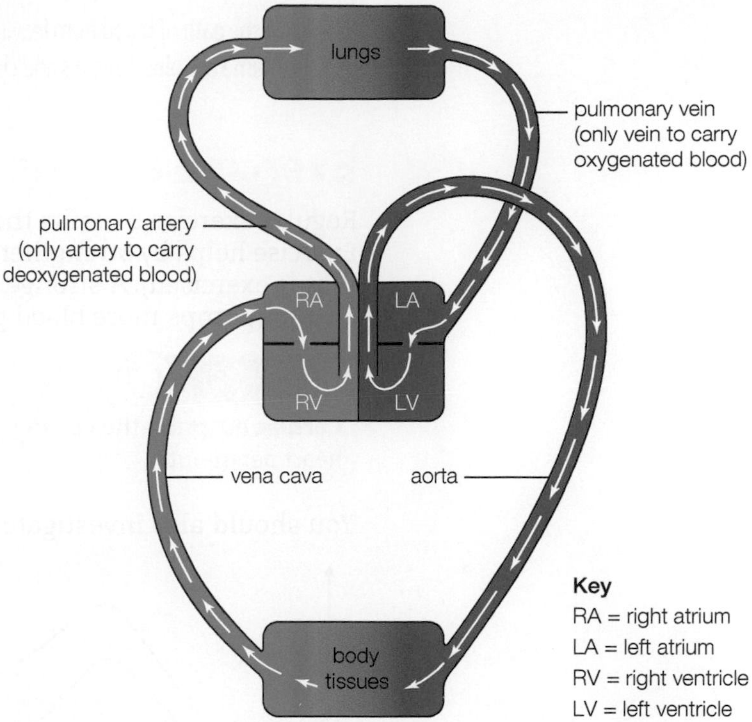

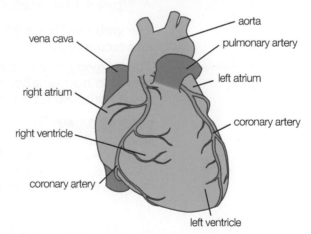

pulmonary vein
(only vein to carry
oxygenated blood)

pulmonary artery
(only artery to carry
deoxygenated blood)

lungs

RA   LA

RV   LV

vena cava   aorta

body
tissues

Key
RA = right atrium
LA = left atrium
RV = right ventricle
LV = left ventricle

▲ **Figure 9.7** Double circulation

The heart itself receives blood from the **coronary arteries**, which branch from the aorta almost immediately after it leaves the heart. These are the fine blood vessels that can be seen running over the outer surface of the heart, shown in Figure 9.8.

vena cava

aorta

pulmonary artery

left atrium

right atrium

coronary artery

right ventricle

coronary artery

left ventricle

▲ **Figure 9.8** External view of the heart showing the coronary arteries

Test yourself 🖉

4 Describe **two** differences between the structure of the aorta and the vena cava.
5 Describe **two** differences between the blood in the aorta and in the vena cava.
6 Name **two** blood vessels that carry blood to the liver.
7 Name the **two** heart chambers that contain deoxygenated blood.

Exercise and the circulatory system

Regular exercise benefits the circulatory system in a number of ways. Exercise helps by strengthening the heart muscle (as with any muscle that is exercised). A stronger heart will have an increased cardiac output (pumps more blood per minute) even when not exercising.

Tip

Cardiac output is the volume (amount) of blood pumped by the heart per minute.

You should also investigate the effect of exercise on pulse rate.

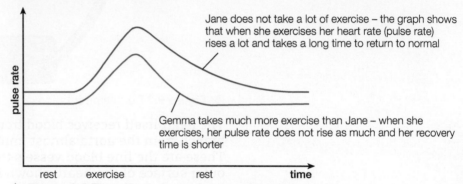

▲ **Figure 9.9** The effect of exercise on pulse rate

The graph in Figure 9.9 shows the effect of exercise on pulse rate and recovery rate.

The recovery rate is the time it takes for the pulse or heart rate to return to normal after exercise and this will usually be shorter for people who exercise regularly or play a lot of sport.

Tip

Heart rate and pulse rate will be the same – heart rate is how often the heart beats and the pulse rate how often a 'pulse' or surge of blood passes round the body – they are the same as each beat causes a new pulse.

Practice questions

1 a) Copy and complete Table 9.3. *(3 marks)*

Table 9.3 The functions of some blood components

Blood component	Function
red blood cells	
	cause blood clotting
white blood cells	

b) Name **three** things that are transported in the plasma. *(3 marks)*

2 a) Give **three** differences between the structures of arteries and veins. *(3 marks)*

b) Figure 9.10 shows how the blood pressure changes as blood flows through the circulatory system.

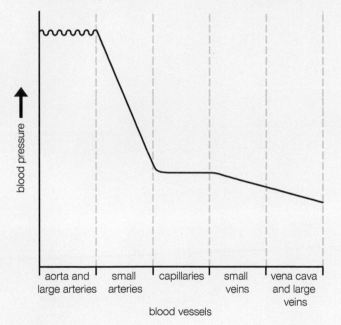

Figure 9.10

i) In which part of the circulatory system is there the largest fall in blood pressure? *(1 mark)*

ii) Suggest what causes the small ripples in blood pressure as the blood flows through the aorta and other large arteries. *(1 mark)*

iii) Suggest **one** reason why it is necessary that blood pressure is low in the capillaries. *(1 mark)*

iv) The blood pressure in the veins is low. Name **one** structure that prevents backflow of blood in the veins. *(1 mark)*

3 Humans are described as having a double circulation system.

a) Describe what is meant by a double circulation system. *(1 mark)*

b) The partially completed Figure 9.11 represents the human circulation. Copy the diagram and use arrows to complete it to show the direction of blood flow between the structures shown. *(4 marks)*

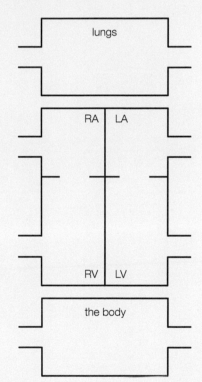

Figure 9.11

4 a) Figure 9.12 shows the effect of exercise on the pulse rate of two girls.

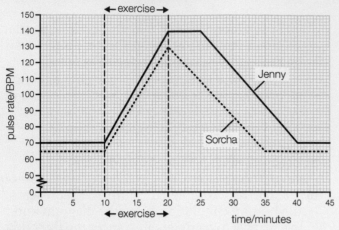

Figure 9.12

i) Calculate the increase in Sorcha's pulse rate during exercise. *(1 mark)*

ii) Calculate the percentage increase in Sorcha's pulse rate during exercise. *(1 mark)*

b) Use the graph to suggest which girl was the fitter. Explain your choice giving **two** pieces of evidence from the graph. *(2 marks)*

c) If the girls continued to exercise for a period of time, suggest the effect this would have, if any, on recovery time. *(1 mark)*

10 The genome, chromosomes, DNA and genetics

Specification points

This chapter covers sections 2.4.1 to 2.4.12 of the specification. It is about the genome, chromosomes, genes, DNA, mitosis and meiosis, genetics, genetic conditions, genetic screening and genetic engineering.

The genome, chromosomes, genes and DNA

Most living cells contain a nucleus (control centre). The nucleus is the control centre because it contains chromosomes that are subdivided into smaller sections called genes. There are hundreds of genes in each chromosome. Figure 10.1 shows a single chromosome. Normally they occur in functional pairs (except sex cells) – we will see why later in this chapter.

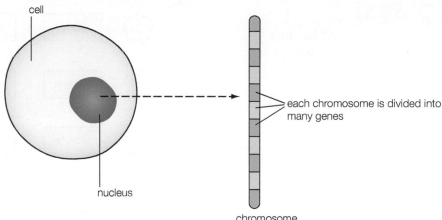

▲ **Figure 10.1** The nucleus, chromosomes and genes

It is the genes in our bodies that control characteristics such as eye and hair colour – the features that make us who we are. Therefore, genes are sections of chromosomes that code for particular characteristics. Inside genes and chromosomes there is a very important molecule that gives them their properties. This molecule is DNA (deoxyribonucleic acid). In effect, genes are short lengths of DNA that code for a particular protein or characteristic.

All the DNA in an individual is referred to as their genome.

The DNA molecule is formed in a double helix with the two outer strands linked as shown.

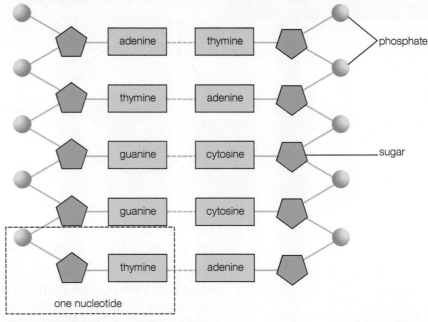

▲ **Figure 10.2** Model of DNA

Tip

It might be easier to remember this diagram if you think of a ladder with interlinking rungs that is twisted round on itself.

Tip

The genome includes all the DNA in all the genes in all the chromosomes in an individual (all the genetic material).

The structure of DNA

DNA consists of three smaller units, which are regularly repeated throughout the length of the molecule. These units are deoxyribose sugar, phosphate and bases. There are four different types of base: adenine, guanine, cytosine and thymine. In the double helix (Figure 10.2), the rungs of the 'ladder' are the bases and the sides are alternating units of deoxyribose sugar and phosphate.

Each repeating unit of DNA (consisting of a phosphate, a sugar and a base) is called a nucleotide. Figure 10.3 shows that bases link the two sides of the molecule together in such a way that adenine only combines with thymine and guanine only combines with cytosine. This arrangement is known as base pairing. The arrangement of the bases along the length of the DNA is what determines how a gene works.

Tip

The letters A, G, C and T are often used to identify the four bases. A only combines with T and C with G.

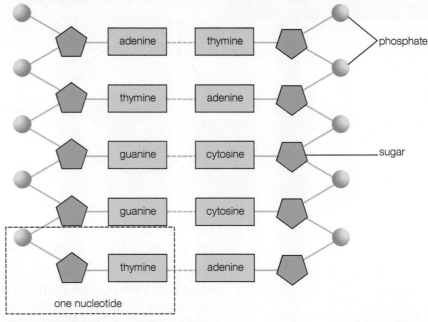

▲ **Figure 10.3** Base pairing in DNA

Test yourself

1 Define the term gene.
2 Name the four DNA bases.
3 What is meant by the unique nature of an individual's DNA?

Show you can ?

If 20% of the DNA bases in an individual are adenine, what percentage will there be of the other bases?

If we map the sequence of bases along each individual's chromosomes we will find that while there will be similarities among different individuals, no two people have the same sequence of bases along the entire length of all their chromosomes (except for identical twins.)

Tip

The sequence of bases in everyone's DNA is unique, apart from identical twins.

How does DNA work?

The DNA works by providing a code to allow the cell to make the proteins that it needs. The DNA determines which proteins, and in particular which enzymes, are made. Enzymes are extremely important proteins that control the cell's reactions. Therefore, by controlling the enzymes, the DNA controls the development of the cell and in turn the entire organism.

The bases along one side (strand) of the DNA molecule – the coding strand – form the genetic code. Each sequence of three bases (a triplet) along this coding strand codes for a particular amino acid – the building blocks of proteins. The sequence of three bases that codes for an amino acid is called a base triplet. As a protein consists of many amino acids linked together it is important that the correct base triplets are arranged in the correct sequence along the coding strand.

Figure 10.4 shows how base triplets code for particular amino acids. In Figure 10.4 the first and fourth base triplets have the same code and this means that amino acid 1 and amino acid 4 are also the same. The model only shows a small section of a gene and a small section of the protein that it produces codes for.

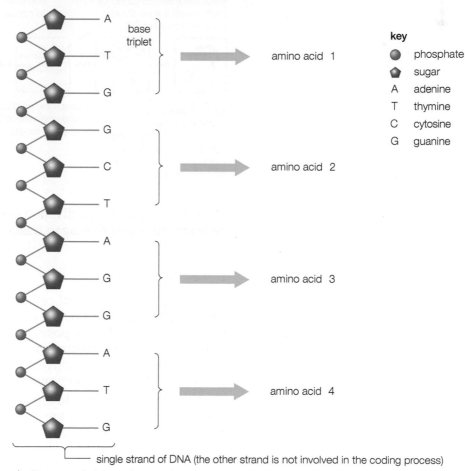

▲ **Figure 10.4** How base triplets code for amino acids

Cell division

Most living organisms grow by increasing their number of cells. Cells double in number by splitting in half. It is important when cells divide during growth that the two new cells (daughter cells) end up with exactly the same genetic makeup as each other and the parent cell – they are clones of the parent cell. This means that every cell in the growing organism has the same number and type of genes and chromosomes, and that these are the same as they were in its very first cell, the zygote. This type of cell division is called mitosis.

Mitosis

Mitosis can be defined as cell division in which the exact duplication of chromosomes takes place to produce daughter cells that are genetically identical to each other and to the parent cell.

Figure 10.5 summarises the process of mitosis and shows that the two daughter cells contain exactly the same chromosomes as each other and also as the parent cell from which they were produced.

Tip

As new (daughter) cells formed by mitosis grow to 'normal' cell size before dividing again, growth of an organism usually involves both cell division and cell growth.

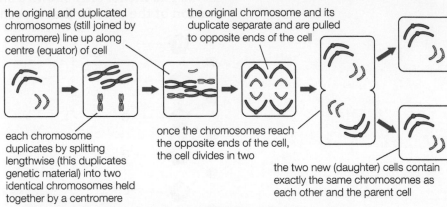

the original and duplicated chromosomes (still joined by centromere) line up along centre (equator) of cell

the original chromosome and its duplicate separate and are pulled to opposite ends of the cell

each chromosome duplicates by splitting lengthwise (this duplicates genetic material) into two identical chromosomes held together by a centromere

once the chromosomes reach the opposite ends of the cell, the cell divides in two

the two new (daughter) cells contain exactly the same chromosomes as each other and the parent cell

▲ **Figure 10.5** Mitosis (only two pairs of chromosomes are shown)

Mitosis is a type of cell division used in growth, to replace worn out cells and to repair damaged tissue.

Meiosis

Meiosis is another type of cell division. It only takes place in the sex organs (the testes and ovaries) during the production of gametes (sperm or eggs). The purpose of meiosis is to produce gametes with half the number of chromosomes of all the other (non-gamete) cells in the body. As meiosis halves the chromosome numbers in the daughter cells it is also known as reduction division.

Most human cells have 46 chromosomes, arranged in 23 pairs, but the sperm and eggs that are produced by meiosis have only 23 chromosomes. It is not just any 23 chromosomes from the 46 but one chromosome from each pair that passes into each gamete. It could be either chromosome of a particular pair that passes into a particular gamete, and as there are 23 pairs of chromosomes in total, there are millions of potential chromosome combinations in any gamete: 2^{23} possibilities.

This random **independent assortment** of chromosomes in meiosis at gamete formation gives unique gametes (the chance of any two gametes being identical is so small that it is virtually impossible) and so helps to produce variation in offspring.

Figure 10.6 summarises the role of meiosis and the random nature of fertilisation itself in producing variation in living organisms.

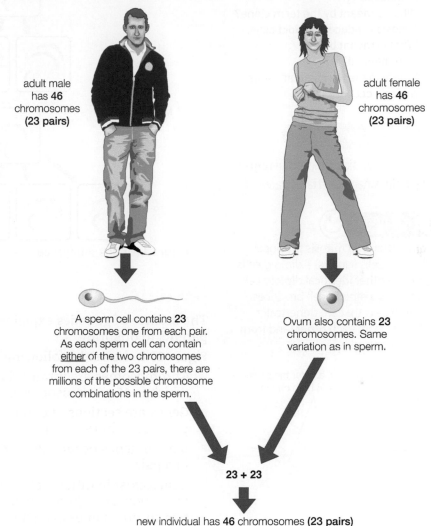

adult male has **46** chromosomes (**23 pairs**)

adult female has **46** chromosomes (**23 pairs**)

A sperm cell contains **23** chromosomes one from each pair. As each sperm cell can contain either of the two chromosomes from each of the 23 pairs, there are millions of the possible chromosome combinations in the sperm.

Ovum also contains **23** chromosomes. Same variation as in sperm.

23 + 23

new individual has **46** chromosomes (**23 pairs**)

▲ **Figure 10.6** Chromosomes in human reproduction

The chromosome number in the gametes is referred to as the haploid number (23 in humans). The normal number in an organism is called the diploid number (46 in humans). The roles of fertilisation include restoring the diploid number in the offspring and combining the different arrangements of chromosomes produced during the process of meiosis.

Show you can

Describe the function of meiosis and explain why it is important that it takes place.

Tip

Remember that mitosis produces identical cells (clones) – diploid cells produce other identical diploid cells. In meiosis a diploid cell produces haploid cells that are genetically different from each other (and from their parent cell).

the form of gene (allele), e.g. for the presence of ear lobes, is the same in both chromosomes

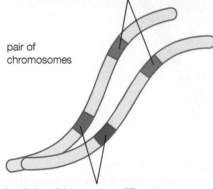

pair of chromosomes

the alleles of the gene are different, e.g. one for brown eyes and one for blue eyes

▲ **Figure 10.8** Chromosomes and genes

Tip

Remember that alleles are different forms of the same gene.

The differences between mitosis and meiosis

Figure 10.7 summarises the differences between mitosis and meiosis.

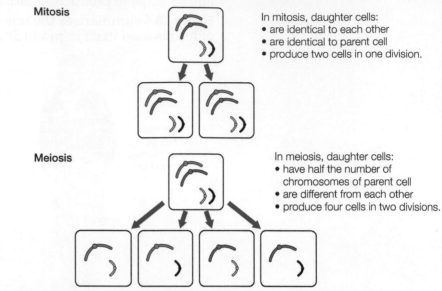

Mitosis

In mitosis, daughter cells:
• are identical to each other
• are identical to parent cell
• produce two cells in one division.

Meiosis

In meiosis, daughter cells:
• have half the number of chromosomes of parent cell
• are different from each other
• produce four cells in two divisions.

▲ **Figure 10.7** The differences between mitosis and meiosis

Genetics

The science of genetics explains how characteristics pass from parents to offspring.

Figure 10.8 shows the following.

▶ Chromosomes are arranged in pairs – humans have 23 pairs, which is 46 chromosomes in total.

▶ Genes are sections of chromosomes that carry the code for particular characteristics such as eye colour.

▶ Similar genes occupy the same position on both chromosomes in a pair.

▶ Genes exist in different forms, called alleles, and the alleles may be homozygous (the same), or heterozygous (different) on the two chromosomes of a pair.

Genetic crosses

A monohybrid genetic cross is a cross between two individuals where the genetics of one characteristic (for example height in peas or eye colour in humans) is considered.

Pea plants occur in their normal tall form or in a much shorter variety. One genetic cross that can be carried out is a cross between tall and short plants. Before carrying out this cross, the tall plants can be allowed to breed with each other for a period of time to ensure they always produce tall plants. The same can be done with the short plants by allowing only short plants to breed together until it is certain they could only produce short offspring. The parent plants to be used are then referred to as pure breeding.

When pure breeding tall plants are crossed with pure breeding short plants (the parental generation), all the plants in the first, or F_1 generation, (the offspring) are tall. However, if these F_1 plants are crossed with other F_1 plants, their offspring (the second or F_2 generation) are a mixture of tall and short plants in the ratio of 75% tall to 25% small.

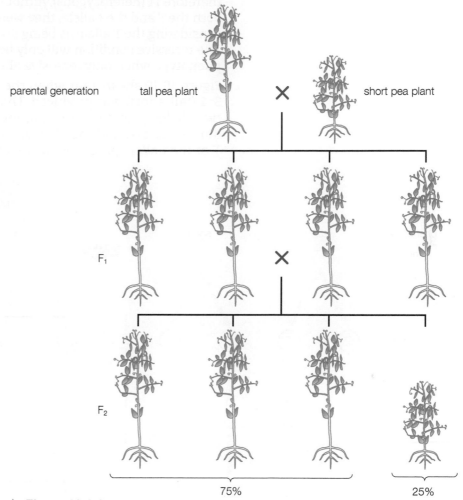

parental generation   tall pea plant   ✕   short pea plant

F_1 ✕

F_2

75%   25%

▲ **Figure 10.9** Crossing pure breeding tall and short pea plants

Explanation of this monohybrid cross

The results of this (and other) genetic crosses can be explained by understanding the gene involved – in this case, the gene that determines height. There are two forms of the height gene – a tall and a short gene. The different forms of a gene are called alleles. In the example of height in peas, there is an allele for tallness and an allele for shortness. In genetic crosses, alleles are represented by a single letter, for example T for tall and t for short.

As the parental plants were pure breeding, the tall plants only carried the tall alleles and short plants only carried the short alleles. These plants containing only one type of allele are homozygous (TT or tt). When both types of allele are present, the individual is heterozygous (Tt).

The paired symbols (TT, Tt or tt) used in genetics are referred to as the genotype and the outward appearance (tall or short) is the phenotype.

When gametes are produced, only one allele (from each gene) from each parent passes on to the offspring. This is fully explained by our understanding of meiosis, as we know that only one chromosome, and therefore one allele, of each pair can pass into a gamete.

The F_1 plants in our cross must have received one T allele from their tall parent and one t allele from the short parent. The F_1 plants were therefore Tt (heterozygous). Although all these plants contained both the T and the t allele, they were tall. This can be explained by considering the T allele as being dominant over the recessive t allele. The recessive condition will only be expressed, or visible, in the phenotype when only recessive alleles are present in the genotype (tt).

Figure 10.10 shows that when the F_1 plants were interbred a ratio of 3:1 (tall:short) was produced. This ratio was achieved because the two alleles (T and t) of one parent were produced in equal numbers during meiosis and they had an equal chance of combining with the T or the t allele produced by the other parent during fertilisation.

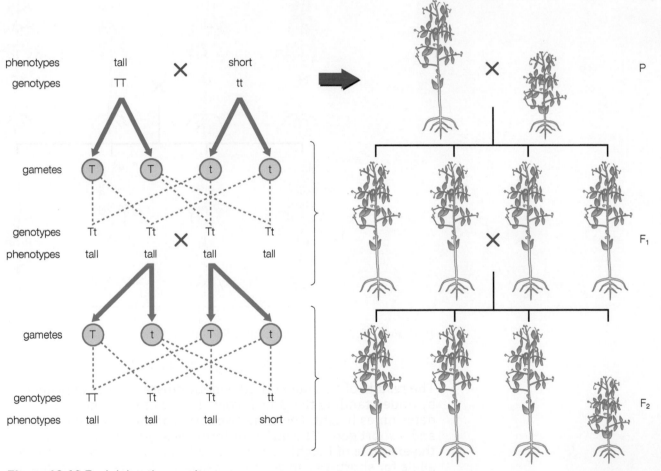

▲ **Figure 10.10** Explaining the results

When completing genetic diagrams, it is helpful to use a small grid called a Punnett square.

Figure 10.11 shows how to use a Punnett square. This is a way of setting out a genetic cross in table format. In this example, using height in peas as before, a heterozygous pea (Tt) is crossed with a homozygous recessive pea (tt). There are extra notes on the diagram to help your understanding of genetic crosses.

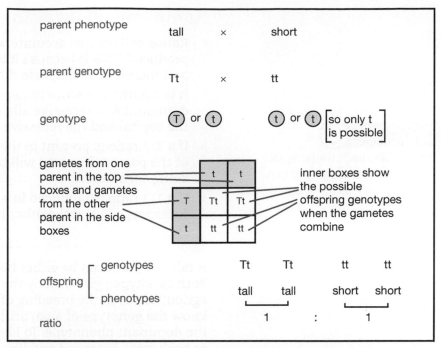

Figure 10.11 A Punnett square showing the genetics of crossing a heterozygous individual with a homozygous recessive individual

Figure 10.12 uses Punnett squares to show examples of some monohybrid crosses that can occur in humans. In these examples of eye colour, brown eye colour (B) is dominant to blue eye colour (b).

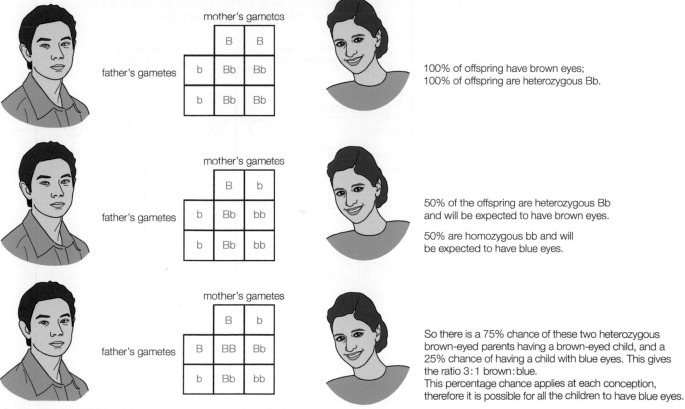

100% of offspring have brown eyes; 100% of offspring are heterozygous Bb.

50% of the offspring are heterozygous Bb and will be expected to have brown eyes.

50% are homozygous bb and will be expected to have blue eyes.

So there is a 75% chance of these two heterozygous brown-eyed parents having a brown-eyed child, and a 25% chance of having a child with blue eyes. This gives the ratio 3:1 brown:blue.
This percentage chance applies at each conception, therefore it is possible for all the children to have blue eyes.

Figure 10.12 The inheritance of eye colour in humans

Some important points about genetic crosses

▶ Ratios will only be accurate when large numbers of offspring are produced. This is because it is totally random which gametes, and therefore alleles, fuse during fertilisation.

▶ It is common practice to use the same letter for both the dominant and recessive alleles, with the dominant allele being the capital and the recessive allele written in lower case.

▶ If a 3:1 ratio is present in the offspring of a particular cross, both of the parents involved will be heterozygous for the characteristic being considered.

▶ If a 1:1 ratio is produced in a cross, one parent will be heterozygous and the other homozygous recessive.

The test cross (back cross)

A tall pea plant can be either homozygous (TT) or heterozygous (Tt). Both genotypes give exactly the same phenotype. Sometimes, in agriculture or in the breeding of domestic animals, it is important to know the genotype of a particular animal or plant that is showing the dominant phenotype. To identify the unknown genotype a test or back cross is carried out, like the one shown in Figure 10.13.

The animal or plant in question is crossed with a homozygous recessive individual. If the offspring are produced in sufficient numbers it is possible to identify the unknown genotype.

> **Tip**
>
> It is useful to learn the parental genotypes that produce 3:1 and 1:1 offspring ratios as these genetic ratios are common in exam questions.

H

> **Tip**
>
> In a test cross if any offspring are homozygous recessive then the unknown parent must have been heterozygous.

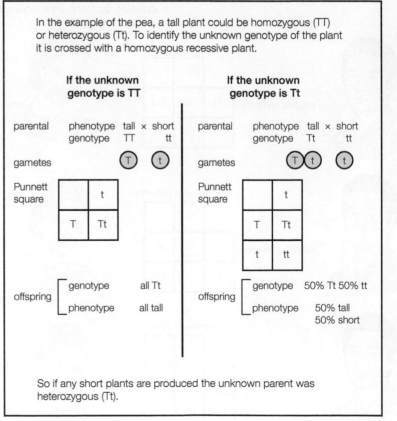

▲ **Figure 10.13** The test cross

Pedigree diagrams

A pedigree diagram shows the way in which a genetic condition is inherited in a family or group of biologically related people. Figure 10.14 is an example of a pedigree diagram showing how the condition cystic fibrosis is inherited. Cystic fibrosis is a medical condition caused by having two recessive alleles of a particular gene.

In this diagram, one of the grandchildren (7) has cystic fibrosis. It is possible to use the information provided to work out the probability of other children having the condition. Genetic counsellors often construct pedigree diagrams and use them to advise parents who have a genetic condition or who may be carriers.

Pedigree diagrams can be used in any type of genetic cross but they are obviously very valuable in tracing and predicting harmful genetic conditions.

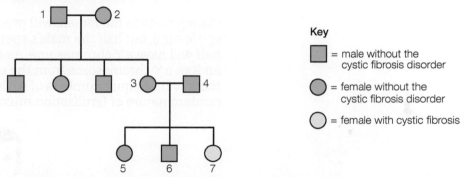

Key

■ = male without the cystic fibrosis disorder

● = female without the cystic fibrosis disorder

○ = female with cystic fibrosis

▲ **Figure 10.14** Pedigree diagram showing the inheritance of cystic fibrosis

Test yourself

Use Figure 10.14 and your knowledge to answer the following questions.
Let F = normal allele; f = cystic fibrosis allele (note: the normal allele is dominant and the cystic fibrosis allele recessive).

8 What is the genotype of the child (7) with cystic fibrosis?

9 What are the genotypes of the parents of child 7 (3 and 4)?

10 What is the probability that the next child of these parents will have cystic fibrosis?

Show you can

What can you say about the genotypes of the grandparents of child 7 (1 and 2) in terms of carrying the cystic fibrosis allele?

Sex determination in humans

Sex in humans is another characteristic that is genetically determined. Humans have 46 chromosomes in each cell (except gametes) consisting of 22 pairs of normal chromosomes and one pair of sex chromosomes. The sex chromosomes determine the sex of each individual. Males have one X and one Y sex chromosome whereas females have two X chromosomes.

An image of a complete set of chromosomes is known as a karyotype. Figure 10.15 shows the karyotypes of the human male and female.

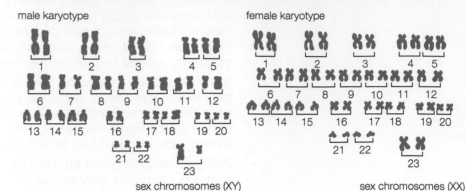

▲ **Figure 10.15** Male and female karyotypes

During meiosis the female will provide one X chromosome for each egg (ovum), but half the male's sperm will have an X chromosome and half will have a Y chromosome. As there will be an equal chance of an X or a Y chromosome from the male being involved in fertilisation there will be equal numbers of males and females produced. Again, the random nature of fertilisation must be emphasised.

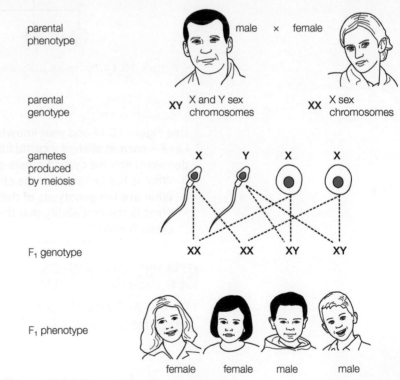

▲ **Figure 10.16** Human sex chromosomes

Sex linkage

The X and Y chromosomes are not only responsible for sex determination, they also have genes that code for a number of body functions. Each of the 22 normal (non-sex) pairs of chromosomes has the same gene present on both chromosomes and in the same position. The alleles may be different (alleles for blue or brown eyes) but the gene (gene for eye colour) is present on both. However, in the sex chromosomes the X is much larger than the Y and carries genes that are not present on the Y.

This is particularly important in males as they only have one X chromosome. Therefore any recessive allele carried on an X chromosome in a male will show its effect in the phenotype – there is no dominant allele to mask its effect, as is the situation with females who have two X chromosomes.

Haemophilia is a sex-linked condition that is almost exclusively found in males. Females seldom show sex-linked conditions but they are often carriers. In sex-linked conditions, carriers are females who have one dominant and one recessive allele on their X chromosomes. In the female the recessive allele usually does not affect the phenotype as it is masked by the dominant allele.

Haemophilia is a condition whereby individuals who are only carrying recessive alleles are unable to make all the products required to clot their blood.

Tip

In genetic conditions caused by a recessive allele, individuals who are heterozygous for the condition (they 'carry' the harmful allele but don't show the condition) are referred to as carriers.

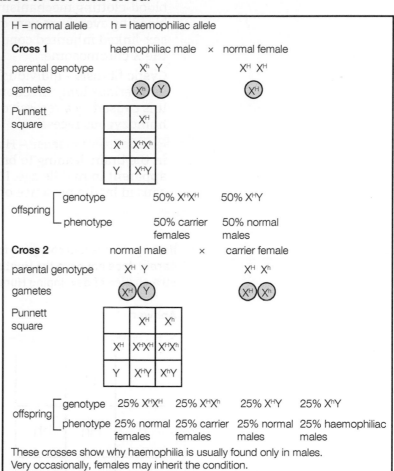

▲ **Figure 10.17** The inheritance of haemophilia

Tip

When carrying out crosses involving sex linkage, it is important to use symbols that represent both the type of sex chromosome (X or Y) and any allele carried, for example X^hY.

Show you can

Although haemophilia is almost always found in males only, it is possible for females to have haemophilia. Use a Punnett square to show how this is possible.

Test yourself

11 What are the sex chromosomes in a human male?
12 On which types of chromosomes are sex-linked conditions such as haemophilia found?

Genetics and health

Genetic conditions

Genetic conditions are caused by a fault with genes or chromosomes (a genetic fault). Some genetic conditions (but not all) are inherited; this means they are passed from parent to child.

Haemophilia, cystic fibrosis, Huntington's disease and Down's Syndrome are all genetic conditions that affect humans. However, they are each caused by a different type of genetic problem.

▶ Haemophilia – this condition is caused by a problem with the blood-clotting mechanism. Sufferers are at risk of excessive bleeding even from very small wounds or bruising. It is a sex-linked inherited condition caused by a recessive allele on the X chromosome.

▶ Cystic fibrosis – individuals with cystic fibrosis have frequent and serious lung infections and problems with food digestion. It is caused by a recessive allele, so affected individuals must be homozygous recessive.

▶ Huntington's disease – Huntington's disease affects nerve cells in the brain, leading to brain damage, which usually becomes apparent in middle age. It is fatal and there is no cure. It is caused by the presence of a dominant allele.

Example

If one parent is heterozygous for Huntington's disease and the other parent does not have the Huntington's disease allele, then each child has a 50% chance of developing Huntington's disease. Explain.

Answer

let H = Huntington allele, h = not affected

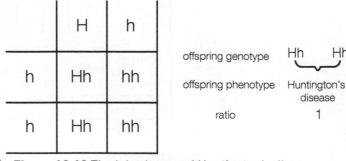

▲ **Figure 10.18** The inheritance of Huntington's disease

Tip

Haemophilia, cystic fibrosis and Huntington's disease are inherited (the faulty genes are passed from parent to child) and can pass through many generations. Down's Syndrome is a mistake during gamete formation – it is not inherited as such (it is not the passing of a faulty gene or chromosome from parent to child).

▶ Down's Syndrome – this condition is caused by the presence of an extra chromosome, so that affected individuals have 47 rather than 46. Humans normally have 23 chromosomes in each gamete (sperm or egg). Occasionally gametes are formed with 24 chromosomes so if one of these gametes is involved in fertilisation with a 'normal' gamete then the child produced will have 47 chromosomes. Individuals with Down's Syndrome have reduced muscle tone and reduced cognitive development.

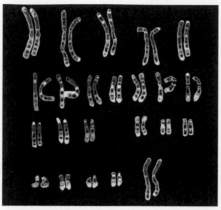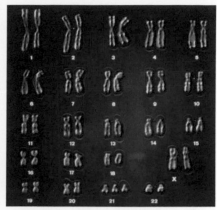

▲ **Figure 10.19** Karyotypes of a normal individual (left) and an individual with Down's Syndrome (right). Note that with Down's Syndrome there is an extra chromosome 21

Genetic screening

Genetic screening may be used to reduce the incidence of diseases or conditions caused by problems with our chromosomes or genes. It involves testing people for the presence of a particular allele or other genetic abnormality. Whole populations can be tested, or testing can be targeted at selected groups or individuals where the probability of having (or passing on) a particular condition is high.

One type of genetic screening is amniocentesis. This involves inserting a needle into the amniotic fluid surrounding the foetus and withdrawing some of the fluid. Foetal cells in the fluid are then examined for the presence of genetic abnormalities.

Figure 10.20 shows amniocentesis taking place. The structure to the left of the needle shows ultrascanning taking place (with the image of the foetus in the background).

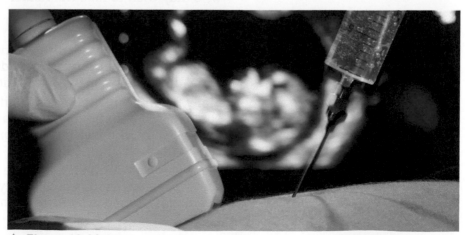

▲ **Figure 10.20** Amniocentesis

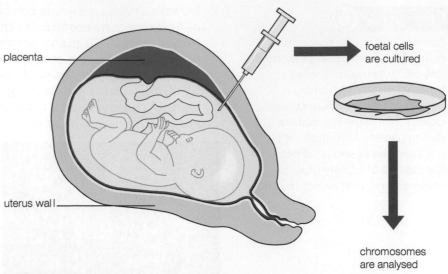

placenta

uterus wall

foetal cells are cultured

chromosomes are analysed

▲ **Figure 10.21** An amniocentesis test

Amniocentesis tests can be carried out for a range of genetic conditions including Down's Syndrome and cystic fibrosis (see Figure 10.21). However, as amniocentesis carries a risk of miscarriage (around 1%), it is usually only routinely used with those pregnant women who have a greater risk of having a child with a genetic abnormality.

Mothers with a greater risk include:

▶ those who have previously carried a foetus with a genetic abnormality

▶ those who have a family history of a genetic condition

▶ those where possible problems have been identified in an earlier medical examination, for example a blood test

▶ older mothers.

When screening for Down's Syndrome, pregnant mothers are offered a blood test between 10–14 weeks of pregnancy. The blood test will determine if the possibility of having a Down's Syndrome child is raised. The amniocentesis is normally then offered to those mothers who appear to have a higher risk based on the earlier screening. While not as accurate as amniocentesis, it carries no risk to the foetus or the mother, but it can help identify those women who may wish to take the riskier amniocentesis procedure.

Genetic screening – ethical and moral issues

If a foetus is diagnosed with a genetic condition the potential parents have some very difficult decisions to make and this creates a real dilemma for many. Is abortion the best thing to do?

Many parents will argue 'yes', as it prevents having a child who could have a poor quality of life; a lot of time may need to be spent caring for the child with the abnormality at the possible expense of time with their other children.

Many parents will argue 'no' as the unborn child doesn't have a say, or they will argue that it is not morally right to 'kill' a foetus. Additionally, abortion is banned in some religions and in some countries.

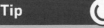

Tip

The blood test for Down's Syndrome is less precise than amniocentesis testing but it poses no risk.

Some other issues arising from genetic screening are listed below.

▶ Who decides on who should be screened?

▶ Is there an acceptable risk associated with genetic screening? For example, amniocentesis for Down's Syndrome screening has a small risk of miscarriage.

▶ Costs of screening compared to the costs of treating individuals with a genetic condition – should cost be a factor?

▶ Should genetic screening be extended to more than just serious genetic conditions? What if it can predict life expectancy?

It is now possible to screen everyone (whether before birth, in childhood or as an adult) for many different alleles. The information obtained is referred to as a genetic profile. Should this information be available to life insurance companies and employers?

If someone is identified with a genetic condition that is not obvious as yet, but may shorten their life span, then life insurance companies may not insure them and if they do then the insurance could be very expensive.

There are arguments for making genetic information publically available – it could help with medical research for example.

> ## Tip
>
> It is possible for anyone to pay for their genetic profile. Many companies carry out genetic profiles and provide a wide range of feedback including the risk of developing a range of diseases.

> ## Test yourself
>
> 13 Which of the following conditions is caused by the presence of a dominant allele?
>
> haemophilia   cystic fibrosis   Huntington's disease
>
> 14 Give the genotype of someone who is a carrier for cystic fibrosis (use F for the dominant allele and f for the recessive allele).
>
> 15 Which of the following conditions is caused by a chromosome (rather than a gene) abnormality?
>
> cystic fibrosis   Down's Syndrome   Huntington's disease

> ## Show you can
>
> When screening for Down's Syndrome explain why the blood test is given before taking an amniocentesis test, rather than the other way around.

Genetic screening is the process of identifying genetic issues that are present in someone's genome. Genetic engineering is the process of actually manipulating the genome in an organism for medical or other reasons.

Genetic engineering

Genetic engineering involves taking a piece of DNA, usually a gene, from one organism (the donor) and adding it to the genetic material of another organism (the recipient).

Genetic engineering is defined as a process that modifies the genome of an organism to introduce desirable characteristics.

Commonly, DNA that codes for a desired product is incorporated into the DNA of bacteria. This is because bacterial DNA is easily manipulated and also because bacteria reproduce so rapidly that large numbers can quickly be produced with the new gene.

△ Figure 10.22 Making human insulin by genetic engineering in giant bioreactors (fermenters)

As a result the bacteria will produce a valuable product coded for by the added gene, such as a drug or hormone that may be difficult or expensive to produce by other means. Once the new genetic material is built into them, the bacteria are added to special fermenters or bioreactors where they reproduce rapidly in suitable growing conditions that maximise the production of the desired product.

One of the best examples of genetic engineering providing essential products for humans is the production of genetically engineered human insulin, as shown in Figure 10.23. Diabetes is becoming increasingly common and as a result many more people require insulin than in the past. Before the development of genetic engineering, insulin was obtained from the pancreases of domestic animals such as pigs and cattle.

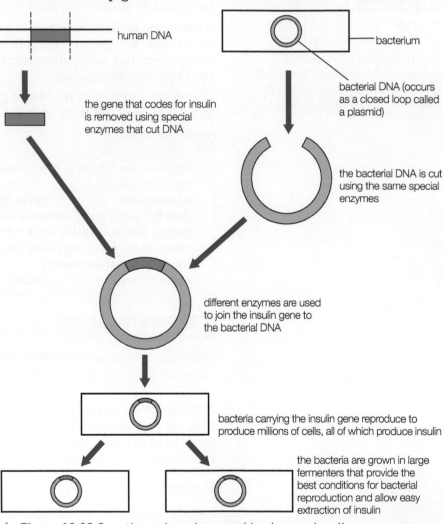

human DNA

the gene that codes for insulin is removed using special enzymes that cut DNA

bacterium

bacterial DNA (occurs as a closed loop called a plasmid)

the bacterial DNA is cut using the same special enzymes

different enzymes are used to join the insulin gene to the bacterial DNA

bacteria carrying the insulin gene reproduce to produce millions of cells, all of which produce insulin

the bacteria are grown in large fermenters that provide the best conditions for bacterial reproduction and allow easy extraction of insulin

△ Figure 10.23 Genetic engineering – making human insulin

The enzymes that cut and isolate the human insulin gene and cut the bacterial plasmid are called restriction enzymes. These cut the DNA in such a way that one of the two strands extends further than the other one. The longer strand will have 'free' exposed bases that are not paired. The key is that each restriction enzyme will leave complementary sections of exposed bases in both the plasmid and the human insulin gene so that they can join by base pairing between each other. Not surprisingly, the exposed strands of DNA and their bases are called 'sticky ends' (Figure 10.24).

Tip

Restriction enzymes are often described as 'molecular scissors' as their function is to cut the molecule of DNA in a particular way.

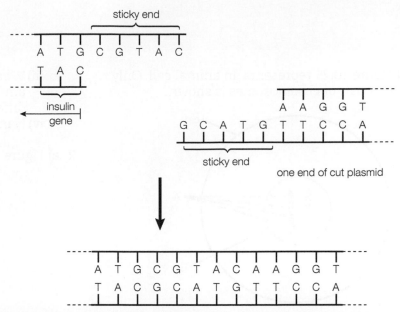

▲ **Figure 10.24** Sticky ends allow base pairing to take place and link the human DNA into the bacterial plasmid

Following the production of the insulin by genetic engineering (also called recombinant DNA technology) the insulin needs to be extracted from the fermenter, purified and packaged before it can be used for medical purposes. These processes that take place after the insulin is produced by the genetically engineered bacteria are referred to as downstreaming.

Many other medical and non-medical products are now made by genetic engineering. In general, pure forms can be produced more quickly, more cheaply and in greater quantities than by older extraction methods.

Advantages of producing human insulin by genetic engineering include:

▶ before genetically engineered insulin, the amount of insulin available was limited by the number of animals brought to the abattoirs for slaughter

▶ the extraction process was time consuming and there was the risk of transferring infections

▶ using animal insulin creates ethical issues for some people

▶ an additional complication is the fact that non-human insulin differs in structure to human insulin and is therefore not quite as effective for humans.

1 a) Figure 10.25 represents an animal cell. Only one pair of chromosomes is shown.

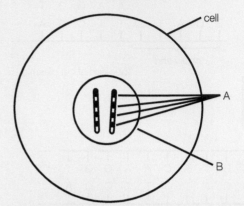

Figure 10.25

i) Name the structures labelled A on the chromosome. *(1 mark)*

ii) Name the part of the cell labelled B where chromosomes are found. *(1 mark)*

b) Chromosomes are formed of DNA.

i) Name the **two** components in the DNA backbone. *(2 marks)*

ii) Explain what is meant by the 'unique nature of an individual's DNA'. *(1 mark)*

H

c) Figure 10.26 outlines the role of DNA in cells.

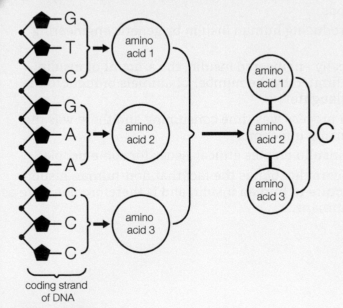

Figure 10.26

i) How many base triplets are present in the diagram? *(1 mark)*

ii) How many bases would be needed to code for a sequence of 45 amino acids? *(1 mark)*

iii) What evidence is there that the three amino acids coded for in the diagram are different from each other? *(1 mark)*

iv) Name structure C. *(1 mark)*

2 a) Figure 10.27 represents the process of mitosis.

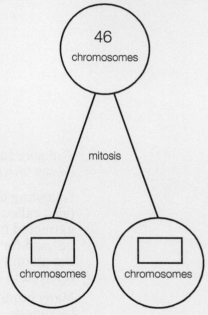

Figure 10.27

i) Copy and complete the diagram to show the number of chromosomes in each daughter cell. *(1 mark)*

ii) Name **one** function of mitosis. *(1 mark)*

b) i) Figure 10.28 below represents the process of meiosis. Only two pairs of chromosomes are shown.

Copy and complete the diagram to show the four possible chromosome arrangements in the daughter cells. *(2 marks)*

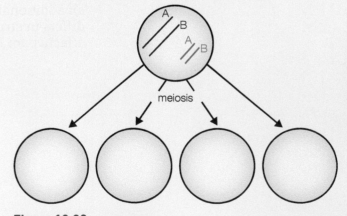

Figure 10.28

ii) Name **one** part of the human body where meiosis occurs. *(1 mark)*

iii) What name is given to the cells that are produced by meiosis? *(1 mark)*

3 a) The allele that causes albinism is recessive to the normal allele.

i) Copy and complete the Punnett square below to show the offspring of a cross between one parent who is heterozygous for albinism and the other parent who has albinism. (Let A = normal; a = albinism) *(2 marks)*

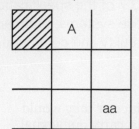

Figure 10.29

ii) From the Punnett square, what is the probability of these parents having a child with albinism? *(1 mark)*

b) Figure 10.30 shows a pedigree diagram for the inheritance of albinism in a family. (Let A = normal; a = albinism)

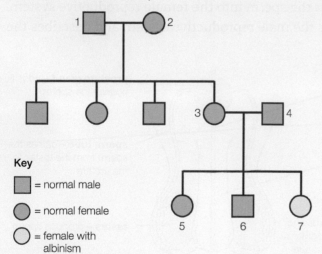

Key
- = normal male
- = normal female
- = female with albinism

Figure 10.30

i) What are the genotypes of the parents of child 7 (3 and 4)? *(1 mark)*

ii) What are the possible genotypes for the brother and sister of the child with albinism (5 and 6)? *(1 mark)*

4 Presence or absence of a widow's peak in humans is genetically determined by the two alleles of one gene.

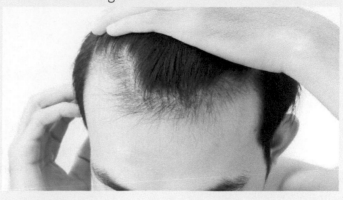

Figure 10.31

The allele for having a widow's peak is dominant. Let W = widow's peak; w = widow's peak absent.

a) State the genotype(s) that will result in the presence of a widow's peak. *(1 mark)*

b) Use a Punnett square to show the possible genotypes of children from two heterozygous parents. *(2 marks)*

c) What proportion of the offspring are homozygous? *(1 mark)*

d) What is the probability of these heterozygous parents' next child being a girl with a widow's peak? *(1 mark)*

5 Figure 10.32 represents a human insulin gene and a plasmid from a bacterium.

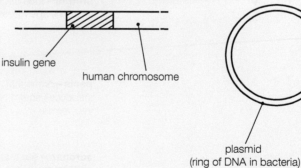

insulin gene

human chromosome

plasmid (ring of DNA in bacteria)

Figure 10.32

a) Describe how the insulin gene can be incorporated into the plasmid. *(3 marks)*

b) Once the plasmid is placed back into the bacterium, describe the stages which follow to enable the bacteria to produce large quantities of insulin. *(2 marks)*

c) State the term that is used to refer to the extraction and purification of the insulin. *(1 mark)*

d) Give **two** advantages of producing insulin by this method. *(2 marks)*

11 Reproduction, fertility and contraception

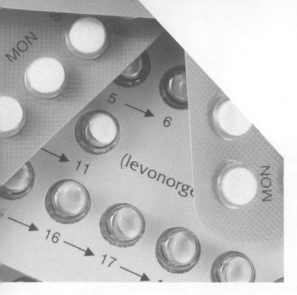

Specification points

This chapter covers sections 2.3.1 to 2.3.7 of the specification. It is about the male and female reproductive systems, pregnancy, sex hormones, the menstrual cycle, infertility and contraception.

Reproduction

Living organisms need to be able to reproduce or they would no longer exist. Humans, like most animals, carry out sexual reproduction. Sexual reproduction involves the joining together of two gametes – the sperm and the egg (ovum).

The male and female reproductive systems

The male reproductive system

The male reproductive system makes sperm (the male gamete) and is adapted to deliver the sperm into the female reproductive system. Figure 11.1 shows the male reproductive system and describes the role of each part.

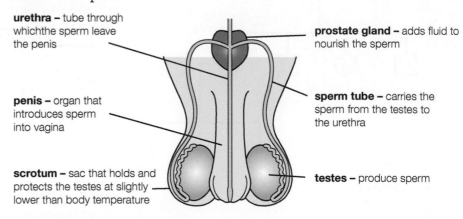

urethra – tube through whichthe sperm leave the penis

prostate gland – adds fluid to nourish the sperm

penis – organ that introduces sperm into vagina

sperm tube – carries the sperm from the testes to the urethra

scrotum – sac that holds and protects the testes at slightly lower than body temperature

testes – produce sperm

▲ **Figure 11.1** The male reproductive system

When a man and a woman have sex they are in intimate contact and as a consequence the man's penis increases in size and becomes firmer. This enables him to place his penis into the vagina of the woman. During ejaculation, sperm is released by reflex action into the female.

Sperm are cells that are highly adapted for their function. They have a flagellum (tail) that allows the sperm to swim to meet the egg. Sperm (and egg cells) are also adapted in being haploid.

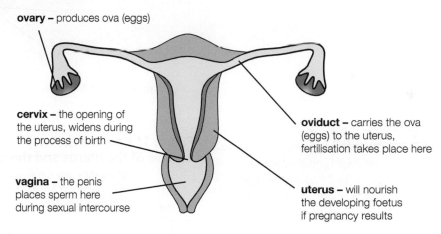

flagellum (tail)

haploid nucleus

▲ **Figure 11.2** A sperm cell

Sperm also have many mitochondria for energy production.

The female reproductive system

The female reproductive system is the part of the body that makes and releases eggs (ova). Additionally, if a sperm joins with an egg and pregnancy results, the embryo and foetus are protected and nourished within the female reproductive system until birth.

Figure 11.3 shows the female reproductive system and describes the role of each part.

ovary – produces ova (eggs)

cervix – the opening of the uterus, widens during the process of birth

vagina – the penis places sperm here during sexual intercourse

oviduct – carries the ova (eggs) to the uterus, fertilisation takes place here

uterus – will nourish the developing foetus if pregnancy results

▲ **Figure 11.3** The female reproductive system

Following sexual intercourse, the male sperm cell is able to swim out of the vagina, through the cervix and uterus and into the oviduct where the sperm and egg can fuse (join).

Fertilisation and pregnancy

If a sperm and an egg (ovum) meet and fuse (join) in an oviduct, fertilisation will result. Fertilisation involves the haploid nuclei of the sperm and egg fusing and restoring the diploid (normal chromosome number) condition. The fertilised egg is the first cell (zygote) of the new individual.

The zygote divides by mitosis and grows into a ball of cells referred to as an embryo that develops further as it travels down the oviduct.

Test yourself

1 What is the role of the prostate gland in the male reproductive system?
2 Name the part of the female reproductive system that produces eggs.
3 In which part of the female reproductive system does fertilisation take place?

Show you can

Describe the path of a sperm from where it is produced until it leaves the male body.

Tip

It is important that all gametes are haploid – if they weren't then every time fertilisation takes place the chromosome numbers in a cell would double!

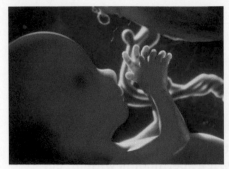

▲ **Figure 11.4** Human foetus at four months, showing the umbilical cord and the placenta

When the embryo enters the uterus, it sinks into the thick lining that has developed inside the uterus wall, becomes attached and receives nourishment (implantation).

At the point where the embryo begins to develop in the uterus lining, the placenta and umbilical cord form. A protective membrane, the amnion, develops around the embryo, as shown in Figure 11.5. It contains a fluid, the amniotic fluid, within which the growing embryo develops. This fluid cushions the delicate developing embryo, which increasingly differentiates into tissues and organs. The embryo is referred to as a foetus after a few weeks when it begins to become more recognisable as a baby.

Obviously the foetus cannot breathe when in the amniotic fluid (its lungs will not be developed enough anyway in the early stages), so during pregnancy useful materials including oxygen and dissolved nutrients (for example amino acids and glucose) pass from the mother to the foetus through the placenta and umbilical cord. Waste excretory materials (including carbon dioxide and urea) pass from the foetus back to the mother.

oviduct
placenta
foetus
umbilical cord
uterus wall
amnion
amniotic fluid
cervix
vagina

▲ **Figure 11.5** A foetus in the uterus

The placenta and umbilical cord in more detail

Figure 11.6 shows the very close relationship between the blood vessels of the uterus and the blood vessels in the placenta.

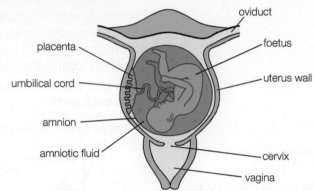

placenta – very rich in maternal blood vessels; mother's blood is rich in oxygen and other nutrients

uterus wall

umbilical vein

umbilical artery

umbilical cord contains:
- umbilical artery, which carries urea, carbon dioxide and other wastes back to the mother
- umbilical vein, which carries oxygen, glucose, amino acids and other nutrients from the mother to the foetus

amnion – contains amniotic fluid

boundary between maternal blood and foetal blood
The blood systems are not joined but are close together and separated by thin membranes to allow diffusion of gases and nutrients to take place in either direction.
The large surface area between the uterus wall and the placenta helps diffusion of materials between the mother and the foetus.

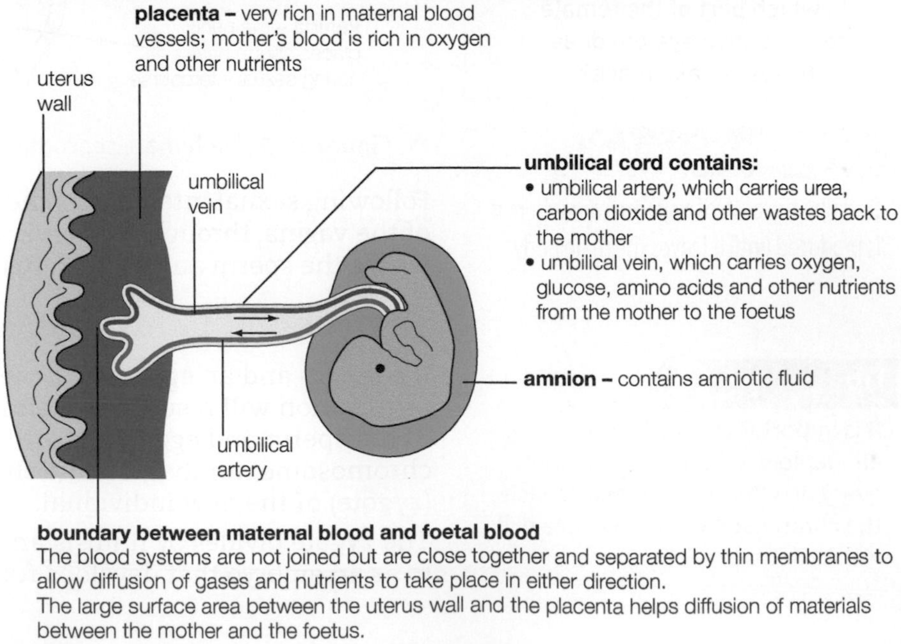

▲ **Figure 11.6** The functions of the placenta

The surface area between the uterine wall and the placenta is further increased by small villi (extensions) in the placenta that extend into the uterus wall.

Sex hormones and secondary sexual characteristics

Testosterone (produced by the testes in males) and oestrogen (produced by the ovaries in females) are important hormones in overall sexual development. One effect they have is the development of the secondary sexual characteristics that are a feature of puberty. The changes that occur in males and females are different but in both sexes they serve to prepare the body for reproduction, both physically and by increasing sexual awareness and drive.

The main secondary sexual characteristics produced by testosterone and oestrogen are summarised in Table 11.1.

Table 11.1 The main secondary sexual characteristics in males and females

Males	Females
body hair and pubic hair develops	hair grows in pubic regions and in the armpits
the sexual organs (genitals) enlarge	the sexual organs enlarge and the breasts develop
the body becomes more muscular	the pelvis and hips widen
the voice deepens	menstruation begins
sexual awareness and drive increase	sexual awareness and drive increase

The menstrual cycle

The menstrual cycle, shown in Figure 11.7, occurs in females from puberty until the end of reproductive life (usually sometime between the ages of 45 and 55). Each menstrual cycle lasts about 28 days. It is a cyclical event with the release of an ovum, the development of a thick lining on the uterus wall, and the breakdown of this lining (menstruation) occurring in each cycle. The purpose of the menstrual cycle is to prepare the reproductive system for pregnancy by controlling the monthly release of an egg and renewing and replacing the uterine lining.

The menstrual cycle is controlled by a number of female hormones. One of the most important hormones is oestrogen. At the start of each menstrual cycle (the onset of bleeding, which we call day 1), the level of oestrogen is low. As the cycle progresses the level of oestrogen rises. It peaks in the middle of the cycle, causing the release of an egg (ovulation). Another very important hormone is progesterone. The level of progesterone is also low during menstruation and peaks in the days following ovulation. The role of the progesterone is to build up and maintain the thick uterine lining (and the subsequent development of the placenta and other structures associated with pregnancy) should pregnancy occur. Oestrogen is also important in the initial buildup of the uterine lining.

If pregnancy does not occur, the levels of oestrogen and progesterone drop towards the end of the cycle and this causes menstruation to occur. Then the cycle begins again.

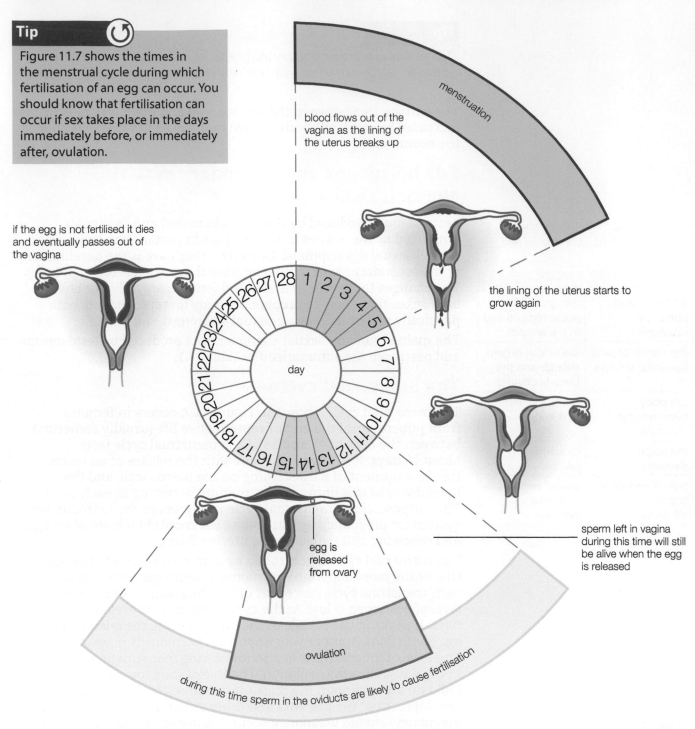

Tip

Figure 11.7 shows the times in the menstrual cycle during which fertilisation of an egg can occur. You should know that fertilisation can occur if sex takes place in the days immediately before, or immediately after, ovulation.

menstruation

blood flows out of the vagina as the lining of the uterus breaks up

if the egg is not fertilised it dies and eventually passes out of the vagina

the lining of the uterus starts to grow again

day

sperm left in vagina during this time will still be alive when the egg is released

egg is released from ovary

ovulation

during this time sperm in the oviducts are likely to cause fertilisation

▲ **Figure 11.7** The menstrual cycle

Test yourself

4 Give **two** functions of oestrogen in the menstrual cycle.
5 What is the function of progesterone in the cycle?

Show you can

Use information provided in earlier sections to draw a diagram to show how the level of oestrogen changes throughout the menstrual cycle.

Fertility problems (infertility) and their treatment **H**

Some people have problems that prevent them having children (fertility problems). Reasons include:

▶ the failure of ovaries to produce eggs

▶ the oviducts may be blocked or twisted, possibly due to infection

▶ complications of some sexually transmitted infections

▶ the lining of the uterus does not develop properly to enable implantation to occur

▶ the vagina may be hostile to sperm entering, for example the lining may be too thick or too acidic

▶ males may not produce enough sperm or the sperm may not be healthy – this can be affected by smoking or drinking alcohol in excess

▶ impotence in males.

There are a number of treatments that can be used to improve the chances of having a baby in someone with a fertility problem. The actual treatment used depends on the nature of the problem.

Fertility drugs (hormone treatment)

These are given to the woman to increase the production of eggs. This may solve the problem if low egg production is the issue but if there are other problems such as blocked oviducts then *in-vitro* fertilisation may be necessary.

In-vitro fertilisation (test-tube babies)

Normally women opting for *in-vitro* fertilisation are given fertility drugs so that several eggs are produced. The eggs are collected from the ovaries surgically. Sperm from a donor (usually the husband or partner) is collected and the sperm and eggs are mixed in the laboratory. Following fertilisation, embryos are placed in the mother's uterus (she will have undergone hormonal treatment to ensure her uterus lining is ready). If the process is successful, an embryo (or possibly more than one) will implant in the uterus lining. Usually only a small number of embryos are placed in the mother's uterus to give a balance between ensuring a successful pregnancy and avoiding multiple births.

Fertility research and treatment is a controversial area. It is now possible to screen embryos to check for abnormalities and even to check the sex of embryos before they are placed in the woman's uterus.

> ### Tip ⟳
>
> Two or more eggs (rather than one) are often used in each cycle of *in-vitro* fertilisation. This is because the process is very complicated and expensive and the use of more than one egg increases the probability of success.

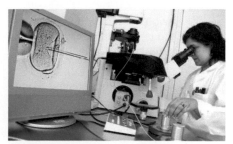

▲ **Figure 11.8** *In-vitro* fertilisation

> ### Test yourself ✎
>
> 6 Give **two** causes of infertility in men.
> 7 Why are fertility drugs used to treat female infertility?

> ### Show you can ❓
>
> Explain why women undergoing fertility treatment may need two different types of hormone treatment.

Contraception – preventing pregnancy

Many people want to have sex but do not want to have children at that particular time. Pregnancy can be prevented by contraception.

 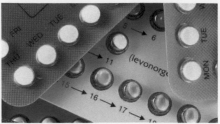

▲ **Figure 11.9** Methods of contraception

Methods of contraception

There are three main types of contraception: mechanical, chemical and surgical. Examples of each and an explanation of how they work, together with their main advantages and disadvantages, are given in Table 11.2.

Table 11.2 Methods of contraception

Type	Example	Method	Advantage	Disadvantage
Mechanical (physical)	male condom	acts as a barrier to prevent the sperm entering the woman	easily obtained and also protects against sexually transmitted infections such as HIV leading to AIDS, chlamydia and gonorrhoea; some STIs can lead to infertility if untreated, for example chlamydia	unreliable if not used properly
	female condom	acts as a barrier to prevent the sperm passing up the female reproductive system	easily obtained and protects against STIs (see above)	unreliable if not used properly
Chemical	contraceptive pill	taken regularly by the woman and prevents the ovaries from releasing eggs by changing hormone levels	very reliable	can cause some side effects such as weight gain, mood swings and may increase the risk of blood clots woman needs to remember to take the pill daily for around 21 consecutive days in each cycle
	implants	small tubes about 4 cm long that are inserted just under the skin in the arm and release hormones slowly over a long period of time to prevent the development and release of an egg	very reliable can work for up to 3 years	do not protect against STIs can prevent menstruation taking place
Surgical	vasectomy	cutting of sperm tubes, preventing sperm from entering the penis	virtually 100% reliable	very difficult or impossible to reverse
	female sterilisation	cutting of oviducts, preventing ova from moving through the oviduct and being fertilised	virtually 100% reliable	very difficult or impossible to reverse

Tip

The actual hormones used in different types of contraceptive pill and implants vary but they are almost always oestrogen and/or progesterone.

Tip

Do not confuse contraceptive implants with contraceptive patches. Patches work for a much shorter time than implants and they work in a slightly different way.

You need to be able to explain how each method of contraception works, its main advantage(s) and disadvantage(s).

Example

Explain how female sterilisation works and give **one** advantage and **one** disadvantage.

Answer

Both oviducts are cut and this prevents eggs (and sperm coming in the other direction) from passing beyond this point so fertilisation cannot occur. The main advantage is that it is virtually 100% reliable. The disadvantage is that it is very difficult or impossible to reverse.

Some people are opposed to contraception but may want to reduce their chances of having children – often because they have a large family already. They can do this by avoiding having sex around the time when the woman releases an ovum each month – this has been called the rhythm or natural method of contraception.

Some people choose this method for religious or ethical reasons but it is much less effective than contraception. In many women the menstrual cycle is irregular, making it difficult to know exactly when an egg is being released.

Test yourself

8 Describe how mechanical contraceptives work.
9 Give **one** advantage of using a mechanical contraceptive.
10 Give **one** disadvantage of using a mechanical contraceptive.

Show you can

Explain why doctors may be unwilling to sterilise an unmarried 25-year-old who has no children.

1 Figure 11.10 represents the female reproductive system.

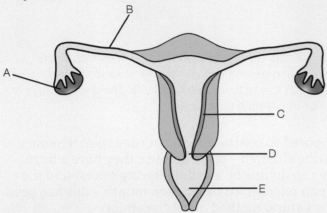

Figure 11.10

a) Identify parts A and B. *(2 marks)*

b) Give the letter on the diagram that represents where:
   i) fertilisation can take place
   ii) an embryo (ball of cells) can implant
   iii) male sperm is deposited during sex.
   (3 marks)

2 a) Sperm have a haploid nucleus.
   i) Describe what is meant by the term haploid. *(1 mark)*
   ii) Why is it important that sperm have a haploid nucleus? *(1 mark)*

   b) Figure 11.11 summarises what happens to the zygote following fertilisation.

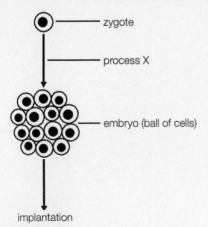

Figure 11.11

   i) Name process X that leads to a zygote forming a ball of cells. *(1 mark)*
   ii) Describe what happens to the ball of cells during and after the process of implantation. *(2 marks)*

3 Figure 11.12 represents the menstrual cycle.

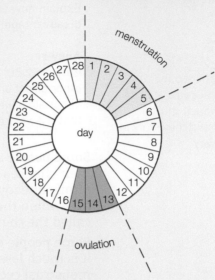

Figure 11.12

a) Describe **one** change that would occur in the uterus between day 6 and day 12. *(1 mark)*

b) Name the hormone responsible for this change. *(1 mark)*

c) Explain fully why it would be possible for a female to get pregnant if sex took place on day 16. *(2 marks)*

4 a) Suggest the effect a blockage in one of the oviducts will have on the chances of a female becoming pregnant. Explain your answer. *(3 marks)*

   b) Figure 11.13 represents the male reproductive system.

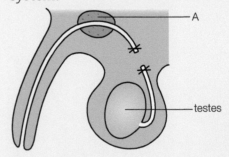

Figure 11.13

   i) Name the gland labelled A. *(1 mark)*
   ii) Name the method of contraception shown. *(1 mark)*
   iii) Using the diagram, explain fully how this method prevents pregnancy. *(2 marks)*
   iv) Give **one** disadvantage of this method of contraception. *(1 mark)*

12 Variation and natural selection

Variation

Living organisms that belong to the same species usually vary from each other in many ways. There are two main types of variation.

Types of variation – continuous and discontinuous variation

Variation in a particular characteristic can be either continuous or discontinuous. In continuous variation there is gradual change in a characteristic across a population. Height is an example of continuous variation in humans. While people can be described as being tall or short, there is not a distinct boundary that separates short and tall people.

Figure 12.1 shows a typical set of values for height in human males. Note that the histogram produced shows a normal distribution. A normal distribution is where most individuals are around the average or mean value and relatively few are found at either extreme.

Mass in humans and length, for example of hand span, will also show continuous variation.

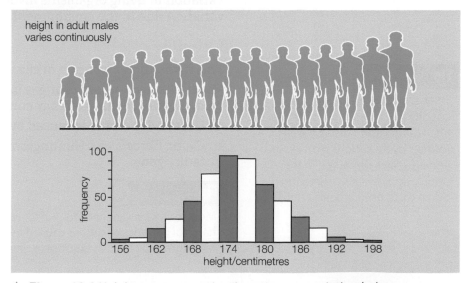

▲ **Figure 12.1** Height as an example of continuous variation in humans

In **discontinuous variation** the population can be clearly divided into discrete groups or categories. A common example used is the ability, or inability, to roll the tongue in humans. In this example, individuals will fit into one of two categories – there are no intermediates

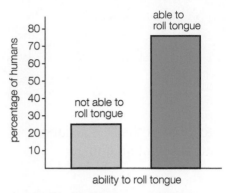

▲ **Figure 12.3** Tongue rolling as an example of discontinuous variation in humans

▲ **Figure 12.2** This man and child can roll their tongues

Hand dominance – being right or left-handed – is another example of discontinuous variation in humans. In other examples of discontinuous variation, there can be more than two categories, for example ABO blood groups, but again all individuals can be clearly identified as belonging to a particular blood group (A, B, AB or O).

You should be able to represent information on variation graphically. Continuous variation should be represented as a histogram and discontinuous variation as a bar chart (with spaces between the bars).

Causes of variation

Variation in living organisms involves genetic variation and/or variation due to the environment.

Genetic variation

This is variation as a result of changes to chromosomes or genes (DNA).

Mutations are random changes in the number of chromosomes or type of gene. We have already come across a number of examples:

▶ Down's Syndrome is caused by an extra chromosome.
▶ Cystic fibrosis and Huntington's disease are each caused by a faulty gene.

Tip

If you see an exam question that has variation shown as a histogram you should be able to identify it as continuous variation. Similarly, if it is represented as a bar chart, it will be discontinuous variation.

Tip

The harmful alleles in genetic conditions such as cystic fibrosis and Huntington's disease were formed by mutations many years ago and have remained in the population ever since, passing down through the generations.

Tip

Most mutations are harmful – in Chapter 10, you learnt about a number of harmful genetic conditions caused by mutations. However, they are not all harmful – look back at question 4 on page 97 about widow's peak.

Tip

Environmental variation can produce variation (different phenotypes) even if the genotypes are the same – think about how plant cuttings (clones) can differ if placed in different light levels.

Test yourself

1 State the **two** different types of variation.
2 State the **two** causes of variation.

Example

Using the photograph, give **two** adaptations to the environment in polar bears.

▲ **Figure 12.4** Polar bear and cub

Answer

They have thick fur to reduce heat loss and their white fur gives them camouflage and allows them to blend in with the snow.

Genetic variation is also caused by the process of sexual reproduction. Meiosis mixes up the chromosome arrangements in gamete formation – this results in all gametes being genetically different. Therefore when any two gametes fuse, the genes in that zygote will be different to the genes in any other zygote. This is why siblings in any one family often show considerable variation in phenotypes.

Environmental variation

The environment can also provide variation. While to a large extent height in humans is genetically controlled, the actual height a human grows to will depend on the quality of nutrition.

Show you can

Table 12.1 gives information about five students in a class.

Table 12.1

Student	Characteristic			
	Able to roll tongue	Height/cm	Hand span/cm	Hand dominance
Kyle	yes	144	17	right
Jonny	no	151	20	right
Eve	no	137	13	right
Mary	yes	146	16	right
Sorcha	yes	153	17	left

a) Which **two** characteristics show discontinuous variation?
b) What percentage of students are able to roll their tongues?
c) What type of graph would you use to represent the data for hand span?

Natural selection and evolution

Natural selection

In nature, adaptations in living organisms are essential for survival and success in all habitats. You should be able to work out the main adaptations in organisms if provided with enough information.

Adaptations in organisms are even more important when organisms compete with each other for resources. This **competition** ensures that the best adapted individuals will survive. For example, the larger seedlings growing in a clump of plant seedlings will be able to obtain vital resources such as light, nutrients and water more easily than the smaller seedlings. As a result of this competition, the stronger individuals will survive, often at the expense of the weaker ones.

This competition for survival, with the result that the better equipped individuals survive, summarises Charles Darwin's theory of natural selection.

Charles Darwin and the theory of natural selection

Charles Darwin (1809–1882) was a naturalist who devoted much of his life to scientific research. As part of his research he spent five years as a ship's naturalist on the HMS Beagle as it travelled to

 Figure 12.5 Charles Darwin

South America. Darwin was greatly influenced by the variety of life he observed on his travels and, in particular, by the animals of the Galapagos Islands. Darwin's famous account of natural selection, *On the Origin of Species*, was published in 1859.

Darwin's main conclusions about natural selection can be summarised as:

▶ There is variation among the phenotypes (individuals) in a population.
▶ If there is competition for resources, there will be a struggle for existence.
▶ The better-adapted phenotypes survive this struggle or competition. This leads to survival of the fittest and these (fittest) individuals are more likely to pass their genes on to the next generation.

It is useful to look at a modern day example to highlight the key features of natural selection.

> **Tip**
>
> Natural selection is the process in which the better-adapted individuals survive (at the expense of the less well adapted individuals) and pass on their genes.

> **Tip**
>
> Natural selection depends on the environment (context). If antibiotics were not used, then the antibiotic resistant bacteria wouldn't be better adapted and therefore wouldn't increase in number.

Antibiotic resistance in bacteria

When bacteria are treated with an antibiotic such as penicillin, most of them are killed. However, as Figure 12.6 shows, a small number (the best adapted phenotypes) may survive, probably because they have a gene (caused by a mutation) that provides resistance. Very soon, the resistant bacteria are the only ones remaining, as they are the only ones surviving and passing their beneficial mutations on to their offspring.

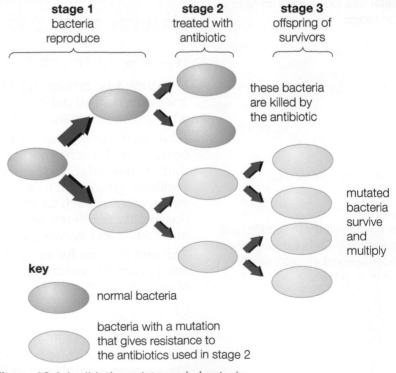

▲ **Figure 12.6** Antibiotic resistance in bacteria

Tip

Evolution is change in species over time as a consequence of natural selection favouring different phenotypes over time and also the formation of new species.

Tip

Species can become extinct if they fail to adapt to environmental change.

▲ **Figure 12.7** Woolly mammoths

The link between natural selection and evolution

Darwin used the theory of natural selection to explain the process of evolution. Evolution shows that species have changed gradually through time in response to changes in the environment and that evolution is a continuing (ongoing) process. As well as species changing over time, evolution can result in the formation of new species.

Extinction

Species are extinct if there are no living examples left. Many species have become extinct and often we only know they did exist in the past due to the discovery of fossils. Examples of extinct species include the dodo, dinosaurs and the woolly mammoth.

Species that are not quite extinct but are at risk are called **endangered species**. For example, a number of big cat species are endangered.

Species can become extinct for many reasons. These include:

▶ climate change or natural disasters, for example, the dinosaurs probably became extinct when a meteor hit the Earth and caused the climate to change

▶ hunting by humans – the dodo was hunted until it became extinct

▶ hunting by animals introduced by humans to areas where they are not normally found

▶ the spread of diseases

▶ loss of habitat – this is causing a lot of species to become extinct today. The loss of habitat is often caused by human activities, for example deforestation and clearing land for towns and cities.

Tip

The list above shows that man is responsible for many of the extinctions that take place today – in the context of most extinctions, the activities of man bring about 'environmental change'.

We can slow down the rate of extinctions in a number of ways. These include:

▶ legislation preventing the hunting of endangered species

▶ international agreements, such as those that plan to limit climate change

▶ special programmes such as creating nature reserves to protect habitats

▶ education to encourage people to do their part in protecting the environment.

Show you can

Explain what biologists mean by natural selection.

Test yourself

3 Describe what is meant by antibiotic resistance.
4 What is evolution?
5 State **two** ways in which man has caused extinctions of other species.

H

Selective breeding

For centuries, people have manipulated the course of natural selection by deliberately selecting particular characteristics or traits in many plants and animals that are of use to us. This is the process of selective breeding (artificial selection).

Characteristics that are advantageous to us include increased crop yield or quality, appearance, hardiness, disease resistance and longer shelf life.

The selective breeding of cereals such as wheat in Figure 12.8, show the principles of selective breeding. Wheat has been bred over many years to produce a shorter stalk length which is less likely to suffer wind damage and the uniform size is easier to harvest. The modern wheat plant has a much larger head of grain (higher yield) than the ancient variety too.

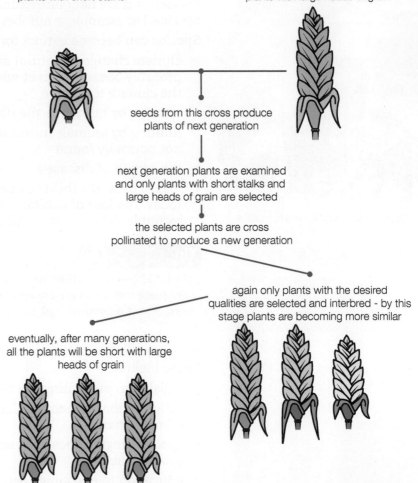

plants with short stalks plants with large heads of grain

seeds from this cross produce plants of next generation

next generation plants are examined and only plants with short stalks and large heads of grain are selected

the selected plants are cross pollinated to produce a new generation

again only plants with the desired qualities are selected and interbred - by this stage plants are becoming more similar

eventually, after many generations, all the plants will be short with large heads of grain

▲ **Figure 12.8** Selective breeding in wheat

Selective breeding has been very important in the development of farm animals and pets. For example, cattle can be bred to have excellent beef properties when used to provide beef for human consumption, for example Aberdeen Angus, or to produce large quantities of high quality milk, for example Friesians. Dogs have been bred for appearance, 'personality', their ability to act as guard dogs and for many other features.

Tip

Selective breeding in wheat shows one of the key characteristics of selective breeding: it takes many generations (reproductive cycles) and a long time to reach the stage where all the animals or plants have the desired characteristics.

Tip

The selective breeding of dogs shows some of the problems of selective breeding if it is carried out to extremes (excessive inbreeding). For example, many modern breeds have limited life spans, joint issues and other problems just because they have been bred to have appearances and 'personalities' that are attractive to pet owners.

Test yourself

6 Name **one** desirable characteristic that could be selectively bred into tomato plants.

Show you can

Suggest the main stages involved in selectively breeding cattle for high milk yield.

Practice questions

1 Figure 12.9 shows some Friesian cattle grazing in a field.

Figure 12.9

   a) Name the type of variation shown by coat colour in the cattle.
   Choose from:
   discontinuous   continuous   environmental
   (1 mark)
   b) Define this type of variation. *(1 mark)*
   c) Suggest **one** other example of variation shown by the cattle in the field. *(1 mark)*

2 a) Figure 12.10 shows how leaf width varies in a particular type of plant leaf.

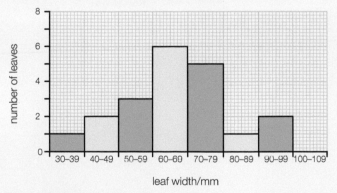

Figure 12.10

   i) What is the most common leaf width in these plant leaves? *(1 mark)*
   ii) How many leaves were sampled in total? *(1 mark)*
   iii) Name the type of variation shown by leaf width. *(1 mark)*

b) In a particular class of Year 11 students, 12 could roll their tongue and 8 could not.
   i) What percentage of students could roll their tongue? *(2 marks)*
   ii) Why is tongue rolling described as discontinuous variation? *(1 mark)*
   iii) Name the type of graph used to show discontinuous variation. *(1 mark)*

3 Scientists cultured two types of bacteria (A and B) in a beaker. Figure 12.11 shows how the numbers of the two types changed after an antibiotic was added to the beaker.

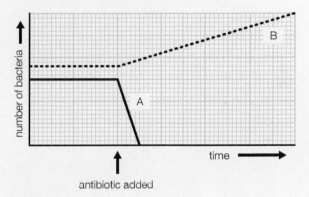

Figure 12.11

a) Describe fully the changes in numbers of bacteria A and B after the antibiotic was added. *(2 marks)*

b) Explain the change in number of bacteria B. *(3 marks)*

c) Name the process that this investigation demonstrates. *(1 mark)*

4 Table 12.2 shows the number of extinctions that have occurred in a country over the last 100 years.

Table 12.2 Extinctions

Year	Number of extinctions
1920	1
1940	2
1960	15
1980	22
2000	46

a) Copy the graph axis in Figure 12.12 below and complete a line graph of the information in Table 12.2.

(3 marks)

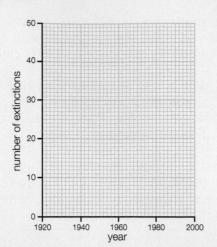

Figure 12.12

b) In which 20 year period was there the largest increase in extinctions? *(1 mark)*

c) Predict how many extinctions there will be by the year 2020. *(1 mark)*

13 Health, disease, defence mechanisms and treatments

Specification points

This chapter covers sections 2.6.1 to 2.6.15 of the specification. It is about communicable diseases, aseptic techniques, the body's defence mechanisms, antibiotics, antibiotic resistant bacteria, vaccinations, non-communicable diseases, heart attacks, strokes and cancer.

Diseases can be described as being communicable or non-communicable.

A communicable disease is a disease that can be passed from one organism (person) to another. A **non-communicable disease** is a disease that is not passed from one organism (person) to another.

Health is defined as being free from communicable and non-communicable disease. If we are healthy, we are free from disease.

Being healthy is important to us and our families. Health is also very important to society. Very unhealthy people cannot work and need care. Billions of pounds are spent each year by the National Health Service (NHS) on keeping people as healthy as possible and on treating and looking after people who are ill.

Tip

The NHS spends money on the salaries of doctors and nurses and other staff, the upkeep of hospitals and health centres, and the drugs and medicines used to treat people.

Communicable diseases

Most communicable diseases are caused by microorganisms such as bacteria, viruses and fungi.

Table 13.1 shows some diseases caused by microorganisms and how they can be spread (what makes them communicable). The table also shows how they can be avoided or treated.

Tip

Communicable diseases are also described as infectious diseases.

Table 13.1 Communicable (infectious) diseases – their cause, spread and prevention or treatment

Microbe	Type	Spread	Control/prevention/treatment
HIV (which leads to AIDS)	virus	exchange of body fluids during sex infected blood	using a condom will reduce risk of infection, as will drug addicts not sharing needles; currently controlled by drugs
Colds and flu	virus	airborne (droplet infection)	flu vaccination for targeted groups
Human papilloma virus (HPV)	virus	sexual contact	HPV vaccination offered to 12–13 years old girls to protect against developing cervical cancer
Salmonella food poisoning	bacterium	from contaminated food	always cooking food thoroughly and not mixing cooked and uncooked foods can control spread; treatment by antibiotics
Tuberculosis	bacterium	airborne (droplet infection)	if contracted, treated with drugs including antibiotics
Chlamydia	bacterium	sexual contact	using a condom will reduce risk of infection; treatment by antibiotics
Athlete's foot	fungus	contact	reducing infection risk by avoiding direct contact in areas where spores are likely to be present, for example by wearing flip flops in changing rooms/swimming pools
Potato blight	fungus	spores spread in the air from plant to plant, particularly in humid and warm conditions	crop rotation and spraying plants with a fungicide

Tip

The only non-human disease in Table 13.1 is potato blight. This is a disease that affects the potato and some other types of plant.

Figure 13.1 shows how the spray of moisture and particles spreads through the air when a woman sneezes. If she has a cold or the flu this is how 'droplet' infection can occur.

▲ **Figure 13.1** A woman sneezing and possibly spreading disease-causing microorganisms

The body's defence mechanisms against communicable diseases

The human body is well adapted to protect us against infection. The body is very successful at preventing most microorganisms from gaining entry and it has very effective defences against those microorganisms that do enter.

Preventing microorganisms gaining entry

The skin itself is an excellent barrier to microorganisms. The openings in the body such as the nose and the respiratory system are protected by mucous membranes that trap the microorganisms and prevent them going any further. Clotting of blood is also important as a defence mechanism. It stops more blood escaping but it also acts as a barrier against infection.

If a microorganism does enter the body it is the blood system that usually helps combat the invader. The blood system is very effective in this role but we are often ill for a period of time before our defence system allows us to recover.

Antigens and antibodies

Invading microorganisms have chemicals on their surface that the body can recognise as being foreign. These chemicals are called antigens and they cause special white blood cells called lymphocytes to produce antibodies.

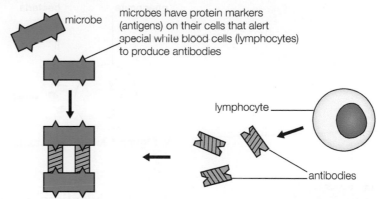

microbe

microbes have protein markers (antigens) on their cells that alert special white blood cells (lymphocytes) to produce antibodies

lymphocyte

antibodies

antibodies immobilise the microbes

▲ **Figure 13.2** How antibodies work

As Figure 13.2 shows, these antibodies have a shape that matches the shape of the antigens on the microorganisms. The antibodies join with the microorganisms (like a jigsaw puzzle) and cause them to clump together. Once clumped (or immobilised) they are easily destroyed by other white blood cells called phagocytes in a process known as phagocytosis.

As the antigens on a particular microorganism and the antibodies used to combat that microorganism are complementary in shape, it is possible to work out the shape of one from the other. Look at the examples shown in Figure 13.3.

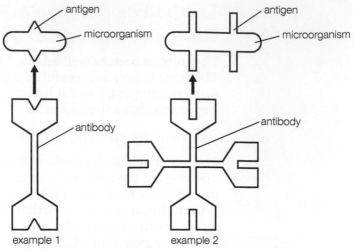

▲ **Figure 13.3** Antigens and their complementary antibodies

Phagocytosis in action

Some white blood cells move around in the blood and destroy microorganisms trapped by antibodies, or destroy them directly without antibody action. This type of white blood cell is called a phagocyte. Phagocytes surround the microorganisms and engulf ('eat') them, as shown in Figure 13.4. Eventually chemicals (enzymes) inside the phagocyte digest the microorganisms and destroy them.

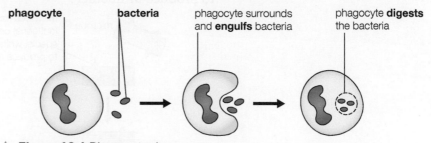

▲ **Figure 13.4** Phagocytosis

Primary and secondary responses

The antibody/antigen reaction described above is a typical response to being infected by a bacterium or a virus. The infected individual is often ill for a few days before the antibody numbers are high enough to provide immunity. This is described as the primary response. However, once infected, the body is able to produce memory lymphocytes that remain in the body for many years. This means that if infection by the same type of microorganism occurs again, the memory lymphocytes will be able to produce antibodies very fast to stop the individual catching the same disease again. This is known as the secondary response.

Immunity

Individuals who are protected against a particular infection or disease are described as being immune to that disease. Most people will be immune to a number of diseases. If someone is immune this means that his or her antibody levels are high enough (or high enough levels can be produced quickly enough) to combat the microorganism should it gain entry to the body again.

There are two types of immunity.

Active immunity is where the body produces the antibodies used to combat the infectious microorganism. This type of immunity is slower acting but usually lasts for a very long time.

Passive immunity is when antibodies from another source (such as those produced by pharmaceutical companies) are injected into the body. These are fast acting but only last for a short period of time.

Test yourself

1 What is a communicable disease?
2 Name **two** human diseases spread by droplet infection.
3 Name the **two** types of white blood cell that help defend against disease.
4 Give **two** features of passive immunity.

Vaccinations

Vaccinations involve the use of dead or modified pathogens (microorganisms which cause diseases) that are injected into the body. The dead or modified microorganisms still have the antigens on their surfaces that cause the body to produce antibodies at a high enough level to prevent the individual becoming ill later (to provide active immunity). Crucially, a vaccination leads to memory lymphocytes being produced that will bring about a rapid immune response if a further infection occurs.

Sometimes we need more than one vaccination to make sure that we remain immune for a reasonable period of time. This is known as a follow-up **booster vaccination**. Figure 13.5 shows what happens following a vaccination that involves a booster.

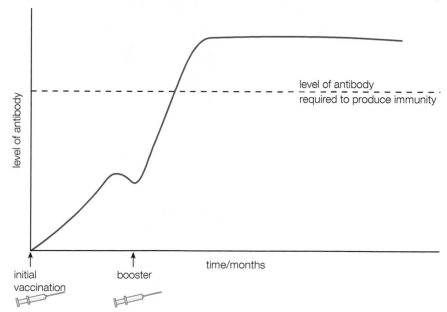

▲ **Figure 13.5** The effect of a vaccination on antibody level

You need to be able to interpret graphs showing the antibody levels typically produced in active and passive immunity. Examples of these are shown in the graphs in Figures 13.6, 13.7, and 13.8.

H

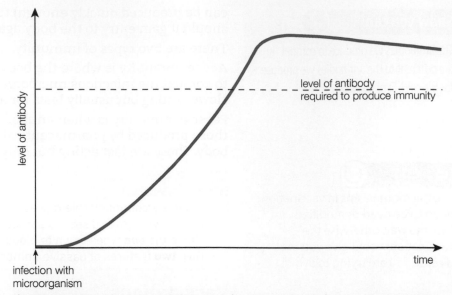

▲ **Figure 13.6** The effect of infection (active immunity) on antibody level

Figure 13.6 represents active immunity in action. The antibodies are produced by the body in response to infection. The relatively slow increase in antibody number is typical of the primary immune response – gaining immunity following infection by a type of microorganism for the first time.

However, in passive immunity the level of antibody increases rapidly but it quickly falls too (Figure 13.7).

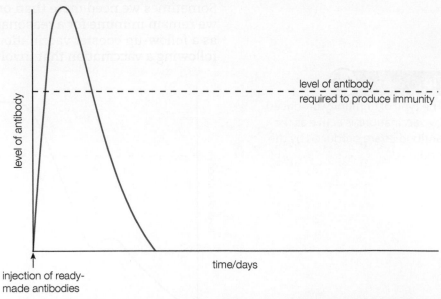

▲ **Figure 13.7** The effect of passive immunity on antibody level

The differences in the speed of the body producing antibodies following a first infection (primary response) and being reinfected by the same pathogen (secondary response) are shown in Figure 13.8.

Tip

The secondary response both produces antibodies more quickly and also produces many more antibodies than the primary response.

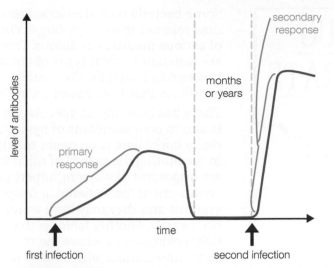

▲ **Figure 13.8** Primary and secondary responses to infection

Test yourself

5 What is a booster injection?
6 Explain why booster injections are given.

Antibiotics

We have come across antibiotics earlier in Unit 2 when looking at natural selection in Chapter 12. Antibiotics, such as penicillin, are chemicals produced by fungi that are used against bacterial diseases to kill bacteria or reduce their growth.

Most people have had antibiotics at some time in their lives to defend against bacterial conditions such as septic throats or infected wounds in the skin. The effect of an antibiotic can be seen in Figure 13.9.

Antibiotics are not as specific as antibodies in that they are not designed to combat only one type of bacteria – they usually act against a range of bacteria and they act in a different way to antibodies. Antibiotics do not all act in the same way and, for this reason, a GP may not prescribe the same antibiotic when a patient is being treated for different infections.

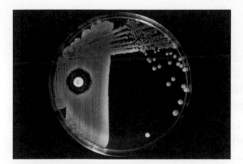

▲ **Figure 13.9** Agar plate showing a clear area (around the white circle on the left) where an antibiotic has killed bacteria growing on the agar

Tip

Antibiotics cause harm to bacteria – they have no effect on viral diseases, such as colds and flu.

Antibiotic-resistant bacteria

Bacterial resistance to antibiotics was used as an example of natural selection in Chapter 12. Bacterial resistance to antibiotics is becoming a major problem and is making many antibiotics ineffective against various bacteria. The overuse of antibiotics is largely responsible and it is very important that antibiotics are only used when they are really necessary. The overuse of antibiotics has allowed many types of bacteria to become resistant to the main antibiotics, and so it is the mutated resistant forms that are now common.

▲ **Figure 13.10** 'Superbugs' in the media

Some bacteria have developed resistance to the extent that they are now referred to as 'superbugs'. They are responsible for a number of serious medical conditions. These superbugs, such as MRSA, are resistant to most types of antibiotic and can be a very serious problem in hospitals. The headline in Figure 13.10 is typical of many headlines that have appeared in the media in recent years.

There has been media speculation that the problem with superbugs is due to poor standards of hygiene in some hospitals, but is this really fair? There is no doubt that good hygiene is very important in preventing the spread of microbes in hospitals but other factors are important in allowing superbugs to flourish in this type of environment too. Patients in hospital often have weak immune systems and they may have wounds that allow microbes to enter the body easily. Another factor is that hospitals provide an 'antibiotic-rich' environment where the microbes have every opportunity to come into contact with a range of antibiotics. This ensures that the non-resistant microbes are eliminated and a high proportion of the surviving microbes are antibiotic resistant.

There is no doubt that 'superbugs' are extremely difficult to eradicate. Nonetheless, new measures in hospitals include increased levels of hygiene (such as the immediate cleaning of spillages of body fluids and the wearing of gloves) and greater care in the administering of antibiotics. Additionally, patients who contract a 'superbug' are often isolated from other patients to reduce the possibility of infecting others.

Aseptic techniques

When working with bacteria and fungi in the laboratory, it is very important that great care is taken to avoid contamination and also the growth of unwanted, pathogenic (harmful) microorganisms. The procedures used to avoid this are referred to as aseptic techniques.

In a school environment there are important health and safety precautions that need to be used when growing or culturing microbes. These include:

▶ not eating or drinking in the laboratory

▶ wiping down lab benches with a disinfectant

▶ not culturing microbes at body temperature

▶ using sterile loops for transferring cultures

▶ flaming the necks of culture bottles to prevent contamination

▶ sterilising (using an autoclave) or disposing of all equipment after use

▶ washing hands thoroughly after each part of the experiment.

Tip

An autoclave is a type of pressure cooker that uses high pressure steam at 120 + °C to kill all living organisms. It sterilises agar dishes, culture media and other apparatus.

When culturing or transferring microorganisms it is important to use aseptic techniques as in the procedure described below.

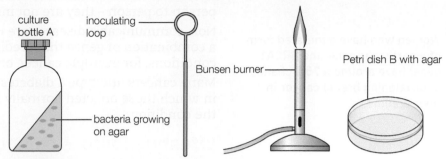

▲ **Figure 13.11** Using aseptic techniques

The apparatus in Figure 13.11 is normally used when inoculating and plating bacteria in the laboratory. All the apparatus used, for example agar plates, should be sterilised and disposed of as indicated by your teacher. A typical sequence is:

1 Pass the metal loop through the flame of the Bunsen burner to sterilise it.

2 Allow the metal loop to cool, as if it is too hot it will kill any microorganisms that it comes into contact with.

3 Remove the lid of the culture bottle (A) and glide the loop over the surface of the agar without applying any pressure. This will ensure that the metal loop has bacteria from the culture bottle over its surface (inoculation).

4 You should 'sweep' the neck of the bottle through the flame to destroy any airborne microbes. Replace the lid of the culture bottle to prevent contamination.

5 Spread the microbes over the surface of the agar in the Petri dish (B) by gently gliding the metal loop over the nutrient agar surface (plating). It is important to hold the Petri dish lid open at an angle rather than completely removing it, as this will reduce the chance of unwanted microbes from the air entering the dish.

6 The metal loop should then be heated again to a high temperature to ensure that any microbes remaining on the loop are destroyed.

7 The Petri dish should be taped in four places and then incubated in an oven at 25 °C. This temperature will allow the microorganisms to grow and form colonies but is below body temperature, meaning pathogenic microbes that could harm humans will not grow.

8 When carrying out the transfer it is important to work close to a Bunsen burner so that the Bunsen flame can kill microorganisms in the air.

Note: instead of using a metal loop it is possible to use sterile disposable plastic loops that do not require heating.

Tip

Nutrient agar is a type of agar enriched with minerals and nutrients essential for the growth of bacteria.

Tip

When incubating Petri dishes (agar plates) in an incubator it is important to store the taped dish upside down to avoid condensation dripping on to the culture.

Non-communicable diseases

Non-communicable diseases are diseases that are not passed from person to person – they are not infectious diseases.

Non-communicable diseases are usually a consequence of inheriting a combination of genes that predispose us to developing some conditions, for example cancer, or are due to lifestyle.

Many cancers and type 1 diabetes are non-communicable diseases in which those affected normally have a genetic predisposition to the condition.

Lifestyle factors and non-communicable disease

Lifestyle factors that are linked to disease include a poor diet, lack of exercise, overexposure to the Sun and the misuse of drugs (smoking and drinking too much alcohol).

Diet and disease

Diet has a major effect on our health. There are a number of features of our diet which contribute to many people not being as healthy as they could be. As a population we are eating too much sugar and fat.

Eating too much of foods high in sugar and fat has two main effects on the body. It means that:

▶ the individual can become overweight and obesity can result

▶ the individual will not be getting a balanced diet and probably not enough fruit and vegetables, meaning that they miss out on essential vitamins and minerals.

Exercise and health

If the energy we use in exercise is less than the amount we eat then we run the risk of becoming overweight or obese. Lack of exercise and eating too much fat and sugar are the two main reasons why so many people are obese in the UK.

Exercise is good for the body in many ways and not just as a means of using energy. The right type of exercise can strengthen our bones, help our circulatory system and can even help our mood.

▲ **Figure 13.12** Exercise brings many benefits

Example

Women who have a mutated form of a particular gene (the BRCA1 gene) have around a 75% chance of developing breast cancer in their lifetime.

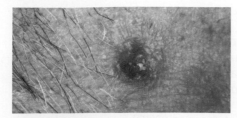

▲ **Figure 13.13** Skin cancer

 Tip

Mutations are random changes to DNA, chromosomes or genes.

▲ **Figure 13.14** A consequence of binge drinking

Tip

Bronchitis can leave affected individuals struggling for oxygen as not enough may reach the alveoli due to the airways (bronchi and bronchioles) being constricted.

Tip

Carbon monoxide combining with red blood cells reduces the number of red blood cells available to carry oxygen. This can result in a shortage of oxygen reaching the body tissues and therefore less oxygen available for respiration.

Overexposure to the Sun

Too much ultraviolet (UV) radiation from the Sun or sunbeds can cause skin cancer (Figure 13.13). The UV light can cause mutations in the skin that can lead to cancer.

In the UK, the number of people getting skin cancer is rapidly rising but the possibility of getting it can be reduced by reducing the time spent in strong sun, covering up and using sun lotion.

Misuse of drugs – alcohol

Many people drink alcohol in moderation and are unlikely to suffer serious harm. However, many people, including many teenagers, drink too much alcohol and can cause harm to themselves and others.

Long-term excessive drinking of alcohol can damage the liver as well as many other parts of the body. Drinking heavily during pregnancy can cause serious damage to the foetus including brain damage (foetal alcohol syndrome).

Binge drinking is a particular problem. This occurs when a large amount of alcohol is drunk over a short period of time, for example, on one night out (Figure 13.14).

Misuse of drugs – tobacco

Smoking can seriously damage health, as summarised in Table 13.2.

Table 13.2 The effects of tobacco smoke on the body

Substance in cigarette smoke	Harmful effect
Tar	causes bronchitis (narrowing of the bronchi and bronchioles), emphysema (damage to alveoli that reduces the surface area for gas exchange) and lung cancer (caused by abnormal cell division in lung cells)
Nicotine	addictive and affects heart rate
Carbon monoxide	combines with red blood cells to reduce the oxygen-carrying capacity of the blood

The introduction of smoking bans in many countries has proved very effective. It both encourages smokers to stop and significantly reduces the chances of people being affected by passive smoking. The use of E-cigarettes is enabling many people to stop smoking tobacco. However, not everyone is in favour of E-cigarettes as many contain nicotine and some people suggest that they encourage people to take up smoking tobacco.

Test yourself

7 What is meant by a non-communicable disease?
8 Describe the link between overexposure to the Sun and skin cancer.
9 State **one** harmful effect of alcohol on the body.
10 State **two** harmful effects of tar (in tobacco smoke) on the body.

Show you can

Explain why people who smoke a lot of tobacco can have low levels of energy.

Heart attacks and strokes (cardiovascular diseases)

Heart disease is caused by cholesterol and other fatty substances being present at such high levels that they build up in the walls of the arteries. Over time this leads to a narrowing of the arteries, making it more difficult for blood to flow through them. This is particularly likely to happen in the very narrow coronary arteries that supply the heart, hence the term coronary heart disease (CHD).

Eventually a coronary artery may become so narrow that a blockage forms (a clot) and stops the blood from flowing in this particular artery, shown in Figure 13.15. This prevents the heart muscle that the artery serves from receiving oxygen and glucose. Respiration can no longer happen in these cells, causing them to die and the heart to stop beating. This is a heart attack.

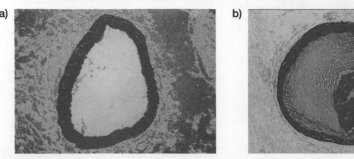

▲ **Figure 13.15 a)** A healthy coronary artery; **b)** A coronary artery that has become blocked by cholesterol (orange)

If the blockage is in the brain, a stroke can result. Again, the cells deprived of oxygen and glucose die and the affected part of the brain stops functioning properly. This often causes paralysis of parts of the body.

Some of the main lifestyle factors that can increase the risk of having a heart attack or a stroke are included in Figure 13.16.

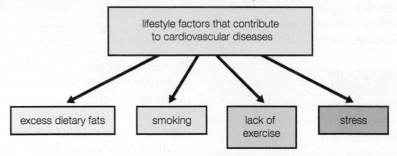

▲ **Figure 13.16** Factors that increase the risk of cardiovascular diseases

Adopting a healthy lifestyle (for example taking exercise and not smoking) can help reduce the risk of developing cardiovascular diseases. In recent decades, cardiovascular disease was often treated by surgery. Now there is a greater emphasis on other less invasive methods.

For example:

▶ Angioplasty and stents are often used to increase the space for blood in arteries. An angiograph is a medical imaging technique that allows doctors to see inside blood vessels. Dye is added to the blood through a thin tube that is inserted into a blood vessel that is close to the skin. The dye helps provide the contrast necessary when viewing the

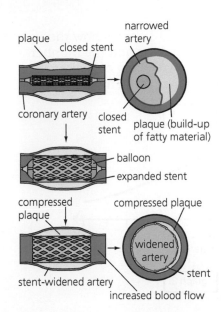

▲ **Figure 13.17** Keeping blood vessels open using stents

affected blood vessel. Balloon-like structures are used to hold the affected artery open for stents (small mesh like structures) to be inserted into the blood vessels to keep them open (see Figure 13.17).

▶ Drugs such as statins and aspirin can help protect against cardiovascular disease. Statins help reduce blood cholesterol and therefore the rate at which blood vessels can become clogged up with fatty deposits. Aspirin has similar effects. Low doses of aspirin are often given to people who have had a heart attack or stroke or are at risk of having one. The aspirin helps 'thin' the blood and makes it less 'sticky', therefore reducing the risk of a clot forming in the narrowed blood vessels and a heart attack or stroke occurring.

Some of the diseases covered in this section are closely linked. Consequently, many people who are affected by one condition (such as obesity) often suffer from one or more other conditions; obese people often develop cardiovascular disease and type 2 diabetes.

Test yourself

11 Give **three** lifestyle factors that can reduce the risk of having a heart attack.
12 Describe how stents can be used to treat someone with cardiovascular disease.

Show you can ?

Draw a flow chart to show how high cholesterol levels can lead to a heart attack.

Cancer

Cancer is uncontrolled cell division and this uncontrolled division can lead to the development of tumours.

There are two types of tumour:

▶ Benign tumours remain in one place and do not spread throughout the body. They may be surrounded by a distinct boundary or capsule (encapsulated).

▶ In malignant tumours groups of cancer cells may break off from the main (primary) tumour and spread around the body, where they can grow into other (secondary) tumours. Malignant tumours are less likely to have a distinct boundary or capsule around them. Malignant tumours are usually much more dangerous.

As with most of the non-communicable diseases discussed in this chapter, lifestyle choices can affect the risk of developing cancer.

In Northern Ireland girls in Year 9 are offered the HPV (human papilloma virus) vaccine. This vaccine helps protect against cervical cancer. The link between smoking and lung cancer is well known, as is the link between high levels of UV radiation and skin cancer.

Tip ↻

It is expected that around one in two people will get cancer at some point during their lives. Thankfully, the treatments are getting better all the time!

Practice questions

1 a) Copy and complete Table 13.3 about communicable diseases and how they are spread. *(5 marks)*

Table 13.3

Disease	Type of microorganism	Method of spread
Salmonella		in contaminated food
Flu	virus	
HPV		sexual contact
Potato blight		

b) Describe how tuberculosis can be:
  i) prevented *(1 mark)*
  ii) treated. *(1 mark)*

2 The flu is a communicable virus disease that can make people ill for a number of days.
  a) What is meant by the term communicable? *(1 mark)*

  b) Each year scientists make a vaccination against the flu. When they do this they have to predict which flu strain is likely to infect the most people. Some targeted groups such as the elderly and people with diabetes are now given the vaccination for free.
    i) Suggest why elderly people are offered the flu vaccination. *(1 mark)*
    ii) Explain how the vaccination enables people to be immune to the disease. *(3 marks)*
    iii) Use the information provided to suggest why it is possible for someone who has been given the vaccination to still get the flu. *(1 mark)*

3 a) Describe and explain **two** aseptic techniques followed when transferring bacteria from one agar plate to a second sterile agar plate. *(4 marks)*
  b) Why are Petri dishes containing microorganisms incubated at 25 °C or lower? *(1 mark)*

4 a) The graphs in Figure 13.18 show how the number of people smoking and the number of people diagnosed with lung cancer have changed over a 50 year period in a large city.

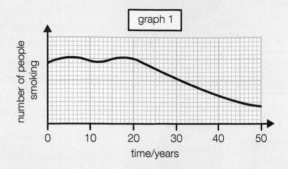

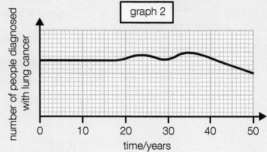

Figure 13.18

  i) What is the evidence from the graphs to suggest that smoking is a cause of lung cancer? *(1 mark)*
  ii) What is the evidence that suggests that it takes a number of years for lung cancer to develop? *(1 mark)*
  b)i) Name the component in tobacco smoke that causes lung cancer. *(1 mark)*
  ii) Name **one** other harmful effect that this component causes. *(1 mark)*

14 Atomic structure

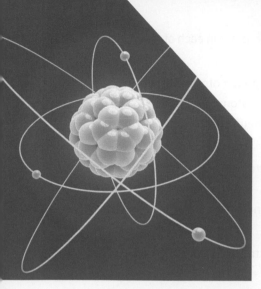

Specification points

This chapter covers specification points 1.1.1 to 1.1.11. It covers the structure of atoms and isotopes. Some aspects of maths – using decimal places – are referred to in terms of calculating the relative atomic masses of different atoms.

What are atoms? And what's inside them? In this chapter you will find the answers to these questions, and learn how to calculate the number of each type of particle found in an atom of different elements.

Protons, neutrons and electrons

An atom is the smallest particle of an element that can exist on its own in a stable environment. Each atom is made up of three types of particle: protons, neutrons and electrons. These are sometimes called subatomic particles. Table 14.1 shows the relative mass and relative charge of each particle, together with their position in the atom. The mass is given relative to the mass of a proton.

Tip

Try to remember that **p**rotons are **p**ositive and **neu**trons are **neu**tral, leaving electrons as negative.

Tip

The nucleus is the central part of the atom containing protons and neutrons.

Tip

The smaller of the two numbers given for each atom in the Periodic Table is the atomic number, and the larger number is the mass number, except for hydrogen where both numbers are the same!

Table 14.1

Particle	Relative mass	Relative charge	Position in the atom
proton	1	+1	nucleus
neutron	1	0	nucleus
electron	$\frac{1}{1840}$	-1	shells

Calculating the number of protons, neutrons and electrons in an atom

Look at the section of the Periodic Table shown in Figure 14.1. You will notice that each element has two numbers. These are called the atomic number and the mass number. **The atomic number is the number of protons in an atom.** The number of protons determines the identity of an atom – for example, all atoms with 7 protons (atomic number 7) are nitrogen atoms. The atomic number also equals the number of electrons in an atom.

▲ **Figure 14.1** A section of the Periodic Table

The heavier part of the atom is the nucleus, which contains protons and neutrons. **The mass number is the total number of protons and neutrons in an atom.** For example, the mass number of oxygen is 16.

The number of protons, neutrons and electrons in any atom can be worked out using the atomic number and mass number.

Number of protons = number of electrons = atomic number

Number of neutrons = mass number – atomic number

Tip

Note that the atomic number and mass number may not be given in the question. Use the Periodic Table in your data leaflet to locate the atom, and write down its mass and atomic number.

Tip

Note that hydrogen has no neutrons.

Example

How many protons, neutrons and electrons are in each of the following atoms?

a) $^{23}_{11}$Na

Number of protons = number of electrons = atomic number = 11

Number of neutrons = mass number – atomic number = 23 – 11 = 12

Answer: 11 protons, 11 electrons, 12 neutrons

b) $^{80}_{35}$Br

Number of protons = number of electrons = atomic number = 35

Number of neutrons = mass number – atomic number = 80 – 35 = 45

Answer: 35 protons, 35 electrons, 45 neutrons

c) $^{1}_{1}$H

Number of protons = number of electrons = atomic number = 1

Number of neutrons = mass number – atomic number = 1 – 1 = 0

Answer: 1 proton, 1 electron, 0 neutrons

Why are atoms electrically neutral?

An atom as a whole has no electrical charge because the number of protons is equal to the number of electrons.

For example, for the atom $^{27}_{13}$Al:

Number of protons = 13, each proton has a +1 charge, so this is a charge of +1 × 13 = +13

Number of neutrons = 14, each neutron has 0 charge so this is a charge of 0 × 14 = 0

Number of electrons = 13, each electron has a –1 charge, so this is a charge of –1 × 13 = –13

Total charge of the aluminium atom = + 13 + 0 – 13 = 0

Test yourself

1 State three differences between an electron and a proton.
2 State two similarities between a neutron and a proton.
3 If an atom has 7 electrons, how many protons does it have?
4 Name the element that has:
   a) 4 protons
   b) 14 protons.
5 Explain why atoms are electrically neutral.
6 How many protons, neutrons and electrons are there in an atom of $^{65}_{30}$Zn?

Show you can

Copy and complete Table 14.2 for each of the elements listed.

Table 14.2

Element	Atomic number	Mass number	Number of protons	Number of electrons	Number of neutrons
$^{7}_{3}Li$					
$^{11}_{5}B$					
$^{24}_{12}Mg$					
$^{39}_{19}K$					
$^{108}_{47}Ag$					
$^{127}_{53}I$					

Electronic configuration

Imagine a hockey team goes on tour with 13 players and stays at a hotel with two rooms on the ground floor, eight on the second and eight on the third. At the hotel, the manager assigns a room to each player using the rule that the rooms on each floor must be filled before a room on the next floor is allocated. As a result, 2 players have a room on the ground floor, 8 on the first and the remaining 3 on the second floor – 2,8,3 as shown in Figure 14.2. How would a netball team of 9 players be allocated?

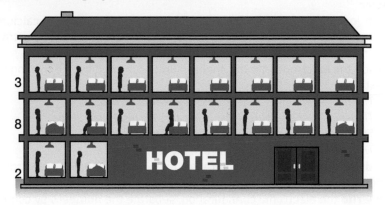

▲ **Figure 14.2** It can be useful to think of the shells in an atom as floors in a hotel, with each person occupying each room on a floor as an electron occupies a position in a shell

In an atom, electrons are arranged in a similar fashion but in shells or energy levels rather than on floors. Electrons occupy the lowest energy levels first. The first shell (lowest energy level), which is closest to the nucleus and can hold up to two electrons, is filled first. The second shell, which can hold up to eight electrons, is filled next. The third shell can also hold up to eight electrons.

Draw and write the electronic configuration for a phosphorus atom.

Answer

- First look at the Periodic Table and find the atomic number of phosphorus, this gives the number of protons, which is equal to the number of electrons.
  - *The atomic number is 15, so there are 15 electrons.*
- Then divide the electrons into the shells, filling each shell before moving to the next one.
  - *The first shell holds 2 electrons, this leaves 13 electrons.*
  - *The second shell holds 8 electrons, this leaves 5 electrons.*
  - *The third shell has 5 electrons.*
- To **draw** the electronic configuration, use a cross (×) or dot (•) to to represent the electrons as shown in Figure 14.3.
- To **write** the electronic configuration, separate the numbers by commas.
  - *The electronic configuration of phosphorus is written as 2,8,5.*

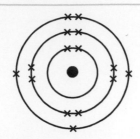

▲ **Figure 14.3** Drawn electronic configuration of phosphorus

Tip

In an examination you may be asked to draw or write the electronic configuration, electronic structure or electronic arrangement for an atom – these three terms all mean the same thing.

Draw and write the electronic configuration for a calcium atom.

Answer

- First look at the Periodic Table and find the atomic number of calcium. This gives the number of protons, which is equal to the number of electrons.
  - *The atomic number is 20, so there are 20 electrons.*
- Divide the electrons into the shells, filling each shell before moving to the next one.
  - *The first shell holds 2 electrons, this leaves 18 electrons.*
  - *The second shell holds 8 electrons, this leaves 10 electrons.*
  - *The third shell has 8 electrons, this leaves 2 electrons.*
  - *These two electrons are placed in the fourth shell.*
- To **draw** the electronic configuration, use × to represent the electrons as shown in Figure 14.4.
- To **write** the electronic configuration, separate the numbers by commas.
  - *The electronic configuration of calcium is written as 2,8,8,2.*

Note that the electronic configuration of calcium is the most difficult one needed for GCSE.

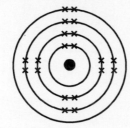

▲ **Figure 14.4** Electronic configuration of a calcium atom

Diagrams of atoms

▲ **Figure 14.5** The electronic configuration of hydrogen is written as 1. This is the simplest electronic configuration

Table 14.3 shows the electronic configuration of a sodium and a neon atom. It also shows a labelled diagram of a sodium and a neon atom. Note the difference. If asked to draw a labelled diagram of an atom you must show the position and number of protons and neutrons in the nucleus of the atom, as well as the electronic configuration. It is best not to draw each individual proton and neutron, but simply to write the number present and show their position in the nucleus with an arrow.

Table 14.3 The difference between drawing an electronic configuration and a labelled diagram of an atom

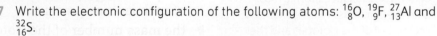

	Atom	
	sodium	neon
electronic configuration		
labelled diagram of the atom	11 protons + 12 neutrons in nucleus	10 protons + 10 neutrons in nucleus

---

Show you can

Figure 14.6 shows an atom of an element X, where:

e represents an electron
n represents a neutron
p represents a proton.

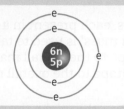

▲ **Figure 14.6**

1 Name the element X.
2 Write the electronic structure of X.
3 What is the mass number of this atom of element X?
4 Name the part of the atom shaded red.

Test yourself

7 Write the electronic configuration of the following atoms: $^{16}_{8}O$, $^{19}_{9}F$, $^{27}_{13}Al$ and $^{32}_{16}S$.
8 Draw the electronic configuration of chlorine and of argon.
9 Beryllium atoms contain 4 electrons and have the electronic structure 2,2. Explain why the electrons are not all in the first shell.
10 Draw a labelled diagram of an atom of:
  a) nitrogen
  b) potassium.

Isotopes

Most elements have atoms which have different mass numbers, for example chlorine has atoms with mass numbers 35 and 37.

Table 14.4 shows that, due to the different mass numbers, the numbers of neutrons are different. These atoms of chlorine are called isotopes.

Table 14.4 The isotopes of chlorine

Atom	Atomic number	Mass number	Number of protons	Number of neutrons	Number of electrons
^{35}Cl	17	**35**	17	**(35 – 17) = 18**	17
^{37}Cl	17	**37**	17	**(37 – 17) = 20**	17

Isotopes are atoms of an element with the same atomic number but a different mass number, indicating a different number of neutrons.

Tip

You will learn in Chapter 2 that when elements react, bonds involving electrons form. All isotopes of an element react in the same way, as they have the same number of electrons in the outer shell.

Hydrogen has three isotopes. Each atom of the isotope has the same atomic number (1), so each also has the same number of protons and electrons. However, each isotope has a different mass number (1, 2 and 3), and so the number of neutrons is different in each isotope. The isotopes of hydrogen are shown in Figure 14.7.

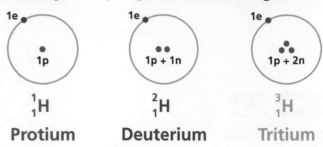

| $^{1}_{1}H$ | $^{2}_{1}H$ | $^{3}_{1}H$ |
| Protium | Deuterium | Tritium |

▲ **Figure 14.7** Isotopes of hydrogen

Calculating relative atomic mass for elements

Most elements contain a mixture of isotopes, each present in a different amount. The term 'weighted mean mass' is used to take into account the contribution made by the isotope to the overall mass of an element. The contribution made by each isotope to the overall mass depends on:

▶ the percentage abundance of the isotope – this gives the amount of each isotope present

▶ the mass number of the isotope.

The contributions of all the isotopes are then combined to give the relative atomic mass. **The relative atomic mass is a weighted mean of the mass numbers.**

$$\text{Relative atomic mass } (A_r) = \frac{(\text{mass number 1} \times \text{abundance}) + (\text{mass number 2} \times \text{abundance})}{\text{total abundance}}$$

Tip

Relative atomic mass is not the same as mass number. Mass numbers are whole numbers because they are the number of protons plus the number of neutrons. Relative atomic mass is an average mass which takes into account all of the isotopes and their abundance, and so is not always a whole number. However, on your data leaflet, apart from chlorine, the relative atomic masses have been rounded to whole numbers.

Example

A sample of chlorine is found to contain 75% of atoms with mass number 35 and 25% of atoms with mass number 37. Calculate the relative atomic mass of chlorine to one decimal place.

Answer

Relative atomic mass $(A_r) = \dfrac{(35 \times 75) + (37 \times 25)}{(75 + 25)} = \dfrac{3550}{100} = 35.5$

Sometimes in calculations you are asked to present your answer to one or two decimal places. Rounding a number to one decimal place means there is only one digit after the decimal point. Rounding a number to two decimal places means there are two digits after the decimal point.

The rules for rounding are:

▶ if the next number is 5 or more, round up

▶ if the next number is 4 or less, do not round up.

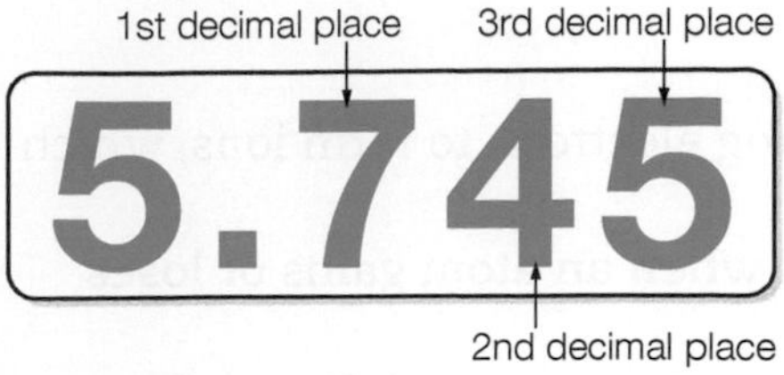

1st decimal place   3rd decimal place

5.745

2nd decimal place

▲ **Figure 14.8** 5.745 has three decimal places

For example, 5.745 rounded to one decimal place is 5.7, because the next number is 4 so it is not rounded up. If 5.745 is rounded to two decimal places, because the next number is 5 it is rounded up to 5.75.

Example

A sample of copper contained 69% ^{63}Cu and 31% ^{65}Cu. Calculate the relative atomic mass of copper to one decimal place.

Answer

$$\text{Relative atomic mass } (A_r) = \frac{(63 \times 69) + (65 \times 31)}{(69 + 31)} = \frac{6362}{100} =$$

$$= 63.62 = 63.6 \text{ to one decimal place}$$

Tip

Relative atomic mass is not a mass in grams, do not include units in your answer.

Example

Table 14.5 shows the abundances of three isotopes of magnesium.

Table 14.5

Isotope	Abundance
^{24}Mg	16
^{25}Mg	2
^{26}Mg	2

Calculate the relative atomic mass of magnesium to one decimal place.

Answer

$$\text{Relative atomic mass } (A_r) = \frac{(24 \times 16) + (25 \times 2) + (26 \times 2)}{(16 + 2 + 2)}$$

$$= \frac{486}{20} = 24.3$$

Test yourself

11 Hydrogen has three isotopes ^{1}H (protium), ^{2}H (deuterium) and ^{3}H (tritium). Calculate the number of protons, neutrons and electrons in each isotope of hydrogen.
12 State the similarities and differences between the isotopes ^{207}Pb and ^{208}Pb.
13 A sample of bromine contains 53% bromine with mass number 79 and 47% bromine with mass number 81. Calculate the relative atomic mass of bromine and give your answer to one decimal place.
14 Calculate the relative atomic mass of silicon using the data in Table 14.6. Give your answer to one decimal place.

Table 14.6

Isotope	^{28}Si	^{29}Si	^{30}Si
% abundance	92.2	4.7	3.1

Ions

Many atoms react by losing or gaining electrons to form ions, which have a full outer shell.

An ion is a charged particle formed when an atom gains or loses electrons.

Ions are charged because the number of protons is different from the number of electrons.

A negative ion is called an anion, for example Cl^-.

A positive ion is called a cation, for example Na^+.

A sodium atom ($^{23}_{11}Na$) has 11 electrons and electronic configuration 2,8,1. If the atom loses an electron, then the sodium ion formed has 10 electrons arranged 2,8. The ion formed has a positive charge as it has one more proton than electrons, as shown in Table 14.7.

Table 14.7 Comparing a sodium atom and sodium ion

	sodium atom, Na	sodium ion, Na+
atomic number	11	11
mass number	23	23
charge due to the protons	(+1 × 11) = 11+	(+1 × 11) = 11+
charge due to the neutrons	(0 × 12) = 0	(0 × 12) = 0
charge due to the electrons	(−1 × 11) = 11−	(−1 × 10) = 10−
overall charge	+11 + 0 − 11 = 0	+ 11 + 0 − 10 = 1+

Tip

When writing the charge of an ion, there is no need to put in the digit 1. Also, the number comes before the sign.

An aluminium atom ($^{27}_{13}Al$) has 13 electrons arranged 2,8,3. If the atom loses three electrons, then the aluminium ion formed has 10 electrons arranged 2,8. The ion formed has a 3+ charge as it has three more protons than electrons.

Table 14.8 Comparing an aluminium atom and aluminium ion

	aluminium atom, Al	aluminium ion, Al³⁺
atomic number	13	13
mass number	27	27
charge due to the protons	(+1 × 13) = 13+	(+1 × 13) = 13+
charge due to the neutrons	(0 × 14) = 0	(0 × 14) = 0
charge due to the electrons	(−1 × 13) = 13−	(−1 × 10) = 10−
overall charge	+13 + 0 − 13 = 0	+13 + 0 − 10 = 3+

An oxygen atom ($^{16}_{8}O$) has 8 electrons arranged 2,6. If the atom gains two electrons, then the oxide ion formed has 10 electrons arranged 2,8. The ion formed has a 2– charge, as it has two more electrons than protons.

Table 14.9 Comparing an oxygen atom and an oxide ion

	oxygen atom, O	oxide ion, O²⁻
atomic number	8	8
mass number	16	16
charge due to the protons	(+1 × 8) = 8+	(+1 × 8) = 8+
charge due to the neutrons	(0 × 8) = 0	(0 × 8) = 0
charge due to the electrons	(−1 × 8) = 8−	(−1 × 10) = 10−
overall charge	+8 + 0 − 8 = 0	+ 8 + 0 − 10 = 2−

Table 14.10 compares an anion and a cation.

Table 14.10 The differences between anions and cations

Cation	Anion
positively charged	negatively charged
formed when electrons are lost from an atom	formed when electrons are gained by an atom
more protons than electrons	more electrons than protons
name of the ion is the same as the atom	name of the ion ends in -ide

During a chemical reaction compounds form, many of which contain anions and cations.

A compound is two or more elements chemically combined. You will learn more about this in Chapter 2.

Drawing diagrams of ions

If asked to draw a labelled diagram of an atom you must show the position and number of protons and neutrons, as well as the electronic configuration. When drawing a labelled diagram of an ion you must also show the position and number of protons and neutrons in the nucleus, and the electronic configuration. Also show the charge on the ion.

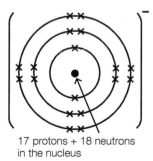

17 protons + 18 neutrons in the nucleus

▲ **Figure 14.9** A labelled diagram of a chloride ion

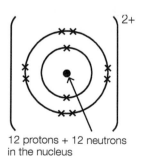

12 protons + 12 neutrons in the nucleus

▲ **Figure 14.10** A labelled diagram of a magnesium ion

Test yourself

15 Copy and complete Table 14.11.

Table 14.11

	Symbol	Mass number	Atomic number	Number of protons	Number of neutrons	Number of electrons	Electronic structure
calcium atom	Ca	40	20				
calcium ion		40	20				
oxygen atom		16	8				
oxide ion		16	8				
nitride ion	N^{3-}	14	7				
sodium atom	Na		11				
sodium ion	Na^+	23	11				

16 An ion has 8 protons and 9 electrons. What is the charge of the ion?

17 An atom of aluminium loses 3 electrons, what is its charge?

18 How many electrons does a potassium ion have? What is its charge?

19 What is the charge of a particle with 7 protons and 10 electrons?

20 How many protons, neutrons and electrons are there in the $^{19}_{9}F^-$ ion?

Show you can

Table 14.12 gives some information about some different particles: A, B, C, D, E and F.
Some particles are **atoms** and some are **ions**. (The letters A, B, C, D, E and F are not the chemical symbols for the elements.)

Table 14.12

Particle	Atomic number	Mass number	Number of protons	Number of neutrons	Number of electrons	Electronic structure
A	18	40				2,8,8
B		27	13			2,8
C			20	20	20	
D		35	17			2,8,7
E	16	32			18	
F	17			20	17	

1 Copy and complete the table.

2 Particle C is an atom. Explain, using the information in the table, why particle C is an atom.

3 Particle E is a negative ion. What is the charge on this ion?

4 Which two atoms are isotopes of the same element?

Practice questions

1 An aluminium atom contains three types of particle.

a) Copy and complete Table 14.13 to show the name, relative mass and relative charge of each particle in an aluminium atom. *(4 marks)*

Table 14.13

Particle	Relative charge	Relative mass
proton		1
		$\dfrac{1}{1840}$
neutron	0	

b) Complete the sentences about an aluminium atom below by choosing **one** of the words in bold. *(4 marks)*

i) In an aluminium atom, the protons and neutrons are in the **nucleus/shells**.

ii) The number of protons in an aluminium atom is the **atomic number/group number/ mass number**.

iii) The sum of the number of protons and neutrons in an aluminium atom is the **atomic number/group number/mass number**.

iv) The number of electrons in an aluminium atom is **13/14/27**.

c) Look at the following atoms and ions.

^{12}C ^{14}C $^{16}O^{2-}$ $^{19}F^-$ ^{20}Ne

Which of these atoms and ions, if any:
i) are isotopes of the same element?
ii) have 9 protons?
iii) have 10 electrons?
iv) have 10 neutrons?
v) have more protons than electrons? *(5 marks)*

2 The structure of the atom has caused debate for thousands of years. In the late 19th century the plum pudding model of the atom was proposed. This was replaced at the beginning of the 20th century with the model of the atom we use today.

a) The diagram below represents an atom of an element. The electrons are missing from the diagram.

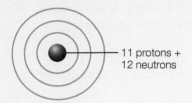

Figure 14.11

i) State the atomic number of this element. *(1 mark)*

ii) State the mass number of this element. *(1 mark)*

iii) Name the part of the atom in which the protons and neutrons are found. *(1 mark)*

iv) Copy and complete the diagram to show the electronic configuration of the atom, using a cross to represent an electron. *(1 mark)*

b) Table 14.14 shows some information for several atoms and simple ions. Copy and complete the table. *(6 marks)*

Table 14.14

Atom/ion	Number of protons	Electronic configuration
	7	2,5
S^{2-}		
Ca^{2+}		
	12	2,8

3 Potassium is a very reactive element. An isotope of potassium has mass number 39 and atomic number 19.

a) What is meant by the term mass number? *(1 mark)*

b) What is meant by the term atomic number? *(1 mark)*

c) Draw a labelled diagram of an atom of potassium. *(4 marks)*

d) Potassium reacts with chlorine to form potassium chloride, which contains potassium ions and chloride ions.

  i) Calculate the number of protons, and number of electrons, in a chloride ion. *(1 mark)*

  ii) Write down the charge and the electronic configuration of the chloride ion. *(1 mark)*

  iii) Potassium chloride is a compound. What is meant by the term compound? *(1 mark)*

4 Silver has two isotopes ^{109}Ag and ^{107}Ag. In a sample of silver, 53% has a mass of 109 and 47% has a mass of 107.

  a) Calculate the relative atomic mass of silver to one decimal place. *(2 marks)*

  b) Define the term isotopes and explain why ^{109}Ag and ^{107}Ag are isotopes. *(3 marks)*

  c) Silver can form Ag$^+$ ions. How many protons, neutrons and electrons are present in a ^{109}Ag$^+$ ion? *(2 marks)*

  d) Silver oxide contains Ag$^+$ ions and oxide ions. What is the formula of an oxide ion? *(1 mark)*

5 Table 14.15 shows some information about atoms and ions.
Copy and complete the table. *(5 marks)*

Table 14.15

Atom/ion	Al	Sn^{2+}	Ba	H$^+$	Se^{2-}	Cl
Atomic number	13	50		1	34	17
Mass number	27	119	137	1		
Number of protons			56			
Number of neutrons					45	20
Number of electrons						

15 Bonding, structures and nanoparticles

Specification points

This chapter covers specification sections 1.2, 1.3 and 1.4 and is called bonding, structures and nanoparticles. It covers ionic, molecular, giant covalent and metallic substances and their properties, and includes an explanation of the three main types of bonding. The idea of nanoparticles is introduced.

In addition to showcasing the best athletes in the world, the Olympic Games also showcase many new materials – including clothing to help reduce the aerodynamic drag of the athlete and different types of plastics, alloys, lighting and adhesives – all used in the design of the Olympic stadium, velodrome and swimming pools. To design new materials chemists need information about their structure – how the atoms or ions are arranged, and the bonding – how they are held together. You will learn about bonding, structure and new materials in this chapter.

Why do atoms bond?

Only six elements exist naturally as single unbonded atoms. These monatomic elements are called the **noble gases** and are listed in Figure 15.1. In the atoms of a noble gas the outer shell is filled – for example, helium has two electrons in the outer shell, which fills this shell, and neon has 8 electrons in the outer shell, which fills it (Figure 15.2). Hence the noble gas atoms are stable and exist on their own.

▲ **Figure 15.1** The noble gases

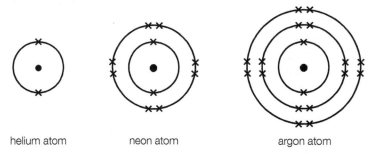

helium atom   neon atom   argon atom

▲ **Figure 15.2** Full outer shells in the noble gases

Noble gases are unreactive due to their **filled outer shell** which is very stable. Most other atoms in the Periodic Table react and bond with other elements in order to obtain a noble gas (stable) electronic structure.

There are three types of bonding studied at GCSE:

▶ ionic bonding
▶ covalent bonding
▶ metallic bonding.

Ionic bonding

Ionic bonding occurs in compounds that contain a metal and a non-metal, usually a compound of a Group 1 or 2 element, with a Group 6 or 7 element. The atoms of these elements do not have filled outer shells, and need to bond to obtain full outer shells and become stable.

When an ionic bond is formed between two atoms:

▶ electrons are transferred from the metal atom to the non-metal atom

▶ oppositely charged ions are formed – the **metal ion** has a **positive** charge, the **non-metal ion** has a **negative** charge – and the ions are held together by attraction between opposite charges – this is the ionic bond. **An ionic bond is the attraction between oppositely charged ions.**

▶ Each ion formed has a filled outer shell, and hence a stable noble gas electronic structure.

Ionic bonding can be represented by dot and cross diagrams. These are diagrams in which all electrons are represented by dots (•) and crosses (×). To draw dot and cross diagrams for ionic bonding, it is useful to follow the steps below:

1 Draw the electronic configuration of each atom, one element with dots and the other with crosses, and work out how many electrons need to be transferred.

2 Show, with arrows, the transfer of electrons.

3 Draw the electronic configuration of each ion.

4 Write the charge of each ion.

▲ **Figure 15.3** Sodium reacting vigorously with chlorine to form the ionic compound sodium chloride

 Example

The bonding in sodium chloride

When sodium reacts with chlorine, each sodium atom loses one electron and each chlorine atom gains one electron, in order to gain full outer shells and become stable. This produces sodium ions with a 1+ charge and chloride ions with a 1– charge. The ionic bond is the attraction between the oppositely charged ions and the compound sodium chloride is formed. The formula of sodium chloride is NaCl.

 Tips

▶ To explain the ionic bonding of a compound, always draw a dot and cross diagram similar to the one shown in Figure 15.4.

▶ Remember that electrons are all the same, but the dots and crosses show which atoms the electrons come from.

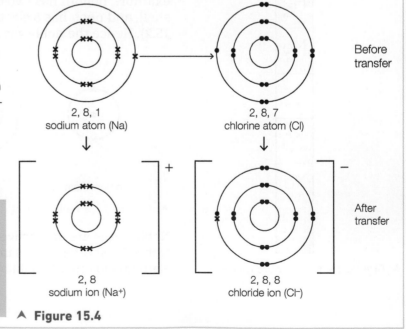

2, 8, 1
sodium atom (Na)

2, 8, 7
chlorine atom (Cl)

Before transfer

2, 8
sodium ion (Na⁺)

2, 8, 8
chloride ion (Cl⁻)

After transfer

▲ **Figure 15.4**

Example

The bonding in magnesium oxide

When magnesium reacts with oxygen, each magnesium atom loses two electrons and each oxygen atom gains two electrons, in order to gain full outer shells and become stable. This produces magnesium ions with a 2+ charge and oxide ions with a 2– charge. The ionic bond is the attraction between the oppositely charged ions, and the compound magnesium oxide is formed. The formula of magnesium oxide is MgO.

Tips

▶ Remember that charges are written with the number first, e.g. 2– not –2, and 2+ not +2.
▶ The metal ion is always positive and the non-metal ion is always negative.

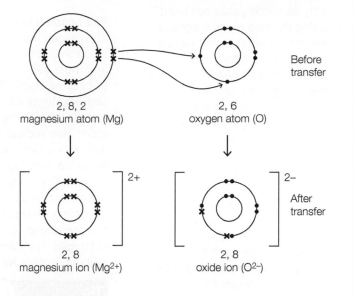

▲ **Figure 15.5**

Example

▲ **Figure 15.6** Calcium fluoride is a white solid. It occurs as the mineral fluorite, which is shown here, and is often coloured due to impurities

The bonding in calcium fluoride

A calcium atom has electronic configuration 2,8,8,2 and that of fluorine is 2,7. The calcium atom must lose two electrons to become stable, and so two fluorine atoms are required to accept the two electrons. Each fluorine atom gains one electron and this produces two fluoride ions, each with a 1– charge, and the calcium atom forms a calcium ion with 2+ charge. The ionic bond is the attraction between the oppositely charged ions, and the compound calcium fluoride is formed. The formula of calcium fluoride is CaF_2.

Tip

Note that the outer two calcium electrons (represented by ×) transfer. You must show that each fluoride ion has 7 electrons (·) of its own and one electron (×) from calcium.

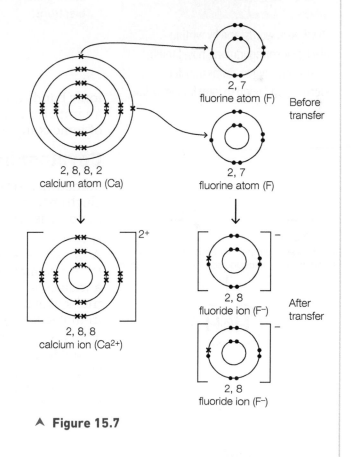

▲ **Figure 15.7**

Example

The bonding in sodium oxide
The oxygen atom has the electronic configuration 2,6 and must gain two electrons to become stable. Two sodium atoms each transfer one electron to the oxygen atom. Two sodium ions, each with a 1+ charge, and an oxide ion with charge 2−, are formed. The ionic bond is the attraction between the oppositely charged ions, and the compound sodium oxide is formed. The formula of sodium oxide is Na_2O.

Tip

Remember that an ion is a charged particle formed when an atom gains or loses electrons.

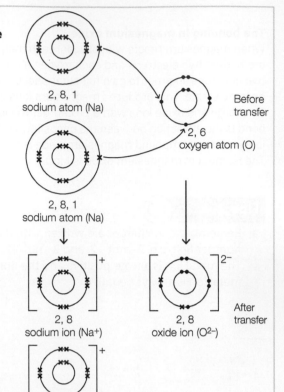

▲ **Figure 15.8**

Ionic bonds are **strong** and need **substantial energy** to **break**. This is an important fact to remember when learning about the properties of ionic compounds on page 149.

Covalent bonding

Sometimes atoms join together to form molecules by sharing electrons, in order to obtain a noble gas configuration. Covalent bonding is typical of non-metallic elements and compounds.

A covalent bond is a shared pair of electrons.
A molecule is two or more atoms covalently bonded together.
Diatomic means that there are two atoms covalently bonded in a molecule. For example, NH_3, H_2O, **N_2**, **HCl**, CO_2 and **Cl_2** are all molecules but only those in bold are diatomic molecules. The atoms in molecules are held together by covalent bonding.

A hydrogen atom (H) has the electronic configuration 1 and is unstable. It requires one more electron in its outer shell to be stable. Two hydrogen atoms each share one electron in a covalent bond and form H_2. Dot and cross diagrams can be used to show the covalent bonding in a molecule (Figure 15.9).

needs one electron so shares one electron   needs one electron so shares one electron

shared pair of electrons – covalent bond

▲ **Figure 15.9** Formation of a covalent bond in a hydrogen molecule

A chlorine atom (Cl) has the electronic configuration 2,8,7 and is unstable. Two chlorine atoms each share one electron to form Cl_2 (Figure 15.10).

2,8,7
needs one electron
so shares one electron

2,8,7
needs one electron
so shares one electron

covalent bond

▲ **Figure 15.10** Formation of a covalent bond in a chlorine molecule

In water (H_2O), the oxygen atom has the electronic configuration 2,6 and requires two more electrons to become stable. Two hydrogen atoms each share one electron with the oxygen atom, and two covalent bonds are formed. In a dot and cross diagram the covalent bond is the shared pair of electrons (•×) and a lone pair **is an unbonded pair of electrons** (××) (Figure 15.11).

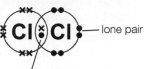

  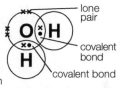 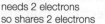

lone pair

covalent bond

covalent bond

2,6
needs 2 electrons
so shares 2 electrons

1
needs one electron
so shares one electron

1
needs one electron
so shares one electron

▲ **Figure 15.11** Formation of covalent bonds in water

A covalent bond may also be represented by a line. A single covalent bond is represented by a single line (–), a double covalent bond by two lines (=) and a triple covalent bond by three lines (≡). Diagrams showing covalent bonds represented as lines are called **structural formulae**. Table 15.1 shows the dot and cross diagrams and the structural formula for each of the molecules on your specification.

Table 15.1 Covalent bonding

Molecule	Dot and cross diagram (a covalent bond is represented as •×)	Number of lone pairs	Number of covalent bonds	Structural formula
H_2	H×H	0	1	H–H
Cl_2	×Cl×Cl×	6	1	Cl–Cl
HCl	H×Cl×	3	1	H–Cl
H_2O	H×O× H	2	2	H–O | H
NH_3	H×N×H H	1	3	H–N–H | H
CH_4	H H×C×H H	0	4	H | H–C–H | H
O_2	×O××O×	4	2	O=O
N_2	×N×××N×	2	3	N≡N
CO_2	×O××C××O×	4	4	O=C=O

Covalent bonds are **strong** and **substantial energy** is required to **break** them.

Test yourself ✐

6 Draw a dot and cross diagram for each of the following molecules.
  a) HF
  b) F_2
  c) SiH_4
  d) H_2S
  e) CCl_4
  f) OCl_2
  g) SiF_4
7 The dot and cross diagram shows nitrogen trichloride.

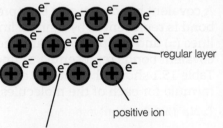

▲ **Figure 15.12**

  a) Copy the diagram and label one lone pair and one covalent bond.
  b) Using a line to represent a covalent bond, draw the structural formula of nitrogen trichloride.
  c) How many lone pairs are there in NCl_3?

Show you can   ?

Phosphorus bonds with hydrogen to from phosphine (PH_3), a colourless gas which has an unpleasant, rotting fish odour. Phosphorus also bonds with chlorine to form phosphorus trichloride, which is a toxic colourless liquid.

1 Draw a dot and cross diagram to show the bonding in PH_3.
2 Suggest the formula of phosphorus trichloride.
3 Draw a dot and cross diagram to show the bonding in phosphorus trichloride.
4 Using your diagram from (3), explain what is meant by a covalent bond and a lone pair.

Metallic bonding

Metallic bonding is the attraction between the positive ions in a regular lattice and the delocalised electrons.

The positive ions in a metal are due to the atoms losing their outer shell electrons. These electrons are delocalised. **Delocalised electrons are electrons that are free to move throughout the whole structure.** The bonding in a metal is shown in Figure 15.13.

delocalised electrons
positive ion
regular layer

▲ **Figure 15.14** All metals have metallic bonding

▲ **Figure 15.13** Metallic bonding

Tip ↻

If asked to draw the bonding in any metal (e.g. copper, aluminium), draw the diagram shown in Figure 15.13. You do not have to adapt it for a specific metal. Try and have the same number of positive ions and negative electrons.

Determining the type of bonding present

It is important that you can determine which type of bonding – ionic, covalent or metallic – is present in a substance.

Table 15.2 Determining the type of bonding present

Description	Type of bonding	Example
a compound of a metal and a non-metal	ionic	magnesium sulfide (MgS)
a compound of non-metals	covalent	carbon dioxide (CO_2)
a diatomic non-metallic element	covalent	chlorine (Cl_2)
a metal	metallic	copper

Test yourself ✐

8 What is a metallic bond?
9 What are delocalised electrons?
10 Decide if the following substances contain ionic, covalent or metallic bonding.
  a) calcium oxide
  b) magnesium
  c) CCl_4
  d) O_2
  e) NaF
  f) Br_2
  g) aluminium
  h) potassium oxide
  i) SiH_4
  j) Ti
  k) OF_2
  l) CaF_2

Structures

Bonding describes how atoms or ions are held together.

Structure describes how the atoms or ions are arranged in space.

There are four different types of structure, as shown in Table 15.3.

Table 15.3 Different types of structure and the bonding present

Structure	Type of bonding in the structure
metallic	metallic (attraction between positive ions and delocalised electrons)
giant ionic lattice	ionic (attraction between oppositely charged ions)
molecular covalent	covalent (atoms are joined by sharing pairs of electrons)
giant covalent	covalent (atoms are joined by sharing pairs of electrons)

Metallic structures

Structure description

The structure of a metal is a giant lattice of positive ions arranged in regular layers, with delocalised electrons free to move throughout the structure. A metallic lattice is a three-dimensional structure of positive ions and delocalised electrons bonded by metallic bonds. The physical properties of a metal can be explained by examining the structure.

Physical properties

1 Metals are good conductors of electricity

Metals are good conductors of electricity because the delocalised electrons can move and carry charge.

2 Metals usually have a high melting point

For metals to melt, the metallic bonds must be broken. Metals have high melting points because the metallic bond (the attraction between the positive ions and delocalised electrons) is strong and takes substantial energy to break.

3 Metals are malleable and ductile

Malleable means the metal can be hammered into shape without breaking.

Ductile means the metal can be drawn into a wire.

Metals are malleable and ductile because the layers of ions can slide over each other, yet the delocalised electrons still attract the ions and hold the structure together – the metallic bonding is not disrupted (see Figure 15.16).

▲ Figure 15.15 Copper is used in electrical cables because it is an excellent conductor of electricity, has a high melting point and is ductile

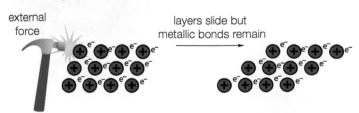

external force

layers slide but metallic bonds remain

▲ Figure 15.16 Metals are malleable

Alloys

An alloy is a mixture of two or more elements, at least one of which is a metal, and the resulting mixture has metallic properties.

> **Tip**
>
> An alloy is a mixture, not a compound – the elements in it are not chemically combined.

Test yourself

11 a) Name the type of bonding in silver.
   b) Name the type of structure in silver.
   c) Why is silver a good conductor of electricity?
   d) Why is silver ductile?
   e) What is meant by the term malleable?

Giant ionic lattices

Structure description

In ionically bonded substances, the oppositely charged ions attract each other from all directions, forming a giant ionic lattice structure containing millions of ions, as shown in Figure 15.18.

A giant ionic lattice is a three-dimensional structure of oppositely charged ions held together by ionic bonds.

In a giant ionic lattice, each ion is surrounded by oppositely charged ions. In sodium chloride, each sodium ion is surrounded by six chloride ions, and each chloride ion is surrounded by six sodium ions, as shown in Figure 15.18.

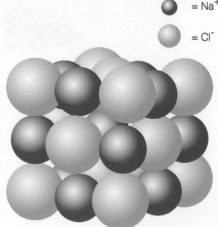

$= Na^+$

$= Cl^-$

▲ **Figure 15.17** Sodium chloride crystals are cubic. They have a giant ionic lattice structure

▲ **Figure 15.18** A three-dimensional model of the giant ionic lattice structure of sodium chloride

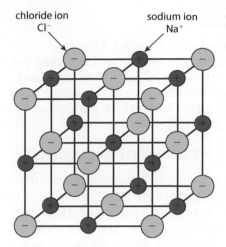

chloride ion
Cl⁻

sodium ion
Na⁺

▲ **Figure 15.19** Part of a sodium chloride giant ionic lattice

Tips

► Note that in Figure 15.19, the ions are shown with lines joining them. The lines are not covalent bonds, they simply help show how the ions are arranged in the lattice. Look at the central positive ion – can you see 6 lines, showing the 6 negative ions attracted to it?

► You should be able to recognise, but do not need to be able to draw, the diagrams shown in Figure 15.18 and Figure 15.19.

Physical properties

1 High melting point and boiling point

When an ionic solid melts, the lattice breaks down and the ions become free to move. Ionic solids have high melting points and boiling points because it takes a substantial amount of energy to break the strong ionic bonds that hold the ions together (see Figure 15.20).

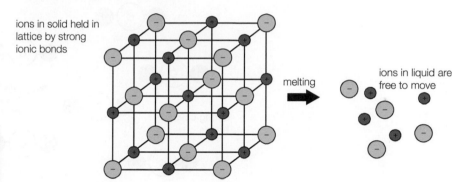

ions in solid held in lattice by strong ionic bonds

melting

ions in liquid are free to move

▲ **Figure 15.20** Strong ionic bonds require substantial energy to break

Tip

Refer to the amount of energy needed to break the bonds – do not say a high temperature is needed.

Tip

You do not need to explain why ionic solids are soluble in water.

Tip

When ionic substances conduct electricity it is because the ions move, it is not due to the movement of electrons.

2 Soluble in water

Most ionic compounds are soluble in water and form aqueous solutions. **An aqueous solution is a solution in which the solvent is water.**

3 Good conductors of electricity when molten or dissolved

An electric current is a flow of charged particles such as ions or electrons. Ionic compounds are made of ions but in the solid state the ions cannot move and so ionic solids do not conduct electricity. If the ionic solid is melted, the **ions can move and carry charge.** Ionic compounds conduct electricity when molten. If an ionic compound is dissolved in water to form a solution then the ions can move and it can conduct electricity.

Test yourself

12 Explain why magnesium chloride has a high melting point.
13 Why does sodium chloride solution conduct electricity?
14 Why does solid sodium chloride not conduct electricity?
15 What is a giant ionic lattice?
16 Which of the following substances have a giant ionic lattice structure?

   CO_2, PH_3, $MgBr_2$, NaF, SiO_2, K_2S

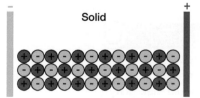

Solid

Does not conduct electricity
The ions are held tightly in the lattice and cannot move to carry the charge.

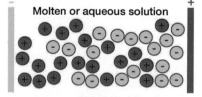

Molten or aqueous solution

Conducts electricity
The ions can now move and carry the charge. The positive ions move to the negative terminal and the negative ions to the positive terminal.

▲ **Figure 15.21**

Table 15.4

Molecules	
Elements	Compounds
Cl_2	H_2O
H_2	CO_2
I_2	NH_3

Molecular covalent structures

A molecule is two or more atoms covalently bonded. Many substances are composed of molecules. The atoms in the molecules may be the same, for example some non-metal elements are made of molecules, or the atoms may be different, for example some non-metal compounds are made of molecules (Table 15.4). Molecular covalent structures contain simple molecules.

Description of a molecular covalent structure

Within each molecule in a covalent molecular structure the atoms are joined by strong covalent bonds, for example in I_2 there is a covalent bond between each iodine atom in the molecule. However, the **molecules** are **not** bonded to each other. There are only weak intermolecular forces between the molecules. These weak forces are called **van der Waals'** forces (see Figure 15.22).

Tip

Inter means 'between', so intermolecular forces are forces between the molecules.

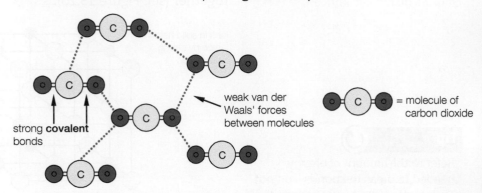

▲ **Figure 15.22** The van der Waals' forces between carbon dioxide molecules

Physical properties

Tip

The covalent bonds in molecules do not break when covalent molecular substances change state.

1 Low melting points and boiling points
Little energy is needed to break the weak van der Waals' forces between the molecules, so covalent molecular structures have low melting and boiling points. As a result, covalent molecular substances are gases, liquids or low melting point solids at room temperature. Generally, the bigger the molecules, the stronger the van der Waals' forces and the higher the melting and boiling points.

2 Do not conduct electricity
Covalent molecular substances do not conduct electricity because the molecules are neutral and there are no charged particles (no ions or electrons) to move and carry charge.

▲ **Figure 15.23** Butter contains a mixture of simple molecular substances. What property does it have that shows this?

Tip

Pure water does not conduct electricity because it is made of molecules. If there are any ionic substances dissolved in water, those ionic substances will conduct electricity.

Tip

Exceptions to the insoluble rule include hydrogen chloride and ammonia, which *are* soluble in water as they react with it.

3 Insoluble in water
Many simple covalent molecular substances, like oxygen gas, are insoluble in water or have very low **solubility** in water (see Figure 15.24).

▲ **Figure 15.24** A bottle of solid iodine (molecular covalent) next to a beaker containing iodine dissolved in hexane (top layer, purple) and iodine dissolved in water (bottom layer). Solid iodine is seen at the bottom of the beaker. Iodine does not dissolve well in water, explaining the faint brown colour observed. It does dissolve well in hexane, forming a strong purple colour

Practical activity

Testing the electrical conductivity of ionic and molecular covalent substances

To investigate the conduction of electricity by a number of compounds in aqueous solution, the apparatus was set up as shown in the diagram.

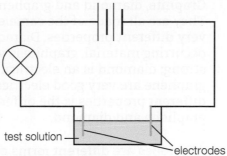

test solution — electrodes

▲ **Figure 15.25**

1 Describe the experimental method which you would use to carry out a fair test on the solutions using the apparatus shown.

2 Copy and complete the results table (Table 15.5).

3 Using the results from columns three and four of the table, write a conclusion for this experiment, stating and explaining any trends shown in the results.

4 Would the results be different if solid copper(II) sulfate was used instead of copper(II) sulfate solution? Explain your answer.

5 Predict and explain the results you would obtain for calcium nitrate solution.

6 Predict and explain the results you would obtain for bromine solution (Br_2).

Table 15.5

Test solution	Does the bulb light?	Does the substance conduct electricity?	Does the substance contain ionic or covalent bonding?
copper(II) sulfate	yes		
ethanol (C_2H_5OH)	no		
magnesium sulfate	yes		
potassium iodide	yes		
glucose ($C_6H_{12}O_6$)	no		
sodium chloride	yes		

Test yourself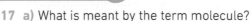

17 a) What is meant by the term molecule?
   b) Classify the substances below as atoms, molecules of elements, or molecules of compounds.
   HF, H_2, F, H_2O, CH_4, Cl_2, O, $SiCl_4$, O_2
18 Explain why sulfur has a low melting and boiling point.
19 What happens to:
   a) the covalent bonds and
   b) the van der Waals' forces in a molecular substance when it boils?
20 Explain why iodine does not conduct electricity.
21 Which of the following substances are covalent molecular structures?
   H_2S, Na_2O, KNO_3, $ZnBr_2$, CO, N_2H_4

Show you can

Metal oxides and non-metal oxides have different properties. Sulfur dioxide, a non-metal oxide has a melting point of −72 °C, and calcium oxide, a metal oxide, has a melting point of 2613 °C. Explain why the melting point of sulfur dioxide is low and that of calcium oxide is high.

Giant covalent structures

A giant covalent structure is a three-dimensional structure of atoms that are joined by covalent bonds.

Such structures are large continuous networks of covalently bonded atoms.

Graphite, diamond and graphene have giant covalent structures. They are all made of the same element, carbon, but they have very different properties. Diamond is the hardest known naturally occurring material, graphite is very soft and graphene is incredibly strong; diamond is an electrical insulator and graphite and graphene are very good electrical conductors. The reason for these different properties is the different structures found in graphite, graphene and diamond.

Allotropes are different forms of the same element, in the same state.

Graphite, graphene and diamond are allotropes. They are different forms of the same element (carbon) in the same state (solid).

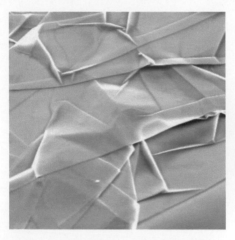

▲ **Figure 15.26** Diamond, graphite and graphene are three very different giant covalent substances, but are all composed of the same element. The photo of graphene is an electron micrograph and shows the thin sheets

Tips

▶ Carbon has six electrons and the electronic configuration 2,4. It has four electrons in the outer shell, which can form four covalent bonds with four other carbon atoms in diamond but with three other carbon atoms in graphite, leaving one electron free per carbon atom.
▶ You should be able to draw and label diagrams of all three allotropes of carbon (Figures 15.27, 15.28, 15.29).

Table 15.6 compares the diagram, structure, physical properties and uses of the three carbon allotropes.

Table 15.6

	Diamond	Graphite	Graphene
Diagram of structure	▲ **Figure 15.27**	▲ **Figure 15.28**	▲ **Figure 15.29**
Structure description	Each carbon atom is covalently bonded to four others in a **tetrahedral** three-dimensional structure.	Layers of carbon atoms are arranged in **hexagons**, with covalent bonds between the atoms and weak forces between the layers. Each carbon atom is covalently bonded to three others, so one electron per carbon atom is unbonded and delocalised and free to move between the layers. Three-dimensional.	A single-atom thick layer of graphite with strong covalent bonds between each carbon atom. The atoms are arranged in **hexagons**. Two-dimensional.
Melting point and boiling point	High The many covalent bonds are strong and substantial energy is needed to break them.	High The many covalent bonds are strong and substantial energy is needed to break them.	High The many covalent bonds are strong and substantial energy is needed to break them.
Electrical conductivity	Does not conduct There are no free ions or delocalised electrons to move and carry the charge.	Good conductor One electron per carbon atom is unbonded. The unbonded electrons are delocalised electrons, which are free to move and carry charge.	Good conductor One electron per carbon atom is unbonded. The unbonded electrons are delocalised electrons, which are free to move and carry charge.
Hardness/ strength	Hard This is due to the three-dimensional tetrahedral structure with strong covalent bonds.	Soft This is because there are weak forces between the layers, which allow the layers to slide.	Very strong Graphene is 100 times stronger than the strongest steel. This is due to the strong covalent bonds in the layer. It is also very light, the thinnest material possible and transparent, due to the one-atom thick layer.
Use	Cutting tools Diamond-tipped tools are used to cut through hard rock, metals and glass.	Lubricants for machinery In **pencil leads**	It is a strong, light and relatively inexpensive electrical conductor and will have many future uses, for example in **solar cells and batteries**.

▲ **Figure 15.30** Pencil makes a mark on paper because the layers of graphite are only held by weak forces and slip off easily, leaving a mark

▲ **Figure 15.31** Graphite in a pencil being used to lubricate a zip fastener

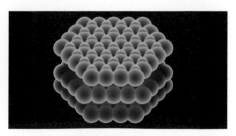

▲ **Figure 15.32** Layered structure of graphite

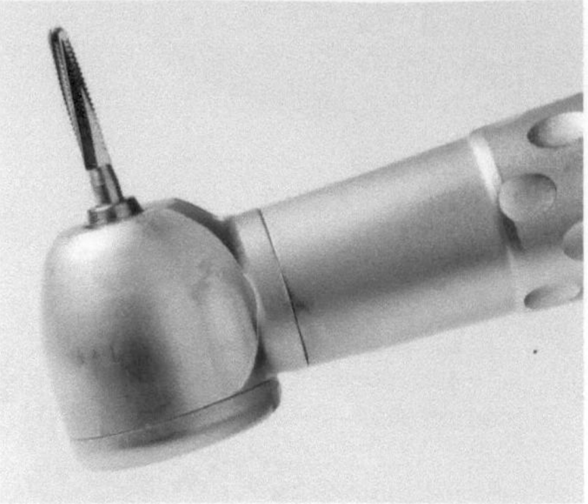

▲ **Figure 15.33** A dental drill often has a diamond coating

Show you can ❓

Copy and complete Table 15.7 to give information about the different structures and uses of carbon.

Table 15.7

	Graphite	Diamond	Graphene
Description of structure			A single-atom thick layer of graphite with strong covalent bonds between each carbon atom, arranged in hexagons.
Uses			

Test yourself ✏️

22 Explain why diamond and graphite both have a very high melting point.
23 Explain why diamond is hard but graphite is soft.
24 Explain why graphite conducts electricity but diamond does not.
25 What is graphene?
26 State two differences and two similarities between graphene and diamond.
27 Draw a labelled diagram of the structure of graphene and the structure of graphite, and state one difference between them.
28 What is an allotrope? Name three allotropes of carbon.

▲ **Figure 15.34** Graphene is a new substance discovered by Andre Geim (above) and Konstantin Novoselov at the University of Manchester. In their initial work they removed flakes from graphite using sticky tape! They won a Nobel Prize in 2010 for their work on graphene

Using data to classify structures

Often in examination questions data is given about different substances, and you may be asked to classify the substance as a giant ionic lattice, molecular covalent, metallic or giant covalent substance. The data may include melting point, boiling point, conductivity or solubility. It is important that you know the different physical properties of each different type of structure. Table 15.8 is a helpful summary.

Table 15.8 Physical properties of different structures

Structure	Melting point: high or low?	Soluble in water?	Conducts electricity?
giant ionic lattice	high	yes	yes, when molten or dissolved
molecular covalent	low	no	no
metallic	usually high	no	yes
giant covalent (diamond)	high	no	no
giant covalent (graphite, graphene)	high	no	yes

Example

Use the data in Table 15.9 to answer the questions.

Table 15.9

Substance	Electrical conductivity as solid	Electrical conductivity as liquid	Electrical conductivity as aqueous solution	Melting point /°C	Boiling point /°C
A	poor	good	good	776	1500
B	good	good	not soluble	327	1760
C	poor	poor	not soluble	−95	69
D	poor	poor	not soluble	3550	4827
E	good	poor	not soluble	3720	4837

1 Which one of A−E is an ionic substance? Explain your answer.
Ionic substances do not conduct electricity when solid, only when dissolved in solution or melted into a liquid, so A is ionic. It also has a high melting point and boiling point.

2 Which one of A−E is simple covalent molecular substance? Explain your answer.
A simple covalent molecular substance does not conduct electricity and is not soluble in water, so options are C and D. However, covalent molecular substances have low melting points and boiling points, so the answer is C.

3 Which one of A−E is diamond? Explain your answer.
Diamond does not conduct electricity and has a high melting point and boiling point, so D is diamond.

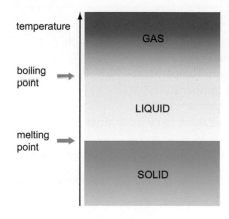

▲ **Figure 15.35** A substance will change states with increasing temperature

It is also useful to be able to interpret data about melting point and boiling point and determine if a substance is solid, liquid or gas at room temperature (20 °C). As illustrated in Figure 15.35, a substance is:

▶ a solid at temperatures below its melting point
▶ a liquid at temperatures between its melting and boiling points
▶ a gas at temperatures above its boiling point.

Example

Use the data in Table 15.10 to answer the questions.

Table 15.10

Substance	Melting point/°C	Boiling point/°C
A	45	137
B	595	984
C	−30	56
D	−189	−186
E	186	302

1 Which substance(s) is/are gases at room temperature (20 °C)?
If a substance is a gas at room temperature, the boiling point must be less than room temperature, so the answer is D.

2 Which substance(s) is/are liquids at 100 °C?
If a substance is a liquid at 100 °C then its melting point must be below 100 °C (so the answer could be A, C or D), but its boiling point must be above 100 °C. So the answer is A.

3 Which substance(s) is/are solids at 100 °C?
If a substance is a solid at 100 °C then its melting point must be above 100 °C and its boiling point must be above 100 °C. So the answer is B and E.

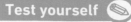

Test yourself

29 Copy and complete Table 15.11.

Table 15.11

Substance	Type of bonding	Type of structure
copper(II) oxide (CuO)		
diamond (C)		
lead carbonate (PbCO$_3$)		
phosphorus oxide (P$_4$O$_{10}$)		
copper (Cu)		
graphene (C)		
ammonia (NH$_3$)		

30 The melting points and boiling points of six substances are shown in Table 15.12.

Table 15.12

Substance	Melting point/°C	Boiling point/°C	Type of bonding present
N$_2$	−210	−196	
CS$_2$	−112	46	
NH$_3$	−78	−34	
Br$_2$	−7	59	covalent
LiCl	605	1137	
Cu	1084	2562	

a) Copy and complete the table by deciding if the bonding in each substance is ionic, covalent or metallic.

b) Which element(s) is/are solid at room temperature (20 °C)?

c) Which compound is liquid at room temperature?

d) Which compound is a gas at room temperature?

e) Which element will condense when cooled to room temperature from 100 °C?

f) Which compound will freeze first on cooling from room temperature to a very low temperature?

Show you can

Substances may be classified in terms of their physical properties. Use Table 15.13 to answer the questions.

Table 15.13

Substance	Melting point/°C	Boiling point/°C	Electrical conductivity as solid	Electrical conductivity as liquid
A	3720	4827	good	poor
B	−95	69	poor	poor
C	327	1760	good	good
D	3550	4827	poor	poor
E	776	1500	poor	good

1 Which substance could be sodium chloride? Explain your answer.

2 Which substance consists of small covalent molecules? Explain your answer.

3 Explain why substance A could not be diamond.

4 Which substance is a metal?

Tip

One nanometre is 0.000 000 001 m
(1×10^{-9} m).

▲ **Figure 15.36** These polymer nanoparticles can pass through vein walls and are used in drug delivery. The drug can be released into tissues in exactly the area needing treatment

Nanoparticles

Few people had heard of the term nanoparticles before the year 2000. Today, many materials are being developed and used in the form of nanoparticles.

What is a nanoparticle?

Lumps and powders are examples of **bulk** materials and consist of huge numbers of atoms. Nanoparticles consist of only a few hundred atoms.

A nanoparticle is a structure that is 1–100 nm in size and contains a few hundred atoms.

Nanoparticles have very different properties and uses from the bulk material. This is due to their high surface area to volume ratio.

Use of nanoparticles in sun creams

▲ **Figure 15.37** Zinc oxide nanoparticles are used in sun cream

Sun cream is a product that absorbs some of the sun's ultraviolet radiation and helps protect against sunburn. Traditional sun creams contain zinc oxide, a white solid that absorbs ultraviolet radiation but is difficult to rub in. Nanoparticles of zinc oxide are now often used in sun creams, and have several **benefits**:

▶ they give better skin coverage to the sun cream
▶ they give more effective protection from the sun's ultraviolet rays
▶ they are clear and colourless, which makes the sun cream invisible on the skin
▶ they do not degrade on exposure to the sun.

There are, however, risks with the use of nanoparticles. It is often assumed that because the bulk material is safe to use that nanoparticles of the same material are also safe. However, nanoparticles have different properties from the bulk material, and so it is reasonable to assume they may have harmful effects on humans or the environment. The risks are difficult to determine because nanoparticulate materials have not been in use for long.

The **risks** of using nanoparticles in sun cream include:

▶ potential cell damage in the body – nanoparticles are so small they may be able to penetrate cell membranes, or be breathed in. In the body they may be more reactive or more toxic than the bulk material.

▶ harmful effects on the environment.

Test yourself

31 What are nanoparticles?
32 Explain why nanoparticles have different properties from the bulk materials.
33 A gold nanoparticle is 21 nm in diameter. Calculate the diameter of the nanoparticle in metres. Use standard form.
34 Explain why there are fears about the use of nanoparticles.

Show you can

Sun creams should show on the label that they contain nanoparticles. Why should this information be included? Pick two statements from the list below.

• Nanoparticles do not occur in nature.
• Nanoparticles have a smaller surface area than larger particles.
• Nanotechnology increases the cost of the sun creams.
• Not all the effects of nanoparticles are fully understood.
• Creams containing nanoparticles are easy to apply.
• Nanoparticles can occur naturally.
• Nanoparticles may be harmful.

Practice questions

1 A dot and cross diagram is given below.

Figure 15.38

a) Name and write the formula for this compound. *(2 marks)*

b) Copy the diagram above and use an arrow to label the following features:
  - a covalent bond
  - a lone pair of electrons. *(2 marks)*

c) Using a line to represent a single covalent bond, redraw the diagram shown above. *(1 mark)*

d) What is meant by the term 'single covalent bond'? *(2 marks)*

2 The diagram below shows three different forms of the element carbon, and carbon dioxide.

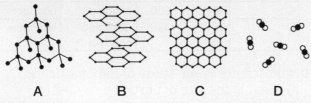

Figure 15.39

a) Name A, B, C and D. *(4 marks)*

b) Name the type of bonding that occurs between the atoms in the substances A–D. *(4 marks)*

c) Name of the type of structure for each substance A–D. *(4 marks)*

d) Explain why carbon dioxide has a low boiling point. *(3 marks)*

e) Which two substances conduct electricity? *(1 mark)*

3 Graphite is used in pencils and for electrodes in the electrolysis of sodium chloride solution.

a) With reference to the structure of graphite, explain why it is used in pencils. *(3 marks)*

b) Graphite electrodes conduct electricity. Explain why graphite is a good conductor of electricity. *(2 marks)*

c) Describe, as fully as you can, what happens when sodium atoms react with chlorine atoms to produce sodium chloride. You may use a diagram in your answer. *(6 marks)*

d) Explain why sodium chloride solution will conduct electricity but sodium chloride solid will not. *(2 marks)*

e) Copy and complete Table 15.14. *(4 marks)*

Table 15.14

	Type of bonding	Type of structure
Graphite		
Sodium chloride		

4 Table 15.15 gives some of the properties of the element magnesium and one of its compounds, magnesium chloride.

Table 15.15

Property	magnesium	magnesium chloride
melting point/°C	649	714
electrical conductivity when solid	conducts	does not conduct
electrical conductivity when molten	conducts	conducts

Use ideas about structure and bonding to explain the similarities and differences between the properties of magnesium and magnesium chloride. *(6 marks)*

5 a) Chlorine is a green-yellow gas which exists as diatomic molecules.
  i) What is meant by the term diatomic? *(1 mark)*
  ii) Use a dot and cross diagram to clearly show how atoms of chlorine combine to form chlorine molecules. *(2 marks)*

b) Chlorine can form a range of compounds with both metals and non-metals.
  i) Describe, as fully as you can, what happens when calcium atoms react with chlorine atoms to produce calcium chloride. You may use a diagram in your answer. *(6 marks)*
  ii) Name the type of bonding found in calcium chloride. *(1 mark)*
  iii) Name the type of structure for calcium chloride. *(1 mark)*
  iv) Explain why calcium chloride solution can conduct electricity. *(3 marks)*
  v) Use a dot and cross diagram to show how atoms of chlorine combine with atoms of carbon to form tetrachloromethane CCl_4. *(3 marks)*
  vi) Name the type of bonding found in CCl_4. *(1 mark)*
  vii) Name the type of structure for CCl_4. *(1 mark)*

c) The properties of compounds depend very closely on their bonding. In Table 15.16 choose the correct word to show some of the expected properties of calcium chloride and tetrachloromethane. *(2 marks)*

Table 15.16

Compound	Formula	Solubility in water	Relative melting point
calcium chloride	$CaCl_2$	soluble/ insoluble	low/high
tetrachloromethane	CCl_4	soluble/ insoluble	low/high

H

d) The bonding and structure of the elements calcium and carbon is very different.
i) Name and describe the bonding in calcium *(3 marks)*
ii) Draw a labelled diagram to show the structure of calcium. *(3 marks)*
iii) Describe the structure of graphite. *(3 marks)*
e) Both calcium and graphite can conduct electricity.
i) Why can calcium conduct electricity? *(2 marks)*
ii) State two physical properties of calcium that are different from those of graphite. *(2 marks)*

6 Carbon dioxide, oxygen and nitrogen are gases found in the air.
a) Copy and complete Table 15.17. *(7 marks)*

Table 15.17

	carbon dioxide	nitrogen	oxygen
formula		N_2	
dot and cross diagram			
structural formula			O=O

b) What is a lone pair of electrons? *(1 mark)*
c) Why do these gases not conduct electricity? *(1 mark)*

7 The table below gives information about the melting point, boiling point and electrical conductivity of four substances A, B, C and D. Use the information in Table 15.18 to answer the questions that follow.

Table 15.18

Substance	Melting point/ °C	Boiling point/ °C	Electrical conductivity when solid	Electrical conductivity when molten
A	−182	−151	does not conduct	does not conduct
B	660	2500	conducts	conducts
C	808	1465	does not conduct	conducts
D	3550	4800	does not conduct	does not conduct

Identify the substance A, B, C or D which: **H**
a) is a gas at room temperature (20 °C)
b) exists as oppositely charged ions in a giant ionic lattice
c) exists as small molecules
d) could be diamond. *(4 marks)*

8 Name the type of structure found in:
a) zinc (Zn)
b) ethane (C_2H_6)
c) diamond (C)
d) magnesium oxide (MgO)
e) iodine trifluoride (IF_3)
f) potassium carbonate (K_2CO_3). *(6 marks)*

9 Use the data to state the type of structure found **H** in A, B, C and D. *(4 marks)*

Table 15.19

Substance	Melting point/ °C	Boiling point/ °C	Electrical conductivity as solid	Electrical conductivity as liquid
A	838	1239	does not conduct	conducts
B	89	236	does not conduct	does not conduct
C	678	935	conducts	conducts
D	1056	1438	does not conduct	conducts

10 Nanoscience is the study of nanoparticles.
a) What is the size of particles studied in nanoscience? *(1 mark)*
b) State two benefits of using nanoparticles in sun cream. *(2 marks)*
c) A silver nanoparticle has a diameter of 19 nm. State the diameter of the particle in metres. Do not use standard form. *(1 mark)*

16 The Periodic Table

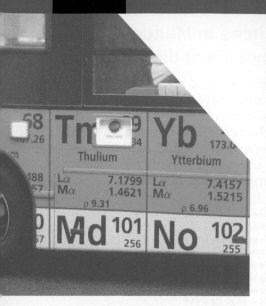

The Periodic Table hangs on the wall of most chemistry laboratories, and can be found in other unusual places, for example on this bus advertising a science park! The table is a powerful icon that summarises much of our knowledge of chemistry. In this chapter you will look at the development of the Periodic Table and the chemistry of Groups 1, 7 and 0 and the transition metals.

Mendeleev and the Periodic Table

Dmitri **Mendeleev** developed the Periodic Table in 1869, while working as a professor of chemistry at St Petersburg University in Russia. He made a set of 63 cards – one for each of the known 63 elements – on which he had written the symbol, atomic weight and properties of the element. He then arranged them in order of increasing atomic weight, and grouped together elements having similar properties. Mendeleev's work resulted in the production of a Periodic Table based on atomic weights but arranged periodically. His work was different from other scientists in two main ways:

▶ Mendeleev listed the elements **in order of increasing atomic weight** but he was prepared to alter slightly the order of the elements if he felt the properties matched better. For example, he swapped the order of iodine (atomic weight 127) and tellurium (atomic weight 128) so that iodine was in the same group as chlorine and bromine, which had similar properties. He believed the atomic weight of tellurium was measured incorrectly.

▶ He left **gaps** where no element fitted the repeating pattern. The gaps were for **undiscovered elements**, and he also predicted the properties of these undiscovered elements.

Some years after Mendeleev produced his table, elements were discovered that he had predicted, and their properties closely matched his predictions. One such element was eka-silicon, which was renamed germanium – its properties and predicted properties are shown in Table 16.1.

▲ **Figure 16.1** T-shirts have different properties (e.g. size and colour), and can be organised in different ways. Similarly, different chemists, over time, organised the elements, which have different properties, in different ways

Table 16.1

	Element name	Appearance	Atomic weight	Density (g/cm³)
Mendeleev's predictions	eka-silicon	grey metal	72	5.5
actual properties of the element	germanium (Ge)	grey metal	74	5.4

Similarities and differences in Mendeleev's Periodic Table and today's Periodic Table

Development of the Periodic Table continued after Mendeleev. For example, an entire group of very unreactive elements – the noble gases – was added after they were discovered at the end of the 19th century. There are now over 100 elements in the Periodic Table. Elements 113, 115, 117 and 118 were discovered in 2016. The Periodic Table we have today is similar to, but also markedly different from, Mendeleev's table. Look at the comparison in Table 16.2.

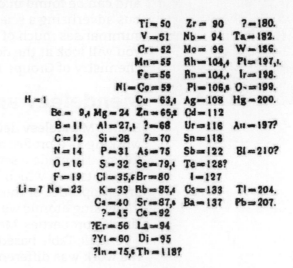

▲ **Figure 16.2** Mendeleev's table. Note the question mark after Te, showing that he thought it was incorrect. Note also the ? and predicted atomic weight for missing elements

Table 16.2 Comparison of Mendeleev's table and the modern Periodic Table

Mendeleev's table	Modern Periodic Table
elements arranged in order of atomic weight	elements arranged in order of atomic number
no noble gases present – they had not been discovered	noble gases present
gaps were left for undiscovered elements (fewer elements in this table)	no gaps (more elements)
no block of transition metals	a block of transition metals is present
actinides and lanthanides not present	actinides and lanthanides are present (a block of elements at the bottom of the table)

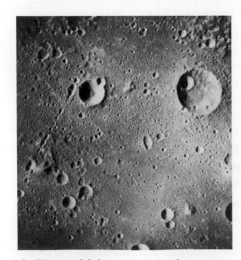

▲ **Figure 16.3** A crater on the moon was named after Mendeleev, and in 1955 a new element was named mendelevium after the famous scientist

PERIODIC TABLE OF ELEMENTS

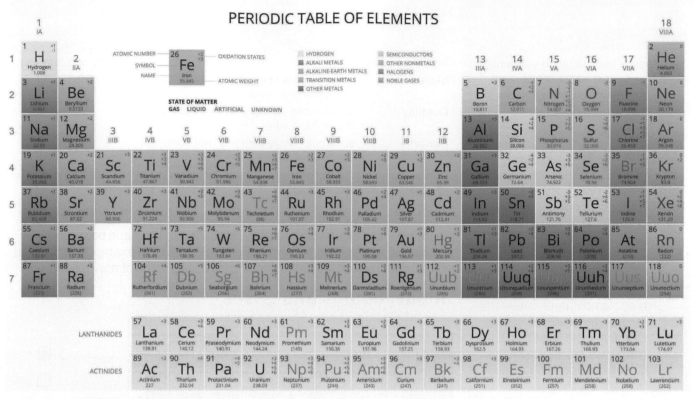

▲ **Figure 16.4** Today's Periodic Table

▲ **Figure 16.6** Metals are sonorous and are used in bells

Arrangement of elements in the Periodic Table

An element is a substance that consists of only one type of atom. Elements cannot be broken down into simpler substances by chemical means.

Elements can be classified as metals and non-metals.

Metals and non-metals

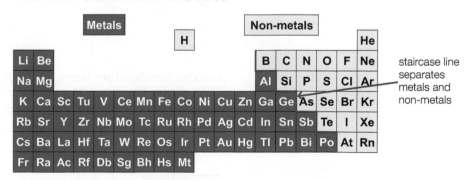

▲ **Figure 16.5** The Periodic Table is separated into metals and non-metals

Figure 16.5 shows the position of metals and non-metals in the Periodic Table. There are more metals than non-metals – over three quarters of the elements are metals. The differences between the properties of metals and non-metals are shown in Table 16.3.

Table 16.3

Test yourself

5 Classify the following elements as metals or non-metals:
   a) chromium   f) sodium
   b) potassium   g) argon
   c) hydrogen   h) bromine
   d) boron   i) manganese
   e) phosphorus   j) nickel.
6 Is each of the following elements a metal or non-metal?
   a) Element 1 makes a ringing sound when struck and is used in electrical cabling.
   b) Element 2 is a gas which does not conduct electricity.
   c) Element 3 is a yellow solid at room temperature that easily melts when warmed.
   d) Element 4 is a solid that conducts heat very well and can be hammered into shape to make saucepans.
7 'Only metals conduct electricity.' Explain if this is a true statement.

Show you can

Carbon (graphite and diamond) are non-metals and aluminium is a metal. State whether each of these substances conducts electricity, and explain why it does or does not.

▲ **Figure 16.8** Mercury expands and contracts quickly and is used in some thermometers

Table 16.3

	Metals	Non-metals
Melting points	high	low (except for graphite and diamond)
Malleability	malleable – can be easily hammered into shape	not malleable – the solids are brittle
Ductility	ductile – can be easily drawn into a wire	not ductile
Sonority	sonorous – make a ringing sound when struck	not sonorous
Conduction of heat and electricity	good conductors of heat and electricity	poor conductors of heat and electricity (except for graphite and graphene)

There are a few elements around the staircase dividing line between metals and non-metals that are difficult to classify. For example, silicon and germanium have some properties of metals and some of non-metals.

Solids, liquids and gases

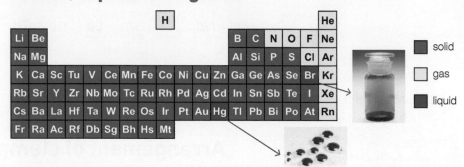

▲ **Figure 16.7** Mercury and bromine are the only two elements that are liquid at room temperature

Figure 16.7 shows the elements in the Periodic Table classified as solids, liquids and gases at room temperature and pressure. The majority of elements are solids. There are two liquids, bromine and mercury, and these are described in Table 16.4.

Table 16.4 Liquids in the Periodic Table

Element	Symbol	Metal or non-metal	Description
bromine	Br	non-metal	red–brown liquid
mercury	Hg	metal	silvery–white liquid

At room temperature there are 11 gases in the Periodic Table, six of which are noble gases. The gases are listed in Table 16.5. Note that all the gases apart from the noble gases are diatomic.

All the other elements are solids at room temperature and pressure.

Table 16.5 Gases in the Periodic Table

Gas	Symbol/ formula	Appearance	Gas	Symbol/ formula	Appearance
hydrogen	H_2	colourless	helium	He	colourless
nitrogen	N_2	colourless	neon	Ne	colourless
oxygen	O_2	colourless	argon	Ar	colourless
fluorine	F_2	yellow	krypton	Kr	colourless
chlorine	Cl_2	yellow–green	xenon	Xe	colourless
			radon	Rn	colourless

Organisation of elements

In the Periodic Table elements are arranged in groups and periods, as shown in Figure 16.9.

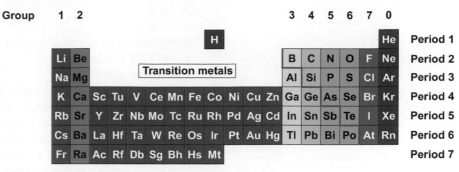

▲ **Figure 16.9** The groups and periods of the Periodic Table

A group is a vertical column in the Periodic Table.

All elements in the same group have **similar chemical properties (reactions)** because they have the **same number of electrons** in their outer shell. **The number of electrons in the outer shell** of the atom is the **same as the group number**. For example, the element bromine is in Group 7 and has 7 electrons in the outer shell of an atom. The element magnesium is in Group 2 and each magnesium atom has 2 electrons in its outer shell. **The exception to this is Group 0**. It is not that the atoms of the elements in this group have no electrons in their outer shell, rather the 0 refers to the fact that these elements are non-reactive because their atoms have full outer shells and are stable. The Group 0 elements are often called the noble gases – some other groups in the Periodic Table have names, which you need to remember. These are shown in Table 16.6.

Table 16.6

Group	Name	Metals or non-metals	Reactive or non-reactive
1	alkali metals	metals	reactive
2	alkaline earth metals	metals	reactive
7	halogens	non-metals	reactive
8	noble gases	non-metals	non-reactive

Group 1 metals are very reactive because the outer shell of the atoms contains one electron which is lost easily to form ions with a 1+ charge. Group 7 elements are also very reactive as they have 7 electrons in their outer shell and can easily gain one electron to form ions with a 1– charge.

Between Group 2 and Group 3 there is a block of metals called the **transition metals**.

A period is a horizontal row in the Periodic Table.

Note that the first period is very short and contains only two elements – hydrogen and helium.

Tip

Do not write 'alkaline metals' for Group 1 – the correct name is 'alkali metals'.

Show you can

An atom of element A has the electronic structure 2,8,1.

1 Explain why element A is not found in Group 5.

2 Determine the atomic number of A.

▲ **Figure 16.10** Alkali metals are stored under oil

▲ **Figure 16.11** Sodium metal is easily cut with a knife, revealing a shiny surface which quickly tarnishes

▲ **Figure 16.12** The alkali metals

Group 1 (alkali metals)

Group 1 elements have similar physical and chemical properties because they all have one electron in the outer shell of their atoms. They are very reactive metals and are **stored under oil** (Figure 16.10) to prevent them reacting with air or water vapour in the air.

Physical properties

1 All alkali metals are **soft** and are easily cut with a knife, as shown in Figure 16.11.

2 Alkali metals are **shiny** when freshly cut but tarnish (go dull) rapidly when exposed to air.

3 Alkali metals have **low density** and the first three – lithium, sodium and potassium – are less dense than water, and thus float.

4 All alkali metals are **grey solids** and have relatively low melting points. The melting points decrease as the atoms get bigger going down the group.

Reaction with water

Group 1 elements all react with water, releasing hydrogen gas and forming a metal hydroxide.

Group 1 metal + water → metal hydroxide + hydrogen

For example:

Word equation: lithium + water   →   lithium hydroxide + hydrogen

Symbol equation: $2Li(s) + 2H_2O(l)$ → $2LiOH(aq) + H_2(g)$

Tip

The hydroxides produced are alkalis. A few drops of indicator can be added to show this.

Lithium, sodium and potassium can all be reacted with water, by a teacher, in the school laboratory. A risk assessment must first be carried out.

Risk assessment for reaction of alkali metals and water

▶ Use tweezers when lifting alkali metals. This prevents handling the metal directly – it could react with water in the skin, causing a burn.

▶ Use a safety screen. This prevents the corrosive hydroxide solution from splashing out, or the metal sparking out and causing damage to the skin.

▶ Wear safety glasses. This prevents the corrosive hydroxide produced from splashing into the eye and causing damage.

▶ Use small pieces of metal to avoid excessive splashing out, and use a large volume of water in a trough.

Preparation of alkali metals for reaction

1. Cut a small piece of metal.
2. Remove the oil layer using filter paper.
3. Use tweezers to drop the metal into a large trough of water, behind a safety screen and wearing safety glasses.

Table 16.8

Metal	Photo	Observations for reaction with water	Word and symbol equation
lithium		• moves on the water surface • bubbles • heat released • Li disappears at the end of the reaction • colourless solution produced.	lithium + water → lithium hydroxide + hydrogen $2Li(s) + 2H_2O(l) \rightarrow 2LiOH(aq) + H_2(g)$
sodium		• Na melts into a tiny ball • moves on the water surface • bubbles • heat released • Na disappears at the end of the reaction • colourless solution produced.	sodium + water → sodium hydroxide + hydrogen $2Na(s) + 2H_2O(l) \rightarrow 2NaOH(aq) + H_2(g)$
potassium		• lilac flame • K melts into a tiny ball • moves on the water surface • bubbles • heat released • K disappears quickly • colourless solution produced • crackles at end.	potassium + water → potassium hydroxide + hydrogen $2K(s) + 2H_2O(l) \rightarrow 2KOH(aq) + H_2(g)$

▲ **Figure 16.13** Rubidium reacting with water

Reactivity increases as the group descends. The least reactive alkali metal is lithium. Rubidium and caesium are not reacted with water in the school laboratory as they react explosively, as shown in Figure 16.13.

Word equation:	rubidium + water	→ rubidium hydroxide + hydrogen
Symbol equation:	$2Rb(s) + 2H_2O(l)$	→ $2RbOH(aq) + H_2(g)$
Word equation:	caesium + water	→ caesium hydroxide + hydrogen
Symbol equation:	$2Cs(s) + 2H_2O(l)$	→ $2CsOH(aq) + H_2(g)$

Francium is at the bottom of the group. The chemistry of francium is not often studied as it is radioactive.

What happens when Group 1 metals react?

Group 1 metals all react in the same way, as their atoms have one electron in their outer shell. When these metals react, each atom loses one electron to form a positive ion with a charge 1+. The ion has a full outer shell and is stable.

A half equation is one that includes electrons. A half equation can be written for the formation of a Group 1 metal ion from its atom.

General half equation: **Group 1 metal atom → Group 1 metal ion + electron**

$$X → X^+ + e^-$$

For example:

Half equation for sodium: $Na → Na^+ + e^-$

Half equation for potassium: $K → K^+ + e^-$

Why does the reactivity of Group 1 metals increase down the group?

When a Group 1 atom reacts it **loses an electron** from its outer shell to form a **positive ion** with a full outer shell (see Figure 16.14).

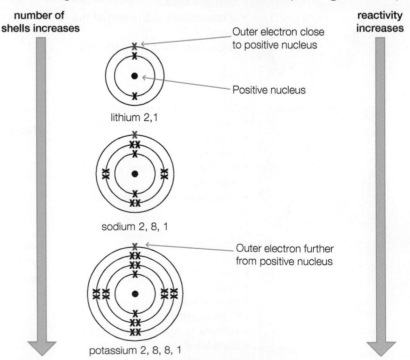

number of shells increases

reactivity increases

Outer electron close to positive nucleus

Positive nucleus

lithium 2,1

sodium 2, 8, 1

Outer electron further from positive nucleus

potassium 2, 8, 8, 1

▲ **Figure 16.14** The further down the group, the further the outer electron is from the positive nucleus and the less attracted it is

Tip

For practice writing half equations see page 249 and the Appendix.

Example

Compare the stability of a potassium atom with a potassium ion, in terms of electronic configuration.

Answer
- The potassium atom has 19 electrons, so its electronic configuration is 2,8,8,1.
- The potassium ion is formed when a potassium atom loses one electron. The electronic configuration of the ion is 2,8,8.
- The potassium ion has a full outer shell and is more stable than the potassium atom.

Tip

You must remember to *compare* the stability.

Going down the group the atoms get bigger due to there being more shells of electrons. As a result the outer electron is further away from the positive charge of the nucleus. This means that the outer negative electron is less strongly attracted to the positive nucleus, and so is easier to lose. The easier the electron is to lose, the more reactive the alkali metal.

Group 1 compounds

Most Group 1 compounds are **white** and dissolve in water to give **colourless** solutions. Figure 16.15 shows white sodium chloride and a colourless solution of sodium chloride.

▲ **Figure 16.15** White sodium chloride dissolves to give a colourless solution

 Tip

Do not use the word 'clear' to describe colour – clear means 'transparent', or 'see-through'. The solution shown in the photograph has no colour and is colourless – it is also clear.

Test yourself

12 Why are the alkali metals reactive?

13 Write a word and a balanced symbol equation for the reaction of potassium with water.

14 Potassium reacts with chlorine to form an ionic compound. Explain why this reaction happens.

15 Explain why potassium is more reactive than sodium.

16 Francium is the last element in Group 1. Predict the physical properties of francium.

Show you can **?**

This question gives information about the reaction of Group 2 elements with water (which is not on the specification) and tests your ability to interpret data.

Table 16.9

Element	Reactivity with water	Name of product
Be	no reaction	no products
Mg	reacts very slowly with cold water	magnesium hydroxide and hydrogen
Ca	reacts moderately with cold water	calcium hydroxide and hydrogen
Sr	reacts rapidly with cold water	strontium hydroxide and hydrogen
Ba	reacts very rapidly with cold water	barium hydroxide and hydrogen

Use the information in Table 16.9 and your own knowledge of Group 1 elements to compare and contrast the reactions of Group 1 and Group 2 elements with water.

In your answer compare:

- the products formed
- the reactivity of the Group 1 elements compared with the Group 2 elements
- the trend in reactivity down both groups.

Group 7 (the halogens)

The halogens are a group of reactive non-metals. A halogen atom has 7 electrons in its outer shell and is unstable, so it can bond covalently with another halogen atom to form a diatomic halogen molecule. Alternatively, the halogen atom may gain an electron from a metal and form a halide ion. For example, a fluorine atom will covalently bond with another fluorine atom to form a fluorine molecule, or it will bond ionically with a metal ion and form a fluoride ion. The halogens are listed in Table 16.10.

Table 16.10

Halogen atom	Halogen molecule	Halide ion
F	F_2	fluoride: F^-
Cl	Cl_2	chloride: Cl^-
Br	Br_2	bromide: Br^-
I	I_2	iodide: I^-
At	At_2	astatide: At^-

Physical properties

Table 16.11 shows the colour and state of the halogens at room temperature and pressure.

Table 16.11

Halogen	Formula		Colour	State at room temperature and pressure
fluorine	F_2		yellow	gas
chlorine	Cl_2		yellow–green	gas
bromine	Br_2		red–brown	liquid
iodine	I_2		grey–black	solid
astatine	At_2		—	—

What is the trend down the group?

You will notice as that as Group 7 is descended, the elements become darker in colour and change from gas to liquid to solid due to the melting point increasing.

Why is there no actual photograph for astatine?

Astatine is a rare element, and no one has ever seen it because a mass of astatine large enough to be seen would immediately be vaporised by the heat of its own radioactivity. Following the trend in the group, astatine may be black and may have a higher melting point than iodine.

Toxicity

The halogens are all toxic. In general, the toxicity decreases down the group. All reactions using halogens must be carried out in a fume cupboard.

Table 16.12

chlorine	iodine

Chlorine is a very toxic gas and was used as a weapon by German troops in the First World War. It is denser than air and thus sank down into the trenches, killing many Allied troops.	Iodine is not as toxic as chlorine and is toxic to humans only in high doses. A low concentration of iodine kills bacteria and can be used to sterilise external wounds.

Heating iodine

Iodine is a grey–black solid at room temperature and pressure. If it is heated it sublimes. Sublimation **is the change of state from solid directly to gas on heating**, without passing through the liquid state.

▲ **Figure 16.16** Iodine subliming

Observations

The grey–black solid changes into a purple vapour (Figure 16.16). On cooling, the purple vapour changes back to grey–black crystals.

What happens when atoms of Group 7 react?

Atoms of Group 7 are reactive, and all react in the same way as they have seven electrons in their outer shell and only need to gain one electron to gain a full outer shell. When they gain an electron, a negative ion with a stable electronic configuration is formed.

A half equation can be used to represent this:

General half equation: **halogen atom + 1 electron → halide ion**

$$X + e^- \rightarrow X^-$$

For example: chlorine atom + electron → chloride ion

$$Cl + e^- \rightarrow Cl^-$$

When a halogen **molecule** reacts two halide ions are formed, as shown in the half equation:

General half equation: **halogen molecule + 2 electrons → 2 halide ions**

$$X_2 + 2e^- \rightarrow 2X^-$$

For example: $Cl_2 + 2e^- \rightarrow 2Cl^-$

Example

Compare the stability of a fluorine atom with a fluoride ion in terms of their electronic configurations.

Answer
- The fluorine atom has 9 electrons so its electronic configuration is 2,7.
- The fluoride ion is formed when a fluorine atom gains one electron. The electronic configuration of the ion is 2,8.
- The fluoride ion has a full outer shell and is more stable than the fluorine atom.

H

Test yourself

17 Table 16.13

Halogen	Melting point/°C	Solubility in water
fluorine	−220	soluble
chlorine	−101	soluble
bromine	−7	slightly soluble

a) Use the data in Table 16.13 to state the trend in the melting points of the halogens.
b) Use the data in the table to state the trend in the solubilities of the halogens.
18 State the colour of the halogen which is found in Period 3.
19 a) How many electrons are in the outer shell of an atom of bromine?
b) Draw a dot and cross diagram of the bonding in a fluorine molecule.
c) What is sublimation?
d) What is observed when iodine is heated?
20 Why is fluorine more reactive than bromine?
H 21 Write half equations for:
a) the formation of an iodide ion from an iodine atom
b) the formation of bromide ions from a bromine molecule
c) the formation of a fluoride ion from a fluorine atom.

What is the trend in reactivity in Group 7?

As Group 7 is **descended** the reactivity of the elements **decreases** (see Figure 16.17). Fluorine is the most reactive halogen.

When the halogens react they gain one electron. Further down the group the atoms have more shells of electrons, so the outermost shell containing electrons is further from the positive attraction of the nucleus. This means the force of attraction between the positive nucleus and an incoming electron decreases. Therefore, it is more difficult to attract an electron, and the reactivity decreases.

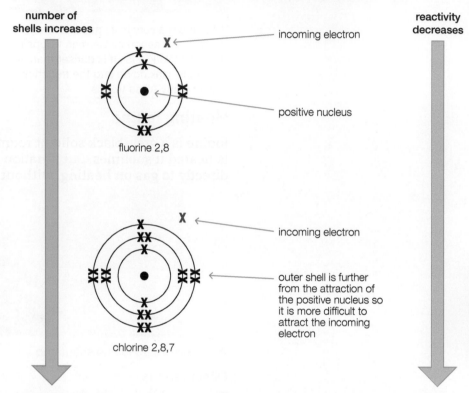

▲ **Figure 16.17** The further down the group, the further the electron gained is from the nucleus

Displacement reactions

In a displacement reaction a more reactive **halogen** (Cl_2, Br_2 or I_2) is added to a **halide solution**. The more reactive halogen pushes out and replaces the less reactive halogen from the solution of its ions.

For example, chlorine is more reactive than the iodine in potassium iodide solution, and displaces it from solution, forming iodine and potassium chloride.

Word equation: chlorine + potassium iodide → iodine + potassium chloride

Symbol equation: $Cl_2(g) + 2KI(aq) \rightarrow I_2(aq) + 2KCl(aq)$

Ionic equation: $Cl_2(g) + 2I^-(aq) \rightarrow I_2(aq) + 2Cl^-(aq)$

Tip

In equations, you must always write the halogens as diatomic molecules: Cl_2, Br_2, I_2.

Chlorine is also more reactive than the bromine in potassium bromide solution, and displaces it from solution, forming bromine and potassium chloride.

Word equation: chlorine + potassium bromide → bromine + potassium chloride

Symbol equation: $Cl_2(g) + 2KBr(aq) → Br_2(aq) + 2KCl(aq)$

Ionic equation: $Cl_2(g) + 2Br(aq)^- → Br_2(aq) + 2Cl^-(aq)$

However, if bromine is added to potassium chloride solution the bromine is less reactive than the chloride in the potassium chloride solution, and no reaction occurs.

Tip

For further information on ionic equations see page 503.

Observations

You need to know the observations for all displacement reactions. Note that the halogen produced dissolves in the halide solution, and produces an aqueous solution. When halogens dissolve they give a slightly different colour, as shown in Table 16.14.

Table 16.14

Halogen	Colour of halogen in normal state	Colour of solution of halogen
chlorine	yellow–green gas	pale green solution
bromine	red–brown liquid	orange solution
iodine	grey–black solid	brown solution

⋏ **Figure 16.18** Chlorine solution, bromine solution, iodine solution

Displacement reactions are often carried out using Group 1 and 2 metal salt solutions, which are colourless. Hence the overall colour change **in the solution** is from colourless to the halogen solution colour. For example, for the reaction:

bromine + potassium iodide → iodine + potassium bromide

the solution changes colour from **colourless** to **brown** (due to the iodine formed).

Table 16.15 Summary of the displacement reactions of halogens

	sodium chloride solution	sodium bromide solution	sodium iodide solution
Cl_2	no reaction	chlorine + sodium bromide → bromine + sodium chloride $Cl_2 + 2NaBr → Br_2 + 2NaCl$	chlorine + sodium iodide → iodine + sodium chloride $Cl_2 + 2NaI → I_2 + 2NaCl$
observation in the solution	—	colourless solution changes to orange	colourless solution changes to brown
Br_2	no reaction	no reaction	bromine + sodium iodide → iodine + sodium bromide $Br_2 + 2NaI → I_2 + 2NaBr$
observation in the solution	—	—	colourless solution changes to brown
I_2	no reaction	no reaction	no reaction

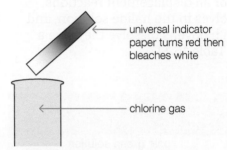

universal indicator paper turns red then bleaches white

chlorine gas

▲ **Figure 16.19** Test for chlorine

Test yourself

22 A halogen is bubbled into a solution of sodium bromide and the solution changes colour from colourless to orange.
   a) Write a balanced symbol equation for this reaction.
   b) Explain why a reaction occurs.

23 Write balanced symbol equations and give observations if a reaction occurs between:
   a) sodium chloride + iodine
   b) potassium iodide + bromine
   c) iodine + potassium chloride
   d) chlorine + calcium bromide
   e) bromine + magnesium iodide
   f) iodine + potassium bromide.

Test for chlorine gas

Damp universal indicator paper changes to **red** and then **bleaches white** in the presence of chlorine gas (see Figure 16.19).

Practical activity

Reactions of the halogens

Figure 16.20 shows chlorine gas being passed through a dilute solution of potassium iodide. The upper layer is a hydrocarbon solvent. A colour change occurs in the potassium iodide solution due to the displacement reaction that occurs.

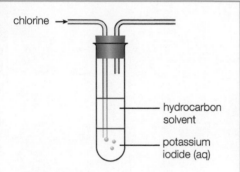

chlorine

hydrocarbon solvent

potassium iodide (aq)

▲ **Figure 16.20**

1 a) What is the most important safety precaution, apart from wearing safety glasses, which must be taken when carrying out this experiment?
   b) Explain why a displacement reaction occurs between chlorine and potassium iodide.
   c) Name the products of the displacement reaction that occurs.
   d) What colour change occurs in the potassium iodide solution?
   e) Write a balanced symbol equation for the reaction between chlorine and potassium iodide.
   f) If this experiment was repeated using bromine, instead of chlorine, explain if the observations would be different.

2 The halogens are more soluble in hydrocarbon solvents than in water and produce coloured solutions, as shown in Table 16.16.

Table 16.16

Halogen	Colour of hydrocarbon solvent when halogen is dissolved
chlorine	pale green
bromine	orange
iodine	purple

After the reaction of chlorine with aqueous potassium iodide, the aqueous layer is shaken with the hydrocarbon solvent and most of the displaced halogen dissolves in the upper layer.
a) Explain the meaning of the word 'solvent'.
b) Use the information in the table to suggest what happens to the colour of the hydrocarbon solvent after shaking.

Show you can ?

Copy and complete Table 16.17.

Table 16.17

	Group 1	Group 7
Number of electrons in outer shell of atom		
Reactive or non-reactive element?		
Trend in reactivity down the group		
Metal or non-metal?		
In reactions, do atoms of the element gain electrons or lose electrons?		

Group 0

Atoms of Group 0 (the noble gases) have a **filled outer shell of electrons** and are **stable**. As a result they are unreactive. They exist as single atoms. Helium has 2 electrons in its filled outer shell, all the other noble gases have 8 electrons.

Physical properties

The noble gases, listed in Figure 16.21, are all colourless.

The noble gases all have low boiling points. This is a typical property of non-metals. You can see from Table 16.18 that helium, at the top of Group 0, has the lowest boiling point in the group. The boiling points then increase as you go down the group because the strength of van der Waals' forces between atoms increases.

Table 16.18

Element	Appearance	Relative atomic mass	Boiling point/°C	Trend in boiling point
helium	colourless gas	4	−269	
neon	colourless gas	20	−246	
argon	colourless gas	40	−190	
krypton	colourless gas	84	−150	
xenon	colourless gas	131	−111	
radon	colourless gas	222	−62	increase in boiling point

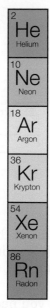

▲ **Figure 16.21** The noble gases

▲ **Figure 16.22** Although unreactive, the noble gases glow with distinctive colours in gas discharge lamps and are used in neon lights, such as these ones in Times Square, New York

Test yourself

24 Why are Group 0 elements unreactive?
25 Why are the noble gases referred to as being in Group 0 rather than Group 8?
26 Some atoms of element 118 have been produced. Element 118 is in Group 0. Predict the appearance and reactivity of this element.

Show you can (?)

Copy and complete Table 16.19.

Table 16.19

Element	Reactive or unreactive?	Metal or non-metal?	Solid, liquid or gas at room temperature?	Electronic structure
He				
Ar				

Transition metals

The transition metals are in the block in the middle of the Periodic Table between Groups 2 and 3. They are all metals, and include many common metals such as chromium (Cr), iron (Fe), nickel (Ni) and copper (Cu).

Some coloured transition metal compounds that you need to know the colours of are shown in Table 16.20.

Table 16.20

Copper(II) oxide	Copper(II) carbonate	Hydrated copper(II) sulfate	Any copper(II) salt in solution, e.g. copper(II) sulfate solution
black solid	green solid	blue crystals	blue solution

Tip

A hydrated salt contains water of crystallisation and will appear crystalline.

Test yourself

27 A metal reacts vigorously with water and fizzes on the surface. Explain if it is a transition metal or an alkali metal.

28 Name a transition metal that forms a green carbonate and a blue hydrated sulfate.

Writing a risk assessment

Whenever experiments and investigations are carried out in the laboratory, you need to decide if the experiment is safe by carrying out a risk assessment. A risk assessment is a judgement of how likely it is that someone might come to harm if a planned action is carried out and how the risks of harm occurring could be reduced.

A good risk assessment includes:

1 a list of all the **hazards** in the experiment

2 a list of the **risks** that the hazards could cause

3 suitable control measures you could take that will reduce or prevent the risk.

For chemicals there should be a hazard warning sign on the container. These are shown on page 210.

The hazards for each chemical can be found by looking up the CLEAPSS student safety sheets, or Hazcards, and should be recorded in your risk assessment. For example, 'When using pure ethanol it should be labelled highly flammable (Hazcard 40 A).' Hazcards tell you about control measures that need to be taken, and state when a fume cupboard needs to be used, when to wear gloves, and the suitable concentrations and volumes of solutions for undertaking an experiment safely.

Table 16.21

Hazard	Risk	Control measure
concentrated sulfuric acid	corrosive	wear protective gloves; use smallest volume possible; wear eye protection
ethanol	highly flammable liquid and vapour	keep away from flames – no naked flames to be used, use a water bath to heat
bromine	toxic and corrosive	handle with gloves; use dilute solutions; wear eye protection; use in fume cupboard
cracked glassware	could cause cuts	check for cracks before use
hot apparatus	could cause burns	allow apparatus to cool before touching
bags and stools	could be a tripping hazard	tuck stools under benches; leave bags in the bag store
chemicals being heated in test tubes	chemicals could spit out of test tubes	wear eye protection; point test tubes away from people
beaker being heated on a tripod and gauze	could fall over and spill hot liquid	keep apparatus away from edge of bench; work standing up; wear eye protection
long hair	could catch fire	tie back long hair

Test yourself

29 Figure 16.23 shows the apparatus used to heat some hydrated copper(II) sulfate in a test tube. Copy and complete Table 16.22 to give a risk assessment for this experiment.

Table 16.22

Hazard	Risk	Control measure

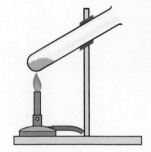

▲ **Figure 16.23**

30 To prepare a new compound, 5 cm³ of ethanol and 5 cm³ of vinegar were mixed in a test tube with 5 drops of concentrated sulfuric acid, and warmed. Write a risk assessment for this experiment.

31 The instructions to react sodium with water were:
   'Place a piece of sodium in a beaker of water and observe the reaction.'
   Rewrite these experimental instructions in view of health and safety precautions.

32 What hazards and risks are associated with carrying out an experiment using a power pack to determine if sodium chloride solution conducts electricity?

Practice questions

1 Use the letters from the table below to answer the questions. The letters do not represent symbols for the elements.

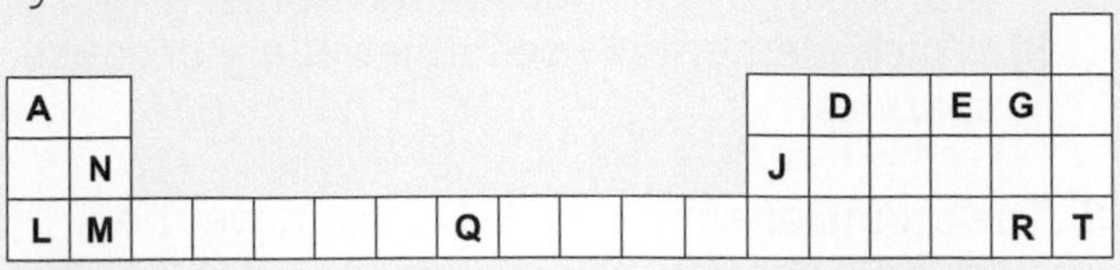

A							D	E	G
	N							J	
L	M			Q				R	T

Figure 16.24

a) Give the letters of any two elements in the same group. *(1 mark)*

b) What is the name for a row of elements? *(1 mark)*

c) What is the letter of:
  i) a transition metal
  ii) a noble gas
  iii) an alkali metal? *(3 marks)*

d) Which is more reactive, A or L? *(1 mark)*

e) Which is more reactive, G or R? *(1 mark)*

f) Give the letter of an element that belongs to an unreactive group. *(1 mark)*

g) Give the formula of the compound which forms between:
  i) A and G
  ii) N and R
  iii) J and G. *(3 marks)*

h) Give the letter of an element that forms an ion with a 2+ charge. *(1 mark)*

2 Give the symbol for:
a) the first element in Group 7 *(1 mark)*
b) the least reactive alkali metal *(1 mark)*
c) the most reactive halogen *(1 mark)*
d) an alkali metal in Period 2 *(1 mark)*
e) a transition metal that forms ions which are blue coloured in solution *(1 mark)*
f) a halogen that is red brown. *(1 mark)*

3 Table 16.23 is a shortened form of the Periodic Table.

Table 16.23

Group 1	2	3	4	5	6	7	0
	Li	Be	B	C	N	O	F	Ne
	Na	Mg	Al	Si	P	S	Cl	Ar
	K	Ca					Br	Kr

Use this table to answer the questions below.
a) What is the general name given to Group 7 elements? *(1 mark)*
b) How many of the elements listed are gases? *(1 mark)*
c) How many of the elements listed are liquids? *(1 mark)*
d) How many of the elements listed are metals? *(1 mark)*

4 When Group 1 elements react, the atoms form ions. For example, when potassium reacts with water, potassium ions are formed from potassium atoms.
a) i) Using electronic structures, compare the stability of a potassium atom with that of a potassium ion. *(3 marks)*
  ii) Why is potassium stored under oil in the laboratory? *(1 mark)*
b) Before reacting Group 1 elements with water a risk assessment is carried out. Give **two** safety precautions, apart from wearing safety glasses, which must be included in the risk assessment for reacting potassium with water. *(2 marks)*
c) Equal-sized pieces of three Group 1 metals are added to separate troughs of water which contain universal indicator. The observations made are recorded in Table 16.24.

Table 16.24

Group 1 metal	Observation on reacting with water	Colour of universal indicator solution
potassium	• melts • burns with a lilac flame • moves on the surface of the water • disappears quickly.	changes colour from green to purple
lithium	• floats • moves on the surface of the water • eventually disappears.	changes colour from green to purple
sodium	• melts • moves on the surface of the water • disappears.	changes colour from green to purple

Use the information in the table to answer the questions below.
i) What happens to the reactivity of the Group 1 elements as the group is descended? *(1 mark)*
ii) Explain fully why the universal indicator changed colour from green to purple. *(2 marks)*
iii) Give one more observation which could be added to the table for all three reactions. *(1 mark)*
iv) Write a word equation for the reaction between sodium and water. *(1 mark)*
v) Write a balanced symbol equation for the reaction between sodium and water. *(3 marks)*

5 The modern Periodic Table has been in use for over 100 years. Its development included the work of several chemists, including that of Dmitri Mendeleev.

a) Fill in the blanks in the following passage: The modern Periodic Table arranges the elements in order of increasing atomic _____ whereas early versions of the Periodic Table arranged them in order of increasing atomic _____. *(2 marks)*

b) State **two** other differences between the modern Periodic Table and Mendeleev's table. *(2 mark)*

c) Elements in the Periodic Table are arranged in groups. Table 16.25 gives details of some of the groups of the Periodic Table. Copy and complete the table. *(6 marks)*

Table 16.25

Group number	Name of group	Number of electrons in the outer shell of an atom	Reactive or unreactive?
1			
		7	

d) Many trends in reactivity and physical properties are apparent as a group is descended.

i) State the trend in reactivity as Group 7 is descended. *(1 mark)*

ii) Name the least reactive element in Group 1. *(1 mark)*

iii) Astatine is found at the bottom of Group 7. Predict its state and colour at room temperature and pressure. *(1 mark)*

iv) Write a half equation for the formation of a bromide ion from a bromine atom. *(2 marks)*

6 Dmitri Mendeleev produced a table that is the basis of the modern Periodic Table of the elements.

a) Define the term 'element'. *(1 mark)*

b) Describe the key features of Mendeleev's table and explain why his table came to be accepted over time by scientists. *(6 marks)*

In this question you will be assessed on your quality of written communication.

7 Use the elements in the box below to answer the questions:

Li	Be	B	C	N	O	F	Ne

Figure 16.25

a) List the elements which are solid at room temperature. *(1 mark)*

b) Name an element which has allotropes. *(1 mark)*

c) Which element is similar to sodium in its reactions? *(1 mark)*

d) Which element would not be expected to form compounds? *(1 mark)*

e) Which element is found in Group 3? *(1 mark)*

f) Which element(s) exists as diatomic molecules? *(1 mark)*

g) Which element has the smallest number of protons in the nucleus of its atoms? *(1 mark)*

h) Which element is used in jewellery because it sparkles? *(1 mark)*

8 The atoms of elements A–D each have the electronic configurations shown:

A 2,8,7   B 2,4   C 2,8   D 2,8,3

a) Which element is in Group 4? *(1 mark)*

b) Which element is a halogen? *(1 mark)*

c) Which element is in Period 3 and Group 3? *(1 mark)*

9 Potassium reacts with water vigorously, moving on the surface of the water with a lilac flame.

a) Write down three other observations of what occurs when potassium reacts with water. *(3 marks)*

b) Write a half equation for the formation of a potassium ion from a potassium atom. *(2 marks)*

c) Write a balanced symbol equation for the reaction of potassium with water. *(3 marks)*

d) Explain if lithium reacts more or less vigorously with water than potassium. *(2 marks)*

e) Potassium reacts with chlorine to form potassium chloride. Write a balanced symbol equation for this reaction. *(3 marks)*

f) Write a half equation for the formation of a chlorine molecule from chloride ions. *(3 marks)*

17 Quantitative chemistry 1

Specification points

This chapter covers specification points 1.7.1 to 1.7.7 and 2.6.1 to 2.6.4. It covers relative atomic mass and relative formula mass, as well as calculations involving moles and the concept of limiting reactant. Empirical formulae and formulae of solids with water of crystallisation are introduced, and the prescribed practical determination of the mass of water of crystallisation present in a hydrated salt is considered.

When following a recipe to make a chocolate cake it is important to mix the ingredients in the correct amounts. In chemical reactions it is important to mix the reactants in the correct amounts to form the maximum amount of product and avoid waste. In this chapter you will learn how to calculate the amounts of chemicals reacting and being produced.

Relative atomic mass (A_r)

Atoms have a tiny mass. For example, an atom of the carbon-12 (^{12}C) isotope has a mass of $0.000\,000\,000\,000\,000\,000\,000\,02\,g$. This is $2.0 \times 10^{-24}\,g$ when written in standard form. This is too small to measure, so chemists measure the mass of atoms relative to each other.

▲ **Figure 17.1** The elephant has a relative mass of 500 compared to the baby. This means that the elephant is 500 times heavier than the baby. The mass of atoms are measured relative to the mass of carbon-12

Chemists use a scale to measure the relative mass of atoms. On this scale the mass of a ^{12}C atom is exactly 12.

For example, ^{24}Mg has a relative mass of 24 – it is twice as heavy as a ^{12}C atom.

^{1}H has a relative mass of 1 – it is 12 times lighter than a ^{12}C atom.

These masses are called relative atomic masses and are represented by the symbol A_r. The relative atomic mass of an atom is the mass of an atom compared with that of the carbon-12 isotope, which has a mass of exactly 12.

In Chapter 12 you learnt that most elements contain a mixture of isotopes, each present in a different amount. The term 'weighted mean mass' is used to take into account the contribution made by each isotope to the overall mass of an element. For example, 25% of chlorine atoms are ^{37}Cl and 75% of chlorine atoms are ^{35}Cl, and the weighted mean (average) mass is 35.5. **The relative atomic mass of an element is a weighted mean mass of the isotopes of an element compared with that of the ^{12}C isotope, which has a mass of exactly 12.**

Tip

Relative atomic masses are not actual masses in grams, they are comparisons – *do not* give units.

Relative formula mass (M_r)

The relative formula mass (M_r) is the sum of the relative atomic mass of all the atoms present in the formula of a substance. For covalent substances it is often referred to as the **relative molecular mass**.

Tip

Use the Periodic Table at the back of this book to find the relative atomic masses of O and H.
Relative atomic mass is the bigger of the two numbers of each element.

Tip

The number of atoms inside brackets is multiplied by the number outside. In this formula there are $1 \times 2 = 2$ nitrogen atoms and $3 \times 2 = 6$ oxygen atoms.

Example

What is the relative formula mass of H_2O?

Answer

$M_r = (2 \times A_r(H)) + (1 \times A_r(O))$
$= (2 \times 1) + (1 \times 16) = 18$

Example

What is the relative formula mass of H_2SO_4?

Answer

$M_r = (2 \times A_r(H)) + (1 \times A_r(S))$
$\quad + (4 \times A_r(O))$
$= (2 \times 1) + (1 \times 32) + (4 \times 16) = 98$

Example

What is the relative formula mass of $Mg(NO_3)_2$?

Answer

$M_r = (1 \times A_r(Mg)) + (2 \times A_r(N))$
$\quad + (6 \times A_r(O))$
$= (1 \times 24) + (2 \times 14) + (6 \times 16) = 148$

Example

What is the relative formula mass of $(NH_4)_2SO_4$?

Answer

$M_r = (2 \times A_r(N)) + (8 \times A_r(H))$
$\quad + (1 \times A_r(S)) + (4 \times A_r(O))$
$= (2 \times 14) + (8 \times 1) + (1 \times 32)$
$\quad + (4 \times 16) = 132$

Test yourself

You can find relative atomic masses on the Periodic Table at the back of this book to help you answer these questions.

1 State the name and mass of the atom to which the mass of all other atoms is compared.
2 What is the relative atomic mass (A_r) of the following?
   a) Na   c) F   e) Al
   b) Mg   d) S
3 Calculate the M_r of
   a) CO_2 h) $Ca(OH)_2$
   b) NaCl   i) $Ca(NO_3)_2$
   c) NH_3 j) $Fe_2(SO_4)_3$
   d) C_2H_4 k) $Cu(OH)_2$
   e) $Al(NO_3)_3$ l) $Al_2(CO_3)_3$
   f) $Pb(NO_3)_2$ m) Na_2CO_3
   g) $Al_2(SO_4)_3$

▲ **Figure 17.2** A dozen donuts is 12 donuts. A mole of a substance is 6×10^{23} particles

▲ **Figure 17.3** One mole of marbles would cover the entire Earth, oceans included, to a depth of 3 miles. 6×10^{23} is a big number!

Show you can

MBr_4 has a formula mass of 439. What is the relative atomic mass of M? Use your periodic table to identify M.

The mole

What is a mole?

Atoms and molecules are so incredibly small that scientists needed to invent a convenient unit of measure for counting them. Chemists use a quantity called **amount of substance** for counting atoms, and it is measured using a unit called the **mole** (abbreviated to mol).

In everyday life, a pair of socks is two socks, a dozen donuts is 12 donuts, a ream of paper is 500 sheets. A mole of a substance is 6×10^{23} particles (this number is called Avogadro's constant). The mass of one mole (6×10^{23} atoms) of carbon is 12 g. For any substance, **the mass of** one mole **in grams is numerically equal to the relative formula mass or relative atomic mass**.

The relative atomic mass of magnesium is 24, so the mass of one mole of magnesium is 24 g and it contains 6×10^{23} atoms of magnesium.

The relative formula mass of water (H_2O) is 18, so the mass of one mole of water molecules is 18 g and it contains 6×10^{23} molecules of water.

Calculating moles

To calculate the number of moles of any substance use the expression:

$$\text{Moles} = \frac{\text{mass (g)}}{M_r}$$

Example

Calculate the number of moles in 4.9 g of H_2SO_4

Answer

$M_r(H_2SO_4) = (2 \times 1) + (1 \times 32) + (4 \times 16) = 98$

number of moles $= \dfrac{\text{mass (g)}}{M_r} = \dfrac{4.9}{98} = 0.05 \text{ mol}$

You can also rearrange this equation to calculate the mass of a given number of moles of a substance:

$$\text{Mass (g)} = \text{moles} \times M_r$$

Tip

If calculating the number of moles of an element, use A_r instead of M_r.

Example

Calculate the mass, in grams, of 20 moles of CO_2.

Answer

$M_r(CO_2) = (1 \times 12) + (2 \times 16) = 44$

mass = moles × M_r = 20 × 44 = 880 g

Example

Calculate the mass, in grams, of 0.5 moles of calcium hydroxide.

Answer

$M_r(Ca(OH)_2) = (1 \times 40) + (2 \times 16) + (2 \times 1) = 74$

mass = moles × M_r = 0.5 × 74 = 37 g

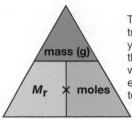

This formula triangle may help you. Cover up the quantity you want to show the equation you need to use.

Thinking of **Mr Moles** with a mass on his head may help you remember this triangle.

▲ **Figure 17.4** Formula triangles can help you easily rearrange equations

Example

Calculate the number of moles in 100 g of $Mg(OH)_2$

Answer

$M_r(Mg(OH)_2) = (24 \times 1) + (2 \times 16) + (2 \times 1) = 58$

number of moles = $\dfrac{mass\ (g)}{M_r} = \dfrac{100}{58}$

= 1.72 mol

Tip

Your calculator gives this value as 1.72413793, but it is best to round your number to one or two decimal places. In this case two decimal places are used.

Test yourself

You can find relative atomic masses on the Periodic Table at the back of this book to help you answer these questions.

4 What is the mass of one mole of the following?
  a) NaCl
  b) Ca
  c) H_2SO_4
  d) $Mg(NO_3)_2$
  e) $(NH_4)_2SO_4$

5 Calculate the number of moles in:
  a) 12 g of H_2SO_4
  b) 16 g of $Mg(OH)_2$
  c) 50 g of $CaCO_3$
  d) 54 g of $Al_2(SO_4)_3$
  e) 4 g of O_2
  f) 5.6 g of $Ca(OH)_2$

6 Calculate the mass in grams of:
  a) 0.5 mole of $FeSO_4$
  b) 0.1 mole of MgO
  c) 0.125 mole of CuO
  d) 2 moles of $Ca(OH)_2$
  e) 4 moles of Na_2CO_3

▲ **Figure 17.5** Some chemists celebrate 'mole day' between 6.00 am and 6.00 pm on 23 October each year. This date represents the number of particles in a mole (6×10^{23})

Show you can ?

0.300 g of a substance was analysed and found to contain 0.0050 moles. Calculate the M_r of the substance.

Molar ratios in equations

A ratio is a way to compare amounts of something. Recipes, for example, are sometimes given as ratios. To make pastry you may need to mix 2 parts of flour to 1 part of fat. This means the ratio of flour to fat is 2:1. Ratios are written with a colon (:) between the numbers, and usually only whole numbers are used.

▲ **Figure 17.6** Pastry is made by mixing the correct ratio of ingredients. In reactions the correct ratio of reactants must be used

In a balanced symbol equation the substances are in a ratio. The balancing numbers (the numbers in front of each species) give the ratio of moles that react together. This is shown in Figure 17.7.

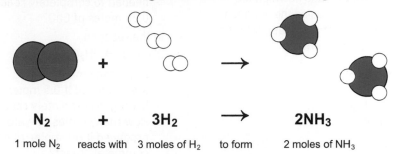

$$N_2 \quad + \quad 3H_2 \quad \longrightarrow \quad 2NH_3$$

1 mole N_2 reacts with   3 moles of H_2 to form   2 moles of NH_3

Ratio: 1 moles N_2 : 3 mole H_2 : 2 moles NH_3

▲ **Figure 17.7** Molar ratio

For example, 2 moles of magnesium react with 1 mole of oxygen to produce 2 moles of magnesium oxide:

$$2Mg + O_2 \rightarrow 2MgO$$

The ratio is 2 moles Mg:1 mole O_2:2 moles MgO. Or in the reaction:

$$2Al + 6HCl \rightarrow 2AlCl_3 + 3H_2$$

The ratio between aluminium and hydrogen is 2Al:3H_2

The ratio between aluminium and hydrochloric acid is 2Al:6HCl, which simplifies to 1Al:3HCl

The ratio can be used to calculate the number of moles that would react and be produced in any one equation, as shown in Figure 17.9 which shows the equation for propane gas (C_3H_8) burning.

C_3H_8	+	$5O_2$	→	$3CO_2$	+	$4H_2O$
1 mole C_3H_8	reacts with	5 moles O_2	to make	3 moles CO_2	and	4 moles H_2O
10 moles C_3H_8	reacts with	50 moles O_2	to make	30 moles CO_2	and	40 moles H_2O
2 moles C_3H_8	reacts with	10 moles O_2	to make	6 moles CO_2	and	8 moles H_2O
0.5 moles C_3H_8	reacts with	2.5 moles O_2	to make	1.5 moles CO_2	and	2.0 moles H_2O

▲ **Figure 17.8** Propane burning

▲ **Figure 17.9** Molar ratio for the complete combustion of propane

Example

In the reaction

$$2Pb(NO_3)_2 \rightarrow 2PbO + 4NO_2 + O_2$$

How many moles of NO_2 are produced from 0.35 moles of $Pb(NO_3)_2$?

Answer

Write down the ratio between the two substances, using the equation. Simplify this ratio and apply to the 0.35 moles of $Pb(NO_3)_2$.

$Pb(NO_3)_2 : NO_2$

2:4

1:2

0.35:?

There is twice as much NO_2 as $Pb(NO_3)_2$, so multiply $Pb(NO_3)_2$ moles by 2:

0.35 : (2 × 0.35) = 0.35 : 0.70

Example

In the reaction

$$4Al + 3O_2 \rightarrow 2Al_2O_3$$

How many moles of aluminium are needed to produce 0.76 moles of aluminium oxide (Al_2O_3)?

Answer

$Al : Al_2O_3$

4:2

2:1

?:0.76

There are twice as many moles of aluminium as aluminium oxide, so multiply by 2:

0.76 × 2 = 1.52 mol

Test yourself

7   In the reaction shown the copper nitrate is heated until it has all decomposed.
$$2Cu(NO_3)_2 \rightarrow 2CuO + 4NO_2 + O_2$$
   a) How many moles of O_2 are produced from 4 moles of $Cu(NO_3)_2$?
   b) How many moles of NO_2 are produced from 0.6 moles of $Cu(NO_3)_2$?

8   In the reaction
$$CaO + 3C \rightarrow CaC_2 + CO$$
   a) How many moles of carbon are needed to completely react with 0.33 moles of CaO?
   b) How many moles of CO are produced when 3.3 moles of carbon react completely?

9   In the reaction
$$2H_2 + O_2 \rightarrow 2H_2O$$
   a) How many moles of H_2O are produced if 0.5 moles of hydrogen completely react?
   b) How many moles of water are produced if 0.1 moles of oxygen completely react?
   c) How many moles of hydrogen react if 0.4 moles of oxygen completely react?
   d) How many moles of oxygen react if 10 moles of hydrogen completely react?

10 In the reaction
$$3Pb + 2O_2 \rightarrow Pb_3O_4$$
   a) How many moles of oxygen are needed to completely react with 0.66 moles of lead?
   b) How many moles of Pb_3O_4 are produced when 2.2 moles of oxygen completely react?
   c) How many moles of Pb_3O_4 are produced when 0.33 moles of lead completely react?

11 Potassium chlorate ($KClO_3$) decomposes to form potassium chloride and oxygen, as shown in the equation:
$$2KClO_3 \rightarrow 2KCl + 3O_2$$
   a) How many moles of potassium chloride are formed when 10 moles of potassium chlorate completely decompose?
   b) How many moles of oxygen are formed when 4 moles of potassium chlorate completely decompose?
   c) How many moles of potassium chloride are formed when 0.5 moles of potassium chlorate completely decompose?

Show you can

1   Select the correct words to complete the sentence below.
   In the equation $C + O_2 \rightarrow CO_2$ one mole of C *atoms/molecules* reacts with one mole of O_2 *atoms/molecules* to form one mole of CO_2 *atoms/molecules*.

2   Write a similar sentence about each equation below.
   a) $C_2H_4 + 2O_2 \rightarrow CO_2 + 2H_2O$
   b) $H_2 + Cl_2 \rightarrow 2HCl$
   c) $C_xH_y + 3O_2 \rightarrow 2CO_2 + 2H_2O$
   d) What are the values of x and y in (c)?

Calculating reacting masses of reactants and products

Chemists need to be able to calculate how much of each reactant to use in a chemical reaction. The general method to do this is:

▶ Write the information (mass) given in the question underneath the equation and underline the substance you need to find the mass of.

▶ Calculate the relative formula mass (M_r) of the substance you have information about and calculate the number of moles. (Remember that the balancing numbers in the equation are *not* part of the formulae.)

▶ Use the ratio from the balanced symbol equation to calculate the number of moles of the substance you need to find the mass of.

▶ Calculate the mass of the substance using mass = moles × M_r.

▲ **Figure 17.10** Technicians working at a pharmaceuticals manufacturing company

Tip

The numbers in front of compounds in a chemical equation give the *ratio* of amounts which react – they do *not* affect the relative formula mass.

Example

What mass of magnesium carbonate reacts with 3.65 g of HCl?

$\underline{MgCO_3} + 2HCl \rightarrow MgCl_2 + H_2O + CO_2$
     3.65 g

Answer

$M_r(HCl) = 1 + 35.5 = 36.5$

moles HCl = $\dfrac{mass}{M_r} = \dfrac{3.65}{36.5} = 0.1$ mol

ratio: 2HCl : 1MgCO₃

     0.1 : ?

     0.1 : $\dfrac{0.1}{2}$

     0.1 : 0.05

$M_r(MgCO_3) = 24 + 12 + (16 \times 3) = 84$

mass MgCO₃ = moles × M_r
        = 0.05 × 84 = 4.2 g

Example

What mass of iron is produced when 320 g of iron(III) oxide is reacted completely with carbon monoxide?

$Fe_2O_3 + 3CO \rightarrow 2Fe + 3CO_2$

Answer

$Fe_2O_3 + 3CO \rightarrow \underline{2Fe} + 3CO_2$
320 g

$M_r(Fe_2O_3) = (2 \times 56) + (3 \times 16) = 160$

moles Fe₂O₃ = $\dfrac{mass}{M_r} = \dfrac{320}{160} = 2$ mol

ratio: 1Fe₂O₃ : 2Fe

     2 : ?

     2 : (2 × 2)

     2 : 4

mass Fe = moles × M_r = 4 × 56 = 224 g

Test yourself

12 What mass of oxygen is needed to completely react with 10 g of hydrogen?
   $2H_2 + O_2 \rightarrow 2H_2O$

13 What mass of calcium oxide is formed from complete decomposition of 25 g of calcium carbonate?
   $CaCO_3 \rightarrow CaO + CO_2$

14 What mass of magnesium oxide would be made from complete reaction of 3 g of magnesium with oxygen?
   $2Mg + O_2 \rightarrow 2MgO$

15 Lithium reacts with water as shown in the equation below. What mass of hydrogen would be produced from complete reaction of 1.4 g of lithium?
   $2Li + 2H_2O \rightarrow 2LiOH + H_2$

16 Copper(II) oxide reacts with aluminium powder as shown below. What mass of aluminium oxide is produced when 10 g of copper(II) oxide reacts?
   $3CuO + 2Al \rightarrow 3Cu + Al_2O_3$

Show you can

4.2 g of sodium hydrogencarbonate is completely decomposed and forms sodium carbonate, as shown in the equation below.

$$2NaHCO_3 \rightarrow Na_2CO_3 + CO_2 + H_2O$$

1 Calculate the relative formula mass of $NaHCO_3$.
2 Calculate the number of moles in 4.2 g of $NaHCO_3$.
3 Calculate the number of moles of sodium carbonate produced.
4 Calculate the mass of sodium carbonate produced.

---

17 What mass of calcium carbonate is needed to completely react with 3.65 g of HCl?
$$CaCO_3 + 2HCl \rightarrow CaCl_2 + H_2O + CO_2$$

18 What mass of NO_2 is obtained from complete decomposition of 33.1 g of lead nitrate?
$$2Pb(NO_3)_2 \rightarrow 2PbO + 4NO_2 + O_2$$

Practical activity

Experiment to find the equation for the action of heat on sodium hydrogencarbonate

A student suggested that there are three possible equations for the thermal decomposition of sodium hydrogencarbonate:

Equation 1: $NaHCO_3 \rightarrow NaOH + CO_2$

Equation 2: $2NaHCO_3 \rightarrow Na_2CO_3 + H_2O + CO_2$

Equation 3: $2NaHCO_3 \rightarrow Na_2O + H_2O + 2CO_2$

In order to find out which is correct, she carried out the following experiment and recorded her results.

- Find the mass of an empty evaporating basin.
- Add approximately 8 g of sodium hydrogencarbonate to the basin and find the mass.
- Heat gently for about 5 minutes.
- Allow to cool and then find the mass.
- Reheat, cool and find the mass.
- Repeat the heating and measurement of mass until constant mass is obtained.

Results

Mass of evaporating basin = 21.05 g
Mass of basin and sodium hydrogencarbonate = 29.06 g
Mass of sodium hydrogencarbonate = 8.01 g
Mass of basin and residue after heating to constant mass = 26.10 g
Mass of residue = 5.05 g

1 Draw a labelled diagram of the experimental set-up for this experiment.
2 Calculate the number of moles of sodium hydrogencarbonate used.
3 From the possible equations, and your answer to question 2, calculate:
   a) the number of moles of NaOH that would be formed in equation 1
   b) the number of moles of Na_2CO_3 that would be formed in equation 2
   c) the number of moles of Na_2O that would be formed in equation 3.

4 From your answers to question 3, calculate:
   a) the mass of NaOH that would be formed in equation 1
   b) the mass of Na_2CO_3 that would be formed in equation 2
   c) the mass of Na_2O that would be formed in equation 3.

5 By comparing your answers in question 4 with the experimental mass of the residue, deduce which is the correct equation for the decomposition of sodium hydrogencarbonate.
6 Why did the mass decrease?

▲ **Figure 17.11** Always check carefully that you have the correct ratio of ingredients before starting to make cookies, or the number you make will be limited!

Limiting reactant

A recipe to make cookies requires 2 cups of flour, 1 cup of chocolate chips and 2 eggs to make 12 cookies. If you have 10 cups of flour, 10 cups of chocolate chips and 4 eggs, how many cookies could you make? The answer is 24 cookies. Why are you limited to just 24 cookies? Because you will run out of eggs after making the 24 cookies. The eggs could be referred to as the 'limiting reactant' because the number of eggs limits the amount of final product (cookies) that can be produced (Figure 17.12). In a chemical reaction the reactant that runs out first in a reaction is the **limiting reactant**. It limits the amount of product that can be produced.

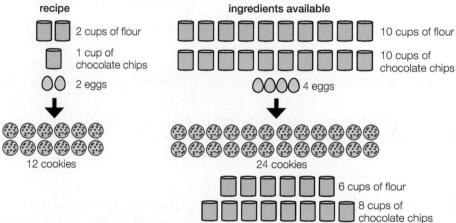

▲ **Figure 17.12** Here, eggs are the limiting reactant as they are used up first. Flour and chocolate chips are in excess

In a chemical reaction between two reactants, if one reactant is in excess the other reactant is the limiting reactant, and is completely used up.

In a chemical reaction between two reactants an excess of one of the reactants can be used to ensure that the other reactant is used up. This is often the case if one of the reactants is in short supply or expensive. For example, when many fuels are burned an excess of oxygen is used. Fuels are expensive and in limited supply, while oxygen is readily available from the air, so using an excess of oxygen ensures that all the fuel burns. The limiting reactant is always used up.

When working out the limiting reactant always use the ratio given in the balanced symbol equation.

H **Example**

5 moles of zinc (Zn) are added to 7 moles of sulfuric acid (H_2SO_4). One of the reagents is in excess.

$$Zn + H_2SO_4 \rightarrow ZnSO_4 + H_2$$

1 Which reactant is in excess?

2 Calculate the number of moles of zinc sulfate and hydrogen formed.

Answer

Table 17.1

	Zn	+	H_2SO_4	$\rightarrow$	$ZnSO_4$	+	H_2
Reacting ratio from the equation	1 mole of Zn	reacts with	1 mole of H_2SO_4	to make	1 mole of $ZnSO_4$	and	1 mole of H_2
Amount provided	5 moles of Zn Limiting reactant		7 moles of H_2SO_4 In excess so it does not all react				
Reaction that takes place	5 moles of Zn	reacts with	5 moles of H_2SO_4	to make	5 moles of $ZnSO_4$	and	5 moles of H_2

1 The reactant in excess is sulfuric acid (2 moles in excess).

2 5 moles of zinc sulfate and 5 moles of hydrogen are formed.

Example

Iron(III) oxide (Fe_2O_3) reacts with carbon monoxide (CO), as shown below, to produce iron. 5 moles of iron(III) oxide are reacted with 25 moles of carbon monoxide.

$$Fe_2O_3 + 3CO \rightarrow 2Fe + 3CO_2$$

1 Which reactant is in excess?

2 Calculate the number of moles of the products formed.

Answer

Table 17.2

	Fe_2O_3	+	3 CO	$\rightarrow$	2 Fe	+	3 CO_2
Reacting ratio from the equation	1 mole of Fe_2O_3	reacts with	3 moles of CO	to make	2 moles of Fe	and	3 moles of CO_2
Amount provided	5 moles of Fe_2O_3 Limiting reactant		25 moles of CO In excess so it does not all react				
Reaction that takes place	5 moles of Fe_2O_3	reacts with	15 moles of CO	to make	10 moles of Fe	and	15 moles of CO_2

1 The reactant in excess is carbon monoxide (10 moles in excess).

2 10 moles of Fe and 15 moles of CO_2 are formed.

Example

0.30 moles of magnesium (Mg) are reacted with 0.20 moles of oxygen (O_2).

$$2Mg + O_2 \rightarrow 2MgO$$

1 Which reagent is in excess?

2 Calculate the number of moles of product formed.

Answer

Table 17.3

	2 Mg	+	O_2	→	2 MgO
Reacting ratio from the equation	2 moles of Mg	reacts with	1 mole of O_2	to make	2 moles of MgO
Amount provided	0.30 moles of Mg _Limiting reagent_		0.20 moles of O_2 _In excess so it does not all react_		
Reaction that takes place	0.30 moles of Mg	reacts with	0.15 moles of O_2	to make	0.30 moles of MgO

1 The reactant in excess is oxygen (by 0.05 moles).

2 0.30 moles of MgO are formed.

Show you can

Calculate the mass of calcium carbide (CaC_2) formed when 28 g of calcium oxide (CaO) is reacted with 48 g of carbon.

$$CaO + 3C \rightarrow CaC_2 + CO$$

Use the following headings:

- number of moles of calcium oxide
- number of moles of carbon
- the reagent in excess is
- number of moles of calcium carbide formed
- mass of calcium carbide formed, in grams.

Tip

Always use the number of moles of the limiting reactant to calculate the number of moles of product formed.

Test yourself

19 Copper(II) oxide reacts with hydrogen according to the equation

$$CuO + H_2 \rightarrow Cu + H_2O$$

How many moles of copper would be made if:

a) 10 moles of copper(II) oxide were reacted with 15 moles of hydrogen

b) 3 moles of copper(II) oxide were reacted with 3 moles of hydrogen

c) 0.6 moles of copper(II) oxide were reacted with 0.3 moles of hydrogen?

20 Copper(II) oxide reacts with methane as shown below.

$$4CuO + CH_4 \rightarrow 4Cu + 2H_2O + CO_2$$

How many moles of copper would be made if:

a) 4 moles of copper(II) oxide were reacted with 4 moles of methane

b) 3 moles of copper(II) oxide were reacted with 1 mole of methane

c) 6 moles of copper(II) oxide were reacted with 2 moles of methane?

21 Magnesium reacts with hydrochloric acid according to the equation

$$Mg + 2HCl \rightarrow MgCl_2 + H_2$$

How many moles of hydrogen would be made if:

a) 4 moles of magnesium were reacted with 4 moles of hydrochloric acid

b) 0.2 moles of magnesium were reacted with 0.5 moles of hydrochloric acid

c) 0.25 moles of magnesium were reacted with 0.75 moles of hydrochloric acid?

22 What mass of calcium sulfide can be made when 6 g of calcium reacts with 8 g of sulfur?

$$Ca + S \rightarrow CaS$$

23 Calculate the maximum mass of titanium produced when 1.9 g of titanium chloride reacts with 6 g of magnesium?

$$TiCl_4 + 2Mg \rightarrow Ti + 2MgCl_2$$

24 23.2 g of tungsten oxide (WO_3, $M_r = 232$) is reacted with 20.0 g of hydrogen (H_2, $M_r = 2$).

$$WO_3 + 3H_2 \rightarrow W + 3H_2O$$

Calculate the mass of tungsten (W, $M_r = 184$) produced from this reaction.

25 What mass of calcium chloride is obtained from 20 g of ammonium chloride and 50 g of calcium oxide? Give your answer to two decimal places.

$$2NH_4Cl + CaO \rightarrow CaCl_2 + H_2O + 2NH_3$$

H

▲ **Figure 17.13** When you make a cake some of the mixture is left behind in the bowl, and so the percentage yield is not 100%. In a chemical reaction some reactants and products may get left behind in the apparatus

Percentage yield

The maximum mass of product that can be obtained in a chemical reaction is called the **theoretical yield**. The mass of product that is actually produced in the laboratory is called the **actual yield**. The percentage yield of any reaction can be calculated using the expression:

Percentage yield = $\dfrac{\text{actual yield}}{\text{theoretical yield}}$ × 100

There are many reasons why the percentage yield is less than 100%:

▶ Some of the product may be lost when it is separated from the reaction mixture. It may be left on the apparatus.

▶ Side reactions may occur – this means that some of the reactants may take part in a different reaction to the desired one.

▶ Some reactions, as you will learn in Chapter 11, do not go to completion. This means they do not finish because they are reversible and some of the products may turn back into reactants.

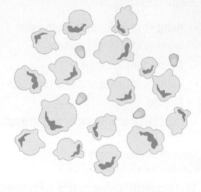

20 unpopped popcorn kernels

16 popped and 4 unpopped popcorn kernels

The theorectical yield is 20, because 20 is the maximum number of pieces of popped corn that can be made from the 20 unpopped kernels.
The actual yield is 16, as 4 kernels did not pop.
The percentage yield is $\dfrac{16}{20}$ × 100 = 80%.

▲ **Figure 17.14** Actual yield is often less than theoretical yield

Example	Example
In a reaction to produce aspirin, the theoretical yield was 40 g but the mass of aspirin produced was 5 g. Calculate the percentage yield of aspirin.	In a reaction, the theoretical yield of ammonia was 24 moles. The number of moles of ammonia produced was 6 moles. Calculate the percentage yield of ammonia.
Answer	**Answer**
percentage yield = $\dfrac{\text{actual yield}}{\text{theoretical yield}}$ ×100 = $\dfrac{5}{40}$ ×100 = 12.5%	percentage yield = $\dfrac{\text{actual yield}}{\text{theoretical yield}}$ ×100 = $\dfrac{6}{24}$ ×100 = 25%

The theoretical yield can be worked out using the method of reacting mass calculations.

Example

Calculate the theoretical yield of magnesium chloride when 8 g of magnesium oxide reacts with hydrochloric acid. If 4 g of magnesium chloride is produced, calculate the percentage yield.

$$MgO + 2HCl \rightarrow MgCl_2 + H_2O$$

Answer

$$MgO + 2HCl \rightarrow \underline{MgCl_2} + H_2O$$
8 g

$M_r(MgO) = 24 + 16 = 40$

moles MgO = $\dfrac{mass}{M_r} = \dfrac{8}{40} = 0.2$ mol

ratio: $1 MgO : 1 MgCl_2$
$\qquad\quad 0.2 : 0.2$

$M_r(MgCl_2) = 24 + (35.5 \times 2) = 95$

mass = moles × M_r = 0.2 × 95 = 19 g

theoretical yield = 19 g

percentage yield = $\dfrac{actual\ yield}{theoretical\ yield} \times 100 = \dfrac{4}{19} \times 100 = 21\%$

When making products in industrial processes it is important that as high a percentage yield as possible is achieved. A high percentage yield means more product is formed and so there is a higher profit for the manufacturing process and/or a lower cost to the consumer.

Test yourself

26 In a reaction to produce paracetamol, the theoretical yield was 200 g but the yield was actually 150 g.
   a) Calculate the percentage yield.
   b) Give three possible reasons for the percentage yield being less than 100%.

27 When 28 g of nitrogen reacts with 6 g of hydrogen, the theoretical yield of ammonia is 34 g. In the reaction the actual yield was 8 g. Calculate the percentage yield for this reaction.

28 In the reaction of iron(III) oxide with carbon monoxide, the theoretical mass of iron that could be produced is 2 kg, but the yield was actually 800 g. Calculate the percentage yield.

29 In the reaction of hydrogen with oxygen, 0.85 moles of water were produced. The theoretical yield for the reaction is 0.95 moles. Calculate the percentage yield of water.

30 a) What is the maximum mass of ammonium chloride that could be formed when 7.3 g of HCl is neutralised by ammonia?
   $$NH_3 + HCl \rightarrow NH_4Cl$$
   b) This experiment produced only 8.3 g of ammonium chloride. What is the percentage yield?

31 a) Calculate the theoretical yield of magnesium oxide when 6 g of magnesium is burned.
   $$2Mg + O_2 \rightarrow 2MgO$$

▲ **Figure 17.15** Salbutamol is an inhaler drug used to treat asthma. There are five reactions needed to produce it, and the overall percentage yield is low (7%). Scientists are continually researching to find ways to improve the percentage yields of the reactions and so decrease the cost of manufacturing the drug

Show you can

Butanol (C_4H_9OH) is used to prepare the chemical bromobutane (C_4H_9Br) according to the following equation:

$$C_4H_9OH + HBr \rightarrow C_4H_9Br + H_2O$$

1 Give an equation to explain the term 'percentage yield'.

2 Assuming a 40% yield, what mass of butanol would be required to produce a yield of 5.48 g of bromobutane?

Tip

For the Double Award Science specification, any calculations from here onwards in this chapter are in C2.

Tip

Sometimes the empirical formula and the molecular formula of a compound are the same, for example C_2H_5N.

▲ **Figure 17.16** Vinegar contains ethanoic acid, sweets contain glucose

194

b) Calculate the percentage yield if 4 g of magnesium oxide were actually produced.

32 Phenol is converted to a trichlorophenol according to the equation below.

$$C_6H_5OH + 3Cl_2 \rightarrow C_6H_2Cl_3OH + 3HCl$$
phenol   trichlorophenol

a) Calculate the theoretical yield of trichlorophenol from 47 g of phenol.

b) If 90.6 g of trichlorophenol were actually produced, calculate the percentage yield.

Calculation of formula of compounds

Empirical formula and molecular formula

There are many different types of formula.

The empirical formula is the simplest whole number ratio of atoms of each element in a compound.

The molecular formula shows the actual number of atoms of each element present in a compound. It will be a simple multiple of the empirical formula. For example:

molecular formula = C_6H_{12} empirical formula = CH_2

molecular formula = C_6H_6 empirical formula = CH

molecular formula = CH_4 empirical formula = CH_4

Figure 17.16 shows vinegar, which contains ethanoic acid (CH_3COOH), and sweets, which contain glucose ($C_6H_{12}O_6$). Both chemicals have different molecular formulae but the same empirical formula (CH_2O).

To find the molecular formula from the empirical formula the relative formula mass is needed.

Example

A compound has a relative formula mass of 28 and the empirical formula CH_2. What is the molecular formula of the compound?

Answer
A molecular formula is a multiple (n) of the empirical formula.

$(CH_2)_n = 28$

$(12 + (2 \times 1))_n = 28$

$14_n = 28$

$n = 2$

$(CH_2)_n = (CH_2)_2 = C_2H_4$
The molecular formula is C_2H_4.

Example

The empirical formula of a compound is CH_2O and its relative formula mass is 60. What is the molecular formula of the compound?

Answer
$(CH_2O)_n = 60$

$(12 + (2 \times 1) + 16)_n = 60$

$30_n = 60$

$n = 2$

$(CH_2O)_n = (CH_2O)_2 = C_2H_4O_2$
The molecular formula is $C_2H_4O_2$.

▲ **Figure 17.17** One type of energy drink contains 0.08 g of caffeine in 50 ml, similar to the amount found in a cup of coffee. Caffeine has the molecular formula $C_8H_{10}N_4O_2$ and the empirical formula $C_4H_5N_2O$

Finding the empirical formula

The general method to follow is:

▶ Find the number of moles of each element in the compound using

Number of moles = $\dfrac{\text{mass}}{M_r}$

▶ Find the simplest ratio of moles – a simple way to do this is to compare the value of moles and find the smallest value, then divide all the molar values by this number.

It is possible to find the empirical formula for a compound using the percentage composition, mass composition or experimental data.

Finding the empirical formula from mass data

Example

In an experiment 128 g of copper combine with 32 g of oxygen. What is the empirical formula of the copper oxide formed?

Answer

Table 17.4

	copper	oxygen
mass in grams	128	32
moles = $\dfrac{\text{mass}}{M_r}$	$\dfrac{128}{64} = 2$	$\dfrac{32}{16} = 2$
simplest ratio	1	1
empirical formula	CuO	

Example

When 128 g of copper combine with oxygen, 143 g of an oxide are formed. What is the empirical formula of the oxide?

Answer

Note that you are not given the mass of oxygen, but the mass of oxide. You need to subtract the mass of copper from the oxide mass to find the mass of oxygen.

Table 17.5

	copper	oxygen
mass in grams	128	$143 - 127 = 16$
moles = $\dfrac{\text{mass}}{M_r}$	$\dfrac{128}{64} = 2$	$\dfrac{16}{16} = 1$
simplest ratio	2	1
empirical formula	Cu_2O	

Finding the empirical formula from experimental data

To find the empirical formula of a substance experimentally, a general method is:

▶ Weigh a crucible and lid.
▶ Weigh a crucible and lid and the solid element.
▶ Place the crucible on a pipeclay triangle and heat, using tongs to raise the lid a little occasionally to allow air in for the solid to react.
▶ Cool and weigh.
▶ Repeat the heating and cooling until the mass is constant.

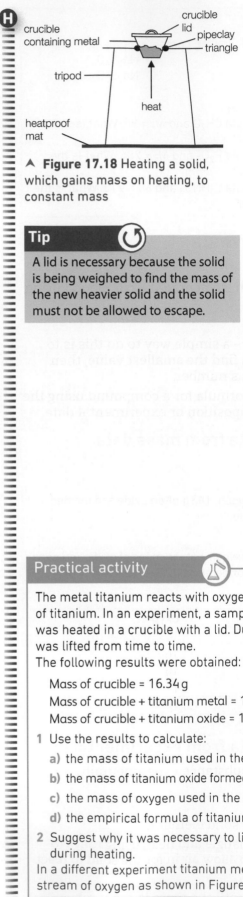

▲ **Figure 17.18** Heating a solid, which gains mass on heating, to constant mass

Tip

A lid is necessary because the solid is being weighed to find the mass of the new heavier solid and the solid must not be allowed to escape.

The experiment is shown in Figure 17.18. Errors that may occur in this experiment include that not all the solid may react and some product may be lost when the lid is lifted.

Example

In an experiment to find the empirical formula of magnesium oxide, some magnesium was heated in a crucible with a lid, which was lifted occasionally to allow air in. The results were:

Mass of crucible + lid = 35.60 g

Mass of crucible + lid + magnesium = 41.12 g

Mass of crucible + lid + contents after heating = 44.82 g

1 Calculate the mass of magnesium used.

2 Calculate the mass of magnesium oxide formed.

3 Calculate the mass of oxygen gained.

4 Calculate the formula of the magnesium oxide.

Answer

1 (Mass of crucible + lid + magnesium) − (mass of crucible + lid)
  = 41.12 − 35.60 = 5.52 g

3 (Mass of crucible + lid + contents after heating) − (mass of crucible + lid)
  = 44.82 − 35.60 = 9.22 g

2 Mass of magnesium oxide − mass of magnesium = 9.22 − 5.52 = 3.7 g

4 **Table 17.6**

	magnesium	oxygen
mass in grams	5.52	3.70
moles = $\frac{mass}{M_r}$	$\frac{5.52}{24}$ = 0.23	$\frac{3.70}{16}$ = 0.23
simplest ratio	1	1
empirical formula	MgO	

Practical activity

The metal titanium reacts with oxygen to form an oxide of titanium. In an experiment, a sample of titanium metal was heated in a crucible with a lid. During heating the lid was lifted from time to time.
The following results were obtained:

Mass of crucible = 16.34 g

Mass of crucible + titanium metal = 17.36 g

Mass of crucible + titanium oxide = 18.06 g

1 Use the results to calculate:

a) the mass of titanium used in the experiment

b) the mass of titanium oxide formed in the experiment

c) the mass of oxygen used in the experiment

d) the empirical formula of titanium oxide.

2 Suggest why it was necessary to lift the crucible lid during heating.

In a different experiment titanium metal was heated in a stream of oxygen as shown in Figure 17.19.

3 Describe a test that could be carried out, before the cylinder is used, to prove that the gas in it is oxygen.

4 What masses should be found before heating to determine the mass of titanium used in the experiment?

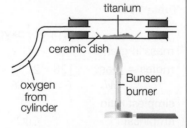

▲ **Figure 17.19**

5 The ceramic container and its contents are repeatedly weighed, heated, reweighed and heated until the mass is constant. State and explain if the mass increases or decreases during this experiment.

6 What safety precautions should be taken in this experiment?

7 How would the reliability of this experiment be checked?

Test yourself

37 In an experiment, 0.96 g of magnesium combine with 2.84 g of chlorine. What is the empirical formula of magnesium chloride?

38 In an experiment, 2.3 g of sodium combine with 0.8 g of oxygen. What is the empirical formula of sodium oxide?

39 In an experiment, 0.72 g of magnesium combine with 0.28 g of nitrogen. What is the formula of magnesium nitride?

40 Calculate the empirical formula of the compound formed when 414 g of lead forms 478 g of a lead oxide.

41 A hydrocarbon (a compound containing carbon and hydrogen only) contains 1.44 g of carbon and 0.36 g of hydrogen. What is the empirical formula of the hydrocarbon?

42 16.25 g of a chloride of iron were produced from the reaction of 5.6 g of iron with an excess of chlorine. Find the empirical formula of the chloride.

43 A compound contains 87.5% silicon and 12.5% hydrogen. Calculate the empirical formula of the compound.

44 A compound contains 14.0% carbon and 44.4% fluorine, and the remainder is chlorine. Calculate the empirical formula of the compound.

Finding the empirical formula from percentage composition data

Example

Find the empirical formula of a compound which contains 22.2% carbon, 3.7% hydrogen and 74.1% bromine.

Answer

In a sample of 100 g of the compound there is 22.2 g of carbon, 3.7 g of hydrogen and 74.1 g of bromine.

Table 17.7

	carbon	hydrogen	bromine
mass in grams	22.2	3.7	74.1
moles = $\frac{mass}{M_r}$	$\frac{22.2}{12}$ = 1.85	$\frac{3.7}{1}$ = 3.7	$\frac{74.1}{80}$ = 0.93
simplest ratio – divide by the smallest number of moles (0.93)	$\frac{1.85}{0.93}$ = 1.989 = 2	$\frac{3.7}{0.93}$ = 3.978 = 4	$\frac{0.93}{0.93}$ = 1
ratio	2	4	1
formula		C_2H_4Br	

Water of crystallisation

▲ **Figure 17.20** Copper(II) sulfate crystals

In 2008 the artist Roger Hiorns made an unusual artwork called 'Seizure', which won the Turner Prize, an annual prize presented to a British visual artist. The artist made an abandoned London flat water tight and flooded it with 700000 litres of hot copper(II) sulfate solution, which then cooled and crystallised, leaving an amazing pattern of hydrated copper(II) sulfate crystals on the walls of the flat, similar to the crystals shown in Figure 17.20.

All crystals are hydrated and contain water of crystallisation.

Water of crystallisation **is water that is chemically bonded into the crystal structure.** Hydrated **means the crystal contains water of crystallisation, and 'anhydrous' means the substance does not contain water of crystallisation.**

The empirical formula of a hydrated compound is written in a unique way:

▶ The water of crystallisation is separated from the rest of the formula by a dot.

▶ The relative number of molecules of water of crystallisation is written after the dot. This is often called the 'degree of hydration'.

For example, hydrated copper(II) sulfate contains five molecules of water of crystallisation for every molecule of copper(II) sulfate, so the degree of hydration is 5 and the empirical formula is written $CuSO_4.5H_2O$. Another example is hydrated barium chloride, which has the formula $BaCl_2.2H_2O$ and has degree of hydration 2, that is, there are two molecules of water for every molecule of barium chloride.

When finding the relative formula mass of compounds containing water of crystallisation, the mass of the molecules of water of crystallisation must be included.

Example

Calculate the relative formula mass of $CoCl_2.6H_2O$.

Answer

$M_r = (1 \times 59) + (2 \times 35.5) + (6 \times 18) = 238$

Example

Calculate the relative formula mass of $CuSO_4.5H_2O$.

Answer

$M_r = (1 \times 64) + (1 \times 32) + (4 \times 16) + (5 \times 18)$
$= 250$

Tip

It is useful for these calculations to remember that the M_r of H_2O is 18 (i.e. $(2 \times 1) + 16$).

Test yourself

45 What is the formula of magnesium sulfate with 7 molecules of water of crystallisation?
46 What is the formula of sodium carbonate with 10 molecules of water of crystallisation?
47 Calculate the M_r of the following compounds:
   a) $BeSO_4.4H_2O$
   b) $Na_2CO_3.8H_2O$
   c) $BaCl_2.2H_2O$
   d) $CaCl_2.2H_2O$
   e) $Ca(NO_3)_2.4H_2O$

Show you can

Calculate the percentage of:

a) oxygen
b) hydrogen
c) water

in $Na_2SO_4.10H_2O$.

Experimental determination of the mass of water of crystallisation in hydrated crystals

Water of crystallisation can be removed by heating to constant mass. The anhydrous salt is formed. This can be represented in an equation. For example, for hydrated copper(II) sulfate the equation is:

Hydrated copper(II) sulfate → anhydrous copper(II) sulfate + water
$$CuSO_4.5H_2O \rightarrow CuSO_4 + 5H_2O$$

To heat to constant mass:

▶ weigh the solid and container

▶ heat for a few minutes, cool then weigh

▶ repeat step 2 until the mass does not change.

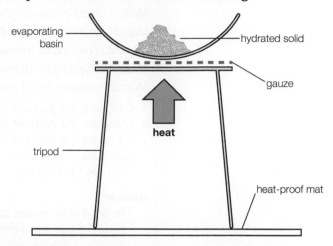

evaporating basin

hydrated solid

gauze

heat

tripod

heat-proof mat

▲ **Figure 17.21** Heating to remove water of crystallisation

The apparatus set-up is shown in Figure 17.21. When heating a hydrated solid to remove water of crystallisation, the mass decreases. The decrease in mass corresponds to the mass of water lost. Heating to constant mass ensures that all the water of crystallisation is removed. The anhydrous solid left, when the water has been removed, is often referred to as the 'residue'.

The solid must not be heated too strongly or it may decompose. Nitrates, in particular, decompose on strong heating.

When a hydrated solid is heated the following **observations** occur:

▶ crystals change to powder

▶ steam may be given off, or beads of colourless liquid (water) may be observed on the side of the apparatus

▶ there may be a colour change, for example blue hydrated copper(II) sulfate changes to white anhydrous copper(II) sulfate, as shown in Figure 17.22.

▲ **Figure 17.22** Blue hydrated copper(II) sulfate changes to white anhydrous copper(II) sulfate on heating

Tip

In an experiment to determine the mass of water in a hydrated salt, an open container, such as an evaporating basin, must be used to allow the water vapour to escape, leaving the anhydrous salt in the basin.

Some **safety rules** to observe in this practical include:

▶ To avoid burning your skin, do not touch the hot evaporating basin but allow it to cool before weighing.

▶ Wear safety glasses, as the solid may spit when being heated.

H

Example

In an experiment to find the mass of water of crystallisation in hydrated magnesium sulfate, ($MgSO_4.xH_2O$) the following results were obtained:

Mass of empty evaporating basin = 12.73 g
Mass of evaporating basin + hydrated magnesium sulfate = 13.96 g
Mass after heating for 5 minutes = 13.56 g
Mass after heating for 10 minutes = 13.33 g
Mass after heating for 13 minutes = 13.33 g

1 Calculate the mass of the anhydrous magnesium sulfate.
2 Calculate the number of moles of anhydrous magnesium sulfate.
3 Calculate the mass of the water of crystallisation removed.
4 Calculate the number of moles of the water of crystallisation removed.
5 Find *x* in the formula.

Answer
1 The solid was heated to constant mass and all the water of crystallisation was removed, so what is left in the basin at the end is anhydrous magnesium sulfate. To calculate the mass of anhydrous magnesium sulfate, subtract the mass of the evaporating basin.

13.33 − 12.73 = 0.6 g

2 First find the M_r of anhydrous magnesium sulfate, and then calculate the number of moles of magnesium sulfate:

M_r = 24 + 32 + (4 × 16) = 120

moles $MgSO_4$ = $\dfrac{mass}{M_r}$ = $\dfrac{0.6}{120}$ = 0.005

3 To find the mass of the water of crystallisation removed you need to subtract the mass of the evaporating basin plus the mass of the anhydrous magnesium sulfate from the mass of the basin plus the mass of the hydrated magnesium sulfate.

13.96 − 13.33 = 0.63 g

4 First find the M_r of water and then calculate the number of moles of water.

M_r = 18

moles H_2O = $\dfrac{mass}{M_r}$ = $\dfrac{0.63}{18}$ = 0.035

5 When working out the value of *x* you are really working out the ratio of anhydrous salt to water. A convenient way of doing this is to use a table.

Table 17.9

	$MgSO_4$	H_2O
Moles	0.005	0.035
Ratio (divide by the smallest number of moles, i.e. 0.005)	$\dfrac{0.005}{0.005}$ = 1	$\dfrac{0.035}{0.005}$ = 7
Formula		$MgSO_4.7H_2O$

The value of *x* is 7.

Example

Calculate the formula of hydrated barium chloride ($BaCl_2.xH_2O$) if 4.88 g of hydrated barium chloride produced 4.16 g of anhydrous barium chloride.

Answer
To find the mass of water, subtract the mass of the anhydrous salt from the hydrated salt.

Table 17.8

	$BaCl_2$	H_2O
Mass in grams	4.16	4.88 − 4.16 = 0.72
Moles = $\dfrac{mass}{M_r}$	$\dfrac{4.16}{208}$ = 0.02	$\dfrac{0.72}{18}$ = 0.04
Ratio (divide by the smallest number of moles, i.e. 0.02)	$\dfrac{0.02}{0.02}$ = 1	$\dfrac{0.04}{0.02}$ = 2
Formula	$BaCl_2.2H_2O$	

Test yourself

48 In an experiment to determine the number of moles of water of crystallisation contained within 1 mole of hydrated zinc sulfate, a pupil heated some zinc sulfate crystals until all the water of crystallisation had been driven out.

Mass of empty evaporating basin = 21.50 g

Mass of evaporating basin + hydrated zinc sulfate = 24.37 g

Mass of evaporating basin + anhydrous zinc sulfate = 23.11 g

a) What is water of crystallisation?

b) How could the pupil make sure that all the water of crystallisation had been removed from the crystals?

c) Calculate the mass of the anhydrous salt.

d) Calculate the mass of 1 mole of $ZnSO_4$.

e) Calculate the number of moles of zinc sulfate in the anhydrous salt.

f) Calculate the mass of water of crystallisation lost.

g) Calculate the number of moles of water lost.

h) What is the number of moles of water of crystallisation contained within 1 mole of hydrated zinc sulfate?

49 Calculate the formula of hydrated lithium sulfate if 3.76 g of hydrated lithium sulfate produces 3.23 g of anhydrous lithium sulfate (Li_2SO_4) on heating.

Prescribed practical

Prescribed practical C5: Determination of the mass of water of crystallisation present in hydrated crystals

The mass of water in $Ni(NO_3)_2.xH_2O$ (hydrated nickel(II) nitrate) may be determined by gently heating a known mass of the solid, to drive off all the water of crystallisation, and reweighing.

1 a) What is meant by the term 'water of crystallisation'?

b) Hydrated nickel(II) nitrate is a green crystalline solid. State one observation which may be made during the heating of the hydrated nickel(II) nitrate.

c) State how you could ensure, using weighings, that all of the water of crystallisation was removed.

2 In an experiment, 2.91 g of hydrated nickel(II) nitrate ($Ni(NO_3)_2.xH_2O$) produced 1.83 g of the anhydrous salt.

a) Write the equation for the reaction that occurs in the experiment.

b) Calculate the mass of water of crystallisation removed.

c) Calculate the number of moles of water of crystallisation removed.

d) Calculate the number of moles of anhydrous nickel(II) nitrate formed.

e) Determine the value of x in $Ni(NO_3)_2.xH_2O$.

f) Suggest one reason why the value of x determined for this experiment may be less than the actual value.

g) Suggest what effect strong heating may have on the salt.

h) Draw a labelled diagram of the assembled apparatus needed to carry out this experiment.

Thermal decomposition by heating to constant mass

Thermal decomposition **is the breakdown of a solid using heat.**

Sometimes a substance thermally decomposes and breaks up completely by the action of heat. Some solids, such as carbonates, thermally decompose to produce a solid and a gas, which is released to the air and the mass decreases. To ensure the thermal decomposition goes to completion, the solid is heated to constant mass. This means the solid is weighed, heated, cooled and reweighed until the mass no longer changes.

Tip

In experiments where carbonates are heated, heat in an open container (e.g. an evaporating basin) to allow the carbon dioxide gas produced to escape.

Practice questions

1 Most metals are found naturally in rocks called ores. Some examples are shown in Table 17.10.

Table 17.10

Metal ore	Formula	Relative formula mass, M_r
galena	PbS	
haematite	Fe_2O_3	
dolomite	$CaMg(CO_3)_2$	

a) Calculate the relative formula mass (M_r) of each metal ore. *(3 marks)*

b) The relative formula mass of another metal ore was calculated to be 102. The molecular formula of this ore can be represented as X_2O_3. Use this information to calculate the relative atomic mass of metal X. *(2 marks)*

c) In an experiment to extract lead from galena, 2.39 g of PbS were burnt in 12.00 g of oxygen (O_2).
  i) Calculate the number of moles in 2.39 g of PbS. *(1 mark)*
  ii) Calculate the number of moles in 12.00 g of oxygen. *(2 marks)*

2 Anaemia is a condition that occurs when the body has too few red blood cells. Anaemia often occurs in pregnant women. To prevent anaemia, iron(II) sulfate tablets can be taken to provide the iron needed by the body to produce red blood cells. One brand of iron(II) sulfate tablets contains 0.2 g of iron(II) sulfate.

a) Fill in the gaps in the following sentence. The masses of all atoms are compared to the mass of a single isotope of the element _____ which has a mass of _____. *(2 marks)*

b) Calculate the relative formula mass (M_r) of iron(II) sulfate ($FeSO_4$). *(1 mark)*

c) What is the mass of one mole of iron(II) sulfate? *(1 mark)*

d) Calculate the number of moles of iron(II) sulfate present in one tablet. *(2 marks)*

3 Glucose has the formula $C_6H_{12}O_6$.
a) What is the empirical formula of glucose? *(1 mark)*

b) Calculate the mass of glucose, in grams, needed to produce 1.08 kg of water in respiration. *(5 marks)*
$$C_6H_{12}O_6 + 6O_2 \rightarrow 6CO_2 + 6H_2O$$

4 Lead is extracted from the ore galena (PbS).
$$2PbS + 3O_2 \rightarrow 2PbO + 2SO_2$$
a) The ore is roasted in air to produce lead(II) oxide (PbO). Calculate the maximum mass of lead(II) oxide produced from 4780 g of galena. *(3 marks)*

b) The lead(II) oxide is reduced to lead by heating it with carbon in a blast furnace. The molten lead is tapped off from the bottom of the furnace.
$$PbO + C \rightarrow Pb + CO$$
Using your answer to a), calculate the maximum mass of lead that would eventually be produced. *(2 marks)*

5 Camping stoves use butane gas (C_4H_{10}) as fuel. The equation for the complete combustion of butane is:
$$2C_4H_{10} + 13O_2 \rightarrow 8CO_2 + 10H_2O$$
a) A camping gas cylinder contains 4.54 kg of butane. Calculate the number of moles of butane (C_4H_{10}) in the cylinder. *(2 marks)*

b) The camping gas is burned in 2.00 kg of oxygen. Calculate the number of moles of oxygen. *(2 marks)*

6 Sodium hydrogencarbonate decomposes when it is heated. 3.36 g of sodium hydrogencarbonate were placed in a test tube and heated over a Bunsen burner for some time.
$$2NaHCO_3 \rightarrow Na_2CO_3 + H_2O + CO_2$$
a) Calculate the number of moles of sodium hydrogencarbonate used. *(2 marks)*

b) Calculate the number of moles of sodium carbonate formed. *(1 mark)*

c) Calculate the mass, in grams, of sodium carbonate expected to be formed. *(2 marks)*

7 Zinc sulfate crystals are prepared in the laboratory by reacting zinc carbonate with sulfuric acid, as shown in the equation below.
$$ZnCO_3 + H_2SO_4 \rightarrow ZnSO_4 + H_2O + CO_2$$

a) What is the maximum mass of zinc sulfate which could be formed when 2.5 g of zinc carbonate are reacted with sulfuric acid? *(3 marks)*

b) A student carried out this experiment and only obtained 2.8 g of zinc sulfate. Calculate the percentage yield. *(2 marks)*

c) Suggest two reasons why the percentage yield is not 100% in this reaction. *(2 marks)*

8 Willow bark contains salicylic acid and was once used as a painkiller. Salicylic acid is now used to manufacture aspirin.

$$C_7H_6O_3 + (CH_3CO)_2O \rightarrow C_9H_8O_4 + CH_3COOH$$

salicylic acid   ethanoic anhydride   aspirin

A student reacted 4.00 g of salicylic acid with 6.50 g of ethanoic anhydride.

a) How many moles of salicylic acid were used? *(2 marks)*

b) How many moles of ethanoic anhydride were present? *(2 marks)*

c) What is the maximum number of moles of aspirin which could be formed? *(1 mark)*

d) Calculate the maximum mass of aspirin which could be formed. *(2 marks)*

e) The student prepared 2.90 g of aspirin. Calculate the percentage yield of aspirin obtained by the student. *(1 mark)*

9 Ammonia is used to make fertilisers such as ammonium nitrate and urea.

a) 7 kg of nitrogen was mixed with 60 kg of hydrogen. What is the limiting reactant? Calculate the maximum mass, in tonnes, of ammonia formed. *(5 marks)*

$$N_2 + 3H_2 \rightarrow 2NH_3$$

b) Urea contains 20.00% carbon, 6.66% hydrogen, 46.67% nitrogen and 26.67% oxygen. Determine the empirical formula of urea. *(5 marks)*

10 Hydrogen peroxide decomposes to produce water and oxygen.

$$2H_2O_2 \rightarrow 2H_2O + O_2$$

a) What is the empirical formula for hydrogen peroxide? *(1 mark)*

b) Calculate the mass of oxygen produced from 5.1 g of hydrogen peroxide. *(3 marks)*

11 Hydrated aluminium oxide is found in toothpastes.

a) What is meant by the term 'hydrated'? *(1 mark)*

To determine the degree of hydration in hydrated aluminium oxide 3.12 g of hydrated aluminium oxide were heated to constant mass. 2.04 g of anhydrous aluminium oxide remained.

b) Find the value of n in $Al_2O_3.nH_2O$. *(4 marks)*

c) Draw a labelled diagram of the apparatus set-up used to determine the mass of water in hydrated aluminium oxide. *(3 marks)*

12 The degree of hydration in samples of hydrated sodium carbonate ($Na_2CO_3.xH_2O$) can be determined by heating. When 14.3 g of a sample was heated to constant mass, 5.3 g of anhydrous sodium carbonate was obtained.

a) Explain the term 'anhydrous'. *(1 mark)*

b) Calculate the number of moles of anhydrous sodium carbonate obtained. *(2 marks)*

c) Calculate the mass of water present in the sample. *(1 mark)*

d) Calculate the number of moles of water present. *(1 mark)*

e) Calculate the value of x in the carbonate. *(1 mark)*

18 Acids, bases and salts

Specification points

This chapter covers specification points 1.8.1 to 1.8.17. It covers indicators and reactions of acids. The prescribed practical activity on the reactions of acids is included.

Strawberries and blueberries contain five different acids including citric acid and ascorbic acid. Many other foods contain acids, and often they make food taste sour – the word 'acid' comes from the Latin word *acere*, which means 'sour'. In this chapter you will explore the different reactions of acids and understand how salts are prepared from acids.

Indicators

Around 1300 AD a Spanish scientist first used blue dye from species of lichens to test for acidity. It was found that this dye was an indicator and it was given the name litmus, which means 'coloured moss'. **An indicator is one colour in acid and a different colour in alkali.**

▲ **Figure 18.1** The dye extracted from red cabbage can be used as an indicator. Boiling flowers (poppy or rose petals) or berries (blueberries or blackcurrants) with water also extracts dyes that are indicators

Litmus paper is a useful indicator for deciding if a substance is acidic or alkaline. Both colours of litmus paper may need to be used. The effects of acid and alkali on litmus paper are shown in Table 18.1.

Table 18.1

	Colour in acid	Colour in alkali	Colour in neutral	
Red litmus paper	red	blue	red	The first solution is acidic, the second is alkaline.
Blue litmus paper	red	blue	blue	

Table 18.2

	Colour in acidic solution	Colour in alkaline solution	Colour in neutral solution
Phenolphthalein	colourless	pink	colourless
Methyl orange	red	yellow	orange

Figure 18.2 The left hand solution of each pair is acid, the right is alkali

What are acids and alkalis?

An acid is a substance that dissolves in water producing hydrogen ions (H^+(aq)).

Table 18.3 shows three common laboratory acids and the ions they contain. It is useful to learn the formula of these acids.

Table 18.3

Acid	Formula	Positive ion	Negative ion
hydrochloric acid	HCl	hydrogen (H^+)	chloride (Cl^-)
sulfuric acid	H_2SO_4	hydrogen (H^+)	sulfate (SO_4^{2-})
nitric acid	HNO_3	hydrogen (H^+)	nitrate (NO_3^-)

An alkali is a substance that dissolves in water producing hydroxide ions (OH^-(aq)).

Table 18.4 shows common laboratory alkalis and the ions they contain.

Table 18.4

Alkali	Formula	Positive ion	Negative ion
sodium hydroxide	$NaOH$	sodium (Na^+)	hydroxide (OH^-)
potassium hydroxide	KOH	potassium (K^+)	hydroxide (OH^-)

Ammonia (NH_3) is a gas. When it reacts with water it produces hydroxide ions, so it is an alkali.

$$NH_3(g) + H_2O(l) \rightarrow NH_4^+(aq) + OH^-(aq)$$

Test yourself

1 Copy and complete Table 18.5.

Table 18.5

Indicator	Colour of the indicator in a solution of:				
	hydrochloric acid	sodium hydroxide	citric acid	calcium hydroxide	ethanoic acid
red litmus paper					
blue litmus paper					

2 a) Explain, in terms of ions, the difference between an acid and an alkali.
  b) Name the ions present in sulfuric acid.
  c) Name the ions present in calcium hydroxide.
  d) What is the formula of nitric acid?

Strong and weak acids and alkalis

Acids can be classified as strong or weak, depending on the number of hydrogen ions dissolved in water ($H^+(aq)$). Strong acids **are acids that are completely ionised in water** – this means they break up completely into ions and there is a high concentration of hydrogen ions in the solution.

For example, hydrochloric acid is ionised completely into hydrogen ions and chloride ions, and the resulting solution contains only these two species.

$$HCl(aq) \rightarrow H^+(aq) + Cl^-(aq)$$

Sulfuric acid, another strong acid, ionises completely into hydrogen ions and sulfate ions.

$$H_2SO_4(aq) \rightarrow 2H^+(aq) + SO_4^{2-}(aq)$$

Nitric acid, another common laboratory acid, is also a strong acid.

Weak acids **are acids that are partially ionised in water.** This means that only a small fraction of the molecules break into ions when added to water.

For example, ethanoic acid is a weak acid. It is partially ionised into hydrogen ions and ethanoate ions in water. The resulting solution will contain hydrogen ions, ethanoate ions (CH_3COO^-) and non-ionised ethanoic acid molecules. There are fewer hydrogen ions in the solution, and so the acid is weak.

$$CH_3COOH(aq) \rightleftharpoons CH_3COO^-(aq) + H^+(aq)$$

Citric acid and carbonic acid are weak acids. Figure 18.3 shows the difference in ionisation between a weak acid and a strong acid.

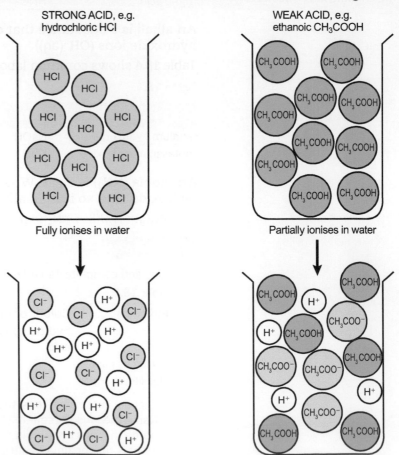

▲ **Figure 18.3** The ionisation of a strong acid and a weak acid

The strength of an alkali (strong or weak) depends on the number of hydroxide ions OH⁻(aq) dissolved in water. **Strong alkalis** are **alkalis that are completely ionised in water** – this means they break up completely into ions. Group 1 hydroxides are strong alkalis.

For example, sodium hydroxide is ionised completely into sodium ions and hydroxide ions, and the resulting solution contains only these two species:

$$NaOH(aq) \rightarrow Na^+(aq) + OH^-(aq)$$

Weak alkalis are **alkalis that are partially ionised in water** – this means they do not completely break up into ions in water, and so there are fewer hydrogen ions, OH⁻, in the solution. Ammonia solution is a weak alkali:

$$NH_4OH(aq) \rightleftharpoons NH_4^+(aq) + OH^-(aq)$$

The pH scale

The pH scale is a measure of how acidic or alkaline a solution is (see Figure 18.5). The scale used in GCSE chemistry runs from 0 to 14. On this scale:

▶ pH 7 is neutral

▶ pH less than 7 is acidic

▶ pH above 7 is alkaline.

> **Tips**
>
> ▶ When writing pH the p is always lower case and the H a capital.
> ▶ pH is an abbreviation for 'power of hydrogen', where the p is short for the German word for power, *potenz*, and H is the element symbol for hydrogen.

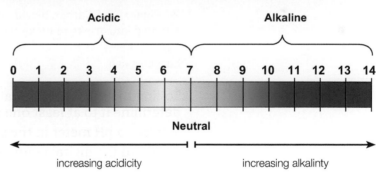

▲ **Figure 18.5** The pH scale

▲ **Figure 18.4** What do Carlsberg beer and pH have in common? Sorenson devised the pH scale when he was director of the Carlsberg laboratory in Copenhagen. This laboratory is supported by the Carlsberg brewing company and is used to research any chemistry related to brewing

Universal indicator solution or paper can be used to find the pH of a solution. Universal indicator paper is often referred to as 'pH paper'. The method used is as follows:

1 dip pH paper into the solution or add 3 drops of universal indicator to the solution

2 compare the colour with the pH colour chart.

The pH value can be used to determine the relative strengths of acidic and alkaline solutions. Table 18.6 gives the pH colours, the associated pH numbers and classifications, and examples from your specification.

Table 18.6 pH classification

	pH	Colour of pH paper	Classification	Examples
	0–2	red	strong acid	sulfuric acid hydrochloric acid nitric acid
	3–6	orange–yellow	weak acid	ethanoic acid (vinegar) carbonic acid
	7	green	neutral	water
	8–11	green-blue–blue	weak alkali	ammonia
	12–14	dark blue–purple	strong alkali	sodium hydroxide, potassium hydroxide

Tip

The pH scale can extend beyond 0–14 and some acids in the chemistry laboratory can have a pH less than 0 (typically −0.3). Fluoroantimonic acid is thought to be one of the world's strongest acids – it can dissolve glass, and has a pH of −25.

▲ **Figure 18.6** Carbon dioxide is bubbled into drinks to make them fizzy. It reacts with water to form carbonic acid

▲ **Figure 18.7** The acids in household substances are generally weak acids. Lemon juice contains citric acid and tea contains tannic acid

To find the pH more accurately, a **pH meter** can be used to determine it to at least one decimal place. The **method** is:
1 place a pH meter in the solution
2 record the digital reading to one decimal place.

▲ **Figure 18.8** A pH meter

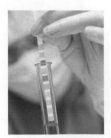

▲ **Figure 18.9** Dipsticks are used by doctors to measure the pH of urine. Urine pH normally ranges between 2.5 and 8.0. High acidity can indicate diabetes, and high alkalinity can indicate a urinary tract infection

Test yourself

3 Classify each of the following solutions as weak or strong, acid or alkali:
   a) a solution with pH 8   c) a solution with pH 9
   b) a solution with pH 3   d) a solution with pH 13.
4 a) Three solutions had pH values of 9, 11 and 14. Which one was the most alkaline?
   b) Three solutions had pH values of 1, 2 and 5. Which one was the most acidic?
   c) A solution has pH 3.5. Describe how the pH was measured.
5 Copy and complete Table 18.7.

Table 18.7

Substance	Acid or alkali?	Strong or weak?	pH range	Colour of pH paper
hydrochloric acid				
potassium hydroxide				
ammonia				
nitric acid				
ethanoic acid				
citric acid				
sodium hydroxide				

6 Using a pH meter, a solution was found to have a pH of 3.6. State what colour you would expect each of the following indicators to be if added to this solution:
   a) red litmus paper   b) blue litmus paper   c) universal indicator.

Show you can ?

In an experiment, a sample of human saliva was removed from a person's mouth every five minutes after a meal and the pH value determined. The graph shows how the pH value of the saliva changed.

1 How were the pH values of the saliva determined in this experiment?
2 When the pH in the mouth is 5.0 or less tooth decay occurs. Use the graph to find the time after the meal at which teeth would start to decay.
3 At what time is the pH of the saliva most acidic?
4 At what time is the pH of the saliva neutral?

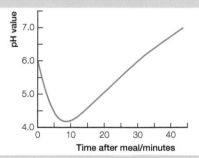

▲ Figure 18.10

Dilute and concentrated acids

A concentrated acid contains a large number of acid particles dissolved per unit volume.

A dilute acid contains a small number of acid particles dissolved per unit volume.

Tip

Do not mix up the terms 'dilute' and 'concentrated' with 'strong' and 'weak'. The terms strong and weak refer to the degree of *ionisation* of the substances in water, but dilute and concentrated refer to the amount *dissolved* in the water.

Example

Solution 1 is a 1.0 mol/dm³ solution of ethanoic acid and solution 2 is a 0.5 mol/dm³ solution of nitric acid.

1 Explain which solution is more concentrated.
2 Explain which solution is a strong acid.
3 Which solution has a lower pH?

Answer

1 A more concentrated solution has more particles of acid per unit volume. The ethanoic acid is more concentrated as it has 1 mole of acid in 1 dm³, while the nitric acid only has 0.5 mole of acid in 1 dm³.
2 A strong acid is one that completely ionises in water. Nitric acid is a strong acid.
3 The nitric acid has a lower pH as it is a strong acid.

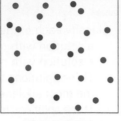 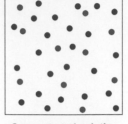

Dilute solution   **Concentrated solution**

● Acid particle
● Water particle

▲ **Figure 18.11** A concentrated solution has more acid particles than a dilute solution

For an acid, the pH is linked to the concentration of H^+ ions in the solution. The **higher the concentration of hydrogen ions** in an acidic solution, the **lower the pH**.

Concentration is measured in moles per cubic decimetre of solution (mol/dm³). One decimetre cubed (dm³) is 1000 cm³. You will learn more about concentration in Chapter 25.

An acid HA has a concentration of 0.2 mol/dm³. It has a higher pH than a 1.0 mol/dm³ solution of HA. This is because 1.0 mol/dm³ is a higher concentration.

Safety

Safety in the laboratory is of prime importance. All risks must be considered and chemicals must be labelled with hazard symbols. There is an international system of labelling chemical hazards. The system uses symbols, which means that there are no language issues, and the hazard symbols are recognised worldwide. The hazard symbols you need to recognise are shown in Table 18.8.

Table 18.8

Hazard symbol	Name	Meaning	Example
	corrosive	burns and destroys living tissue	concentrated acid
	explosive	explodes if exposed to flame, heat or knocked	potassium
	flammable	catches fire easily when in contact with air	ethanol
	toxic	can kill by poisoning	weedkiller, cyanide
	moderate hazard	may be harmful or an irritant	dilute sodium hydroxide

▲ **Figure 18.12** Oven cleaner contains sodium hydroxide, which is corrosive

7 What is the difference between a concentrated alkali and a dilute solution of an alkali?

8 a) Sulfuric acid is a strong acid. Explain what this means.

b) Citric acid is a weak acid. Explain what this means.

c) Comment on the concentration of these two acids: $0.4\,mol/dm^3$ ethanoic acid and $0.4\,mol/dm^3$ nitric acid. Is one more concentrated or less concentrated than the other?

d) Which of the two acids in c) has the lowest pH? Explain your answer.

e) What hazard symbol is used on a bottle of:
   i) concentrated hydrochloric acid
   ii) petrol?

9 Table 18.9 gives some information about three solutions.

Table 18.9

Solution	A	B	C
pH	4	3	2

a) Which solution has the highest concentration of H^+ ions?

b) Describe how the pH was measured.

c) Two solutions of the same acid have concentrations of $0.5\,mol/dm^3$ and $0.8\,mol/dm^3$. Which of the two solutions has the higher pH?

d) A solution E has a pH of 7. Comment on the concentration of H^+ ions compared with the concentration of OH^- ions.

Show you can

Two solutions A and B were tested using a pH meter, red litmus paper, blue litmus paper and universal indicator paper. The results are shown in Table 18.10.

Table 18.10

Test	Result for solution A	Result for solution B
pH meter	1.82	3.85
red litmus	red	red
blue litmus	red	red
universal indicator	red	orange

1 Explain how the results obtained using universal indicator may be converted into a pH value.

2 Are solutions A and B acidic, neutral or alkaline?

3 If the experiment was repeated using a more concentrated solution of A would the results be different? Explain your answer.

4 Explain why universal indicator gives more information than litmus.

Reactions of acids

Salts

All acids produce salts when they react.

A salt is the compound formed when some or all of the hydrogen ions of an acid are replaced by metal or ammonium ions.

For example, zinc sulfate is the salt formed when zinc ions replace the hydrogen ions of sulfuric acid.

H_2SO_4 forms $ZnSO_4$ The two H^+ ions are replaced by a zinc ion (Zn^{2+})

Sodium sulfate is the salt formed when sodium ions replace the hydrogen ions of sulfuric acid.

H_2SO_4 forms Na_2SO_4 The two H^+ ions are replaced by two sodium ions (Na^+)

Ammonium sulfate is the salt formed when ammonium ions replace the hydrogen ions of sulfuric acid.

H_2SO_4 forms $(NH_4)_2SO_4$ The two H^+ ions are replaced by two ammonium ions (NH_4^+)

Each acid produces a differently named salt. Table 18.11 shows the general names of the types of salts formed by some acids.

Table 18.11

Acid	Salt
hydrochloric	chloride
sulfuric	sulfate
nitric	nitrate

A salt is named in two parts. The first part of the name is the name of the metal ion or the ammonium ion. The second part of the name comes from the acid.

For example:

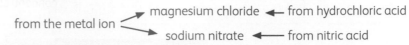

from the metal ion — magnesium chloride ← from hydrochloric acid
— sodium nitrate ← from nitric acid

Observations for acid reactions

There are some general observations that can be made during acid reactions:

▶ If a gas is produced the observation is **'bubbles'**.

▶ If a solid such as a solid metal, base or carbonate reacts with the acid then the observation is **'solid disappears and a solution is produced'**.

▶ Most acid reactions are exothermic – the observation is **'heat released'**. Two notable exceptions to this are the reactions of copper(II) oxide and sodium hydrogencarbonate with acid.

▶ You need to know the colours of the chemicals given in Table 18.12, and from this you can work out if there is a colour change in a reaction or if the solution remains colourless.

Table 18.12

Substance	Colour
copper(II) oxide	black solid
copper(II) carbonate	green solid
hydrated copper(II) sulfate	blue crystals
copper(II) salts in solution	blue solution
group 1, group 2, aluminium and zinc compounds	white solids, and if they dissolve in water they give colourless solutions

Tip

When an acid reacts with a solid, the solid disappears, it does not dissolve.

Example

What are the observations for the reaction below?

copper(II) carbonate + sulfuric acid → copper(II) sulfate + water + carbon dioxide

Answer
- A gas (carbon dioxide) is produced, so the observation is 'bubbles'.
- The copper(II) carbonate is a solid and it is green, so the observation is 'green, solid copper(II) carbonate disappears'.
- The salt copper(II) sulfate is produced in solution, so the observation is 'blue solution produced'.
- Heat released.

Example

What are the observations for the reaction below?

sodium hydroxide + sulfuric acid → sodium sulfate + water

Answer
- Sodium hydroxide is a colourless solution, as is sulfuric acid and sodium sulfate solution, so the solution remains colourless.
- Heat released.

Example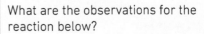

What are the observations for the reaction below?

magnesium + hydrochloric acid → magnesium chloride + hydrogen

Answer
- Bubbles
- Grey solid magnesium disappears
- Colourless solution produced
- Heat released

Test yourself

Write all the observations that can be made for each of the following reactions:

10 magnesium carbonate + hydrochloric acid → magnesium chloride + water + carbon dioxide

11 potassium hydroxide + sulfuric acid → potassium sulfate + water

12 calcium + hydrochloric acid → calcium chloride + hydrogen

13 magnesium oxide + nitric acid → magnesium nitrate + water

14 copper(II) oxide + sulfuric acid → copper(II) sulfate + water

There are four different types of reactions of acids, each of which produce a salt.

1 Acids and metals

Many metals react with acids. The general equation is

metal + acid → salt + hydrogen

Tips

▶ Remember MASH:

metal + acid → salt + hydrogen
    m   a   s   h

▶ Unreactive metals such as copper, silver and gold do not react with acids. You will learn more about this in Chapter 8.

Example

Word equation:

magnesium + hydrochloric acid → magnesium chloride + hydrogen

Balanced symbol equation:

$Mg + 2HCl → MgCl_2 + H_2$

Observations: grey solid magnesium disappears, colourless solution produced, heat released, bubbles.

Example

Word equation:

zinc + sulfuric acid → zinc sulfate + hydrogen

Balanced symbol equation:

$Zn + H_2SO_4 → ZnSO_4 + H_2$

Observations: solid grey zinc disappears, colourless solution produced, heat released, bubbles.

▲ **Figure 18.13** Testing for hydrogen

Tip

Hydrogen is diatomic and its formula is H_2.

The hydrogen produced in these reactions can be tested. The **test for hydrogen** is:

▶ apply a lighted splint

▶ a popping sound results.

In this test the hydrogen burns in oxygen to produce water:

$2H_2 + O_2 → 2H_2O$

2 Acids and bases

The general equation for the reaction of an acid with a base is

base + acid → salt + water

A base is a metal oxide or metal hydroxide that neutralises an acid to produce a salt and water.

An alkali is a soluble base.

▲ **Figure 18.14** A coil of magnesium ribbon reacting with acid

Copper(II) oxide and zinc oxide are examples of bases that neutralise an acid to produce a salt and water. However, they are not soluble so they are not alkalis. Sodium hydroxide and potassium hydroxide are bases, and they are soluble bases (alkalis).

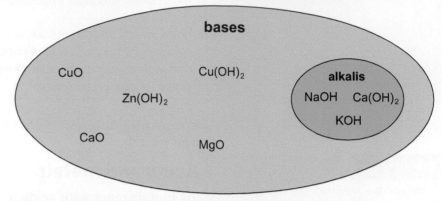

▲ **Figure 18.15** Some bases are soluble (alkalis)

When an acid reacts with an alkali, neutralisation occurs. **Neutralisation is the reaction between the hydrogen ions in an acid and the hydroxide ions in an alkali to produce water.**

The ionic equation for neutralisation is:

$$H^+(aq) + OH^-(aq) \rightarrow H_2O(l)$$

Example

Word equation:

sodium hydroxide + hydrochloric acid → sodium chloride + water

Balanced symbol equation:

$NaOH + HCl \rightarrow NaCl + H_2O$

Observations: colourless solution remains, heat released.

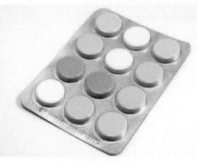

▲ **Figure 18.16** Indigestion tablets use mixtures of carbonates such as calcium carbonate to neutralise excess stomach acid which causes indigestion

Example

Word equation:

copper(II) oxide + sulfuric acid + → copper(II) sulfate + water

Balanced symbol equation:

$CuO + H_2SO_4 \rightarrow CuSO_4 + H_2O$

Observations: black solid copper(II) oxide disappears, blue solution produced.

Neutralisation reactions are exothermic. **An exothermic reaction is one in which heat energy is given out.**

It is possible to carry out an experiment to measure the temperature during a neutralisation reaction. Often the reaction is carried out in a polystyrene cup, which prevents heat loss to the surroundings. The polystyrene cup is placed in a glass beaker for support.

▲ **Figure 18.17** Green copper(II) carbonate reacting with sulfuric acid to produce a blue solution of copper(II) sulfate

Prescribed practical

Prescribed practical C1: Investigating the reaction of acids – temperature changes which occur during a neutralisation reaction.

In an experiment the following method was used:

- 25 cm³ of 1 mol/dm³ sodium hydroxide were measured out and placed in a polystyrene cup.
- The polystyrene cup was placed in a glass beaker and the temperature of the sodium hydroxide was measured and recorded.
- A burette was filled with 1 mol/dm³ hydrochloric acid.
- 5.0 cm³ of hydrochloric acid were added from the burette to the polystyrene beaker, and the temperature recorded. This was repeated until the total volume of hydrochloric acid added was 40.0 cm³.
- The experiment was repeated and the average temperature was calculated.

The results are shown in Table 18.13.

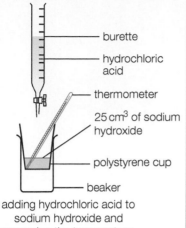

adding hydrochloric acid to sodium hydroxide and measuring the temperature.

▲ **Figure 18.18**

Table 18.13

Volume of acid added/cm³	0.0	5.0	10.0	15.0	20.0	25.0	30.0	35.0	40.0
Temperature/°C	19.4	21.6	24.0	25.2	25.4	26.2	25.4	25.2	25.1
Repeat temperature/°C	19.6	21.8	24.0	25.0	25.6	26.4	25.4	25.0	24.9
Average temperature/°C	19.5	21.7	24.0	25.1	25.5	26.3	25.4	25.1	25.0

Questions

1 In this experiment identify the:
  a) independent variable   b) dependent variable   c) controlled variables.
2 Why was the experiment repeated?
3 Why was a polystyrene cup used?
4 Why was the polystyrene cup placed in a beaker?
5 Why should the solution in the beaker be stirred after each addition of acid?
6 State one possible source of error in this experiment and state how the method could be improved to minimise this source of error.
7 Plot a graph of average temperature (*y* axis) against volume of acid added (*x* axis).
8 Describe the trend shown by the results plotted.
9 It is thought that when complete neutralisation has occurred the highest temperature is reached. Suggest how you would confirm experimentally that the highest temperature reached is the point at which neutralisation has occurred.

A different experiment was carried out by adding different acids of concentration 1 mol/dm³ to 25 cm³ of 1 mol/dm³ sodium hydroxide solution in a polystyrene cup. The highest temperature reached was recorded and presented as a table:

Table 18.14

	hydrochloric acid	sulfuric acid
highest temperature/°C	26.9	27.1
repeat highest temperature/°C	26.9	26.9
average highest temperature/°C	26.9	27.0

10 In this experiment identify the:
  a) independent variable   b) dependent variable   c) controlled variables.
11 State two conclusions you can draw from the results of the experiment.

3 Acid and carbonate

Acids react with metal carbonates and hydrogencarbonates in the same way. The general equation is:

metal carbonate + acid → salt + water + carbon dioxide

or

metal hydrogencarbonate + acid → salt + water + carbon dioxide

Group 1 hydrogencarbonates are solids but Group 2 hydrogencarbonates only exist in solution.

> **Example**
>
> Word equation:
>
> calcium carbonate + hydrochloric acid → calcium chloride + water + carbon dioxide
>
> Balanced symbol equation:
>
> $CaCO_3 + 2HCl → CaCl_2 + H_2O + CO_2$
>
> Observations: solid white calcium carbonate disappears, colourless solution produced, heat released, bubbles.

> **Example**
>
> Word equation:
>
> sodium hydrogencarbonate + nitric acid → sodium nitrate + water + carbon dioxide
>
> Balanced symbol equation:
>
> $NaHCO_3 + HNO_3 → NaNO_3 + H_2O + CO_2$
>
> Observations: solid white sodium hydrogencarbonate disappears, colourless solution produced, bubbles.

> **Tip**
>
> Sodium and potassium hydrogencarbonate react endothermically with acid and the temperature of the solution falls.

The carbon dioxide gas produced in these reactions can be tested. The **test for carbon dioxide** is:

▶ bubble the gas into **colourless limewater** (calcium hydroxide solution)

▶ the solution will change to milky if the gas is carbon dioxide.

If excess carbon dioxide is bubbled into the milky limewater it will eventually go colourless again.

Limewater is calcium hydroxide.

carbon dioxide + limewater → calcium carbonate + water

$CO_2(g) + Ca(OH)_2 (aq) → CaCO_3 (s) + H_2O (l)$

The calcium carbonate produced is insoluble and so a milky white precipitate forms.

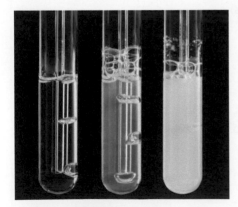

▲ **Figure 18.19** Bubbling carbon dioxide into limewater changes it to a milky colour, due to the production of insoluble calcium carbonate

When excess carbon dioxide is added then soluble calcium hydrogencarbonate is formed, the white precipitate disappears and a colourless solution is formed.

calcium carbonate + water + carbon dioxide → calcium hydrogencarbonate

$$\underbrace{CaCO_3 \text{ (s)} + H_2O \text{ (l)}}_{\text{milky limewater}} + CO_2 \text{ (g)} → \underbrace{Ca(HCO_3)_2 \text{ (aq)}}_{\text{colourless solution}}$$

4 Acid and ammonia

The general equation is:

ammonia + acid → ammonium salt

Example

Word equation:

ammonia + hydrochloric acid → ammonium chloride

Balanced symbol equation:

$NH_3 + HCl → NH_4Cl$

Example

Word equation:

ammonia + sulfuric acid → ammonium sulfate

Balanced symbol equation:

$2NH_3 + H_2SO_4 → (NH_4)_2SO_4$

Test yourself

15 Complete the following word equations:
   a) zinc + nitric acid →
   b) sodium hydroxide + sulfuric acid →
   c) magnesium + sulfuric acid →
   d) hydrochloric acid + sodium carbonate →
   e) copper(II) oxide + sulfuric acid →
   f) potassium hydrogencarbonate + sulfuric acid →
   g) calcium carbonate + hydrochloric acid →
   h) sodium hydroxide + nitric acid →
16 Write observations for each reaction in question 16.
17 Write balanced symbol equations for each reaction in question 16.
18 What is an alkali? Use the solubility table on your data leaflet to decide if the following bases are alkalis:
   a) copper(II) oxide
   b) zinc oxide
   c) magnesium oxide
   d) sodium hydroxide
   e) copper(II) hydroxide
   f) magnesium hydroxide
   g) potassium hydroxide.

Show you can

Sodium sulfate, produced by the reaction between sulfuric acid and sodium hydroxide, is used in washing powders.

1 Write a word equation and a balanced symbol equation for the reaction of sulfuric acid with sodium hydroxide solution.

The compositions of two washing powders, A and B, are shown in Table 18.15.

Table 18.15

	Composition (%)				
	sodium sulfate	sodium carbonate	sodium silicate	soap	detergent
A	29	20	20	0	15
B	35	0	26	6	13

Dilute nitric acid was added to each of the powders. Only one of the powders reacted.

2 Which powder reacted, A or B? Explain your answer.
3 During the reaction, bubbling was noted. Name the gas produced in the reaction.
4 Describe a chemical test for this gas and state the result for a positive test.
5 The silicate ion is SiO_3^{2-}. Suggest the formula for sodium silicate.

Test yourself

19 Copy and complete the following word equations:
   a) lithium oxide + _____ → lithium chloride + _____
   b) _____ + _____ → potassium sulfate + water + carbon dioxide
   c) calcium carbonate + _____ → calcium chloride + _____ + _____
   d) _____ + _____ → aluminium nitrate + hydrogen
   e) sodium hydroxide + _____ → sodium sulfate + _____

20 Suggest two chemicals that could be reacted together to make the following salts:
   a) calcium nitrate
   b) copper(II) chloride
   c) zinc sulfate.

Show you can

A solution of the salt magnesium chloride can be prepared by reactions A to D in Figure 18.20.

1 Write word equations for each of the reactions A to D.
2 State two observations that you would make during reaction D.

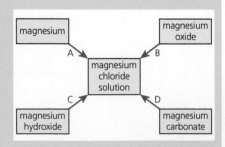

▲ Figure 18.20

Practice questions

1. **a)** Copy and complete Table 18.17. *(7 marks)*

   Table 18.17

Substance	pH	
baking soda	8–10	
ethanoic acid		
caustic soda solution		strong alkali
hydrochloric acid		
lemon juice		weak acid

   b) What is observed when:
   i) red litmus and
   ii) blue litmus
   are dipped into a solution of baking soda? *(2 marks)*
   c) What is an acid? *(1 mark)*
   d) Two solutions of sulfuric acid have concentrations $0.5\,mol/dm^3$ and $1.0\,mol/dm^3$. Explain which acid has the higher pH. *(2 marks)*

2. In an experiment a student slowly added solution Y in $0.5\,cm^3$ portions to solution X in a conical flask and swirled the solution. The apparatus for the experiment is shown below.

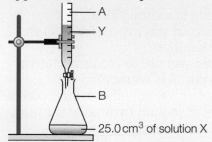

Figure 18.21

The student measured and recorded the pH after each addition of Y, and drew a graph of pH against volume of Y added.

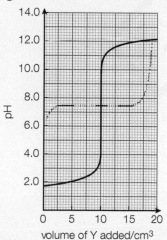

Figure 18.22

a) What is the piece of apparatus labelled A? *(1 mark)*
b) What is the piece of apparatus labelled B? *(1 mark)*
c) What piece of apparatus was used to measure out $25.0\,cm^3$ of solution X? *(1 mark)*
d) How could the pH of the solution be measured? *(2 marks)*
e) Why was the flask swirled after each addition? *(1 mark)*
f) Is X an acidic, alkaline or neutral solution? Use the graph to explain your answer. *(1 mark)*
g) Is Y an acidic, alkaline or neutral solution? Use the graph to explain your answer. *(1 mark)*
h) Use the graph to describe what happens to the pH of the mixture in the conical flask as solution Y is slowly added. *(1 mark)*
i) Describe what would happen to the shape of the graph if solution X was in the burette and solution Y was in the conical flask. *(3 marks)*
j) What volume of Y is needed to react with all of solution X? *(1 mark)*

3. Neutralisation occurs when an acid and an alkali react to form a salt and water.
   a) i) Copy and complete Table 18.18 to give the names and formulae of the ions present in all acids and alkalis. *(4 marks)*

   Table 18.18

	Ion present in all acids	Ion present in all alkalis
Name		
Formula		

   ii) Write an ionic equation for neutralisation, including state symbols. *(3 marks)*
   b) Sulfuric acid solution can be neutralised using an alkali such as sodium hydroxide or by adding a solid oxide such as copper(II) oxide.
   i) Write a balanced equation for the reaction between sodium hydroxide and sulfuric acid. *(3 marks)*
   ii) Write a balanced equation for the reaction between copper(II) oxide and hydrochloric acid. *(2 marks)*
   iii) State two observations for the reaction in **i)** and the reaction in **ii)**. *(4 marks)*
   c) Sulfuric acid also reacts with magnesium metal.
   i) State two observations for this reaction. *(1 mark)*
   ii) Describe a test for the gas produced. State the result for a positive test. *(1 mark)*

4 a) State the approximate pH of:
  i) hydrochloric acid
  ii) ethanoic acid
  iii) sulfuric acid
  iv) potassium hydroxide solution. *(4 marks)*
  b) What hazard symbol should be on a bottle of concentrated acid? *(1 mark)*
  c) Ethanoic acid is a weak acid. What is a weak acid? *(1 mark)*
  d) Sodium hydroxide is a strong alkali. What is a strong alkali? *(1 mark)*
  e) When acids and alkalis react, salts are formed. What is meant by the term 'salt'? *(2 marks)*
  f) Crystalline salts are hydrated. What is meant by the term 'hydrated'? *(1 mark)*
  g) One solution has pH 3.3 and another has pH 11.4.
  i) Describe how you would determine the pH of these two solutions. *(2 marks)*
  ii) Classify each solution as a strong or weak acid or alkali. *(2 marks)*
  h) What colour is:
  i) solid magnesium sulfate *(1 mark)*
  ii) copper(II) nitrate solution *(1 mark)*
  iii) magnesium chloride solution? *(1 mark)*

5 a) Copy and complete Table 18.19.

Table 18.19

Reaction	Balanced symbol equation	Observations
magnesium + hydrochloric acid	*(3 marks)*	*(3 marks)*
calcium carbonate + hydrochloric acid	*(3 marks)*	*(3 marks)*
copper(II) oxide + sulfuric acid	*(3 marks)*	*(3 marks)*

  b) Decide if each of the following substances is a base or a salt:
  i) magnesium oxide
  ii) sodium hydroxide
  iii) aluminium sulfate
  iv) magnesium nitrate
  v) sodium chloride
  vi) magnesium hydroxide
  vii) copper(II) sulfate
  viii) copper(II) oxide
  ix) potassium hydroxide. *(9 marks)*
  c) What is an alkali? *(1 mark)*
  d) Name two alkalis from the list in **b)**. *(2 marks)*

6 Hydrochloric acid can react with sodium hydroxide and with magnesium. Compare and contrast the reaction of hydrochloric acid with calcium hydroxide with the reaction of hydrochloric acid with calcium. In your answer you must include:
  • the names of all products for each reaction
  • the observations for each reaction. *(6 marks)*
In this question you will be assessed on your quality of written communication.

7 Copper(II) chloride is a salt which may be prepared by adding excess copper(II) carbonate to hydrochloric acid.
  a) What is meant by the term 'salt'? *(1 mark)*
  b) What would you observe when solid copper(II) carbonate is added to dilute hydrochloric acid? *(3 marks)*
  c) Write a balanced symbol equation for the reaction of copper(II) carbonate with hydrochloric acid. *(3 marks)*
  d) Describe how you would test for the gas produced in the reaction in **c)**. State the result for a positive test. *(3 marks)*
  e) Hydrated copper(II) chloride contains 2 moles of water of crystallisation. Write the formula of hydrated copper(II) chloride. *(1 mark)*

8 Sodium hydroxide is a strong alkali.
  a) i) What pH would you expect for sodium hydroxide solution?
  ii) What would you use to confirm that your answer to part i) is correct?
  b) A few drops of universal indicator are placed into a sample of sodium hydroxide solution. What colour will be seen?
Bottles containing concentrated sodium hydroxide solution often display the hazard symbol shown below.

Figure 18.23

  c) What does this hazard symbol mean?
  d) Give two reasons why hazard symbols are shown on bottles of chemicals.

e) i) Sodium hydroxide reacts with hydrochloric acid. Write a balanced symbol equation for the reaction.

ii) The reaction is exothermic. What does this mean?

iii) How could you tell that the reaction is exothermic?

9 a) i) Describe the difference between strong acids and weak acids in terms of hydrogen ions dissolved in water.

ii) Give one example of a strong acid and one example of a weak acid.

b) Explain why sodium hydroxide solution is a strong alkali but ammonia solution is a weak alkali.

A strong acid and a strong alkali react together in a test tube. At the end of the reaction all of the acid and all of the alkali have been used up.

c) i) What name is given to this process?

ii) What is the pH of the solution in the test tube at the end of the reaction?

iii) The final solution is tested with pH paper. What colour would you expect to see?

A mysterious white powder, a blood smear, and a half eaten biscuit – completely unrelated items to most, but at a crime scene they are important evidence for forensic scientists to analyse. Forensic scientists use chemical methods to analyse traces of crime scene material – gunshot residues, drugs, hair or blood. Chemical analysis is also used to separate and test the materials in our food and water, and to identify and test the purity of different substances.

Pure substances

When we see the stamp 'pure orange juice', like that shown in Figure 19.1, we consider the orange juice to be a natural substance with nothing added to it during manufacture; just the juice without colourings or sweeteners. A chemist, however, would not consider the orange juice to be pure as it is a mixture of different substances.

In chemistry a pure substance **is a single element or compound not mixed with any other substance.**

Orange juice is not a single element or compound, as it contains water, citric acid, vitamin C and other ingredients. It is a mixture and not a pure substance.

▲ **Figure 19.1** A scientist would not consider orange juice to be pure

Table 19.1 Some pure substances and some mixtures

Pure substances	Mixtures
Diamond only contains the element carbon.	Air is a mixture of oxygen, carbon dioxide, nitrogen and other trace gases.
Water only contains the compound water.	Mineral water is a mixture of water and dissolved salts.
Table salt only contains the compound sodium chloride.	Milk is a mixture of water, lactose, fat and minerals.

▲ **Figure 19.2** Mineral water is not pure as it is not a single compound. It is mostly water but there are other substances mixed with it

Melting and boiling point of pure substances and mixtures

Melting point is the temperature at which a solid changes into a liquid.

Boiling point is the temperature at which a liquid changes into a gas.

Pure substances have **specific** melting points and boiling points. For example, the compound water has a boiling point of 100 °C and a melting point of 0 °C; the element sulfur has a melting point of 115 °C and a boiling point of 444 °C. Substances can be identified by their melting points and boiling points.

A substance must absorb heat energy so that it can melt or boil. The temperature of the substance does not change during melting, boiling or freezing, even though energy is still being transferred. You can see this in the graph in Figure 19.3.

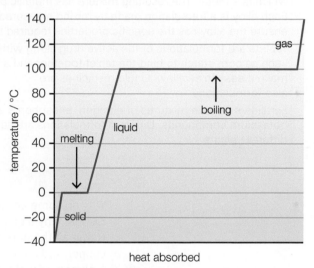

▲ **Figure 19.3** Graph of temperature against heat absorbed for water. Note that the temperature is constant during melting and boiling

Impure substances **do not have sharp** melting or boiling points, but melt or boil over a range of temperatures. The presence of an **impurity** usually **lowers** the melting point of a solid or raises the boiling point of a liquid. The greater the amount of an impurity, the bigger the difference from the true melting point and boiling point.

Table 19.2 The differences in melting and boiling points between pure and impure substances

Pure substances	Mixtures (impure substances)
have sharp, specific melting and boiling points	melt or boil over a range of temperatures
	an impure substance has a lower melting point and a higher boiling point

Tip

The melting point of a substance is the same temperature as the freezing point.

Formulations

Tip

Remember that a formulation is a mixture; no chemical reaction occurs when it is made.

A formulation is a mixture that has been designed as a useful product. It is formed by mixing together several different substances in carefully measured quantities to ensure the product has the required properties.

Many foods are formulations and this means they have consistent properties. Would you buy tomato ketchup if it tasted different each time? An 18-carat-gold alloy used in jewellery is a formulation; it is a mixture of carefully measured quantities of gold, silver, copper and zinc to ensure that it has the correct hardness. Three types of formulation are shown in Table 19.3.

Table 19.3 Examples of formulations

Formulation	Comment	Named example
alloys	An alloy is a mixture of two or more elements, at least one of which is a metal. The resulting mixture has metallic properties. Each alloy is a formulation made by mixing measured quantities to ensure the alloy has the specific properties required for its use.	18-carat gold (gold, silver, copper, zinc)
medicine	Tablets are formulations of the active drug along with ingredients such as corn starch to bind the tablet together and a lubricant to make it easy to swallow. Liquid medicines are also formulations.	Calpol® is a formulation of paracetamol and liquid flavourings.
fertiliser	Fertilisers contain mixtures of nitrogen, phosphorus and potassium compounds. Different formulations are suitable for different plants.	NPK fertilisers

Test yourself

1. a) Why do some people state that a bottle of mineral water is pure water?
   b) Explain why a chemist would not say that mineral water is pure.
   c) Name all the pure substances from the following list: apple juice, copper, sodium chloride, rock salt, air, magnesium, milk.
2. a) A sample of water was found to freeze between −1 °C and −3 °C. Was this water pure? Explain your answer.
   b) Pure water boils at 100 °C. What happens to the boiling point if some salt is dissolved in the water?
3. The melting point of paracetamol is 169 °C. Four students, A–D, made paracetamol in the laboratory and recorded the melting point. Which student(s) made pure paracetamol? Explain your answer.

Table 19.4

Student	A	B	C	D
Melting point/°C	164–166	160–162	169	169–170

4. The melting point of aspirin is 136 °C. A student prepared some aspirin and found the melting point to be 132–134 °C. What can the student deduce about the prepared aspirin?
5. A fertiliser contained 23% ammonium nitrate, 23% magnesium sulfate and 54% ammonium phosphate. Why is this fertiliser a formulation?

Table 19.5 shows the melting points and boiling points of some metallic elements and alloys named by the letters A–C.

Table 19.5

	Melting point/°C	Boiling point/°C
A	−34	356
B	420	913
C	1425–1540	2530–2545

a) In what state is substance A at room temperature and pressure (20 °C)?
b) In what state is substance B at room temperature and pressure (20 °C)?
c) In what state is substance C at room temperature and pressure (20 °C)?
d) What is an alloy?
e) Classify the substances in the table as elements or alloys and explain your answer.

Separating mixtures

A mixture is defined as two or more substances mixed together. The substances in a mixture are quite easy to separate because the substances are not chemically joined to each other. It is useful to understand the following key terms when thinking about separating mixtures.

▶ A solute is the substance that dissolves in a solvent.
▶ A solvent is the liquid in which a solute dissolves.
▶ A solution is a solute dissolved in a solvent.
▶ A soluble substance is one which will dissolve in a solvent.
▶ An insoluble substance is one which does not dissolve in a solvent.

Different methods are used to separate different types of mixtures.

Filtration

Filtration is used to separate an insoluble solid from a liquid. For example, it could be used to separate solid insoluble sand from water.

The mixture is poured through a filter funnel containing a piece of filter paper. The insoluble solid is caught in the filter paper but the liquid passes through. **The liquid which passes through the filter paper during filtration is called the** filtrate. **The solid that remains on the filter paper after filtration is called the** residue.

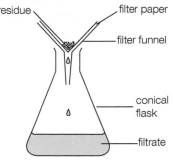

▲ **Figure 19.4** Filtration

Tip
When drawing diagrams, always make sure they are labelled.

Evaporation

To separate a dissolved solid from a solvent, the mixture is placed in an evaporating basin and heated gently until all the solvent has evaporated and the solid is left in the basin. This method is often used to separate salt from a salt solution.

Evaporation **is the change of state from liquid to gas when heated.**

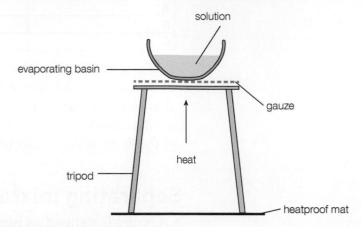

▲ **Figure 19.5** Evaporation

Crystallisation

Another method of separating a dissolved solid from a solvent involves heating the solution to boil off some of the solvent. This creates a saturated solution. **A saturated solution is one in which no more solid can dissolve at that temperature.** The saturated solution is then cooled. The dissolved solid (solute) becomes less soluble and so cannot remain dissolved. Consequently, it crystallises out of solution. The crystals may then be separated from the saturated solution by filtration. Many salts, for example copper(II) sulfate, are separated from their solution by crystallisation.

▲ **Figure 19.6** Crystals of copper(II) sulfate produced by crystallisation

Simple distillation

To separate a solvent from a solution, simple distillation can be used. For example, pure water can be obtained from sea water or from copper(II) sulfate solution using distillation. Distillation is evaporation followed by condensation.

Condensation **is the change of state from gas to liquid when cooled.**

The mixture to be separated is placed in the distillation apparatus, as shown in Figure 19.7, and heated until the solvent boils.

Anti-bumping granules **are added to the mixture in the flask to promote smooth boiling.**

The vaporised solvent passes into the condenser where it cools and condenses and then runs into the collection flask. **The** distillate **is the liquid that is cooled from the vapour and collected during distillation.**

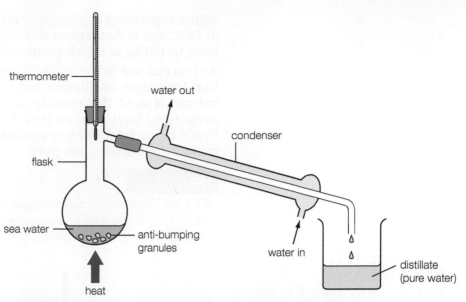

thermometer

thermometer

water out

condenser

flask

sea water

anti-bumping granules

water in

distillate (pure water)

heat

▲ **Figure 19.7** Simple distillation of sea water to obtain pure water

Tip ↻

Remember when labelling that water enters a condenser at the bottom and leaves at the top; this makes sure the outer jacket is completely filled and an effective cooling system is created.

Tip ↻

The bulb of the thermometer is opposite the exit to the condenser so that the temperature of the exiting vapour, which condenses to form the product, can be recorded.

Fractional distillation

Fractional distillation is used to separate miscible liquids which have different boiling points.

Miscible liquids **are liquids that mix together**, such as alcohol and water.

Immiscible liquids **are liquids which do not mix together but form two distinct layers**, for example oil and water.

▲ **Figure 19.8** Gin and tonic water are miscible liquids

▲ **Figure 19.9** Oil and water are immiscible

In fractional distillation, the mixture of miscible liquids is heated and the liquids boil one by one as the temperature rises. Each vapour rises up, is condensed and then collected. The distillate collected at each different temperature is called a fraction and a different receiver is used for each fraction.

When separating a mixture of ethanol and water, the ethanol boils at 79 °C and is condensed and collected first. The temperature then rises to 100 °C, at which point water boils and is collected.

As you can see in Figure 19.10, the apparatus used is similar to that for simple distillation, but a long column called a **fractionating column** is used; this provides a better separation of liquids. Any evaporated liquids below their boiling point condense on the glass beads in the fractionating column and run back into the flask. This means the fractions are pure.

> **Tip** ↻
>
> If the liquids in the mixture are flammable, it is safer to heat the flask with an electric heating mantle or a water bath (if the temperature needed is less than 100 °C), rather than a Bunsen burner.

> **Tip** ↻
>
> You will study fractional distillation of crude oil in industry in Unit 2.

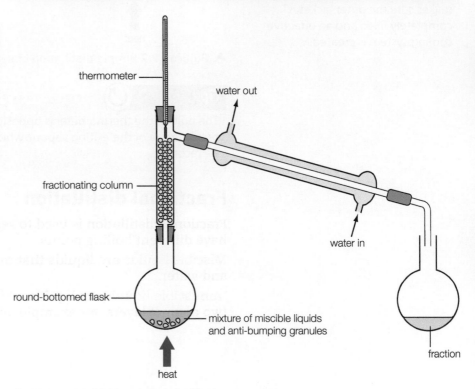

▲ **Figure 19.10** Fractional distillation apparatus

Test yourself 🖉

6 A mixture of insoluble copper(II) carbonate and water was separated by filtration.
  a) Name the filtrate.
  b) Name the residue.
7 Pure water can be obtained by distillation of copper(II) sulfate solution.
  a) Name the solute in copper(II) sulfate solution.
  b) Name the change of state which occurs in the distillation flask.
  c) What is the distillate?
  d) What is added to the distillation flask to ensure smooth boiling?

Show you can

Three common methods of separation are filtration, distillation and fractional distillation.

For each of these separation methods pick two words or phrases from the list below and insert them into a copy of the table with an explanation of their meaning.

condenser, distillate, fractionating column, filtrate, miscible liquids, residue

Then complete Table 19.6 by writing the type of mixture separated by each method.

Table 19.6

	Filtration	Distillation	Fractional distillation
Type of mixture separated			
Important word and definition			
Important word and definition			

Practical activity

Rock salt

Common salt has the chemical name sodium chloride. It is found naturally in large amounts in seawater or in underground deposits. Sodium chloride can be extracted from underground deposits by the process of solution mining, which is shown in Figure 19.11.

1 a) On what physical property of sodium chloride does this process depend?

 b) Suggest one reason why solution mining uses a lot of energy.

 c) Suggest one negative effect that solution mining has on the environment.

 d) How is sodium chloride obtained from the concentrated salt solution?

Rock salt is a mixture of salt, sand and clay. Pure salt can be separated from rock salt using the laboratory method detailed below.

Method

A Wear eye protection.

B Place eight spatulas of rock salt into a mortar and grind using a pestle.

C Place the ground rock salt into a beaker and quarter fill with water.

D Place the beaker on a gauze and tripod and heat gently while stirring with a glass rod. Stop heating when the salt has dissolved. The sand and clay will be left.

E Allow the mixture to cool and then filter.

F Heat until half the volume of liquid is left.

G Place the evaporating basin on the windowsill to evaporate off the rest of the water slowly. Pure salt crystals should be left.

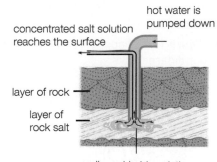

▲ **Figure 19.11** Salt extraction by solution mining

a)

b)

c)

▲ **Figure 19.12**

2 Which step of the method detailed on page 229 (A–G) is best represented by each photograph in Figure 19.12?

3 a) Why is rock salt considered to be a mixture?

  b) What was the purpose of grinding the rock salt?

  c) Why was the mixture heated and stirred?

  d) State what the filtrate contains.

  e) State what the residue contains.

  f) Explain why the salt obtained may still be contaminated with sand and suggest how you would improve your experiment to obtain a purer sample of salt.

Paper chromatography

Chromatography can be used to separate mixtures of soluble substances in a solution. Mixtures that are suitable for separation by chromatography include inks, dyes and colouring agents in food.

Method

1 Using a pencil, draw a base line on the chromatography paper around 1–2 cm from the bottom. Use pencil as it will not dissolve in the solvent.

2 Using a capillary tube, place a spot of the substance to be analysed on the base line. When this is dry, add another spot on top to make it concentrated.

3 Place the paper in a beaker with the solvent at the bottom. The pencil line and spots must be above the level of the solvent so that the spots do not dissolve into the solvent in the beaker.

4 The solvent travels up the paper.

5 When the solvent is near the top, take the paper out of the solvent and mark the level that the solvent reached. This is known as the solvent front.

6 Leave the paper to dry. The mixture should have separated into different components, which are seen as spots on the paper.

▲ **Figure 19.13** Paper chromatography of four different inks

Tip ↻

If the spots are colourless and not visible, they can be sprayed with a chemical developing agent.

The solvent front is the furthest distance travelled by the solvent.

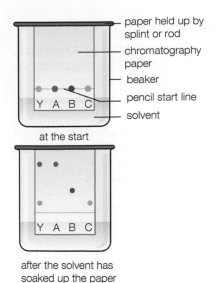

paper held up by splint or rod

chromatography paper

beaker

pencil start line

solvent

at the start

after the solvent has soaked up the paper

▲ **Figure 19.14** Paper chromatography

Tip ↻

In an examination you will need to use a ruler to measure the distance moved by the substance and the solvent. If the spot is large, measure to the middle of the spot.

solvent front

For substance Q:
$$R_f = \frac{4}{5} = 0.8$$

For substance P:
$$R_f = \frac{3}{5} = 0.6$$

5 cm

4 cm

3 cm

base line

P   Q

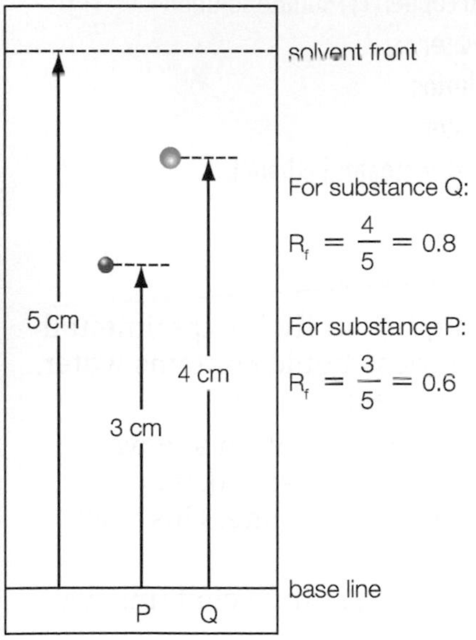

▲ **Figure 19.15** Finding R_f values

A **pure substance** usually produces a **single spot** in chromatography, whatever solvent is used. In Figure 19.14, A, B, and C are pure substances. Mixtures will usually produce more than one spot; one for each substance in the mixture. Y is a mixture of two substances as it produces two spots. By comparing Y with A, B and C, it should be clear that Y is a mixture of substances A and C.

How chromatography works

In paper chromatography, the **stationary phase** is the paper and the **mobile phase** is the solvent.

Each substance in the mixture moves at a different rate up the paper. This depends on its relative attraction to the paper and the solvent. Substances that have a stronger attraction to the solvent, and are more soluble in it, move quickly and travel a long way up the paper. Substances that have a stronger attraction to the paper move slowly and only travel a short distance up the paper (for example substance C in Figure 19.14).

R_f values

The separated components of the mixtures can be identified by calculating the R_f value using the equation:

$$R_f = \frac{\text{distance moved by substance}}{\text{distance moved by solvent}}$$

The R_f value of a particular substance in the same solvent is always the same. An example of how to calculate R_f values for two substances, P and Q, is shown in Figure 19.15.

Test yourself ✎

8 The chromatogram in Figure 19.16 shows the results when a mixture of dyes, S, was compared to substances 1–6.
   a) Explain which of substances 1 to 6 are pure substances.
   b) Identify the substances in S.
   c) Calculate the R_f values for each spot in colouring S. Give your answer to two significant figures.
   d) Which colour spot moved slowest during the experiment?
   e) Explain what is meant by the terms solvent and solvent front.
   f) Name the stationary phase.
   g) When setting up this experiment how would you draw the base line?
9 The same substance was analysed by chromatography using two different solvents, as shown in Figure 19.17.
   a) Calculate the R_f value of the substance in each solvent.
   b) Explain why the substance moved further in solvent 2 than solvent 1.

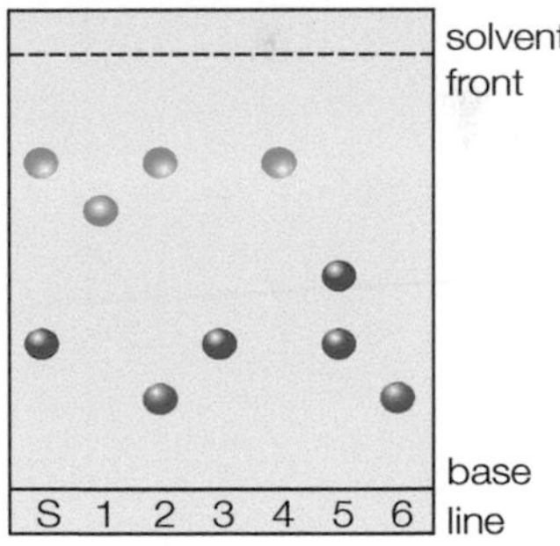

solvent front

base line

S 1 2 3 4 5 6

▲ **Figure 19.16**

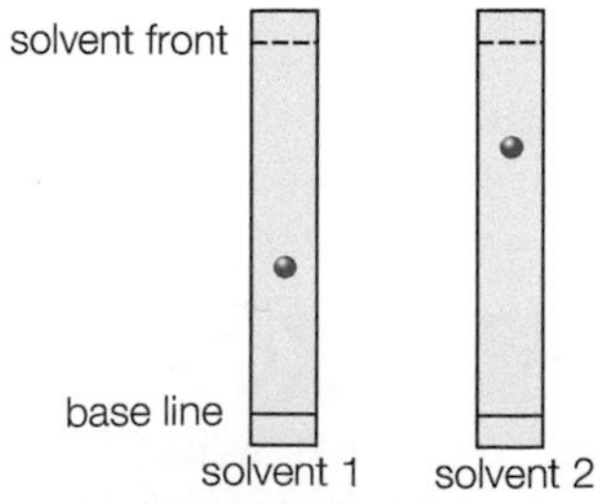

solvent front

base line

solvent 1   solvent 2

▲ **Figure 19.17**

Choosing methods of separation

To decide on the most appropriate method of separating a mixture, it can be useful to first decide the type of mixture. The **summary in Table 19.7** can then help you decide which method is best.

Table 19.7 Summary of separation methods

Type of mixture	insoluble solid and liquid	soluble solid dissolved in a solvent	two miscible liquids	soluble solids dissolved in a solvent (often coloured)
Example	sand and water	sodium chloride solution	ethanol and water	dyes food colourings ink
Method of separation	filtration	▶ evaporation (to obtain solid) ▶ crystallisation (to obtain solid) ▶ simple distillation (to obtain solvent)	fractional distillation	paper chromatography

Show you can

Name the best method of separation in each case:

a) to separate the dyes in red food colouring

b) to separate salt and water from salt solution

c) to produce crystals of copper(II) sulfate from copper(II) sulfate solution

d) to separate a miscible mixture of wine and water

e) to separate water from copper(II) sulfate solution

f) to separate insoluble copper(II) oxide and water

g) to separate a miscible mixture of ethanoic acid, water and ethanol.

Test for water

Chemists often encounter colourless liquids in their experimental work. To find out whether a given colourless liquid contains water, they use the anhydrous copper(II) sulfate test.

Anhydrous copper(II) sulfate ($CuSO_4$) does not contain any water of crystallisation and is powdery. When water is added, blue hydrated copper(II) sulfate forms, which is crystalline. This is an exothermic reaction.

The equation shows that hydrated copper(II) sulfate contains five moles of water of crystallisation.

anhydrous copper(II) sulfate + water → hydrated copper(II) sulfate

$$CuSO_4 + 5H_2O \rightarrow CuSO_4.5H_2O$$

This reaction can be used as a chemical test for the presence of water. The test is:

▶ A few drops of the liquid to be tested are added to **white anhydrous copper(II) sulfate**.

▶ If water is present, the white powder will turn to blue hydrated copper(II) sulfate.

▲ **Figure 19.18** White anhydrous copper(II) sulfate turns blue in the presence of water

Tip

Remember that hydrated means contains water of crystallisation and anhydrous means contains no water of crystallisation.

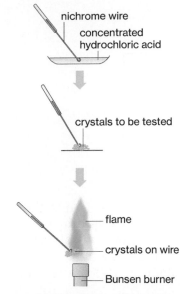

nichrome wire

concentrated hydrochloric acid

crystals to be tested

flame

crystals on wire

Bunsen burner

▲ **Figure 19.19** Procedure for a flame test

▲ **Figure 19.20** The photograph shows a flame test being carried out on a potassium salt

Cation tests

Cations can be identified by carrying out a flame test.

1 Flame tests

Some positive metal ions produce an intense colour in a blue Bunsen burner flame. A flame test can be used to identify them.

Method

▶ Make a loop on the end of a piece of **nichrome** wire.
▶ Dip the loop into **concentrated hydrochloric acid** and then into the salt to be tested.
▶ Place the loop into a **blue** Bunsen burner flame and record the first colour observed.

The flame test colours which you need to learn are shown in Table 19.8.

Table 19.8 Flame test colours

Metal ion present	Flame colour
lithium (Li^+)	crimson
sodium (Na^+)	yellow/orange
potassium (K^+)	lilac
calcium (Ca^{2+})	brick-red
copper (Cu^{2+})	blue-green/green-blue

Tip

Flame colour is produced by the metal ion in the compound. Don't confuse flame colours with the colour of the flame when a metal burns in air. For some metals the colour is the same, but in others it is not. For example, magnesium compounds do not have a flame test colour but magnesium burns with a white flame in air.

Test yourself

10 Describe how you would determine experimentally whether a solid powder contained potassium ions.
11 Write the formula of the metal ion which causes:
   a) a yellow flame colour
   b) a blue-green flame colour.
12 What is a cation?
13 What is observed when a flame test is carried out on:
   a) calcium chloride
   b) lithium chloride?

Practice questions

1 Aspirin tablets contain the active drug aspirin, a binding agent and a sweetener.
   a) Why is an aspirin tablet considered to be a formulation? *(1 mark)*
   b) How would you prove that a sample of the active drug was pure? *(1 mark)*
   c) Apart from drugs, state two other examples of formulations. *(2 marks)*

2 Mixtures may be separated in the laboratory in many different ways. Three different methods of separating mixtures are shown below.

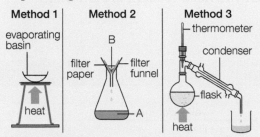

Figure 19.21

   a) Name each method of separation. *(3 marks)*
   b) Which method (1, 2 or 3) would be most suitable for obtaining water from potassium chloride solution? *(1 mark)*
   c) Which method would be most suitable for removing sand from a mixture of sand and water? *(1 mark)*
   d) What general term is used for liquid A and solid B in Method 2? *(2 marks)*
   e) State why Method 2 would not be suitable to separate copper(II) sulfate from copper(II) sulfate solution. *(1 mark)*

3 To determine whether two different orange drinks, X and Y, contained the food colourings E102, E101 or E160, a student put a drop of each orange drink and a drop of each food colouring along a pencil line on a piece of filter paper. The filter paper was placed in a solvent and the coloured components separated out. The results are shown in Figure 19.22.

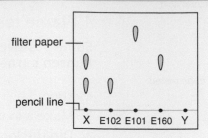

Figure 19.22

   a) What is the name of the process used by the student to analyse the two orange drinks? *(1 mark)*
   b) Orange drink X contains the food colouring E102. How do the results show this? *(1 mark)*
   c) What other food colouring does orange drink X contain? *(1 mark)*
   d) Re-draw the diagram and add a spot to show that orange drink Y contained food colouring E160. *(1 mark)*
   e) The line across the bottom of the filter paper was drawn with a pencil not with ink. Why should the line not be drawn with ink? *(1 mark)*

4 Drinking water has been distilled from sea water since at least 200AD when the process was described clearly by the Greek philosopher Aristotle. The distillation apparatus used to separate pure water from salt solution in the laboratory is shown below.

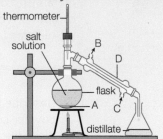

Figure 19.23

   a) Name the solute in salt solution. *(1 mark)*
   b) Name the solvent in salt solution. *(1 mark)*
   c) What is solid A? Suggest its purpose. *(2 marks)*
   d) What labels should be inserted at positions B and C? *(2 marks)*
   e) Name the piece of apparatus labelled D. *(1 mark)*
   f) What is the temperature on the thermometer when distillation is occurring? What does this suggest? *(2 marks)*
   g) What happens to the salt dissolved in the solution during this process? *(1 mark)*
   h) Distillation involves evaporation. What is meant by evaporation? Where in the apparatus does evaporation take place? *(2 marks)*

i) Distillation involves condensation. What is meant by condensation? Where in the apparatus does condensation take place? *(2 marks)*

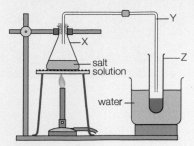

Figure 19.24

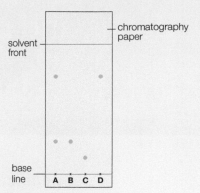

Figure 19.25

j) What advantages does the apparatus shown in Figure 19.23 have over the apparatus in Figure 19.24? *(1 mark)*

k) What is the purpose of the beaker of water in Figure 19.24? *(1 mark)*

l) Name the pieces of apparatus X, Y and Z. *(3 marks)*

m) Describe how you would test the purity of the distillate. *(1 mark)*

n) How would you test and prove that the distillate is water? *(2 marks)*

5 To determine whether a sample of water contained some dissolved metal ions, a chromatography experiment was carried out using the water sample (A) and known metal ion solutions B (containing copper(II) ions), C (containing iron(II) ions) and D (containing iron(III) ions). The method used was:

- Draw a base line on the chromatography paper 1.5 cm from the bottom using a pencil.
- Place a concentrated drop of each solution to be tested on the base line.
- Place the chromatography paper into a chromatography tank containing water at a depth of 1 cm.
- After the water had soaked up the paper, it was dried and sprayed with sodium hydroxide solution and coloured spots appeared on the paper.

a) Why is it necessary to use a pencil line rather than a line drawn with a pen? *(1 mark)*

b) Why is the water solvent at a depth of only 1 cm? *(1 mark)*

c) How is a concentrated drop of each solution added to the chromatography paper? *(2 marks)* The chromatogram obtained is shown in the diagram.

d) Write an equation which is used to calculate R_f value and calculate the R_f value for spot B. *(2 marks)*

e) If the experiment was repeated using a different solvent, would the R_f value be the same? *(1 mark)*

6) Hydrochloric acid reacts with bases to form salts such as sodium chloride and magnesium chloride. An antiseptic mouthwash is thought to contain both of these salts.
Describe how you would confirm that the mouthwash contained sodium ions. *(2 marks)*

20 Metals and reactivity series

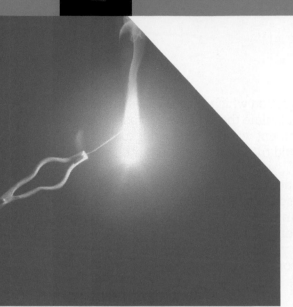

Magnesium is a reactive metal and burns brightly with a white light. It is used in distress flares. Other metals such as gold do not burn and are unreactive. Why are some metals more reactive than others? In this chapter you will learn why metals have different reactivities, and will explore the reaction of metals with water and with other metal ion solutions.

The reactivity series of metals

By comparing the reactivity of metals with water, oxygen and acids it is possible to list the metals in a reactivity series, with the most reactive at the top. Figure 20.1 shows the reactivity series needed for GCSE, with a mnemonic to help you remember it.

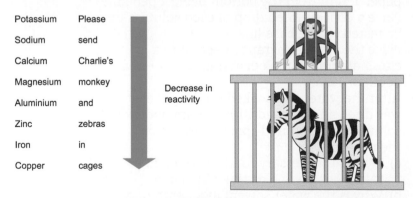

Potassium	Please
Sodium	send
Calcium	Charlie's
Magnesium	monkey
Aluminium	and
Zinc	zebras
Iron	in
Copper	cages

Decrease in reactivity

▲ **Figure 20.1** Mnemonic for remembering the reactivity series

Reaction of metals with water

You learnt in Chapter 16 that potassium and sodium are very reactive metals. They are stored **under oil** to prevent their reaction with oxygen and moisture in the air, and they react vigorously with cold water. These two metals are at the top of the reactivity series. Table 20.1 shows the metals in the reactivity series that react with cold water to produce a metal hydroxide and hydrogen.

metal + water → metal hydroxide + hydrogen

Tip

Check back to page 167 and revise the risk assessment for reacting potassium and sodium with water.

Table 20.1

Metal	What does it react with?	Observations	Word and symbol equation
potassium	reacts very vigorously with cold water	• lilac flame • K melts into a tiny ball • moves on surface/floats • bubbles • heat released • K disappears at end • colourless solution produced.	potassium + water → potassium hydroxide + hydrogen $2K(s) + 2H_2O(l) \rightarrow 2KOH(aq) + H_2(g)$
sodium	reacts vigorously with cold water	• Na melts into a tiny ball • moves on surface • bubbles • heat released • Na disappears • colourless solution produced.	sodium + water → sodium hydroxide + hydrogen $2Na(s) + 2H_2O(l) \rightarrow 2NaOH(aq) + H_2(g)$
calcium	reacts vigorously with cold water	• sinks then rises • bubbles • heat released • Ca disappears • colourless solution with a white precipitate (due to calcium hydroxide which is only slightly soluble).	calcium + water → calcium hydroxide + hydrogen $Ca(s) + 2H_2O(l) \rightarrow Ca(OH)_2(aq) + H_2(g)$
magnesium	reacts very, very slowly with cold water	• a few bubbles of gas over a long time.	magnesium + water → magnesium hydroxide + hydrogen $Mg(s) + 2H_2O(l) \rightarrow Mg(OH)_2(aq) + H_2(g)$

It is possible to test the solution produced in each case, to prove that it is an alkali, by adding an indicator. For example, add phenolphthalein and it turns pink.

In the reaction of calcium (or magnesium) with water it is possible to collect the gas produced using the apparatus in Figure 20.3. The inverted filter funnel helps channel the gas into the test tube. The gas can then be tested – it should give a pop when a lighted splint is applied. With magnesium only a very small amount of hydrogen is produced over a few days.

▲ **Figure 20.2** A chunk of calcium reacting with water. Notice that it sinks and there are bubbles. A white precipitate of calcium hydroxide forms

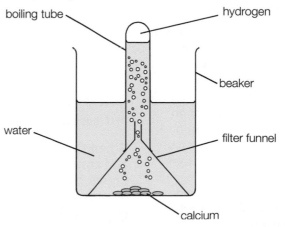

▲ **Figure 20.3** Collecting the gas produced when calcium reacts with water

237

Table 20.2 shows the reactions of the metals that react with steam to produce an oxide and hydrogen.

metal + steam → metal oxide + hydrogen

Table 20.2

Metal	What does it react with?	Observations	Word and symbol equation
magnesium	reacts with steam	• bright white light • white solid formed • heat given out.	magnesium + steam → magnesium oxide + hydrogen $Mg(s) + H_2O(g) \rightarrow MgO(s) + H_2(g)$
aluminium	reacts with steam	• powdered aluminium reacts to give a white solid • heat is given out. note that there is no reaction with aluminium foil unless the protective layer of aluminium oxide is removed.	aluminium + steam → aluminium oxide + hydrogen $2Al(s) + 3H_2O(g) \rightarrow Al_2O_3(s) + 3H_2(g)$
zinc	reacts with steam	• powdered zinc reacts to form a yellow solid (which changes to white on cooling) • heat is given out.	zinc + steam → zinc oxide + hydrogen $Zn(s) + H_2O(g) \rightarrow ZnO(s) + H_2(g)$
iron	reacts with steam	• powdered iron reacts to form a black solid.	iron + steam → iron oxide + hydrogen $3Fe(s) + 4H_2O(g) \rightarrow Fe_3O_4 + 4H_2(g)$ Due to the two different charges of iron, both iron(II) oxide (FeO) and iron(III) oxide (Fe_2O_3) are formed, and together these form a mixed oxide: Fe_3O_4.

Tip

Aluminium often reacts more slowly than its position in the reactivity series would suggest. This is because it is covered in a protective aluminium oxide layer which is unreactive and prevents it reacting. Powdered aluminium shows the element's true reactivity as the aluminium oxide layer is removed.

Tips

When drawing this diagram it is useful to remember the following points:
▶ There must be no blocks in delivery tube – it must go through the stopper.
▶ The delivery tube does not go through the wall of the trough.
▶ The water level in the trough must be higher than the beehive shelf.
▶ Two Bunsen burners are needed, one to heat the metal and one to heat the damp mineral wool and generate steam.

The reaction of metals with steam can be carried out using the apparatus shown in Figure 20.4. The mineral wool is dampened with water, and when it is heated steam is generated. The gas produced is collected under water.

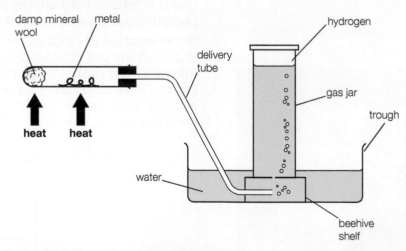

▲ **Figure 20.4** Heating metals with steam

When the heating is stopped in this experiment, the gases cool and contract, so there is a risk that the water in the trough may be drawn back into the hot boiling tube. This is called 'suck back'. It can be prevented by removing the bung from the boiling tube.

Copper metal **does not react** with cold water or steam. It is used in plumbing. Metals below copper in the reactivity series, such as silver and gold, also do not react with cold water or steam.

Reaction of metals with air

Most metals react with air to form a metal oxide. It is the oxygen part of air that reacts.

metal + oxygen → metal oxide

Observations and equations for the reaction of some metals in the reactivity series with air, are shown in Table 20.3.

Table 20.3

Metal	Equation	Observations
potassium	potassium + oxygen → potassium oxide $4K(s) + O_2(g) \rightarrow 2K_2O(s)$	burns with a lilac flame producing a white solid
sodium	sodium + oxygen → sodium oxide $4Na(s) + O_2(g) \rightarrow 2Na_2O(s)$	burns with a yellow flame producing a white solid
calcium	calcium + oxygen → calcium oxide $2Ca(s) + O_2(g) \rightarrow 2CaO(s)$	burns with a red flame producing a white solid
magnesium	magnesium + oxygen → magnesium oxide $2Mg(s) + O_2(g) \rightarrow 2MgO(s)$	burns with white light producing a white solid
aluminium	aluminium + oxygen → aluminium oxide $4Al(s) + 3O_2(g) \rightarrow 2Al_2O_3(s)$	burns only when finely powdered producing a white solid
zinc	zinc + oxygen → zinc oxide $2Zn(s) + O_2(g) \rightarrow 2ZnO(s)$	burns producing a yellow solid which becomes white on cooling
iron	iron + oxygen → iron oxide $3Fe(s) + 2O_2(g) \rightarrow Fe_3O_4(s)$	iron filings burn with orange sparks producing a black solid
copper	copper + oxygen → copper(II) oxide $2Cu(s) + O_2(g) \rightarrow 2CuO(s)$	red brown copper glows red, there is a blue-green flame and it becomes covered with a black layer

The more reactive metals burn completely when heated. However, less reactive metals such as copper only react on the surface and a layer of product forms.

What happens to metal atoms when they react?

When metal atoms react they lose electrons to form positive ions.

In the reaction of metals with water the **metal atoms lose electrons** and **form positive metal ions**. For example, when sodium reacts with water the sodium atoms lose electrons and form sodium ions in the product sodium hydroxide.

Symbol equation: $2Na + 2H_2O \rightarrow 2NaOH + H_2$

Half equation: $Na \rightarrow Na^+ + e^-$

Tip

There is more information on writing half equations on page 506.

The reactivity of metals is related to the tendency of a metal to lose electrons and form a positive ion (Table 20.4). **More reactive metals** have a **greater tendency** to lose electrons and **form positive ions.** Reactive metals such as potassium and sodium easily lose electrons to form ions but less reactive metals such as copper do not lose electrons easily.

Table 20.4 The tendency of metal atoms to lose electrons

Reactivity series	Reactivity	Tendency of metal atoms to lose electrons and form positive ions
potassium		
sodium		
calcium		
magnesium	decreases	decreases
aluminium		
zinc		
iron		
copper		

When a metal reacts with acid, the metal atoms lose electrons and form positive metal ions.

For example, when zinc atoms react with hydrochloric acid they lose electrons and form positive zinc ions.

Symbol equation: $Zn + 2HCl \rightarrow ZnCl_2 + H_2$

Half equation: $Zn \rightarrow Zn^{2+} + 2e^-$

When a metal burns in oxygen, the metal atoms lose electrons and form positive ions.

For example, when magnesium reacts with oxygen, the magnesium atoms lose electrons and form positive magnesium ions.

Symbol equation: $2Mg + O_2 \rightarrow 2MgO$

Half equation: $Mg \rightarrow Mg^{2+} + 2e^-$

Show you can

Magnesium (Mg) and zinc (Zn) both react with sulfuric acid.

1 a) Which metal reacts more vigorously with sulfuric acid?
  b) Explain, in terms of the tendency to form ions, why this metal reacts more vigorously.
2 Write a balanced symbol equation and a half equation for the reaction of magnesium with sulfuric acid.

Test yourself

1 Compare and contrast the reaction of calcium and cold water with that of magnesium and cold water.
2 Write balanced symbol equations, with state symbols for the reactions of:
  a) potassium and cold water
  b) magnesium and steam
  c) aluminium and steam.
3 Calcium reacts with water.
  a) What is observed in this reaction?
  b) Write a word equation for this reaction.
  c) Write a balanced symbol equation for this reaction.
  d) Explain what happens to the calcium atoms in this reaction in terms of electrons.
  e) Copper does not react with water. Explain this in terms of electrons.
4 Potassium and sodium both react with cold water.
  a) Describe two differences in the observations in these reactions.
  b) Write half equations for the metal atom in these reactions.
  c) Why is potassium more reactive than sodium?
5 Compare the reaction of magnesium and air, with that of copper and air. Give observations and balanced symbol equations.

▲ **Figure 20.5** The thermite reaction between iron oxide and aluminium is a displacement reaction. The reaction is very exothermic, and molten iron is produced

 Tip

For information on how to write ionic equations for displacement see page 505.

▲ **Figure 20.6** A copper coin was placed in silver nitrate solution. A displacement reaction occurred, as copper is more reactive than silver. Silver has been displaced and is deposited on the coin. Note the solution becomes slightly bluish due to Cu^{2+} ions forming in solution

Tip

The metal salt solution used in practical work is usually a nitrate solution, because all nitrates are soluble, or a sulfate solution, because most sulfates are soluble.

Displacement reactions

A displacement reaction occurs when a **more reactive metal takes the place of a less reactive metal in a compound**.

Displacement reactions can occur when a solid metal reacts with a solid metal oxide. For example, aluminium will displace iron from iron oxide because aluminium is more reactive than iron:

aluminium + iron oxide → aluminium oxide + iron

$$2Al + Fe_2O_3 → Al_2O_3 + 2Fe$$

This particular reaction is called the thermite reaction and produces a lot of heat. It is used to weld railway lines. A mixture of aluminium and iron oxide is placed over the gap between the railway lines and produces molten iron, which flows into the gap, cools and solidifies to weld the lines together.

Displacement reactions also take place in solution. For example, magnesium will displace copper from a copper ion solution because magnesium is more reactive than copper:

Word equation: magnesium + copper(II) nitrate → magnesium nitrate + copper

Symbol equation: $Mg + Cu(NO_3)_2 → Mg(NO_3)_2 + Cu$

Ionic equation: $Mg + Cu^{2+} → Mg^{2+} + Cu$

Observations for displacement reactions

The most important observations are for reactions with copper(II) sulfate or any copper metal solution, as these give a colour change. In general, for a metal with any copper ion solution the observations are:

▶ **heat is released** (the bigger the difference in the reactivities of the metals the more heat is given out)

▶ **red–brown** solid copper is deposited

▶ **blue colour** of the copper(II) ion solution **fades**.

For example, for the displacement reaction

zinc + copper(II) sulfate → zinc sulfate + copper

the observations are:

▶ red–brown deposit of copper

▶ blue colour of copper(II) sulfate solution fades

▶ heat released.

Analysing experimental data to predict where an unfamiliar element should be placed in the reactivity series

To compare the reactivity of metals, a piece of solid metal is often placed in different metal salt solutions to determine if a displacement reaction occurs.

The results are often recorded in a table similar to Table 20.5. The shaded parts of the table show that a metal should not be placed in a solution of its own salt. Examination questions often give information on displacement reactions for unfamiliar metals such as lead or tin.

Table 20.5

Metal	Solution				
	CuSO$_4$ copper(II) sulfate	ZnSO$_4$ zinc sulfate	FeSO$_4$ iron(II) sulfate	Pb(NO$_3$)$_2$ lead(II) nitrate	SnCl$_2$ tin(II) chloride
Cu		✗	✗	✗	✗
Zn	✓		✓	✓	✓
Fe	✓	✗		✓	✓
Pb	✓	✗	✗		✗
Sn	✓	✗	✗	✓	

✓ – a reaction occurred; ✗ – no reaction occurred

Tip

An easy way to place metals in order of reactivity is to count the ticks for each metal – the more ticks the more reactive it is.

From Table 20.5 an order of reactivity can be worked out:

▶ Zinc displaces copper, iron, lead and tin from solution.
▶ Iron displaces copper, lead and tin from solution.
▶ Tin displaces copper and lead from solution.
▶ Lead displaces copper from solution.
▶ Copper does not displace any other metal from solution and so it is least reactive.

The order of reactivity is:

zinc,   iron,   tin,   lead,   copper

Most reactive ——————————————————→ Least reactive

Show you can ?

To determine the order of reactivity of the metals copper, magnesium, nickel and zinc, each metal was heated with the oxides of the other metals. The results obtained are recorded in Table 20.6.

Table 20.6

	copper	magnesium	nickel	zinc
copper(II) oxide		Reaction	Reaction	Reaction
magnesium oxide	No reaction		No reaction	No reaction
nickel oxide	No reaction	Reaction		Reaction
zinc oxide	No reaction	Reaction	No reaction	

1 Give the order of reactivity of the four metals, from the most reactive to the least reactive.
2 Write a balanced chemical equation for the reaction of nickel oxide (NiO) with magnesium.
3 From the list below, write word and balanced symbol equations for all reactions that occur:
 nickel + hydrochloric acid
 zinc + cold water
 nickel + cold water
 zinc + sulfuric acid
 magnesium + zinc oxide.

Test yourself

6 a) Write a word and balanced symbol equation for the reaction of calcium and copper(II) sulfate.
 b) Explain why this reaction occurs.
 c) Write an ionic equation for this reaction.
 d) Give two observations for this reaction.

7 Decide if a reaction occurs in the following examples. If so, write balanced symbol and ionic equations.
 a) magnesium + zinc nitrate →
 b) iron(III) chloride + aluminium →
 c) sodium chloride + copper →
 d) zinc sulfate + aluminium →
 e) copper + zinc nitrate →
 f) copper(II) nitrate + zinc →

8 The metal chromium can be made in a displacement reaction between aluminium and chromium(III) oxide.

 aluminium + chromium(III) oxide → aluminium oxide + chromium

 a) Why does aluminium displace chromium in this reaction?
 b) Write a balanced symbol equation for this reaction.
 c) Write an ionic equation for this reaction.

9 In Table 20.7 the metal was added to the metal solution. Copy and complete the table: if a reaction occurs, insert a tick; if not, insert a cross.

Table 20.7

Metal solution	Metal			
	Mg	Al	Zn	Cu
magnesium sulfate				
aluminium chloride				
zinc sulfate				
copper(II) sulfate				
iron(II) sulfate				

Prescribed practical

Prescribed practical C3: Investigating reactivity of metals

The reactivity of metals can be studied using displacement reactions. If a displacement reaction occurs there is a temperature rise. In an experiment the following method was used:

- Pour some copper(II) sulfate solution into a polystyrene cup and record the temperature of the solution.
- Add a known mass of metal and stir.
- Record the maximum temperature of the mixture.
- Repeat the experiment.

The results of this experiment are shown in Table 20.8.

Table 20.8

Metal	Temperature increase/°C		Average temperature rise/°C
	Experiment 1	Experiment 2	
magnesium	11.5	16.5	14.0
silver	0.0	0.0	0.0
iron	3.0	4.0	3.5
gold	0.0	0.0	0.0
zinc	7.0	8.0	7.5

1 State two factors which should be kept the same in this experiment to make it a fair test.

2 State and explain which of the metals gave the least reliable temperature rise.

3 Explain why there is no temperature rise when silver is added to copper(II) sulfate solution.

4 State and explain which of the metals used in the experiment is the most reactive.

5 Why do the results make it impossible to decide which of the metals is the least reactive?

6 Write a balanced symbol equation for the displacement reaction between zinc and copper(II) sulfate solution.

7 Suggest two observations which may occur in the reaction in question (6).

8 Why was a polystyrene cup used in this experiment?

9 Suggest one improvement which could be made to the experimental method, to enable more accurate results to be obtained.

Extraction of metals

Very unreactive metals such as gold and platinum are found 'native' on Earth. This means they are found naturally as the element, uncombined with anything else. Most metals are more reactive than gold and are found in compounds in rocks. The process of obtaining a metal from these compounds is called **extraction**.

An ore is a rock that contains a metal compound from which the metal can be extracted.

Most metal ores are oxides, and to extract the metal the oxygen must be removed. Removal of oxygen is called **reduction**. There are two methods of extraction of metals from ores:

▶ electrolysis
▶ reduction using carbon.

▲ **Figure 20.7** Gold is an unreactive metal. It is found uncombined in nature. This photo shows gold which has been found in a river by gold panning

The method of removing the oxygen depends on the position of the metal in the reactivity series. Table 20.9 shows the reactivity series, with carbon inserted. Metals that are less reactive than carbon, for example iron, can be extracted by heating the metal oxide ore with carbon. This is studied in detail in Chapter 21.

Metals that are more reactive than carbon can be extracted using the more powerful method of electrolysis. This is studied in detail in Chapter 26. Electrolysis involves passing electricity through a molten compound to decompose it into its elements. For example, aluminium is produced by electrolysis of aluminium oxide found in the ore bauxite. Much energy is needed to keep the ore molten for electrolysis and it is an expensive method.

Table 20.9

Decrease in reactivity	Metal	Method of extraction
	potassium	electrolysis
	sodium	
	calcium	
	magnesium	
	aluminium	
	carbon	
	zinc	reduction of the ore by heating with carbon
	iron	
	copper	

Test yourself

10 Most metals are extracted from compounds found in rocks called ores. A few metals are found as elements.
   a) Why are some metals, such as gold, found as elements and not in compounds?
   b) What is an ore?
11 Which method is used to extract the following metals:
   a) iron
   b) aluminium
   c) magnesium
   d) zinc?
12 The ore tin(II) oxide is heated with carbon. Carbon dioxide and tin are formed.
   a) Write a balanced symbol equation for this reaction.
   b) Explain why this is a reduction reaction.

Show you can

From the list of elements below:

carbon   copper   gold   hydrogen   sodium   zinc

1 Choose an element that is most likely to be found in rocks as the metal itself.
2 Choose two elements that are likely to be found in ores.
3 Choose an element that can be heated with an ore to extract a metal.
4 Choose an element that is most likely to be extracted from its ore by electrolysis.

Practice questions

1 Potassium metal is removed from under oil, and added to a trough of water. A lilac flame is produced and the potassium melts into a tiny ball.

  a) State two ways in which the potassium metal is prepared for reaction. *(2 marks)*
  b) State three other observations for this reaction. *(3 marks)*
  c) Write a balanced symbol equation for the reaction of potassium and water. *(3 marks)*
  d) Write a half equation for the reaction of potassium atoms in this reaction. *(2 marks)*
  e) Write a balanced symbol equation for the reaction of calcium with water. *(3 marks)*
  f) State three observations for the reaction of calcium with water. *(3 marks)*
  g) Draw a labelled diagram of the apparatus set-up used to react calcium with water and collect the gas produced. *(3 marks)*
  h) Suggest why potassium is more reactive with water than calcium. *(1 mark)*
  i) Write a balanced symbol equation for the reaction of zinc with steam. Include state symbols. *(3 marks)*
  j) Draw a labelled diagram of the apparatus set-up used to react zinc with steam. *(5 marks)*

2 A student investigated some displacement reactions and recorded the observations in Table 20.10.

Table 20.10

Reaction	Reactants	Observations
1	copper + silver nitrate	colourless solution turned blue, solid formed
2	iron + zinc sulfate	none
3	silver + iron(II) sulfate	none
4	zinc + copper(II) sulfate	blue solution turned colourless, solid formed
5	iron + copper(II) sulfate	blue solution turned colourless, solid formed

  a) What colour is the solid formed in reaction 5? *(1 mark)*
  b) Suggest why the colourless solution turned blue in reaction 1. *(1 mark)*
  c) Place the metals in order from most to least reactive. *(1 mark)*
  d) Which metal would you expect to react most quickly with hydrochloric acid? *(1 mark)*
  e) Write a balanced symbol equation for reaction 1. Include state symbols. *(3 marks)*
  f) Write an ionic equation for reaction 1. *(3 marks)* **H**
  g) Explain why reaction 4 occurs. *(1 marks)*

3 a) Use Table 20.11 to write the metals magnesium, nickel, chromium and manganese in order of reactivity. *(1 mark)*

Table 20.11

Metal	Solution			
	magnesium nitrate	nickel(II) nitrate	chromium(III) nitrate	manganese(II) nitrate
Mg		✓	✓	✓
Ni	✗		✗	✗
Cr	✗	✓		✗
Mn	✗	✓	✓	

  b) Write a balanced symbol reaction for the reaction of magnesium with nickel(II) nitrate. *(2 marks)*
  c) Why are nitrates often used in solution displacement reactions? *(1 mark)*

4 a) Strips of four different metals were placed in solutions of the nitrates of the same metals. Any displacement reactions that occurred are represented by a ✓ in Table 20.12.

Table 20.12

Metal	P nitrate	Q nitrate	R nitrate	S nitrate
P	✗	✗	✗	✓
Q	✓	✗	✓	✓
R	✓	✗	✗	✓
S	✗	✗	✗	✗

  What is the order of reactivity of metals P, Q, R and S (put the most reactive first)? *(1 mark)*
  b) When zinc is added to a solution of copper(II) sulfate a displacement reaction occurs.
   i) Write a balanced symbol equation for this reaction. *(2 marks)*
   ii) State two observations for this reaction. *(2 marks)*
   iii) Explain why this reaction occurs. *(1 mark)*
  c) When zinc reacts with steam a gas is produced.
   i) Name the gas. *(1 mark)*
   ii) Describe a test for this gas, giving the observation for a positive test. *(2 marks)*
   iii) Write a half equation for the zinc atoms reacting. *(2 marks)* **H**
  d) Suggest how zinc is extracted from its ore. *(1 mark)*

21 Redox, rusting and iron

The Eiffel tower in Paris is made of iron. Every 7 years it is painted with a specially mixed 'Eiffel tower brown' coloured paint to prevent it from rusting. In this chapter you will explore how iron metal is extracted from its ore, and understand the chemistry of redox reactions, including rusting.

What is redox?

In Figure 21.1 there are many redox reactions occurring. The electric currents from the batteries that power the computer and games consoles are generated by redox reactions. The combustion of the natural gas in the fire is a redox reaction; even the respiration that all the family members are undergoing to keep alive is a redox reaction.

▲ **Figure 21.1** What redox reactions are occuring?

Redox reactions **are reactions in which oxidation and reduction occur at the same time** (reduction–oxidation). Oxidation and reduction can be defined in one of two ways, as shown in Table 21.1.

Table 21.1 Definitions of oxidation and reduction

Oxidation	Reduction
gain of oxygen	loss of oxygen
loss of hydrogen	gain of hydrogen

Recognising oxidation and reduction in terms of loss or gain of oxygen or hydrogen

When identifying and explaining oxidation or reduction in an equation use the following steps:

1 Choose and state the **appropriate** definition of oxidation or reduction – do not give both definitions.

2 Name the species that is being oxidised or reduced.

Example

Why is $S + O_2 \rightarrow SO_2$ an oxidation?

Answer

Step 1 – oxidation is the gain of oxygen or loss of hydrogen. Looking at the equation, the appropriate definition is oxidation is the gain of oxygen.

Step 2 – Sulfur has had oxygen added to it to form SO_2, so it is oxidised.

Example

Why is $H_2 + Cl_2 \rightarrow 2HCl$ a reduction?

Answer

Step 1 – Reduction is the loss of oxygen or the gain of hydrogen. Looking at the equation, the appropriate definition is reduction is the gain of hydrogen.

Step 2 – Chlorine has gained hydrogen to form HCl, so it is reduced.

Example

Explain why this reaction is a redox reaction:

copper(II) oxide + hydrogen →
                     copper + water

Answer

Reduction is loss of oxygen. The copper(II) oxide loses oxygen and forms copper, so the copper(II) oxide is reduced.

Oxidation is gain of oxygen. The hydrogen gains oxygen and forms water, so the hydrogen is oxidised.

This reaction is a redox reaction because both oxidation and reduction occur at the same time.

Tip

This reaction is used in a blast furnace to produce iron. See Figure 21.9.

Tip

It is better to state that 'the iron(III) oxide loses oxygen' rather than that 'the iron loses oxygen'.

In some reactions oxidation and reduction both occur. These are redox reactions.

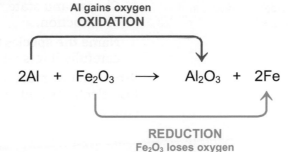

▲ **Figure 21.2** Redox in terms of oxygen content

The reaction of aluminium and iron(III) oxide is a redox reaction (Figure 21.2). Oxidation is the gain of oxygen. The aluminium gains oxygen to form aluminium oxide (Al_2O_3), and is oxidised. Reduction is loss of oxygen. The iron(III) oxide (Fe_2O_3) loses oxygen to form iron, and is reduced.

Example

Why is the following reaction a redox reaction?

$Fe_2O_3 + 3CO \rightarrow 2Fe + 3CO_2$

Answer

Reduction is the loss of oxygen.

The Fe_2O_3 loses oxygen and forms iron, so the Fe_2O_3 is reduced.

Oxidation is gain of oxygen. The CO gains oxygen and forms CO_2, so the CO is oxidised.

This reaction is a redox reaction because both oxidation and reduction occur at the same time.

1 Explain if the following reactions are oxidation or reduction.
   a) $2Ca + O_2 \rightarrow 2CaO$
   b) $2HI \rightarrow H_2 + I_2$
   c) $2Na + H_2 \rightarrow 2NaH$
   d) $N_2 + 3H_2 \rightarrow 2NH_3$
   e) $C + O_2 \rightarrow CO_2$
2 Explain, in terms of oxygen and hydrogen content, why each of the following reactions is a redox reaction.
   a) $Mg + CuO \rightarrow MgO + Cu$
   b) $ZnO + H_2 \rightarrow Zn + H_2O$
   c) $2H_2 + O_2 \rightarrow 2H_2O$

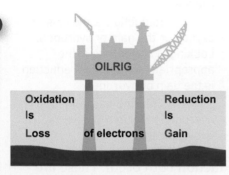

▲ **Figure 21.3** Learn OILRIG to help you remember oxidation and reduction definitions

Oxidation and reduction in terms of electrons

Oxidation **can be defined as a reaction in which a substance loses electrons.**

Reduction **can be defined as a reaction in which a substance gains electrons.**

Remember the word **oilrig** (Figure 21.3) to help you remember these definitions.

When identifying and explaining oxidation or reduction in an equation, again use the following steps:

1 Choose and state the appropriate definition of oxidation or reduction.
2 Name the species that is being oxidised or reduced. Note carefully if it is an atom or an ion.

When metals react with oxygen, water or an acid the metal atoms lose electrons and form metal ions. The metal atoms are oxidised (Table 21.2).

Table 21.2

Reaction	Example	Half equation	Explanation
metal + oxygen	$Mg + O_2 \rightarrow 2MgO$	$Mg \rightarrow Mg^{2+} + 2e^-$	The magnesium atom loses electrons and is oxidised.
metal + acid	$Ca + 2HCl \rightarrow CaCl_2 + H_2$	$Ca \rightarrow Ca^{2+} + 2e^-$	The calcium atom loses electrons and is oxidised.
metal + water	$2Na + 2H_2O \rightarrow 2NaOH + H_2$	$Na \rightarrow Na^+ + e^-$	The sodium atom loses electrons and is oxidised.

Example

Why is $Mg \rightarrow Mg^{2+} + 2e^-$ an oxidation reaction?

Answer
Oxidation is the loss of electrons. The magnesium atom loses electrons, and so is oxidised.

Example

Why is $Cl_2 + 2e^- \rightarrow 2Cl^-$ a reduction reaction?

Answer
Reduction is the gain of electrons. The atoms of the chlorine molecule gain electrons, and so the chlorine molecule is reduced.

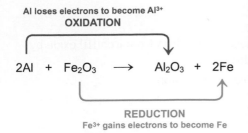

Al loses electrons to become Al^{3+}
OXIDATION

$$2Al + Fe_2O_3 \longrightarrow Al_2O_3 + 2Fe$$

REDUCTION
Fe^{3+} gains electrons to become Fe

▲ **Figure 21.4** Redox in terms of electrons

Displacement reactions involve oxidation and reduction and this can be explained by loss and gain of electrons. For example, in the displacement of iron from iron(III) oxide by aluminium, the aluminium atom loses three electrons to become Al^{3+} and the iron(III) ion in the iron(III) oxide gains electrons to form iron atoms.

In a reaction in which one substance loses electrons, another substance gains those electrons. This means that both reduction and oxidation take place and it is a **redox** reaction.

Writing ionic equations and half equations for displacement reactions

Ionic equations and half equations can be written for displacement reactions.

A general method of explaining why a displacement reaction is a redox reaction is as follows:

1 Write an ionic equation.
2 Write a half equation for the metal **atom** losing electrons and name the metal atom losing electrons.
3 State the definition of oxidation.
4 Write a half equation for the metal **ion** gaining electrons and name the metal ion gaining electrons.
5 State the definition of reduction.
6 State the definition of redox.

Example

Write two half equations for the displacement of silver from silver nitrate by copper and explain why it is a redox reaction.

copper + silver nitrate → copper(II) nitrate + silver

$$Cu + 2AgNO_3 \rightarrow Cu(NO_3)_2 + 2Ag$$

Answer
It is useful to first write the ionic equation by ignoring any spectator ions (in this case NO_3^-).

$$Cu + 2Ag^+ \rightarrow Cu^{2+} + 2Ag$$

From the ionic equation you can see that in this reaction the Cu atoms become Cu^{2+} ions in $Cu(NO_3)_2$ while the Ag^+ ions in $AgNO_3$ become Ag atoms.

The Cu atoms lose electrons to form Cu^{2+} ions:

$$Cu \rightarrow Cu^{2+} + 2e^-$$

This is oxidation. Electrons are lost.
The Ag^+ ions in $AgNO_3$ gain electrons to form Ag atoms:

$$Ag^+ + e^- \rightarrow Ag$$

This is reduction. Electrons are gained.
Reduction and oxidation are occurring at the same time, so this is a redox reaction.

H

Example

Write two half equations for the displacement of iron from iron(III) oxide by aluminium and explain why it is a redox reaction.

aluminium + iron(III) oxide → aluminium oxide + iron

$2Al + Fe_2O_3 \rightarrow Al_2O_3 + 2Fe$

Answer

It is useful to first write an ionic equation, ignoring the spectator ions (O^{2-}).

$2Al + 2Fe^{3+} \rightarrow 2Al^{3+} + 2Fe$

From the ionic equation you can see that the Al atoms become Al^{3+} ions in Al_2O_3 while the Fe^{3+} ions in Fe_2O_3 become Fe atoms.

The Al atoms lose electrons to form Al^{3+} ions:

$Al \rightarrow Al^{3+} + 3e^-$

This is oxidation. Electrons are lost.

The Fe^{3+} ions in Fe_2O_3 gain electrons to form Fe atoms:

$Fe^{3+} + 3e^- \rightarrow Fe$

This is reduction. Electrons are gained.

Reduction and oxidation are occurring at the same time, so this is a redox reaction.

Tip

If you need to revise writing ionic equations, turn to page 503.

Test yourself

3 Explain in terms of electrons why each of the following reactions is a redox reaction. You may use half equations in your answer.
   a) $Fe + CuSO_4 \rightarrow FeSO_4 + Cu$
   b) $2Al + 3ZnSO_4 \rightarrow 3Zn + Al_2(SO_4)_3$
   c) $Zn + CuO \rightarrow ZnO + Cu$

4 Magnesium reacts with copper(II) oxide in a displacement reaction to form copper.

   $Mg + CuO \rightarrow MgO + Cu$

   a) Explain, in terms of oxygen, why the magnesium is oxidised in this reaction.
   b) Write a half equation to show the oxidation of magnesium atoms to magnesium ions in this reaction.
   c) Explain, in terms of electrons, why the magnesium is oxidised in this reaction.
   d) Explain, in terms of oxygen, why the copper(II) oxide is reduced in this reaction.
   e) Write a half equation to show the reduction of copper(II) ions to copper atoms in this reaction.
   f) Explain, in terms of electrons, why the copper(II) oxide is reduced in this reaction.

5 Write an overall ionic equation and two half equations for each of the following displacement reactions:
   a) $Zn + CuSO_4 \rightarrow ZnSO_4 + Cu$
   b) $Zn + 2AgNO_3 \rightarrow Zn(NO_3)_2 + 2Ag$
   c) $2Al + 3CuSO_4 \rightarrow Al_2(SO_4)_3 + 3Cu$

6 Zinc displaces iron from a solution of iron(II) sulfate.
   a) Write a word equation for this reaction.
   b) Write a balanced equation for this reaction.
   c) Write an ionic equation for this reaction.
   d) Write the two half equations for this reaction.
   e) Identify which half equation is a reduction process.
   f) Identify which half equation is an oxidation process.
   g) Explain why this is a redox reaction.

7 Magnesium displaces silver from a solution of silver nitrate.
   a) Write a word equation for this reaction.
   b) Write a balanced equation for this reaction.
   c) Write an ionic equation for this reaction.
   d) Write two half equations for this reaction.
   e) Identify which half equation is a reduction process.
   f) Identify which half equation is an oxidation process.
   g) Explain why this is a redox reaction.

Show you can ❓

Each of the following reactions can be classified as an oxidation or a reduction reaction.

Reaction 1: $CH_4 + 2O_2 \rightarrow CO_2 + 2H_2O$

Reaction 2: $CuO + Mg \rightarrow Cu + MgO$

Reaction 3: $ZnO + H_2 \rightarrow Zn + H_2O$

1 Write the formula of the substance that is oxidised in reaction 1.
2 Explain which substance in reaction 2 is reduced.
3 Explain which substance in reaction 3 is oxidised.
4 Write a balanced ionic equation for reaction 2 and state and use a half equation to explain which species is oxidised.

Rusting

Why are the Statue of Liberty and the dome of Belfast City Hall green? Both structures are made of copper, and due to the copper reacting over time with the atmosphere, a thin layer of green copper(II) carbonate has formed. This layer actually protects the copper underneath from further corrosion.

▲ **Figure 21.5** The Statue of Liberty

▲ **Figure 21.6** Belfast City Hall

Corrosion is the destruction of materials by chemical reactions with substances in the environment. Many metals corrode in air. However, only iron (and steel, an alloy of iron) rusts.

What is rust?

Rust is **hydrated iron(III) oxide** ($Fe_2O_3.xH_2O$). It is formed when iron reacts with oxygen and water.

iron + oxygen + water → hydrated iron(III) oxide
(rust)

Rust is a red–brown flaky solid. When bits of rust flake off an iron object, more iron metal is exposed to air and water, so the iron continues to corrode and weaken.

Rusting is an oxidation reaction. Oxidation is the gain of oxygen. In rusting, the iron gains oxygen and is oxidised.

▲ **Figure 21.7** An iron padlock rusting. Note that rust is a red-brown flaky substance

Experimental investigation of rusting

To investigate rusting experimentally, iron nails can be placed in test tubes under different conditions for a week (Table 21.3).

Table 21.3

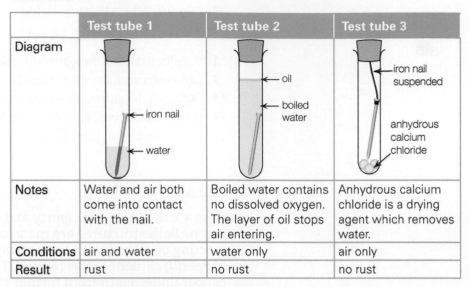

	Test tube 1	Test tube 2	Test tube 3
Notes	Water and air both come into contact with the nail.	Boiled water contains no dissolved oxygen. The layer of oil stops air entering.	Anhydrous calcium chloride is a drying agent which removes water.
Conditions	air and water	water only	air only
Result	rust	no rust	no rust

This experiment shows that the conditions needed for **rusting** are **oxygen (air) and water**.

Preventing rusting

Rusting is a problem because it is unsightly and dangerous, as it weakens the metal.

There are many methods of preventing rusting.

Some are barrier methods, which prevent air and water from coming into contact with the iron (Table 21.4).

Table 21.4

Method	How it works	Example	
Painting	It acts as a barrier, preventing oxygen and water from reaching the iron.	Paint is used to protect cars, bridges and railings.	
Oiling	It acts as a barrier, preventing oxygen and water from reaching the iron.	Moving parts, such as bicycle chains and gate hinges, are oiled to prevent rusting.	
Plastic coating	It acts as a barrier, preventing oxygen and water from reaching the iron.	A plastic coating is used to cover garden fences, fridge shelves, weights, etc.	

Other methods include covering with a suitable metal coating and sacrificial protection (Table 21.5).

Table 21.5

Method	How it works	Example	
Metal covering/ plating	A thin metal covering is applied by electroplating. It acts as a barrier, preventing air and water from coming into contact with the iron. '**Galvanising**' is the term given to plating iron with a layer of **zinc**. The zinc acts as a barrier but also protects by sacrificial protection if the surface is scratched.	Food cans are made from iron and plated with tin. Buckets and chains are often galvanised.	
Sacrificial protection	A metal that is more reactive than iron (higher up in the reactivity series) is attached to the iron, and it reacts (corrodes) instead of the iron.	Zinc or magnesium blocks are attached to the hulls of steel ships to prevent the steel rusting. Magnesium is attached to steel pipelines or oil rigs to prevent rusting. The magnesium reacts instead of the iron, and must be replaced periodically.	

Tip

Note that iron is the only metal that rusts; other metals corrode.

Show you can

Iron can be protected from rusting by sacrificial protection. In an experiment to investigate sacrificial protection, different metals are wrapped around iron nails and left in water for one week.

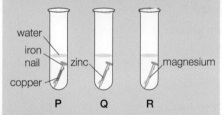

▲ **Figure 21.8**

1 State two ways in which you could ensure that this experiment is a fair test.
2 At the end of the week, in which tube(s) would rusting have occurred? Explain your answer.

Test yourself

8 a) What is the chemical name of rust?
　b) What conditions are needed for rusting?
9 a) Write a word equation to show what happens when iron rusts.
　b) Describe an experiment you could do to show that it is the oxygen rather than the nitrogen or argon in air that is required for iron to rust. Describe the results you would expect.
10 Explain how each of the following prevents steel from rusting:
　a) connecting the steel to some magnesium
　b) painting the steel
　c) coating the steel with a layer of chromium by electroplating
　d) galvanising the steel.

Extraction of iron from haematite

The main ore of iron is **haematite** and it contains iron(III) oxide (Fe_2O_3). The iron is extracted from the ore in a blast furnace. Three raw materials (Table 21.6) are fed into the top of the blast furnace, and are often called the 'charge'.

Table 21.6

Common name	Chemical name	Formula/symbol
haematite	iron(III) oxide	Fe_2O_3
limestone	calcium carbonate	$CaCO_3$
coke	carbon	C

There are three main steps in the extraction of iron from haematite (see Figure 21.9).

1 The production of the reducing agent

Hot air is blasted into the furnace through pipes at the bottom. The hot air heats the furnace and allows the coke to burn rapidly to form carbon dioxide.

carbon + oxygen → carbon dioxide

$$C + O_2 \rightarrow CO_2$$

This is an exothermic reaction, and the high temperature enables more coke (carbon) to react with the carbon dioxide to form carbon monoxide. **Carbon monoxide** is the **reducing agent**.

carbon dioxide + carbon → carbon monoxide

$$CO_2 + C → 2CO$$

2 The reduction of haematite

The carbon monoxide reducing agent reduces the iron ore by removing the oxygen from it, producing iron and carbon dioxide.

iron(III) oxide + carbon monoxide → iron + carbon dioxide

$$Fe_2O_3 + 3CO → 2Fe + 3CO_2$$

Iron sinks to the bottom and stays **molten**.

3 The removal of acidic impurities

Acidic impurities such as **silicon dioxide** from the sand in the haematite rock are present. **Limestone** (calcium carbonate) is added to remove these acidic impurities. The limestone **thermally decomposes** in the heat of the furnace to give calcium oxide and carbon dioxide.

calcium carbonate → calcium oxide + carbon dioxide

$$CaCO_3 → CaO + CO_2$$

Thermal decomposition is the breaking down of a substance using heat.

The basic calcium oxide produced reacts with the acidic silicon dioxide to produce **calcium silicate**, which is known as **slag**. The molten slag floats on top of the molten iron at the bottom of the furnace. The molten iron and slag are tapped off as liquids, separately, at the bottom of the furnace.

calcium oxide + silicon dioxide → calcium silicate

$$CaO + SiO_2 → CaSiO_3$$

Tip ↻

Learn to label the parts of a blast furnace.

Iron is a cheap metal because the ore is abundant. It is used in many structures and bridges due to its strength.

Tip ↻

Make sure you can explain why this is a redox reaction. See page 247.

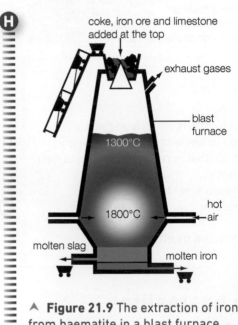

coke, iron ore and limestone added at the top

exhaust gases

blast furnace

1300°C

1800°C

hot air

molten slag

molten iron

▲ **Figure 21.9** The extraction of iron from haematite in a blast furnace

▲ **Figure 21.10** A worker in a protective suit beside a stream of molten metal that is pouring from a blast furnace. Heat radiated from the metal is reflected away by the shiny surface of the protective suit

▲ **Figure 21.11** The Golden Gate Bridge in San Francisco is a 3 mile long structure made of steel, an alloy of iron and carbon

Practice questions

1 a) Many laboratory reactions involve oxidation.
  i) Explain why the burning of calcium in air to produce calcium oxide is an oxidation reaction. *(2 marks)*
  ii) Write a balanced symbol equation for this reaction. *(4 marks)*
  iii) Write a half equation for the oxidation reaction. *(3 marks)*
b) Many industrial reactions also involve oxidation. In the production of sulfuric acid a reaction which occurs is

$$2SO_2 + O_2 \rightarrow 2SO_3$$

Explain why this is an oxidation reaction.
(2 marks)

c) In industry, iron is manufactured in a blast furnace by the following redox equation:

$$Fe_2O_3 + 3CO \rightarrow 2Fe + 3CO_2$$

Explain fully, in terms of change in oxygen content, why this reaction is described as a redox reaction. *(5 marks)*

2 Magnesium displaces copper from a solution of copper(II) nitrate. For this reaction:
  a) write a balanced symbol equation *(2 marks)*
  b) write an ionic equation *(2 marks)*
  c) write two half equations. *(6 marks)*
  d) Explain which half equation is a reduction process. *(2 marks)*
  e) Explain why this is a redox reaction. *(1 mark)*

3 A displacement reaction occurs between zinc and copper(II) chloride.

zinc + copper chloride → zinc chloride + copper
  a) Write a balanced symbol equation, with state symbols, for this reaction. *(3 marks)*
  b) Write an ionic equation for this reaction. *(2 marks)*
  c) Explain in terms of electrons why this is a redox reaction. *(5 marks)*

4 Most metals corrode as they react with oxygen and/or water. The corrosion of iron and steel is called rusting.
  a) Explain how a surface layer such as a layer of paint prevents iron from rusting. *(1 mark)*
  b) Explain how attaching a block of magnesium to the bottom of a steel ship prevents it from rusting. *(2 marks)*
  c) How would the following substances be protected from rusting?
    i) gate hinge *(1 mark)*
    ii) steel food can *(1 mark)*

d) Explain why galvanising stops iron from rusting even when the galvanised surface is scratched or broken. *(2 marks)*

5 a) Reactions 1–3 are reactions of iron.
  1: iron + oxygen + water → hydrated iron(III) oxide
  2: iron + copper(II) sulfate → iron(II) sulfate + copper
  3: iron + hydrochloric acid → iron(II) chloride + hydrogen
  i) What is the common name for hydrated iron(III) oxide? *(1 mark)*
  ii) Explain why iron is oxidised in reaction 1. *(2 marks)*
  iii) Reaction 2 is described as a redox reaction. Explain, in terms of electrons, why it is described as a redox reaction. *(5 marks)*
  iv) Write a balanced symbol equation for reaction 3. *(3 marks)*
  v) Using a half equation, explain if the iron is oxidised or reduced in reaction 3. *(4 marks)*
  vi) What is meant by the word 'hydrated'? *(1 mark)*

b) Iron is extracted from its ore in a blast furnace. Aluminium is extracted from its ore by electrolysis.
  i) Name the ore from which iron metal is extracted. *(1 mark)*
  ii) Explain why aluminium and iron are produced by different methods. *(1 mark)*
  iii) Describe in words how the reducing agent is produced in the blast furnace. *(3 marks)*
  iv) Write two balanced symbol equations to show how calcium carbonate is used to remove acidic impurities in the blast furnace. *(4 marks)*

6 a) The formation of rust is described as an oxidation reaction.
  i) Name a metal that rusts. *(1 mark)*
  ii) Name the gaseous element that is required for the formation of rust. *(1 mark)*
  ii) Name the compound that is required for the formation of rust. *(1 mark)*
  iv) Explain why rusting is an oxidation reaction. *(2 marks)*
  v) Describe the appearance of rust. *(2 marks)*

b) To prevent an iron gate rusting it can be painted and the hinges can be oiled.
  i) Explain why these two methods prevent rusting. *(1 mark)*

ii) A galvanised gate does not rust. Explain how a galvanised gate is different from an iron gate. *(1 mark)*

c) Iron can be made by heating carbon with iron(II) oxide.

$FeO + C \rightarrow Fe + CO$

i) Name the substance that is reduced in this reaction. Explain your answer. *(2 marks)*

ii) Name the substance that is oxidised in this reaction. Explain your answer. *(2 marks)*

d) What is sacrificial protection? *(2 marks)*

7 a) Magnesium reacts with copper(II) sulfate solution. The reaction is described as a redox reaction, as both oxidation and reduction occur.

$Mg + CuSO_4 \rightarrow Cu + MgSO_4$

i) In this reaction, which ion does not undergo any change? *(1 mark)*

ii) Write an ionic equation for the reaction, including state symbols. *(3 marks)*

iii) What is oxidised in this reaction? *(1 mark)*

iv) Write a half equation for the oxidation process that occurs in this reaction. *(3 marks)*

v) What is observed in this reaction? *(2 marks)*

b) Magnesium also reacts with hydrochloric acid in a redox reaction.

i) Write a balanced symbol equation for this reaction. *(3 marks)*

ii) In this reaction, which ion does not undergo any change? *(1 mark)*

iii) Write a half equation for the reduction process that occurs in this reaction. *(3 marks)*

8 Iron is produced in a blast furnace.

a) Name an ore of iron that is added to the blast furnace. *(1 mark)*

b) Name two other solid materials that are added at the top of the furnace. *(2 marks)*

c) Why is hot air added to the furnace? *(1 mark)*

d) Name the reducing agent used in a blast furnace, and give two balanced symbol equations for its production in the furnace. *(5 marks)*

e) Write a balanced symbol equation for the reduction of iron ore in the furnace. *(3 marks)*

f) Explain in words how slag is made in a blast furnace. *(3 marks)*

g) How is slag removed from a blast furnace? *(1 mark)*

h) Suggest the name of an exhaust gas that may leave the top of the furnace. *(1 mark)*

22 Rates of reaction

Specification points

This chapter covers specification points 2.3.1 to 2.3.6. It covers the different factors affecting the rate of a reaction, including catalysts. Interpretation of experimental data is covered, as is a prescribed practical exploring how changing a variable can change the rate of a reaction.

When glow sticks are activated, a small container inside the stick which is filled with hydrogen peroxide breaks, and the peroxide mixes and reacts with the other chemical in the glow stick. Placing a glow stick in a beaker of hot water makes it glow more intensely, and placing it in a beaker of ice causes it to glow less intensely. It appears that temperature affects the rate of this reaction. In this chapter you will look at temperature and other factors that affect the rate of reaction.

What is rate of reaction?

Some reactions are fast and others are slow. The rate of a reaction is a measure of the speed at which reactants are changed into products.

In a **fast** reaction it takes a **short** time to convert reactants to products. Dynamite exploding and potassium reacting with water are examples of fast reactions.

In a **slow** reaction it takes a long time to convert reactants to products. Rusting of iron is a slow reaction.

Measuring rate of reaction

The rate of a reaction may be determined by measuring:

▶ the loss of a reactant over time
▶ the gain of a product over time.

The rate can be calculated using the equation

$$Rate = \frac{1}{time}$$

The **units** of rate are s^{-1}.

▲ **Figure 22.1** Explosives can be used to demolish unwanted buildings such as this hotel. An explosion is a fast reaction – the reactants are converted to products quickly

▲ **Figure 22.2** Fermentation of sugars to produce home-brewed wine is a slow reaction

Test yourself

1 A reaction takes 15 s to complete. Calculate the rate of the reaction and state the units.
2 A reaction takes 60 s to finish. Calculate the rate of the reaction and state the units.
3 Table 22.1 gives the results of an investigation.

Table 22.1

Experiment	Concentration/mol/dm³	Time/s	Rate of reaction/s⁻¹
1	0.4	105	0.0095
2	0.8	79	0.0127
3	1.2	54	0.0185
4	1.6	32	

a) Calculate the rate of the reaction for experiment 4.
b) From the results of the experiments state the effect of increasing the concentration on the rate of the reaction.

Practical methods for measuring rate of reaction

Reactions used in rate of reaction experiments

The four reactions most often used to investigate rate of reaction are described below.

1 Metals with dilute acid

When metals react with dilute acid, hydrogen gas is produced. An example reaction is:

magnesium + hydrochloric acid → magnesium chloride + hydrogen

$$Mg + 2HCl \rightarrow MgCl_2 + H_2$$

To experimentally investigate reactions that produce a gas, the volume of gas produced or the mass of gas lost can be measured over a period of time.

2 Marble chips (calcium carbonate) with dilute acid

When marble chips (calcium carbonate) react with dilute acid, carbon dioxide gas is produced.

calcium carbonate + hydrochloric acid → calcium chloride + carbon dioxide + water

$$CaCO_3 + 2HCl \rightarrow CaCl_2 + CO_2 + H_2O$$

This reaction also produces a gas, and to investigate the reaction experimentally the volume of gas produced or the mass of gas lost could be measured over a period of time.

Tip

The production of gases in these reactions allows the rate to be monitored in several ways.
The mass will decrease as the gas is released. The volume of gas produced may be measured. The time for the reaction to stop (fizzing) may also be measured.

▲ **Figure 22.3** Hydrogen peroxide decomposition can also be catalysed by the enzyme catalase, which is found in raw liver (liver can be seen at the bottom of the flask). Catalase protects cells from the toxic effects of hydrogen peroxide, which is produced as a biochemical by-product

3 Catalytic decomposition of hydrogen peroxide (H_2O_2) to produce oxygen

Hydrogen peroxide will decompose (break up) in the presence of a catalyst of manganese(IV) oxide to produce oxygen gas.

hydrogen peroxide → oxygen + water

$$2H_2O_2 \rightarrow 2H_2O + O_2$$

Tip

This reaction is used to prepare oxygen gas in the laboratory, and is studied again on pages 338–9.

This reaction also produces a gas, and to investigate the reaction experimentally the volume of gas produced or the mass of gas lost could be measured over a period of time.

4 Sodium thiosulfate with acid

Sodium thiosulfate is a colourless solution which reacts with acid to form a precipitate of sulfur. The time taken to form a certain amount of precipitate can be measured.

$$2HCl(aq) + Na_2S_2O_3(aq) \rightarrow 2NaCl(aq) + SO_2(g) + S(s) + H_2O(l)$$

Tip

You are not required to learn the equation for this reaction but it is useful to note that two aqueous solutions mix to form a solid precipitate of sulfur.

Recording results

During experimental activities you must often record results in a table. When drawing tables and recording data ensure that:

▶ the table is a ruled box with ruled columns and rows (in pencil)

▶ there are headings for each column and row (in pen)

▶ there are units for each column and row – usually placed after the heading after a solidus (/) or brackets, for example 'Temperature/°C' or 'Temperature (°C)'

▶ there is room for repeat measurements and averages – remember the more repeats you do, the more reliable the data.

Table 22.2 Example of a table to record experimental results

Time/minutes	Mass of oxygen lost/g	Mass of oxygen lost/g	Average mass of oxygen lost/g
1			
2			
3			
4			
5			

Plotting graphs

When carrying out experiments you must be able to translate data from one form to another. Most often this involves using data from a table to draw a graph. A graph is an illustration of how two variables relate to one another.

When drawing a graph remember that:

▶ The independent variable is placed on the x-axis, while the dependent variable is placed on the y-axis (Figure 22.4).

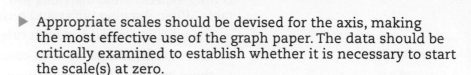

Tip

The independent variable is the factor that you change during an experiment. The dependent variable is the factor that changes as a result of this change and is measured during the reaction.

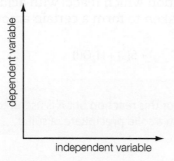

▲ **Figure 22.4** Independent variable is on the x axis and dependent is on the y axis

▶ Appropriate scales should be devised for the axis, making the most effective use of the graph paper. The data should be critically examined to establish whether it is necessary to start the scale(s) at zero.

▶ Axes should be labelled with the name of the variable followed by a solidus (/) and the unit of measurement. For example, the label may be Temperature/°C.

▶ A line of best fit should be drawn. When judging the position of the line, there should be approximately the same number of data points on each side of the line. Resist the temptation to simply connect the first and last points. The line of best fit may be a straight line or curve. Ignore any anomalous results when drawing the line.

▶ A graph must have a title. A title is often 'A graph of y axis quantity against x axis quantity'. For example for Figure 22.5, a suitable title would be 'A graph of mass of flask and contents against time'.

Tip

You need to judge the line of best fit with your eyes. Using a see-through plastic ruler or a flexible curve can help you. If drawing a curve by hand, it is easier to get a smoother line if you have your elbow on the inside of the curve as you draw it.

Show you can

In an experiment some calcium carbonate and acid were placed in a conical flask on a balance and the balance reading recorded every minute. The results were recorded and the graph shown below was drawn.

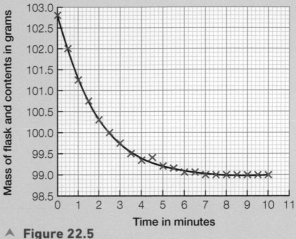

▲ **Figure 22.5**

1 Identify any anomalous results.
2 The *y*-axis is not labelled correctly. Write down the correct label for this axis.
3 The *x*-axis is not labelled correctly. Write down the correct label for this axis.
4 What is the independent variable in this experiment?
5 At time 2 minutes, what is the mass of the flask and contents?

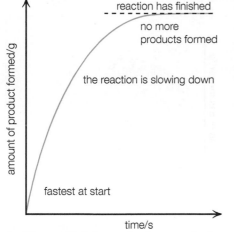

▲ **Figure 22.6** A graph of amount of product against time

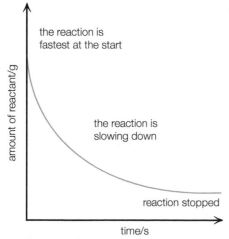

▲ **Figure 22.7** A graph of amount of reactant against time

Rate graphs

To investigate rate of reaction, a graph can be drawn to show how the quantity of reactant used or product formed changes with time. The gradient or slope of the line represents the **rate** of the reaction. A steep slope means a fast reaction.

A typical graph showing how the **amount of product** changes with time for an individual reaction is shown in Figure 22.6. The line is steepest at the start when more reactant particles are present and the reaction is fastest. As time proceeds the line becomes less steep, as the reaction slows down as there are fewer reactant particles. The line eventually becomes horizontal when the reaction stops.

A typical graph showing how the **amount of reactant** changes with time for an individual reaction is shown in Figure 22.7. Again the line is steepest at the start, and the reaction is fastest here. The reaction slows down, and the graph becomes less steep. When one of the reactants is used up the reaction stops and the line is horizontal.

The rate of one reaction can be compared to the rate of another using graphs. For example, in Figure 22.8, graph A (blue line) is steeper than B (red line), and represents a faster reaction. It also finishes at an earlier time. Note that both reactions finish at the same amount of product. This is the case if the same amounts of reactants are used.

A **faster** reaction:

▶ has a steeper slope (the line is higher at every time point)
▶ will be over in a faster time (the line levels off earlier).

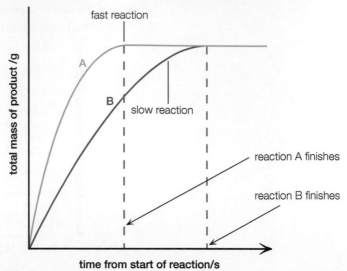

▲ **Figure 22.8** Graph of amount of product against time for reaction A and B

Test yourself ✏

6 The graph shows how the mass of carbon dioxide formed when sodium carbonate reacts with hydrochloric acid varies over time.
  a) At what point on the graph is the rate of reaction greatest? Explain your answer.
  b) At what point on the graph does the reaction stop? Explain your answer.
  c) Is the rate of reaction faster at point B or E? Explain your answer.
  d) Write a balanced symbol equation for the reaction.

7 Look at the graphs for reactions P, Q and R, which all produce a gas.
  a) Which reaction is the fastest?
  b) Which reaction is the slowest?

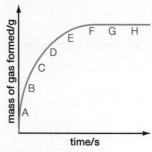

▲ **Figure 22.9**

8 Hydrogen gas is formed when magnesium reacts with sulfuric acid. Table 22.3 shows how the volume of hydrogen changed with time when some magnesium was reacted with sulfuric acid.

Table 22.3

Time/s	0	10	20	30	40	50	60	70	80	90	100
Volume of hydrogen/ cm³	0	30	55	75	88	98	102	104	104	104	104

a) Plot a graph of the volume of hydrogen against time.
b) At what time did the reaction finish?
c) Sketch a graph, on the same axes, of a reaction which is faster but uses the same amounts of magnesium and sulfuric acid.
d) Write a word equation for this reaction.
e) Write a balanced equation for this reaction.

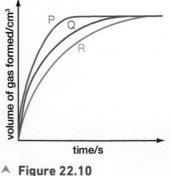

▲ **Figure 22.10**

Experimental methods for finding reaction rate

1 Measuring the volume of gas produced per unit time

Many of the reactions used to study rate of reaction produce a gas, for example the reaction of an acid and metal, or of marble chips and acid or the catalytic decomposition of hydrogen peroxide.

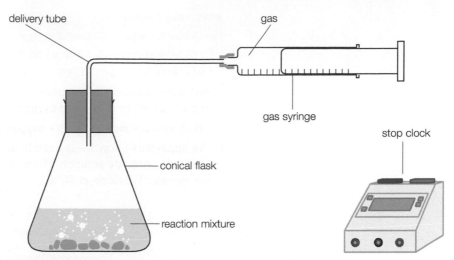

▲ **Figure 22.11** Investigating the rate of reaction by measuring the volume of a gas produced. The reaction mixture could be an acid and metal or carbonate, or hydrogen peroxide with the catalyst manganese(IV) oxide

The apparatus shown in Figure 22.11 can be used to measure the volume of gas produced per unit time. Safety glasses will also need to be worn. A stopwatch is started and the gas is collected in the gas syringe. The volume of gas is measured and recorded at various time intervals. The reaction is over when two readings are the same, showing that no more gas is being produced because one of the reactants has been used up.

A graph of volume of gas against time can be plotted. One variable can be changed and the experiment repeated to determine the effect of this variable on the rate of reaction.

2 Measuring the change in mass in a given time

Using the apparatus shown in Figure 22.12, the change in mass during a reaction can be recorded at various time intervals. If the reaction produces a gas, then the mass decreases because the gas has escaped from the flask into the atmosphere. The cotton wool plug allows the gas out and prevents any loss of liquid from the flask, as bubbling in the reaction may cause the solution to splash out.

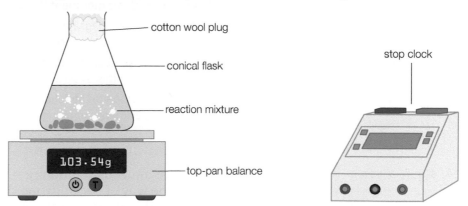

▲ **Figure 22.12** Investigating the rate of reaction by measuring the mass decrease. The reaction mixture could be an acid and metal or carbonate

A graph of loss in mass against time, or simply mass against time, can be plotted.

263

Practical activity

Investigating the rate of decomposition of hydrogen peroxide solution by measuring the loss in mass

Hydrogen peroxide decomposes in the presence of solid manganese(IV) oxide to produce water and oxygen.

hydrogen peroxide → water + oxygen

The apparatus shown was used to investigate the rate of decomposition of hydrogen peroxide solution. 20 cm³ of the solution were added to 1.0 g of manganese(IV) oxide at 20 °C.

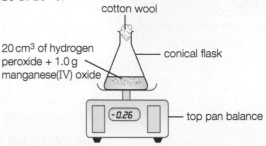

▲ **Figure 22.13**

The following results were obtained.

Table 22.4

Time/min	1	2	3	4	5	6	7
Mass of oxygen lost/g	0.20	0.34		0.45	0.47	0.48	0.48

1 Write a balanced symbol equation for the decomposition of hydrogen peroxide.

2 What is the purpose of the cotton wool plug?

3 Plot a graph of mass of oxygen lost against time.

4 Use the graph to state the mass of oxygen lost after 3 minutes.

5 Suggest an alternative method of measuring the rate of this reaction without measuring the mass of oxygen lost.

6 Sketch, on the same axes, the graph you would expect if the reaction was carried out at a higher temperature, giving a faster reaction.

At the end of this experiment the manganese(IV) oxide can be recovered.

7 Draw a labelled diagram of the assembled apparatus that could be used to recover the manganese(IV) oxide at the end of the experiment.

8 How would you experimentally prove that the manganese(IV) oxide was not used up in this experiment?

9 Copy and complete the following sentence:
In this experiment the manganese dioxide is acting as a _____ .

3 Formation of a precipitate

When two aqueous solutions are mixed, sometimes a precipitate forms. One way of investigating the rate of a precipitation reaction is to place the reaction flask containing one reactant solution on top of a piece of paper with a cross drawn on it, as shown in Figure 22.14. The time is recorded from when the second reactant solution is added until the precipitate forms and obscures the cross from the view of an observer looking down into the flask. The reaction of sodium thiosulfate with acid can be investigated using this method.

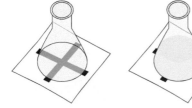

▲ **Figure 22.14** A cross drawn on a piece of paper is placed under the reaction mixture of sodium thiosulfate and hydrochloric acid. The experiment is timed from when the reactants are mixed until a precipitate of sulfur is formed and obscures the cross

Show you can

An antacid tablet containing citric acid and sodium hydrogencarbonate fizzes when added to water due to a reaction taking place that produces carbon dioxide.

In an experiment, one tablet was added to 50 cm³ of water (an excess) in a conical flask at a temperature of 20 °C. The flask was loosely stoppered with cotton wool, placed on a balance and a stopwatch started. The mass was recorded every 20 seconds.

1 Explain why the total mass of the flask and its contents decreased during the experiment.
2 What is the purpose of the cotton wool plug?
3 Sketch a graph of total mass against time for the reaction. Label the line A.
4 On the same axes, sketch a graph of total mass against time for the same reaction using the same conditions but using a crushed tablet, which leads to a faster reaction. Label the line B.

▲ **Figure 22.15** Collision theory states that particles must collide with each other in order to react. This is similar to judo and wrestling, wherein the two competing players must contact one another to score a point

Collision theory

Collision theory states that for a reaction to occur the reacting particles (which may be atoms, ions or molecules) must collide together with enough energy to react.

Activation energy is the minimum energy needed for a reaction to occur.

Successful collisions are ones that result in a reaction, and they take place when reactant particles collide with the activation energy. The activation energy is used to help break bonds so that the atoms can be rearranged to make products (Figure 22.16). Unsuccessful collisions are collisions that do not result in reaction, and occur when the particles collide with less than the activation energy.

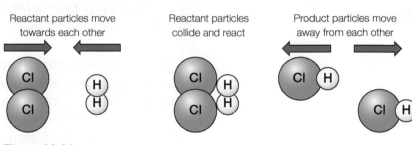

| Reactant particles move towards each other | Reactant particles collide and react | Product particles move away from each other |

▲ **Figure 22.16**

Tip ↻

Frequency is the number of collisions in a certain amount of time.

In general, reaction rates are increased when the energy of the collisions is increased and when the frequency of collisions is increased.

▲ **Figure 22.17** This milk has been kept at room temperature and has curdled. Keeping milk in the fridge slows down the chemical reactions that make it go off

Factors affecting the rate of reaction

Different factors can change the rate of a reaction.

1 Temperature

Change: increasing the temperature increases the rate of reaction.

Explanation:

▶ the particles are moving faster, so there are more frequent collisions

▶ the particles also have more energy, so more of the particles have the activation energy and react when they collide

▶ hence there more successful collisions between particles in a given time (the successful collisions are more frequent)

▶ and so the rate of reaction is faster.

2 Concentration of solution

▲ **Figure 22.18** Increasing the number of dodgems in the same space means there are more frequent collisions

Change: increasing the concentration of a solution increases the rate of reaction (see Figure 22.19).

Explanation:

▶ there are more particles present in the same volume

▶ and so there are more successful collisions (with the activation energy) between particles in a given time (the successful collisions are more frequent)

▶ and so the rate of reaction is faster.

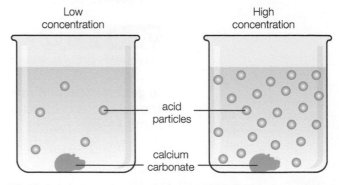

▲ **Figure 22.19** Calcium carbonate in low and high concentrations of acid

3 Surface area of solid reactants

Change: increasing the surface area by breaking a solid reactant into smaller pieces increases the rate of reaction.

Explanation:

▶ there are more particles on the surface exposed to the other reactant

▶ and so there are more successful collisions (with the activation energy) between particles in a given time (the successful collisions are more frequent)

▶ and so the rate of reaction is faster.

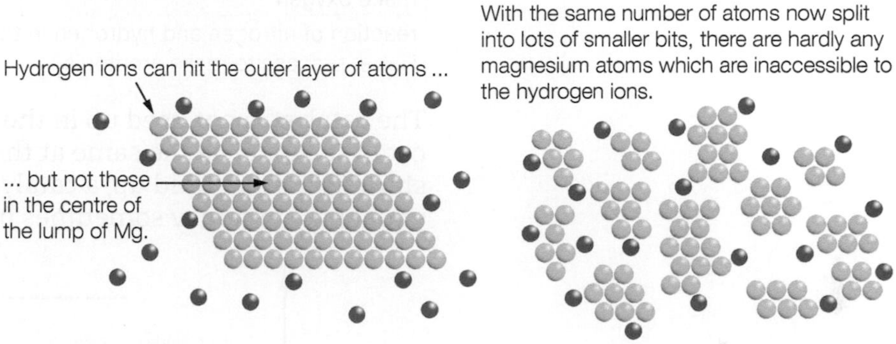

Hydrogen ions can hit the outer layer of atoms ...

With the same number of atoms now split into lots of smaller bits, there are hardly any magnesium atoms which are inaccessible to the hydrogen ions.

... but not these in the centre of the lump of Mg.

▲ **Figure 22.20** The effect of changing surface area on the reaction of acid and magnesium

To determine the effect of breaking a solid into smaller particles consider a cube of sides of length 2 cm and volume 8 cm³. If it is broken down into eight smaller cubes with sides of length 1 cm then the volume is still 8 cm³ but the surface area is increased from 24 cm² to 48 cm², as shown in Table 22.5. The surface area to volume ratio is thus much greater.

Table 22.5

Cube	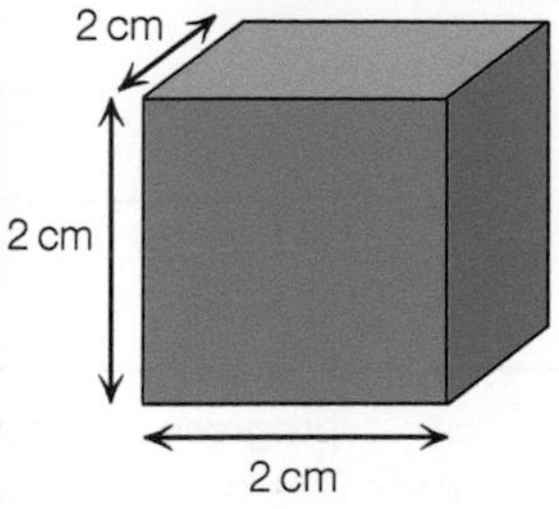 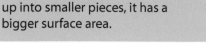 one larger cube	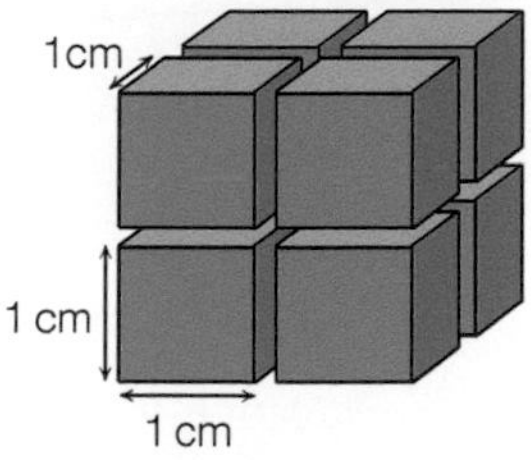 broken up into smaller cubes
Surface area	surface area of each side = $2 \times 2 = 4\,cm^2$ there are 6 sides, so total surface area = $6 \times 4 = 24\,cm^2$	**Each cube:** surface area of each side = $1 \times 1 = 1\,cm^2$ there are 6 sides, so surface area of each cube = $6 \times 1 = 6\,cm^2$ **For all 8 cubes:** total surface area = $8 \times 6 = 48\,cm^2$
Volume	volume = $2 \times 2 \times 2 = 8\,cm^3$	**Each cube:** volume = $1 \times 1 \times 1 = 1\,cm^3$ **For all 8 cubes:** total volume = $8 \times 1 = 8\,cm^3$
Surface area : volume ratio	24 : 8 = 3 : 1	48 : 8 = 6 : 1

4 The presence of a catalyst

A catalyst is a substance that increases the rate of a reaction without being used up.

Different reactions have different catalysts. Many catalysts are transition metals or compounds of transition metals. Some catalysts and their reactions are shown in Table 22.6.

Table 22.6 Some catalysts and their reactions

Reaction	Catalyst	Type
decomposition of hydrogen peroxide to make oxygen	manganese(IV) oxide	transition metal compound
reaction of nitrogen and hydrogen in the Haber process to make ammonia	iron	transition metal

The catalyst is not used up in the reaction, and the mass of the catalyst should be the same at the end of the reaction as at the start. As it is not used up, a catalyst should not appear in a chemical equation, but it may sometimes be included on top of the arrow.

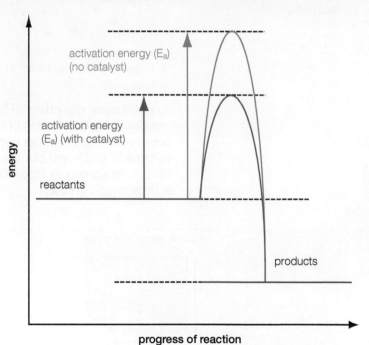

▲ **Figure 22.21** You will study this graph in more detail in Chapter 15. It shows how a catalyst provides an alternative reaction pathway with lower activation energy

▲ **Figure 22.22** Platinum, the catalyst used to make hydrogen by electrolysis of water, is expensive. A new catalyst made from soybeans and a small amount of molybdenum metal has been discovered and is 1500 times cheaper

A catalyst works by:

▶ providing **an alternative reaction pathway of lower activation energy**

▶ and so there are more successful collisions (with the activation energy) between particles in a given time (the successful collisions are more frequent)

▶ and so the rate of reaction is faster (see Figure 22.23).

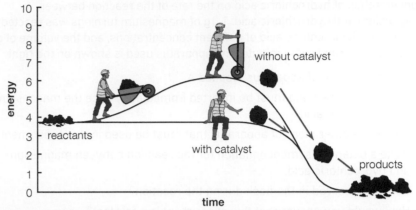

Tip

A catalyst provides an alternative pathway for the reaction which has lower activation energy. It does not lower the activation energy.

▲ **Figure 22.23** When the pathway that is less steep is chosen (with catalyst) the activation energy is lowered and less energy is needed to form the product

Prescribed practical

Prescribed practical C4: Investigating how changing a variable (concentration) changes the rate of a reaction by observing the formation of a precipitate

Sodium thiosulfate solution ($Na_2S_2O_3$) reacts with dilute hydrochloric acid according to the equation:

$$Na_2S_2O_3(aq) + 2HCl\,(aq) \rightarrow 2NaCl(aq) + S(s) + SO_2(g) + H_2O(l)$$

In an experiment, $25\,cm^3$ of sodium thiosulfate was placed in a conical flask on top of a piece of paper with a cross drawn on it. Hydrochloric acid was added and the stopwatch started. A precipitate was produced that caused the solution to become cloudy. The stopwatch was stopped when the experimenter could no longer see the cross on the paper through the solution, due to the precipitate formed.

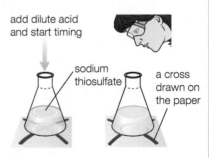

▲ **Figure 22.24**

1 Look at the equation and identify the product which causes the solution to become cloudy.
The experiment was repeated using different concentrations of sodium thiosulfate. The results are recorded in Table 22.7.

Table 22.7

Experiment	Concentration of sodium thiosulfate/mol/dm³	Time taken for cross to disappear/s	Rate of reaction (1/time)/s⁻¹
1	0.4	105	0.0095
2	0.8	79	0.0127
3	1.2	54	0.0185
4	1.6	32	

2 Calculate the value for the rate of reaction for experiment 4.
3 Identify three variables that must be kept constant to make this a fair test.
4 State and explain one change which could be made to this experiment to give more accurate results.
5 From the results of the experiment state the effect of increasing the concentration of sodium thiosulfate solution on the rate of the reaction.
6 Explain in terms of collision theory how increasing the concentration of sodium thiosulfate solution has this effect.

Prescribed practical

Prescribed practical C6: Investigating how changing a variable (concentration) changes the rate of a reaction by measuring the volume of gas produced

An experiment was carried out to determine the effect of changing the concentration of hydrochloric acid on the rate of the reaction between magnesium and hydrochloric acid. 1.0 g of magnesium turnings was reacted with excess hydrochloric acid of different concentrations, and the volume of gas produced recorded every minute. The apparatus used is shown on the right.

1 Name the piece of apparatus labelled A.

2 Explain why the bung must be inserted immediately once the magnesium is added to the acid.

3 Name one other piece of apparatus that must be used in this experiment.

4 Write a balanced symbol equation for the reaction between magnesium and hydrochloric acid.

5 What is observed in the flask during the reaction?

6 How would you ensure that this experiment is a fair test?

7 State one source of error in this experiment.

The results from this experiment are plotted on the axes as shown in Figure 22.26.

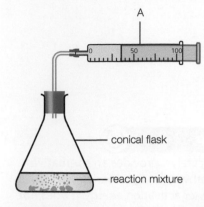

▲ **Figure 22.25**

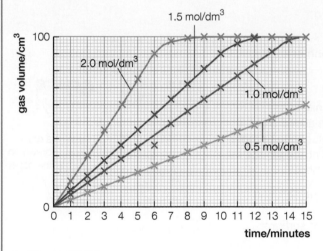

▲ **Figure 22.26**

8 Use the graph to determine the time at which each reaction ended for acid of concentration 2.0 mol/dm³ and 1.5 mol/dm³.

9 Are there any results that are anomalous? Explain your answer.

10 Which concentration of acid produced the slowest rate of reaction?

11 What is the relationship between the concentration of acid and the rate of reaction?

H 12 Explain this relationship in terms of collision theory.

Practice questions

1 Sodium thiosulfate solution reacts with hydrochloric acid to form solid sulfur. As the sulfur is formed, the solution becomes cloudy. The reaction can be carried out in a conical flask on top of a piece of paper with a cross drawn on it. The time it takes for the solution to become so cloudy that the cross can no longer be seen can be measured and used to compare reaction rates under different conditions. Table 22.8 shows the results of some experiments carried out in this way.

Table 22.8

Experiment	Concentration of sodium thiosulfate solution/mol/dm³	Temperature/°C	Time taken for solution to become too cloudy to see cross/s
A	0.10	20	75
B	0.20	20	37
C	0.30	20	25
D	0.20	30	18

a) Which experiment has the fastest rate of reaction? *(1 mark)*

b) What is the effect of changing the concentration on the rate of reaction? Which experiments did you use to work this out? *(2 marks)*

c) What is the effect of changing the temperature on the rate of reaction? Which experiments did you use to work this out? *(2 marks)*

d) Calculate the rate of reaction for experiment A. *(1 mark)*

2 Calcium carbonate reacts with hydrochloric acid and produces carbon dioxide gas. An experiment was carried out to compare the rate of reaction using large pieces, small pieces and powdered calcium carbonate. In each case the same mass of calcium carbonate was used.

Table 22.9

Type of calcium carbonate	Time taken to produce 100 cm³ of carbon dioxide/s
large pieces	75
small pieces	48
powder	5

a) Write a word equation for the reaction between calcium carbonate with hydrochloric acid. *(1 mark)*

b) Write a balanced symbol equation for the reaction. *(3 marks)*

c) Calculate the rate for the powder. *(1 mark)*

d) What is the effect of changing the surface area of the calcium carbonate on the rate of reaction? *(1 mark)*

3 Some reactions have catalysts that are transition metals or transition metal compounds.

a) What is a catalyst? *(2 marks)*

b) Name a transition metal compound used to catalyse the decomposition of hydrogen peroxide. *(1 mark)*

c) Name a transition metal used as a catalyst in the Haber process. *(1 mark)*

d) Explain why a catalyst changes the rate of a reaction. *(3 marks)*

4 Hydrogen peroxide decomposes to form water and oxygen gas according to the equation:
$$2H_2O_2(aq) \rightarrow 2H_2O(l) + O_2(g)$$
The catalyst used in the laboratory is manganese(IV) oxide.

a) What is meant by the term *catalyst*? *(2 marks)*

b) 50.0 cm³ of 0.08 mol/dm³ hydrogen peroxide were decomposed using manganese(IV) oxide as the catalyst at 20 °C.
The volume of oxygen gas was measured. The curve below shows how the total volume of oxygen collected changed with time under these conditions. Four points on the graph are labelled A, B, C and D.

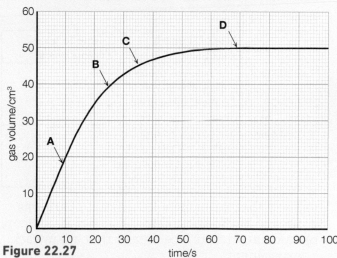

Figure 22.27

i) At which point (A, B, C or D) is the rate of reaction fastest? *(1 mark)*

ii) At which point (A, B, C or D) is the rate of reaction zero? *(1 mark)*

iii) Draw a curve on the graph above to show how the total volume of oxygen will change with time if the reaction is repeated at 60 °C using 50 cm³ of 0.04 mol/dm³ hydrogen peroxide solution. Label this curve Y. *(3 marks)*

c) In a second experiment, the mass of manganese(IV) oxide was measured at various times. Which one of the following graphs best shows the experimental results? *(1 mark)*

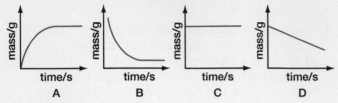

Figure 22.28

5 a) Explain, as fully as possible in terms of particles, how an increase in concentration of hydrochloric acid increases the rate of the reaction between magnesium and HCl. *(3 marks)*

b) Name two factors apart from the concentration of HCl that will affect the rate of this reaction. *(2 marks)*

6 Rhubarb stalks contain some ethanedioic acid. If rhubarb stalks are placed in a purple solution of dilute acidified potassium manganate(VII), the solution goes colourless as the ethanedioic acid reacts with it. The results from two experiments are shown in the table below. In the first experiment one piece of rhubarb stalk was used in the shape of a cuboid measuring 5 cm × 1 cm × 1 cm. In the second experiment, a similar piece of rhubarb stalk was chopped up into five cubes, each measuring 1 cm × 1 cm × 1 cm.

Table 22.10

	Size of rhubarb stalk used	Time to go colourless/s
Experiment 1	one piece of rhubarb stalk in the shape of a cuboid measuring 5 cm × 1 cm × 1 cm	75
Experiment 2	five pieces of rhubarb stalk in the shape of cubes each measuring 1 cm × 1 cm × 1 cm	52

a) Show that the same amount of rhubarb was used in each experiment. *(3 marks)*

b) Calculate the surface area of rhubarb in each experiment. *(2 marks)*

c) Calculate the surface area to volume ratio of rhubarb in each experiment. *(2 marks)*

d) What does this experiment show about the effect of surface area on rate of reaction? *(1 mark)*

e) Explain, in terms of collisions, why changing the surface area has the effect that it does. *(3 marks)*

f) What variables should be controlled between experiments 1 and 2 to ensure that this is a fair test? *(3 marks)*

7 In a laboratory experiment 0.5 g of magnesium ribbon was reacted with excess dilute hydrochloric acid at room temperature. The volume of gas produced was noted every 20 seconds. The results were plotted below as graph C.

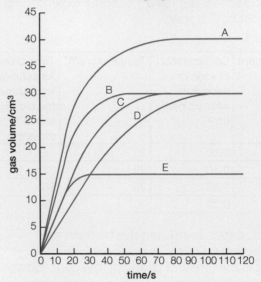

Figure 22.29

a) At what time did reaction C finish? *(1 mark)*

b) What is the total volume of gas produced for reaction C? *(1 mark)*

The experiment was repeated using different conditions and the results obtained given in graphs A, B, D and E.

c) State and explain which of the graphs A, B, D or E would have been obtained if the 0.5 g of magnesium ribbon were replaced by 0.5 g of magnesium powder. *(2 marks)*

d) State and explain which of the graphs A, B, D or E would have been obtained if the 0.5 g of magnesium ribbon were added to excess dilute hydrochloric acid at a temperature below room temperature. *(2 marks)*

e) State and explain which of the graphs A, B, D or E would have been obtained if the 0.25 g of magnesium ribbon were reacted with excess dilute hydrochloric acid at room temperature. *(2 marks)*

f) State and explain, in terms of collision theory, the effect of increasing the concentration of hydrochloric acid on the rate of the reaction between hydrochloric acid and magnesium. *(3 marks)*

23 Equilibrium

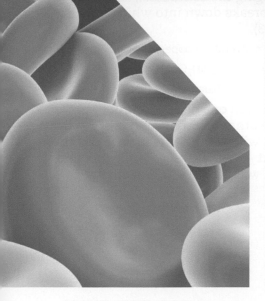

Specification points

In this chapter specification points 2.4.1 to 2.4.2 are covered. Reversible and irreversible reactions and the concept of dynamic equilibrium are explored.

Without chemical equilibrium, life as we know it would not be possible. Haemoglobin is a molecule found in red blood cells. It transports oxygen around our bodies, and without it we would not survive. The transport of oxygen by haemoglobin is a reversible equilibrium reaction. The haemoglobin has to be able to take up oxygen, but also to release it; this is done through changes in the conditions in different places in our bodies. In this chapter we will study the meaning of the term 'equilibrium', and discover how different conditions affect it.

Reversible reactions

Irreversible reactions are reactions where the reactants convert to products and the products cannot convert back to the reactants. For example, you cannot change a baked cake back into its raw ingredients or change the ashes from a piece of burnt newspaper back into a newspaper.

▲ **Figure 23.1** Baking a cake is not a reversible reaction

When a metal such as magnesium reacts with an excess of acid to produce a salt and hydrogen the reaction stops when all of the magnesium has reacted. The products do not go back to reactants once the products have been formed. The reaction is said to have 'gone to completion', and this is indicated by an arrow →.

$$Mg + H_2SO_4 \rightarrow MgSO_4 + H_2O$$

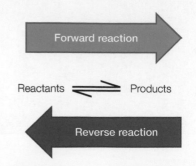

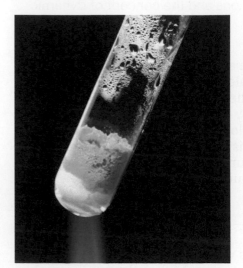

▲ **Figure 23.2** A reversible reaction goes in both directions

▲ **Figure 23.3** Heating hydrated copper(II) sulfate

Some chemical reactions are reversible. **A reversible reaction is one in which the products, once made, can react to reform the reactants.** Reversible arrows (⇌) are used to show that a reaction is reversible (Figure 23.2).

By altering the reaction conditions, the direction of a reversible reaction can be changed. For example, white anhydrous copper(II) sulfate reacts with water to form blue hydrated copper(II) sulfate. This reaction is reversible, and by changing the conditions, i.e. by heating the hydrated copper(II) sulfate, it breaks down into white anhydrous copper(II) sulfate again (Figure 23.3).

anhydrous copper(II) sulfate + water ⇌ hydrated copper(II) sulfate

$$CuSO_4 + 5H_2O \rightleftharpoons CuSO_4.5H_2O$$

white   blue

If a reversible reaction is exothermic in one direction, and gives out heat, then it is endothermic in the other direction, and takes in heat. The amount of energy transferred will be the same in each direction. For example, if a reaction gives out −92 kJ of energy in the forward reaction then the reverse reaction will take in +92 kJ of energy.

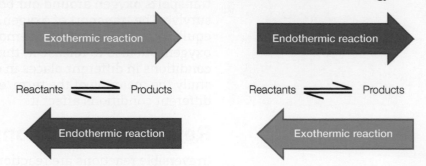

▲ **Figure 23.4** If a reversible reaction is exothermic in one direction it will be endothermic in the other direction

Test yourself

1 a) What is meant by the term 'reversible reaction'?
   b) Decide if the following reactions are reversible or irreversible.
      i) $Mg + O_2 \rightarrow MgO$
      ii) $CH_3COOH + C_2H_5OH \rightleftharpoons CH_3COOC_2H_5 + H_2O$
      iii) $Ca + 2HCl \rightarrow CaCl_2 + H_2$
      iv) $N_2 + 3H_2 \rightleftharpoons 2NH_3$

2 For the reaction
   $$A + B \rightleftharpoons C + D$$
   Write the equation for the:
   a) forward reaction
   b) reverse reaction.

3 Ammonia reacts with hydrogen chloride gas at room temperature to produce ammonium chloride. When heated, ammonium chloride breaks down to form ammonia and hydrogen chloride.
   a) Write a balanced symbol equation to represent this reversible reaction.
   b) Write an equation for the reverse reaction.
   c) Is the reverse reaction exothermic or endothermic?

Dynamic equilibrium

Open and closed systems

A closed system is one where no substances can get in or out. A flask with a stopper or a saucepan of boiling water with a lid is a closed system.

An open system allows entry and exit of substances, for example an open saucepan of boiling water.

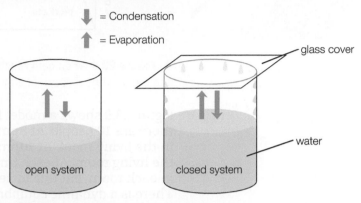

▲ **Figure 23.5** Open and closed systems

▲ **Figure 23.6** An unopened fizzy drinks bottle is a closed system

Figure 23.5 shows an open and a closed system. If equal amounts of water are placed in both beakers, in the open system the rate of evaporation is greater than the rate of condensation, as the vapour can escape. There is less water in the beaker. In the closed beaker, the rate of evaporation and the rate of condensation become equal, and the system is in equilibrium.

What is dynamic equilibrium?

When a system is in equilibrium there is no observable change – nothing seems to be happening. However, the system is *dynamic*, i.e. it is in constant motion. As fast as reactants are changed into products, the products are being converted back into reactants. The reaction is still taking place in both directions. All the species in the system are present (see Figure 23.8).

Dynamic equilibrium occurs:

▶ in a closed system
▶ when the rates of the forward and reverse reactions are equal
▶ when the amounts of reactants and products remain constant.

For example, for the reaction

hydrogen + iodine ⇌ hydrogen iodide

in a closed system, at equilibrium all the species are present – hydrogen, iodine and hydrogen iodide and:

▶ the rate of the forward reaction equals the rate of the reverse reaction
▶ the amounts of hydrogen and iodine remain constant
▶ the amount of hydrogen iodide remains constant.

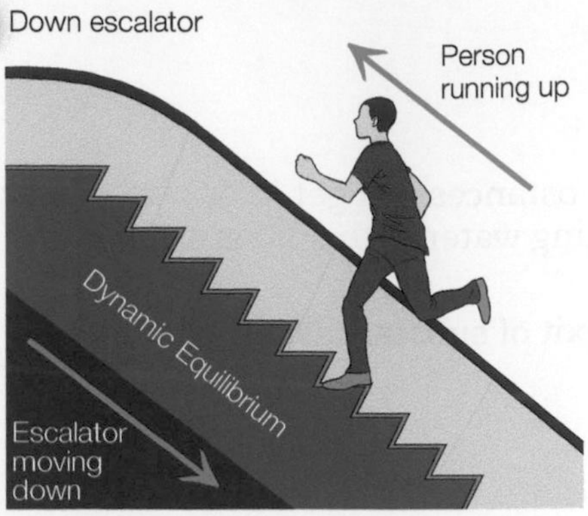

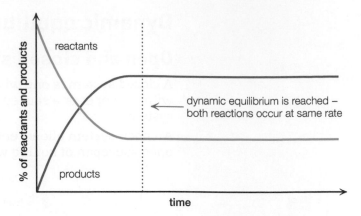

H Down escalator

Person running up

Dynamic Equilibrium

Escalator moving down

▲ **Figure 23.7** If you run up an escalator at the same rate that the escalator is moving down you will stay in the same position. This is dynamic equilibrium, and the rates of the forward and reverse reactions are equal

reactants

% of reactants and products

dynamic equilibrium is reached – both reactions occur at same rate

products

time

▲ **Figure 23.8** When both reactions occur at the same rate the amount of product and the amount of reactant stay the same. Dynamic equilibrium has been reached

Figure 23.9 shows a model for dynamic equilibrium. At a house party there are 10 people. At 8pm there are 6 people in the kitchen and 4 in the living room. At 10pm there are 6 people in the kitchen and 4 in the living room. At each time there are the same numbers of people in each room, but due to people moving, they are different people. There is a dynamic equilibrium between the people in the living room and the kitchen. If there is a spill on the floor in the kitchen and everyone leaves the kitchen so the floor can be mopped, the equilibrium will be disturbed because the conditions have changed.

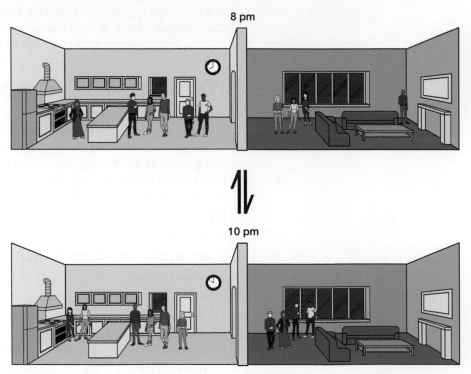

8 pm

10 pm

▲ **Figure 23.9** The movement of people at a house party can help us visualise dynamic equilibrium. There are the same amount of people in each room at 8 pm and 10 pm, however people have moved and there are different ones in each room

Test yourself ✏️

4 Ethanol can be made in the following reaction, which is reversible and reaches a state of dynamic equilibrium in a closed system. The energy change for the forward reaction has the value −43 kJ.

ethene + steam ⇌ ethanol

a) What is the energy change for the reverse reaction?
b) Name all the species present at equilibrium.
c) What can you say about the rate of the forward and reverse reactions at equilibrium?
d) What can you say about the amount of reactants and products present?
e) What is a closed system?

5 Hydrogen can be made by reacting methane with steam, as shown in the equation below:

methane + steam ⇌ hydrogen + carbon monoxide

$$CH_4(g) + H_2O(g) \rightleftharpoons 3H_2(g) + CO(g)$$

This reaction is reversible and a dynamic equilibrium can be reached.

a) What is meant by the term 'equilibrium'?
b) State two necessary conditions for equilibrium to occur in a reversible reaction.

Practice questions

1 Carbon monoxide reacts with steam to form carbon dioxide and hydrogen in a reversible reaction. The energy change for the forward reaction is −42 kJ.

    carbon monoxide + steam ⇌ carbon dioxide + hydrogen

a) How does the word equation show that the reaction is reversible? *(1 mark)*

b) What is the energy change for the reverse reaction? *(1 mark)*

c) The reaction is carried out using a catalyst. What is a catalyst? *(2 marks)*

d) Write a balanced symbol equation for the reversible reaction. *(2 marks)*

e) Write the equation for the reverse reaction. Include state symbols *(2 marks)*

2 Ethene reacts with steam to form ethanol, in a reversible, exothermic reaction. A catalyst of phosphoric acid is used.

$$C_2H_4(g) + H_2O(g) \rightleftharpoons C_2H_5OH(g)$$

a) What is a reversible reaction? *(1 mark)*

b) What is an exothermic reaction? *(1 mark)*

24 Organic chemistry

What is organic chemistry? Most things in the party photograph are made from organic chemicals – the plastic tablecloth, cups and plates, the sugar in the marshmallows and cake, the clothes, even the children themselves. The common factor in each of these products is the element carbon, and it is the element on which organic compounds are based. In this chapter you will study some different organic compounds and how they react.

Organic chemistry

Organic materials are used in all areas of our lives. Pharmaceutical drugs, cosmetics, paints, clothes and plastics are all made of organic compounds. All organic compounds contain carbon. The chemistry of carbon is vast; there are in excess of ten million known carbon compounds. It is estimated that 300 000 new carbon compounds are discovered each year.

Carbon can form so many compounds because:

▶ a carbon atom can form bonds with other carbon atoms to make chains and rings

▶ a carbon atom can form a single or double bond to another carbon atom

▶ a carbon atom can bond with other atoms for example hydrogen, oxygen and chlorine.

Organic chemistry is the study of compounds containing C–H bonds.

▲ **Figure 24.1** Most medicinal drugs are organic compounds. Discovering, producing and trialling new medicines is the business of the pharmaceutical industry

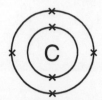

▲ **Figure 24.2** A carbon atom

Bonding in organic compounds

A carbon atom has four electrons in its outer shell (Figure 24.2) and forms four covalent bonds in all of its compounds.

The four bonds can be single bonds or sometimes a mixture of single bonds and double bonds as shown in Figure 24.3.

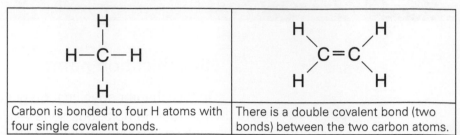

Carbon is bonded to four H atoms with four single covalent bonds.	There is a double covalent bond (two bonds) between the two carbon atoms.

▲ **Figure 24.3** Carbon can form single and double bonds with itself to form organic compounds

Homologous series

To make the study of the large number of carbon compounds more manageable carbon compounds are divided into families called homologous series.

A homologous series is a family of organic molecules that:

▶ have the same general formula
▶ show similar chemical properties
▶ show a gradation in their physical properties and
▶ differ by a 'CH_2' unit.

For GCSE, you will study 4 different homologous series – the alkanes, the alkenes, the alcohols and the carboxylic acids.

Naming organic compounds

In organic chemistry compounds are named according to the number of carbon atoms present. A prefix shows the length of the carbon chain:

▶ **Meth** means the organic compound contains 1 carbon atom
▶ **Eth** means the organic compound contains 2 carbon atoms
▶ **Prop** means the organic compound contains 3 carbon atoms
▶ **But** means the organic compound contains 4 carbon atoms.

The name of a compound also has a suffix which shows the homologous series it belongs to.

▶ The suffix for alkanes is 'ane '.
▶ The suffix for alkenes is 'ene'.
▶ The suffix for alcohols is 'anol'.
▶ The suffix for carboxylic acids is 'anoic acid'.

Monkeys	Eat	Peanut	Butter
Meth	Eth	Prop	But
1C	2C	3C	4C

▲ **Figure 24.4** You can use mnemonics to help you remember the prefixes for organic compounds

▲ **Figure 24.5** One cow can release 500 dm³ of methane a day! Cows and sheep are thought to be responsible for one fifth of global methane production

For example:

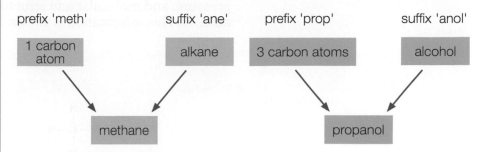

prefix 'meth'   suffix 'ane'   prefix 'prop'   suffix 'anol'

| 1 carbon atom | | alkane | 3 carbon atoms | | alcohol |

methane   propanol

General formula

All homologous series are represented by a general formula. A general formula is one involving a variable number n, which gives the number of carbon atoms present in the chain. It can be used to work out the molecular formula of any compound in the homologous series.

For the alkanes the general formula is C_nH_{2n+2}.

For the alkane propane, which has three carbon atoms, $n = 3$, then there are 8 hydrogen atoms $((2 \times 3) + 2)$ and the molecular formula is C_3H_8.

Show you can

A carboxylic acid **A** which contains 1 carbon atom reacts with an alcohol **B** which contains 4 carbon atoms.

Name **A** and **B**.

Alkanes

The alkanes are a homologous series which can be represented by the general formula C_nH_{2n+2}.

A **structural formula** represents the bonds in a compound as lines. To draw the structural formula of butane, for example, follow the steps below:

▶ 'But' means 4, so draw four carbon atoms bonded in a chain.

$$C—C—C—C$$

▶ Then draw in bonds so that each carbon has four bonds in total.

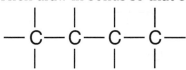

▶ Now draw hydrogen atoms at the end of each bond.

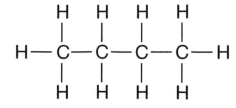

You need to know the names, state at room temperature and pressure, and molecular and structural formula of the first four alkanes. This information is shown in Table 24.1.

Table 24.1

Number of carbon atoms (n)	Name	Molecular formula	Structural formula	Colour and state at room temperature and pressure
1	methane	CH_4	H—C—H (with H above and below)	colourless gas
2	ethane	C_2H_6	H—C—C—H	colourless gas
3	propane	C_3H_8	H—C—C—C—H	colourless gas
4	butane	C_4H_{10}	H—C—C—C—C—H	colourless gas

The alkanes are all hydrocarbons. **A hydrocarbon is a molecule consisting of carbon and hydrogen atoms only.**

The alkanes are also saturated molecules. **Saturated (in the context of organic chemistry) means that all the carbon–carbon bonds are single.**

Tip

A hydrocarbon is a compound.

Show you can ?

An alkane has 5 carbon atoms.

1 Write its molecular and structural formula.
2 How many covalent bonds are in the molecule?

Test yourself

5 What is the general formula of alkanes?
6 What is a hydrocarbon?
7 Name and give the molecular formula of:
  a)
    H—C—C—C—C—H
  b)
    H—C—H

8 How many covalent bonds are there in one ethane molecule?
9 In an alkane how many covalent bonds are attached to every:
  a) carbon atom
  b) hydrogen atom.
10 Hexane is an alkane containing six carbon atoms.
  a) Alkanes are saturated hydrocarbons. What does saturated mean in this context?
  b) Give the molecular formula of hexane.
  c) Draw the structural formula of hexane.

Reactions of alkanes

Some organic molecules have a reactive group which is responsible for its chemical properties. **A** functional group **is a reactive group in a molecule.**

Alkanes do not have a functional group and so they are a less reactive group of organic molecules. The main reaction of alkanes is combustion.

1 Complete combustion

Combustion **is the reaction of fuels with oxygen, forming oxides and releasing heat energy.**

When alkanes burn, they react with oxygen. Complete combustion occurs if there is a **plentiful** supply of oxygen from the air. The products are **carbon dioxide** and **water**, and energy is released. An orange flame is observed in the reaction (the higher the percentage of carbon present, the more orange the flame.)

Example

Write a balanced equation for the complete combustion of methane, CH_4.

Word equation: methane + oxygen $\rightarrow$ carbon dioxide + water

Answer

Unbalanced equation: $CH_4 + O_2 \rightarrow CO_2 + H_2O$

CO_2: there is 1 C atom in CH_4 so there will be $1CO_2$ formed

H_2O: there are 4 H atoms in CH_4 so there will be $2H_2O$ formed

O_2: $CO_2 + 2H_2O$ contains 4 O atoms and so $2O_2$ needed

Balanced equation: $CH_4 + 2O_2 \rightarrow CO_2 + 2H_2O$

Example

Write a balanced equation for the complete combustion of butane, C_4H_{10}.

Word equation: butane + oxygen $\rightarrow$ carbon dioxide + water

Answer

Unbalanced equation: $C_4H_{10} + O_2 \rightarrow CO_2 + H_2O$

CO_2: there are 4 C atoms in C_4H_{10} so there will be 4 CO_2 formed

H_2O: there are 10 H atoms in C_4H_{10} so there will be 5 H_2O formed

O_2: 4 CO_2 + 5 H_2O contains 13 O atoms and so $6\frac{1}{2}$ O_2 needed

As there is a half in the balancing number equation you can double all the balancing numbers.

Balanced equation: $2C_4H_{10} + 13O_2 \rightarrow 8CO_2 + 10H_2O$

Testing for the products of combustion

To test for **carbon dioxide**, bubble the combustion products through **colourless limewater**. If it becomes **milky**, carbon dioxide is present.

To test for water, condense the vapour by cooling, and add to **white anhydrous copper(II) sulfate**. If it changes to **blue**, water is present.

Figure 24.6 Fire breathing performers use a fuel made of alkanes

Tip

If you are asked to include state symbols in combustion equations, all reactants and products are gases – the water is formed as water vapour.

Figure 24.7 Most gas barbecues use propane or butane gas

▲ **Figure 24.8** A carbon monoxide detector

▲ **Figure 24.9** It is thought that the greenhouse effect has caused an increase in the Earth's temperature. This has led to melting of some ice caps, which causes flooding

Tip

Acid rain reacts with limestone buildings. To use the term 'erodes', which implies a physical process, is incorrect.

▲ **Figure 24.10** This limestone statue outside Leeds Town Hall has been damaged by acid rain

2 Incomplete combustion

If there is **limited oxygen** present, then alkanes undergo incomplete combustion to give **carbon monoxide and water,** and release energy. Sometimes soot (carbon) is also produced, particularly with larger alkanes, but you do not need to include soot in equations.

For example:

ethane + limited oxygen → carbon monoxide + water

$$C_2H_6(g) + 2\tfrac{1}{2}O_2(g) \rightarrow 2CO(g) + 3H_2O(g)$$

propane + limited oxygen → carbon monoxide + water

$$C_3H_8(g) + 3\tfrac{1}{2}O_2(g) \rightarrow 3CO(g) + 4H_2O(g)$$

As there is a half in the balancing numbers in the equation it is usual to double all the values:

$$2C_3H_8(g) + 7O_2(g) \rightarrow 6CO(g) + 8H_2O(g)$$

Carbon monoxide is a toxic gas which prevents haemoglobin in the blood carrying oxygen, hence incomplete combustion is dangerous. Symptoms of carbon monoxide poisoning include headache, dizziness, nausea, fatigue, and lack of concentration. Exposure to high concentrations can cause unconsciousness and death.

Common atmospheric pollutants and their sources

The combustion of fuels is a major source of atmospheric pollution. Table 24.2 shows some of the pollutants, how they are formed and the problems they cause.

Table 24.2

Pollutant	How it is formed?	Problems it causes
carbon dioxide	complete combustion of hydrocarbon fuels	Carbon dioxide is a greenhouse gas and leads to the greenhouse effect – it absorbs infrared radiation given off by the Earth and causes the Earth's surface to warm leading to: • sea level rises • flooding • climate change.
carbon monoxide	incomplete combustion of fuels	Carbon monoxide is a toxic gas as it combines with haemoglobin in the blood, reducing its capacity to carry oxygen.
soot (carbon particles)	incomplete combustion of fuels	The carbon particles pollute the air and can cause lung damage and respiratory problems.
sulfur dioxide leading to acid rain	Many fuels contain sulfur impurities which burn and produce acidic sulfur dioxide. $S + O_2 \rightarrow SO_2$ The sulfur dioxide reacts with water in the atmosphere to form sulfurous acid which falls as acid rain. $H_2O + SO_2 \rightarrow H_2SO_3$	Acid rain can: • damage buildings, especially limestone buildings • damage vegetation • kill fish in lakes and rivers.

Test yourself

11 Name the products for:
  a) complete combustion of ethane
  b) incomplete combustion of propane
  c) butane burning in excess oxygen
  d) methane burning in limited oxygen.

12 Write balanced symbol equations for:
  a) ethane burning completely in oxygen
  b) propane burning completely in oxgyen
  c) butane burning in limited oxygen
  d) methane burning in limited oxygen
  e) butane burning completely in oxygen.

13 Complete and balance the equations for complete combustion for:
  a) $C_8H_{18} + O_2 \rightarrow$
  b) $C_{10}H_{22} + O_2 \rightarrow$
  c) $C_6H_{14} + O_2 \rightarrow$

14 What is the test for carbon dioxide?

15 What is the test for water?

Tip

Acid rain can be prevented by removing the sulfur impurity from coal, or passing acidic sulfur dioxide through an alkali.

▲ **Figure 24.11** Acid rain is a weak acid. It destroys trees by damaging the leaves and causing defoliation

▲ **Figure 24.12** The pH of this lake is low due to acid rain, causing the fish living in it to die

Show you can

Burning petrol in car engines can lead to the formation of acid rain.

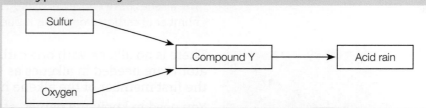

▲ **Figure 24.13**

1 What is the formula of compound Y?
2 How is acid rain formed from compound Y?
3 Suggest one way in which acid rain could be prevented.
4 State two environmental problems caused by acid rain.

Practical activity

The experiment shown in the diagram was carried out to investigate the products of the combustion of a hydrocarbon candle wax. In the experiment the products of the combustion were drawn through the apparatus by the vacuum pump.

a) Explain why a colourless liquid forms at the bottom of the U tube.

b) The colourless liquid in the U tube changes white anhydrous copper(II) sulfate to blue hydrated copper(II) sulfate. Suggest the name of the colourless liquid.

c) Solution A was clear and colourless at the start of the experiment and it slowly became milky. Identify solution A and explain why it became milky.

d) From the results of this experiment identify two of the combustion products formed when the hydrocarbon burns completely.

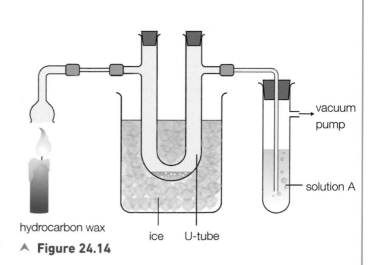

hydrocarbon wax
ice   U-tube
▲ **Figure 24.14**

e) Some solid black particles were found on the apparatus at the end of the experiment. Suggest the identity of these particles and suggest how they are formed.

f) The experiment was repeated using a different fuel, which contained sulfur as an impurity.

   i) Name the gas formed when sulfur is burned in air.

   ii) Sulfur dioxide gas is acidic. Suggest what happens to it as it passes through the apparatus.

   iii) Suggest why burning sulfur-containing fuels is a problem for the environment.

Alkenes

The alkenes are a homologous series which can be represented by the general formula C_nH_{2n}.

The **functional group** for alkenes is a **carbon-carbon double** bond (C=C). Due to this functional group, alkenes are more reactive than alkanes.

> **Tip**
>
> The alkenes have two fewer hydrogen atoms than alkanes, with the same number of carbon atoms. This is due to the carbon-carbon double bond.

There is no alkene with one carbon atom because at least two carbon atoms are needed in alkenes as they all contain a C=C bond. Hence, the first member of the alkene homologous series is ethene.

You need to know the names, state at room temperature and pressure, and molecular and structural formula of the four alkenes shown in Table 24.3.

> **Tip**
>
> Alkenes are hydrocarbons, molecules consisting of carbon and hydrogen only.

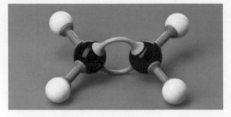

▲ **Figure 24.15** A model of an ethene molecule

Table 24.3

Number of carbon atoms (n)	Name	Molecular formula	Structural formula	Colour and state at room temperature and pressure
2	ethene	C_2H_4		colourless gas
3	propene	C_3H_6		colourless gas
4	but-1-ene	C_4H_8		colourless gas
4	but-2-ene	C_4H_8		colourless gas

Notice that but-1-ene and but-2-ene have the same molecular formula C_4H_8, but their structural formula is different because the carbon-carbon double bond is in a different position. The numbers in the names but-1-ene and but-2-ene give the position of the carbon-carbon double bond. The number 1 means the double bond starts at the first carbon. Note that either end of the molecule could be designated as carbon 1, as shown by the red numbers.

$$H-C_1=C_2-C_3-C_4-H$$

$$H-C_4-C_3-C_2=C_1$$

When drawing organic molecules it does not usually matter what angle you draw the atoms at or which way round you draw the molecules. For example, the following four structures are all correct structures of propene. Notice that the double bond can only be in position one and so numbers are not necessary in the name.

$$H-C-C=C-H \qquad H-C-C=C \qquad H-C=C-C-H \qquad C=C-C-H$$

To draw the structural formula of but-2-ene, for example, follow the steps below:

▶ 'But' means 4 so draw four carbon atoms bonded in a chain:

$$C-C-C-C$$

▶ Then draw in the functional group (C=C) in position 2 (but-2-ene):

$$C-C=C-C$$

▶ Now add bonds so that every carbon has four bonds:

$$-C-C=C-C-$$

▶ Add hydrogen atoms to each bond:

$$H-C-C=C-C-H$$

Show you can

Pentene is an alkene containing five carbon atoms.

1 Give the molecular formula of pentene.

2 Suggest a structural formula for pentene.

Test yourself

16 The general formula of alkenes is C_nH_{2n}.
  a) If $n = 3$ what is the molecular formula and the name of the alkene?
  b) If $n = 8$ what is the molecular formula?
  c) If $n = 2$ what is the molecular formula and the name of the alkene?
  d) If $n = 10$ what is the molecular formula?

17 Name and give the molecular formula of:

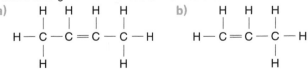

a) $$H-C-C=C-C-H$$ \qquad b) $$H-C=C-C-H$$

18 a) How many covalent bonds are there in one propene molecule?
  b) Why is propene a hydrocarbon?

Ⓗ

Reactions of alkenes

1 Complete combustion

Alkenes, like alkanes, will completely combust in a plentiful supply of oxygen from the air to produce carbon dioxide and water. They burn with an orange flame.

For example:

ethene + oxygen → carbon dioxide + water

$$C_2H_4(g) + 3O_2(g) \rightarrow 2CO_2(g) + 2H_2O(g)$$

propene + oxygen → carbon dioxide + water

$$C_3H_6(g) + 4\tfrac{1}{2}O_2(g) \rightarrow 3CO_2(g) + 3H_2O(g)$$

As there is a half in the balancing numbers it is usual to double all their values.

$$2C_3H_6(g) + 9O_2(g) \rightarrow 6CO_2(g) + 6H_2O(g)$$

2 Incomplete combustion

If there is **limited oxygen** present, then alkenes undergo incomplete combustion to give **carbon monoxide and water**, and release energy. Sometimes soot (carbon) is also produced, particularly with larger alkenes, but again, you do not need to include soot in equations. The flame is smoky due to soot.

For example:

ethene + limited oxygen → carbon monoxide + water

$$C_2H_4(g) + 2O_2(g) \rightarrow 2CO(g) + 2H_2O(g)$$

propene + limited oxygen → carbon monoxide + water

$$C_3H_8(g) + 3\tfrac{1}{2}O_2(g) \rightarrow 3CO(g) + 4H_2O(g)$$

As there is a half in the balancing numbers it is usual to double all their values:

$$2C_3H_8(g) + 7O_2(g) \rightarrow 6CO(g) + 8H_2O(g)$$

Alkenes are not usually used as fuels, as they are very valuable and they can be used to make polymers or as a starting material for many other chemicals.

3 Addition reactions

Alkenes are unsaturated. **Unsaturated (in the context of organic chemistry) means that the molecule contains one or more carbon-carbon double bonds.**

An **addition reaction** is one in which two molecules react to form one product. Alkenes can undergo addition reactions because the C=C bond breaks open and atoms add on to the two carbon atoms. The product contains a C–C single bond and is saturated.

▲ **Figure 24.16** A sooty flame

alkene molecule   molecule that reacts with alkene   the C=C bond opens up and the molecule adds onto these C atoms

Key
○ represent the atoms/groups bonded to the two C atoms in the C=C double bond
● represent the atoms/groups in the molecule adding onto the C=C double bond

▲ **Figure 24.17**

Addition reaction of alkenes with bromine

In this reaction the C=C breaks and one bromine atom adds on to each of the carbon atoms in the double bond. We say the bromine 'adds across the double bond'. The reaction with ethene is shown:

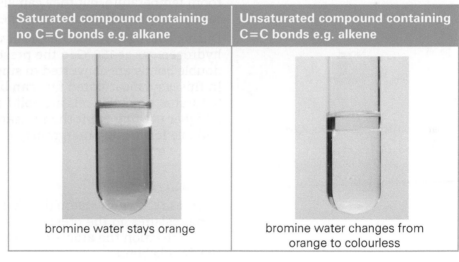

Using molecular formulae: $C_2H_4 + Br_2 \rightarrow C_2H_4Br_2$

Condition: room temperature

This reaction is used to **test for the presence of a C=C bond**.

Test:

▶ If the organic substance is a liquid, shake with bromine water; if a gas, bubble the gas into bromine water.

Result:

▶ **Orange bromine water** changes to **colourless** in the presence of a C=C bond. This is shown in Table 24.4.

Table 24.4

Saturated compound containing no C=C bonds e.g. alkane	Unsaturated compound containing C=C bonds e.g. alkene
bromine water stays orange	bromine water changes from orange to colourless

Bromine water is a solution of bromine in water. It has an orange colour due to the dissolved bromine molecules (Br_2). On reacting with a C=C the bromine atoms add on to the double bond, and are removed from the bromine water, hence it changes to colourless.

Tip

The orange bromine water reacts and becomes colourless. Remember not to confuse clear and colourless – clear means see-through and colourless means there is no colour.

Tip

This reaction should be carried out in a fume cupboard.

Tip

Note that you must state the result in each fat to fully answer this question.

Example

How would you experimentally decide if a sample of liquid fat is saturated or unsaturated?

Answer

• Shake with bromine water.
• If the bromine water changes from orange to colourless the fat is unsaturated.
• If the bromine water stays orange the fat is saturated.

H

Addition reactions with hydrogen

In this reaction the C=C breaks and one hydrogen atom adds on to each of the carbon atoms in the double bond. The hydrogen 'adds across the carbon-carbon double bond'. An alkane is produced.

The reaction with ethene is shown:

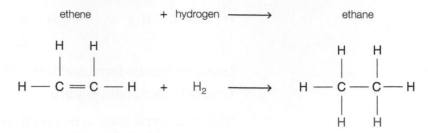

Using molecular formulae: $C_2H_4 + H_2 \rightarrow C_2H_6$

Condition: temperature 150 °C
           catalyst of nickel

This reaction is called **hydrogenation** because hydrogen is added across the carbon-carbon double bond.

Unsaturated vegetable oils (these contain C=C) tend to be liquid at room temperature, but they can also be 'hardened' by hydrogenation, to make them solid at room temperature.

During hydrogenation, the liquid vegetable oils are reacted with hydrogen gas at 150 °C in the presence of a nickel catalyst. The double bonds are converted to single bonds by the hydrogenation. In this way unsaturated fats can be made into saturated fats. Saturated vegetable oils are solid at room temperature, and have a higher melting point than unsaturated oils. This makes them suitable for making margarine, or for commercial use in the making of cakes and pastry.

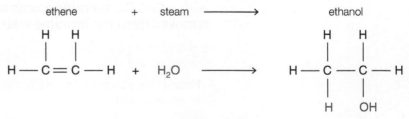

Unsaturated fat (liquid)   Saturated fat (solid)

▲ **Figure 24.18** Unsaturated oils can be hydrogenated to form solids at room temperature

Addition reaction with steam

Alkenes react with steam ($H_2O(g)$) in industry at high temperatures and pressures, in the presence of a catalyst to produce alcohols. In this reaction the atoms of H_2O add on to the carbon atoms in the double bond.

Ethene reacts with steam to produce ethanol.

Using molecular formulae: $C_2H_4(g) + H_2O(g) \rightarrow C_2H_5OH(l)$

Condition: catalyst and high temperature

This reaction is called hydration because water is being added across the C=C double bond.

Tip ↻

Writing an equation including structural formula for the organic molecules helps you understand the reaction, and is acceptable in an examination.

Show you can

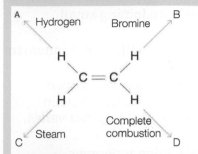

▲ **Figure 24.19**

1 Write the formula of the products of the reaction of ethene with hydrogen, bromine, steam and plentiful oxygen.

2 What is observed in reaction B?

3 Which reaction A, B, C or D is hydration?

4 Which reaction A, B, C or D is hydrogenation?

5 Suggest the structure of the product if propene is reacted with hydrogen.

6 Suggest the structure of the product if propene is reacted with bromine.

Test yourself

19 Name the products for:
   a) complete combustion of but-1-ene
   b) incomplete combustion of but-2-ene.

20 Write balanced symbol equations for:
   a) ethene burning completely in oxygen
   b) propene burning completely in oxgyen
   c) butene burning in limited oxygen
   d) ethene burning in limited oxygen

21 Complete and balance the equations for complete combustion for:
   a) $C_8H_{16} + O_2 \rightarrow$
   b) $C_{10}H_{20} + O_2 \rightarrow$
   c) $C_6H_{12} + O_2 \rightarrow$

22 Alkenes are unsaturated hydrocarbons.
   a) What does the term unsaturated mean?
   b) What does the term hydrocarbon mean?
   c) Explain why alkenes are much more reactive than alkanes.
   d) How would you experimentally decide if a sample is chloroethane or chloroethene?

23 Hexene is an alkene with six carbon atoms.
   a) Give the molecular formula of hexene.
   b) Describe what would be observed if some bromine water was added to some hexene.
   c) Name the product when hexene reacts with hydrogen.

24 a) What is an addition reaction?
   b) Write symbol equations for the following addition reactions of alkenes.
      i) ethene + H_2
      ii) ethene + H_2O
      iii)ethene + Br_2
   c) Name the product in b) i).
   d) Name the product in b) ii).
   e) In which of the reactions in part b) is/are saturated molecules produced?
   f) Which of the reactions in part b) is/are hydration reactions?
   g) Name the catalyst used in b) i).

Polymers

Paper clips can be joined together to make a long chain as shown in Figure 24.20. Simple molecules containing a C=C double bond can also be joined together to make a long chain (polymer).

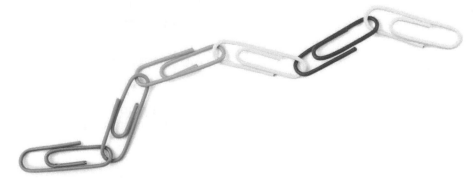

▲ **Figure 24.20** Single paper clips can be joined to make a chain. Single molecules (monomers) can be joined to make a polymer

H

▲ **Figure 24.21** Poly(chloroethene) (PVC) is used to make vinyl LP records

When alkene molecules react with other alkene molecules a polymer is formed.

A **polymer** is a long chain molecule made from joining small molecules together.

A **monomer** is a small molecule that combines with other monomers to make a polymer.

In the reaction between monomers the C=C double bonds open up and the molecules join onto each other to make a long chain molecule. The exact number of molecules that join together varies, but is likely to be several hundred.

Addition polymerisation is the process of joining monomer molecules together to form a long chain molecule.

The word poly means many and the name of a polymer is the word poly followed by the name of the monomer in brackets. For example, lots of ethene molecules join together to make the polymer poly(ethene), better known as polythene and many chloroethene molecules join together to make poly(chloroethene).

Figure 24.22 shows the polymerisation of ethene to form poly(ethene). The double bond breaks and the repeating units add together. You will notice that when monomers with C=C join together in addition polymerisation, only one product, the polymer is made. It does not have a carbon-carbon double bond and is unreactive.

lots of small molecules (monomers)

join together to form a long chain molecule (polymer)

▲ **Figure 24.22** Ethene monomers join together to form poly(ethene)

The chemical equation for this reaction is:

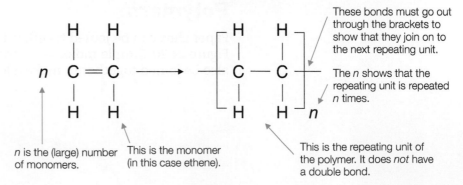

n is the (large) number of monomers.

This is the monomer (in this case ethene).

These bonds must go out through the brackets to show that they join on to the next repeating unit.

The n shows that the repeating unit is repeated n times.

This is the repeating unit of the polymer. It does *not* have a double bond.

The repeating unit is the structure that repeats many times, and the 'open' bonds on the end show this:

The polymer is made up of many repeats of the repeating unit and is represented as:

$$\left[\begin{array}{c} \begin{array}{cc} H & H \\ | & | \\ -C - C - \\ | & | \\ H & H \end{array} \end{array} \right]_n$$

You may be asked for example to draw a polymer showing several repeating units. To do this simply link together the required number of repeating units.

Example

Draw the structure of polythene showing 3 repeating units.

Answer

$$\begin{array}{cccccc} H & H & H & H & H & H \\ | & | & | & | & | & | \\ -C - C - & C - C - & C - C - \\ | & | & | & | & | & | \\ H & H & H & H & H & H \end{array}$$
1 unit   1 unit   1 unit

You must know the structures of the polymers, and the equations for their formation, as shown in Table 24.5.

Table 24.5

Monomer	Repeating unit of polymer	Polymer	Equation for formation of polymer
H H | | C = C | | H H ethene	H H | | —C — C— | | H H	H H ┌| |┐ ┼C—C┼ └| |┘n H H poly(ethene) common name polythene	H H | | n C = C ⟶ | | H H ethene (monomer) ┌H H┐ ┼C — C┼ └H H┘n poly(ethene) (polymer)
H Cl | | C = C | | H H chloroethene (vinyl chloride)	H Cl | | —C — C— | | H H	H Cl ┌| |┐ ┼C—C┼ └| |┘n H H poly(chloroethene)/ poly(vinylchloride) common name PVC	H Cl | | n C = C ⟶ | | H H chloroethene (monomer) ┌H Cl┐ ┼C — C┼ └H H┘n poly(chloroethene) (polymer)

You may be asked to draw the structure of a polymer, or write a polymerisation equation for any given monomer. To help with writing an equation for polymerisation use the general equation:

$$n \quad \begin{array}{c} | \quad | \\ C = C \\ | \quad | \end{array} \quad \longrightarrow \quad \left[\begin{array}{c} | \quad | \\ C - C \\ | \quad | \end{array} \right]_n$$

Adapt this equation by drawing in the different atoms or groups shown in the monomer.

Example

Write an equation for the polymer formed from the monomer shown.

$$\begin{array}{cc} H & CH_3 \\ | & | \\ C & = C \\ | & | \\ H & H \end{array}$$

Answer

First write the general polymerisation equation.

Then add in the different atoms or groups from the monomer – H and CH_3 in the correct positions.

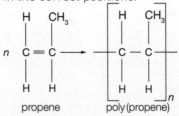

propene   poly (propene)

You must also be able to deduce the structure of a monomer, from the given structure of a polymer.

Example

The structure below shows part of a polymer chain. Draw the monomer and the repeating unit.

$$\begin{array}{cccccc} H & C_2H_5 & H & C_2H_5 & H & C_2H_5 \\ | & | & | & | & | & | \\ -C & -C & -C & -C & -C & -C- \\ | & | & | & | & | & | \\ H & H & H & H & H & H \end{array}$$

Answer

It is important to be able to identify the repeating unit from the polymer, it is circled below.

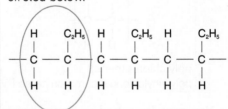

The repeating unit is:

$$\begin{array}{cc} H & C_2H_5 \\ | & | \\ -C & -C- \\ | & | \\ H & H \end{array}$$

To draw the monomer, add a double bond and remove the open bonds at the ends:

$$\begin{array}{cc} H & C_2H_5 \\ | & | \\ C & = C \\ | & | \\ H & H \end{array}$$

Tip

When asked to draw the structure of a repeating unit of a molecule you should not draw brackets around it and you should not put an *n* beside it – it is not the polymer.

Test yourself

25 a) What is a polymer?

   b) What is a monomer?

   c) What functional group do monomers that form addition polymers contain?

   d) Write an equation for the polymerisation of poly(chloroethene).

   e) Write the structure of the repeating unit in poly(ethene).

   f) Suggest the name of the addition polymer formed from styrene.

   g) Suggest the name of the monomer used to make poly(bromoethene).

26 Write an equation for the polymerisation of the monomer:

27 Draw three repeating units of the polymer made from the monomer shown:

28 Copy and complete the table.

Table 24.6

Structure of monomer	Repeating unit of polymer	Structure of polymer
CH₃ H C=C CH₃ H		
		F F C—C H H n

Show you can ?

Part of the structure of a polymer is shown below.

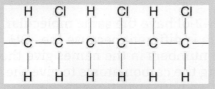

▲ **Figure 24.24**

1 Draw the repeating unit of this polymer.

2 Draw the structure of the monomer used to make this polymer.

3 Explain if this polymer is a hydrocarbon.

Alcohols

The alcohols are a homologous series which can be represented by the general formula $C_nH_{2n+1}OH$.

The **functional group** for alcohols is the **hydroxyl** group (–OH).

> **Tip**
>
> Do not confuse the hydroxyl group (–OH) with the hydroxide ion OH⁻.

The alcohols are not hydrocarbons because they also contain oxygen, not just hydrogen and carbon.

You need to know the names, state at room temperature and pressure, and molecular and structural formula of the four alcohols shown in Table 24.7.

▲ **Figure 24.25** A major use of ethanol is in alcoholic drinks

Table 24.7

Number of carbon atoms	Name	Molecular formula	Structural formula	Colour and state at room temperature and pressure
1	methanol	CH₃OH	H—C—OH with H above and below	colourless liquid
2	ethanol	C₂H₅OH	H—C—C—OH	colourless liquid
3	propan-1-ol	C₃H₇OH	H—C—C—C—OH	colourless liquid
4	propan-2-ol	C₃H₇OH	H—C—C—C—H with OH on middle	colourless liquid

Notice that propan-1-ol and propan-2-ol have the same molecular formula, but their structural formula is different because the –OH group is in a different position. The numbers in the names give the position of the OH. When numbering the carbon atoms to give the position you can start at either end, so both structures shown in Figure 24.26 are propan-1-ol.

$$H-C_3-C_2-C_1-O-H \qquad H-O-C_1-C_2-C_3-H$$

▲ **Figure 24.26** Both of these structures are propan-1-ol

To draw the structural formula of propan-2-ol, for example, follow the steps below:

▶ 'prop' means 3 so draw 3 carbon atoms bonded in a chain.

$$C - C - C$$

▶ It is an alcohol and has the –OH group. Draw the functional group in position 2.

$$\overset{OH}{\underset{}{C - C - C}}$$

Tip

The C always bonds to the O in the OH. It does not bond to the H.

▶ Now add bonds so that every carbon has four bonds.

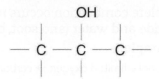

Tip

If you are asked to draw the structure of the alcohol and show ALL the bonds, you must show the bond between the oxygen and hydrogen: for example, Figure 24.27.

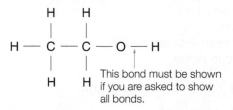

This bond must be shown if you are asked to show all bonds.

▲ **Figure 24.27** Structural formula of ethanol showing all bonds

▶ Add hydrogen atoms to each bond.

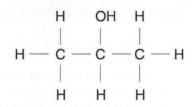

Combustion of alcohols

a) Complete combustion

Complete combustion occurs if there is a **plentiful supply of oxygen** from the air. The products are **carbon dioxide and water**, and energy is released. A clean (non smoky) blue flame with an orange tip is observed.

For example:

ethanol + oxygen → carbon dioxide + water

$C_2H_5OH(l) + 3O_2(g) \rightarrow 2CO_2(g) + 3H_2O(g)$

propanol + oxygen → carbon dioxide + water

$2C_3H_7OH(l) + 9O_2(g) \rightarrow 6CO_2(g) + 8H_2O(g)$

Tip

When balancing equations for the complete combustion of alcohols remember that there is an oxygen atom in the alcohol as well as the atoms that come from O_2.

▲ **Figure 24.28** A methylated spirits camp stove uses alcohol as fuel

b) Incomplete combustion

Incomplete combustion occurs **in limited** oxygen and produces **carbon monoxide** and **water** (and soot, but soot is not needed in equation).

For example:

ethanol + limited oxygen → carbon monoxide + water

$C_2H_5OH \text{ (l)} + 2O_2 \text{ (g)} \rightarrow 2CO \text{ (g)} + 3H_2O \text{ (g)}$

propanol + limited oxygen → carbon monoxide + water

$C_3H_7OH \text{ (l)} + 3O_2 \text{ (g)} \rightarrow 3CO \text{ (g)} + 4H_2O \text{ (g)}$

> **Tip** ↻
>
> When adding state symbols remember that the alcohols are liquid and the water formed will be formed as vapour.

Test yourself ✎

29 a) What is meant by the term functional group?
   b) Name the functional group in propan-1-ol.
   c) Explain if ethanol is a hydrocarbon.
   d) What homologous series does propan-2-ol belong to?
   e) State four features of a homologous series.

30 Name the following:

a) 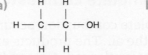 b) 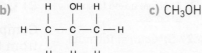 c) CH_3OH

31 When ethanol is burnt in excess oxygen, the products are bubbled into limewater.
   a) What is observed in the limewater?
   b) Name the products when ethanol is burnt in excess oxygen.
   c) Write a balanced symbol equation.
   d) Complete the balanced symbol equation for the complete combustion of butanol.

     $C_4H_9OH + O_2 \rightarrow$

   e) Name three products which may be formed when propan-2-ol undergoes incomplete combustion.

▲ **Figure 24.29** Wine has been produced by fermentation for 9000 years

Making ethanol

1 By fermentation

Ethanol is made by fermentation. **Fermentation is the breakdown of sugars to produce ethanol and carbon dioxide** under the conditions:

▶ the sugars are dissolved in solution

▶ in the presence of yeast

▶ at a warm temperature (not above 37 °C)

▶ in the absence of air.

The reaction takes several days and the ethanol formed must be separated from the other substances in the flask by distillation.

Fermentation produces an ethanol solution of approximately 12-15%. A higher concentration of ethanol would kill the yeast.

Fractional distillation can be used to produce a more concentrated solution of ethanol such as those in spirits such as vodka.

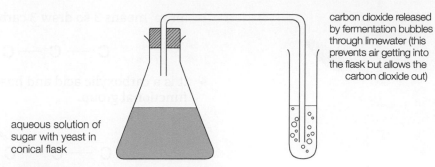

▲ **Figure 24.30** Fermentation in the laboratory

aqueous solution of
sugar with yeast in
conical flask

carbon dioxide released
by fermentation bubbles
through limewater (this
prevents air getting into
the flask but allows the
carbon dioxide out)

2 From ethene

Ethanol can be produced by reacting ethene and steam as shown on page 290.

Test yourself

32 **a)** What is fermentation?
 b) State three conditions for fermentation.
 c) Write an equation for the production of ethanol from ethene.

Show you can ?

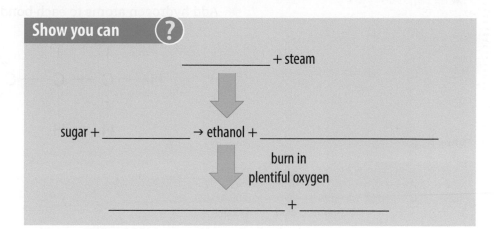

Carboxylic acids

The carboxylic acids have the **carboxyl (–COOH)** functional group which is written:

They are not hydrocarbons because they contain oxygen, and not just hydrogen and carbon only.

You need to know the names, state at room temperature and pressure, and molecular and structural formula of the four carboxylic acids shown in Table 24.8.

To draw the structural formula of propanoic acid for example, follow the steps listed on the next page:

▲ **Figure 24.31** Butanoic acid is responsible for the smell of rancid butter and also gives vomit its unpleasant smell. Perhaps not surprisingly, this acid is used in stink bombs

Tip

If you are asked to show all the bonds in a structure, show the bond in the –O–H group.

▶ 'prop' means 3 so draw 3 carbon atoms bonded in a chain.

$$C — C — C$$

▶ It is a carboxylic acid and has the –COOH group. Draw the functional group.

▶ Now add bonds so that every carbon has four bonds.

▶ Add hydrogen atoms to each bond.

Table 24.8

Number of carbon atoms	Name	Molecular formula	Structural formula	Colour and state at room temperature and pressure
1	methanoic acid	HCOOH		colourless liquid
2	ethanoic acid	CH_3COOH		colourless liquid
3	propanoic acid	C_2H_5COOH		colourless liquid
4	butanoic acid	C_3H_7COOH		colourless liquid

Salts of carboxylic acids

Carboxylic acids are weak acids.

This means that they are acids that are only partially ionised in solution.

For example, for ethanoic acid:

$$CH_3COOH \rightleftharpoons CH_3COO^- + H^+$$

Carboxylic acids react with bases, metals and carbonates to form salts. For example, ethanoic acid forms salts called ethanoates and propanoic acid forms propanoate salts. Your data leaflet gives the following information about the ions found in carboxylic acid salts.

Table 24.9

Name	Symbol
methanoate	$HCOO^-$
ethanoate	CH_3COO^-
propanoate	$C_2H_5COO^-$
butanoate	$C_3H_7COO^-$

▲ **Figure 24.32** Vinegar is a dilute solution of ethanoic acid

When a salt is formed from an acid, the hydrogen of the COOH group is replaced by a metal or ammonium ion. For example:

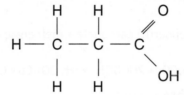

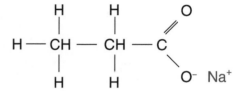

Sodium propanoate

You must be able to write the formulae of carboxylic acid salts. (See page 496 for more detail on writing formulae).

Tip

There is an ionic bond between the sodium and the propanoate group. Don't draw a line between the sodium and the oxygen. That would represent a covalent bond.

Tip

The charges do not have to be shown, but if they are *both* plus and minus must be given.

Example

Write the formula of sodium propanoate.

Answer

sodium propanoate

Symbol	Na	C_2H_5COO
Charge	+1	-1

Same

NaC_2H_5COO

Salts are named with the metal part first, however, by convention when the formula of carboxylic acid salts are written the metal comes last.

The formula is C_2H_5COONa, showing more clearly where the metal ion is bonded.

Example

Write the formula of magnesium ethanoate.

Answer

magnesium ethanoate

Symbol	Mg	CH_3COO
Charge	+2	-1

SWOP DROP

1   2

$Mg(CH_3COO)_2$

However, by convention it is written $(CH_3COO)_2Mg$.

Tip

Check out page 212 for more information on writing observations for acid reactions.

Reactions of carboxylic acids

Carboxylic acids take part in typical acid reactions – with carbonates, metals and bases to form salts.

With carbonates

carbonate + acid → salt + water + carbon dioxide

Carboxylic acids react with carbonates to release carbon dioxide, which changes colourless limewater to milky. This reaction can be used to **test for carboxylic acids**.

Example

sodium carbonate + ethanoic acid → sodium ethanoate + carbon dioxide + water

$Na_2CO_3 + 2CH_3COOH \rightarrow 2CH_3COONa + CO_2 + H_2O$

Observations

white solid sodium carbonate disappears

colourless solution produced

bubbles

heat released

Example

copper(II) carbonate + methanoic acid → copper(II) methanoate
$\qquad\qquad\qquad\qquad\qquad$ + carbon dioxide + water

$CuCO_3 + 2HCOOH \rightarrow (HCOO)_2Cu + CO_2 + H_2O$

Observations

green solid copper(II) carbonate disappears

bubbles

heat released

blue solution formed

heat released

With metals

Carboxylic acids react with metals to release hydrogen, which gives a pop when tested with a lighted splint.

metal + acid → salt + hydrogen

Example

magnesium + ethanoic acid → magnesium ethanoate + hydrogen

$Mg + 2CH_3COOH \rightarrow 2(CH_3COO)_2Mg + H_2$

Observations

bubbles

grey solid magnesium disappears

colourless solution produced

heat released

▲ **Figure 24.33** Magnesium reacting with ethanoic acid

With bases

base + acid → salt + water

Example

sodium hydroxide + ethanoic acid → sodium ethanoate + water

$NaOH + CH_3COOH → CH_3COONa + H_2O$

Observations
solution remains colourless
heat released

Example

magnesium hydroxide + propanoic acid → magnesium propanoate + water

$Mg(OH)_2 + 2C_2H_5COOH → (C_2H_5COO)_2Mg + 2H_2O$

Observations
solution remains colourless
heat released

Example

ammonia + acid → ammonium salt

ammonia + ethanoic acid → ammonium ethanoate

$CH_3COOH + NH_3 → CH_3COONH_4$

Observations
There is release of heat and the colourless solution remains.

Test yourself

33 Write formula for:
   a) sodium ethanoate
   b) calcium ethanoate
   c) magnesium methanoate
   d) aluminium methanoate
   e) copper(II) propanoate
   f) sodium propanoate.

34 Complete the word equations for:
   a) calcium + ethanoic acid →
   b) magnesium carbonate + methanoic acid →
   c) sodium hydroxide + propanoic acid →
   d) magnesium + methanoic acid →
   e) copper(II) carbonate + propanoic acid →
   f) sodium carbonate + propanoic acid →

35 Write balanced symbol equations for the reactions in question 35.
36 Give observations for the reactions in question 35.
37 How do you test for the gas produced in 35 a)?

Making and recording observations

Making and recording observations is an important skill in Chemistry. Important types of observations in chemistry, and notes on how to record these are shown in Table 24.10.

Table 24.10

Type of observation	Notes on recording observations	Examples
colour change	Always state the colour of the solution before the reaction and after.	• Bubbling an alkene into bromine water – the colour change is orange solution to a colourless solution.
bubbles produced	If a gas is produced, then bubbles are often observed in the liquid and the solid reactant disappears.	• When sodium carbonate reacts with an acid the observation is bubbles and the solid reactant disappears – note: writing that carbon dioxide is formed is not an observation.
heat produced	Often in a reaction the temperature changes – in an exothermic reaction it increases and in an endothermic reaction it decreases.	• When water is added to anhydrous copper sulfate the temperature increases. • When acids react with alkalis, the temperature increases.
precipitate produced	When two solutions mix, often an insoluble precipitate forms. Ensure you use the word precipitate in your observation – a common mistake is to write that the solution becomes cloudy. Also state the colour of the precipitate and the colour of the solution before adding the reagent.	• When barium chloride solution is added to a solution containing sulfate ions a white preciptate is formed in the colourless solution.
solubility of solids	When a spatula of a soluble solid is added to water, the observation is often that the solid dissolves to form a solution. Make sure you state the colour of the solution formed.	• Copper(II) sulfate crystals dissolve in water to produce a blue solution.
solubility of liquids	When a liquid is added to water always record if it is miscible or immiscible with water.	• Ethanol and water are miscible. The liquids in the right hand test tube are immiscible and form two layers. Those in the left are miscible, hence there are no layers.

Test yourself

38 In a laboratory experiment a student added some magnesium metal to copper(II) sulfate solution and recorded the observation that 'the magnesium became covered in copper and magnesium sulfate was formed.'
   a) Write a word equation for the reaction occurring.
   b) State and explain if the observations given by the student are correct.
39 Some hydrochloric acid was placed in a conical flask and a spatula of calcium carbonate powder was added.
   a) Write a balanced symbol equation for this reaction.
   b) State the observations which were noted.
40 Copy and complete Table 24.11:

Table 24.11

Reaction	Observations
ethanoic acid + sodium carbonate	
potassium iodide solution + silver nitrate solution	
bromine water + alkene	
hydrochloric acid + magnesium	
Acidified barium chloride + sulfuric acid	

Crude oil

What is crude oil?

The dark yellow black sticky liquid shown in Figure 24.34 is crude oil. It contains over 150 carbon based compounds and is the world's main source of hydrocarbons.

Crude oil (petroleum) **is a mixture containing mainly alkane hydrocarbons.**

Petrochemicals **are chemicals made from petroleum (crude oil) and natural gas.**

The petrochemical industry makes chemicals from the raw materials crude oil and natural gas. Crude oil is a **feedstock** (raw material) for the petrochemical industry.

Crude oil is a finite resource. **A finite resource is a resource that once used, cannot be replaced in a human lifetime.**

▲ **Figure 24.34** Crude oil

How is crude oil separated?

For crude oil to be useful, the hydrocarbons it contains have to be separated. The hydrocarbons have **different boiling points** and this difference is used to separate them by **fractional distillation** at an oil refinery (Figure 24.35).

This process separates the hydrocarbons into fractions. **A** fraction **is a mixture of molecules with similar boiling points**. In each fraction the hydrocarbons contain a similar number of carbon atoms.

In fractional distillation:

▶ The crude oil is heated and vaporised.

▶ The vaporised crude oil enters a fractionating tower which is hotter at the bottom and cooler at the top.

▶ The hydrocarbons cool as they rise up the tower and condense at different heights because they have different boiling points.

▶ Hydrocarbons with large molecules are collected near the bottom of the tower while those with small molecules are collected at the top.

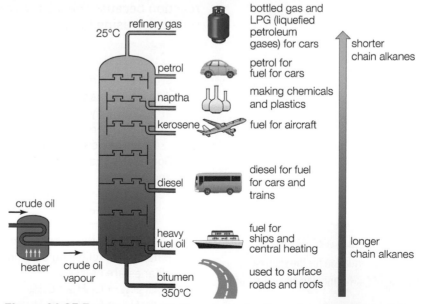

▲ **Figure 24.35** Fractional distillation of crude oil produces many useful products

The fractions from crude oil have different uses because they have different properties. Some names and uses of the fractions which you need to remember are:

▶ refinery gas for bottled gases
▶ petrol used as a fuel for cars
▶ naptha for manufacture of chemicals and plastics
▶ kerosene as a fuel for aircraft
▶ diesel as a fuel for cars and trains
▶ fuel oils used for fuels for ships
▶ bitumen used to surface roads and roofs.

Cracking

Shorter chain alkanes are in very high demand as fuels but longer chain alkanes are in less demand as they do not burn well and are less flammable. This means that there is a surplus of the longer chain alkanes, such as those in the bitumen fraction, from the fractional distillation of crude oil.

These longer alkanes can be broken down into the shorter, more useful alkanes by a process called cracking.

Cracking is the breakdown of larger saturated hydrocarbons (alkanes) into smaller more useful ones, some of which are unsaturated (alkenes).

The alkenes produced can be used as a starting material to make many other substances such as polymers.

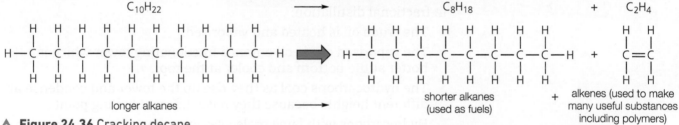

▲ **Figure 24.36** Cracking decane

Cracking can be carried out by heating. It is a thermal decomposition reaction because the alkanes are broken down into smaller molecules using heat.

Show you can

Balance the equation for the thermal cracking of $C_{10}H_{22}$ and give the values of y:

$$C_{10}H_{22} \rightarrow C_6H_{14} + \underline{\quad\quad} C_2H_y$$

Test yourself ✎

41 Crude oil is a mixture of hydrocarbons. It is separated by fractional distillation at an oil refinery.
  a) What is a hydrocarbon?
  b) What property of the hydrocarbons allows them to be separated in this way?
  c) Describe how fractional distillation separates the hydrocarbons.
42 Some long chain alkanes are cracked to produce shorter chain alkanes and alkenes.
  a) Give a reason why longer alkanes are in less demand as fuels than shorter alkanes.
  b) What is cracking?
  c) What does the term thermal decomposition mean?
43 Balance the following equations for reactions that can take place when alkanes are cracked.
  a) $C_{18}H_{38} \rightarrow C_{12}H_{26} + C_3H_6$
  b) $C_{20}H_{42} \rightarrow C_{12}H_{26} + C_2H_4$
  c) $C_{18}H_{38} \rightarrow C_{10}H_{22} + C_4H_8 + C_2H_4$
  d) $C_{25}H_{52} \rightarrow C_{11}H_{24} + C_4H_8 + C_3H_6$

Practice questions

1 Hexane and propane are members of the alkane homologous series. All members of a homologous series have the same general formula.
   a) State the general formula of the alkanes. *(1 mark)*
   b) State three other features of a homologous series. *(3 marks)*
   c) Hexane contains six carbon atoms. Draw its structural formula. *(1 mark)*
   d) Why are the alkanes hydrocarbons? *(1 mark)*
   e) Draw the structural formula of propane. *(1 mark)*
   f) Write the balanced symbol equation of the complete combustion of propane. *(3 marks)*

2 Ethene is an alkene.
   a) What is the state of ethene at room temperature and pressure? *(1 mark)*
   b) Write the molecular formula of ethene. *(1 mark)*
   c) Ethene burns in oxygen, and also reacts with other chemicals as shown in the diagram below, forming products A, B, C, and D.

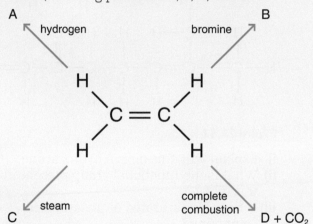

Figure 24.37
   i) Draw the structure of A. *(1 mark)*
   ii) Draw the structure of B. *(1 mark)*
   iii) Draw the structure of C. *(1 mark)*
   iv) Name substance D. *(1 mark)*
   v) Describe how you would test for carbon dioxide. State the result for a positive test. *(3 marks)*
   vi) Name the type of reactions which produce A, B, and C. *(1 mark)*

3 Polythene is described as a hydrocarbon polymer whereas polyvinylchloride (PVC) is not.
   a) Why is polyvinyl chloride not a hydrocarbon? *(1 mark)*
   b) Explain what you understand by the term polymer. *(1 mark)*

   c) Name the monomer from which polythene is formed. *(1 mark)*
   d) Write a structural equation to show the polymerisation of vinyl chloride to form polyvinylchloride (PVC). *(4 marks)*

4 Ethene is used in many industrial reactions such as the one represented below.

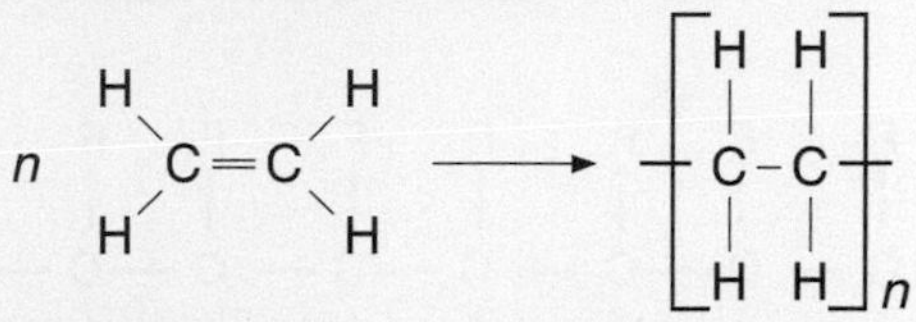

$$n \quad \begin{array}{c} H \\ \diagdown \\ C = C \\ \diagup \\ H \end{array} \begin{array}{c} H \\ \diagup \\ \diagdown \\ H \end{array} \longrightarrow \begin{array}{c} H \ H \\ | \ | \\ C - C \\ | \ | \\ H \ H \end{array}_n$$

Figure 24.38
   a) State fully the type of chemical reaction shown above. *(2 marks)*
   b) Name the product formed in this reaction. *(4 marks)*
   c) What does *n* represent in the equation above? *(1 mark)*

5 Candles, which are made of hydrocarbons, have been used for around 2000 years to provide light. The diagram below shows the apparatus used to test for the products of complete combustion of candle wax.

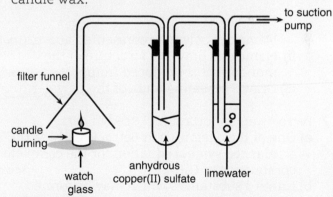

Figure 24.39
   a) Copy and complete Table 24.12 describing the appearance of the anhydrous copper(II) sulfate and limewater at the beginning of the experiment and five minutes after it began. *(4 marks)*

Table 24.12

	At the beginning of experiment	Five minutes after the start of the experiment
limewater		
anhydrous copper(II) sulfate		

b) Carbon monoxide and soot can be formed in the incomplete combustion of a candle.
  i) What condition is required for incomplete combustion to occur? *(1 mark)*
  ii) Why is carbon monoxide dangerous? *(1 mark)*
  iii) Why is it difficult to detect carbon monoxide? *(1 mark)*
  iv) What element is soot? *(1 mark)*
  v) What problems can soot in the environment cause to humans? *(1 mark)*

H **6 a)** The structure below represents part of a polymer.

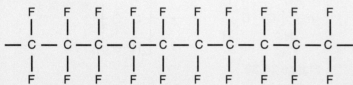

Figure 24.40

  Draw the structural formula of the monomer from which this polymer is formed. *(1 mark)*

  b) i)

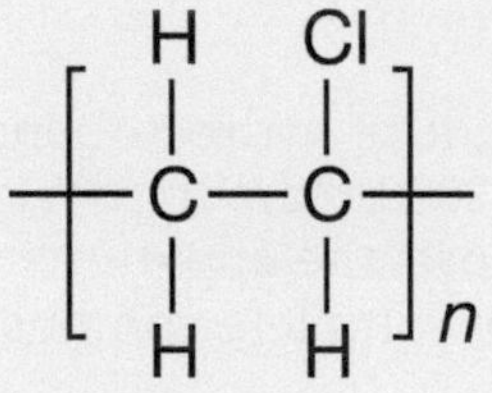

  ![Figure 24.41 polymer structure]

  Figure 24.41

  Name the polymer represented above. *(1 mark)*
  ii) Name and draw the structure of the monomer it is produced from. *(2 marks)*
  iii) Draw 3 repeating units of the polymer *(1 mark)*

7 Petrol is an important fuel used in cars.
  a) One of the chemicals in petrol is C_8H_{18}. Write a balanced symbol equation for the complete combustion of C_8H_{18}. *(3 marks)*
  b) Name 3 substances which may be formed when octane incompletely combusts. *(3 marks)*
  c) Explain why some sulfur dioxide may be formed when petrol burns. *(2 marks)*
  d) Acid rain may form from sulfur dioxide. State three environmental problems of acid rain. *(3 marks)*

8 Crude oil is the source of hydrocarbons such as alkanes.
  a) i) What is crude oil? *(1 mark)*
  ii) How are alkanes obtained from crude oil? *(1 mark)*
  iii) Name two fractions which are obtained from crude oil. *(2 marks)*

b) Octane, C_8H_{18}, is an alkane which is a constituent of petrol.
  i) Octane is a saturated hydrocarbon. What is meant by the terms **saturated** and **hydrocarbon**? *(2 marks)*
  ii) What is the general formula of the alkanes? *(1 mark)*
c) Alkenes, such as propene, can be obtained from large alkanes such as octane.
  i) What name is given to the process of forming alkenes from alkanes? *(1 mark)*
  ii) Write an equation for the formation of propene and an alkane, from octane. *(2 marks)*
  iii) Describe a chemical test for an unsaturated hydrocarbon such as propene. *(3 marks)*
  iv) Write an equation for the reaction involved in the test described in part **iii)**. *(2 marks)* **H**

9 Perfume is a mixture of essential oils dissolved in a solvent.
  a) One of the essential oils used in making perfume is shown below.

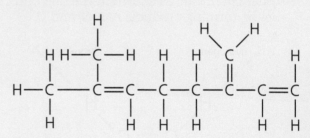

Figure 24.42

  i) Explain why this oil is a hydrocarbon. *(1 mark)*
  ii) What is the functional group present in this oil? *(1 mark)*
  iii) Is the oil a saturated or unsaturated compound? *(1 mark)*
  iv) What is the observation when bromine water was mixed with this oil? *(2 marks)*
  b) Another essential oil with formula $C_{10}H_{17}OH$ is used in perfume making. It has a sweet lavender like smell. **H**

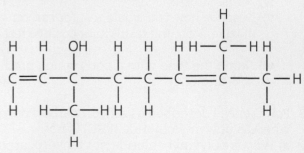

Figure 24.43

i) Identify the two functional groups in this oil. *(2 marks)*

ii) Complete and balance the symbol equation below for the complete combustion of this oil. *(2 marks)*

$C_{10}H_{17}OH + O_2 \rightarrow$

c) Ethanol is an alcohol which is often used as a solvent in perfumes.

i) Draw the structural formula of ethanol. *(1 mark)*

ii) Ethene can be used to manufacture the ethanol used in perfumes. Write a balanced symbol equation for the reaction of ethene to produce ethanol and state the conditions necessary. *(3 marks)*

iii) What type of reactions do alkenes undergo? *(1 mark)*

d) Ethanoic acid is a carboxylic acid which can be used to make other solvents. These solvents are also used in perfumes.

i) Draw the structural formula of ethanoic acid. *(1 mark)*

ii) State two observations you would make when sodium carbonate reacts with ethanoic acid. *(2 marks)*

iii) Write a balanced symbol equation for this reaction. *(3 marks)*

iv) Ethanoic acid is a weak acid. What is a weak acid? *(1 mark)*

10. Copy and complete Table 24.13. *(21 marks)*

Table 24.13

Name	Molecular formula	Structural formula	State at room temperature and pressure
		H H | | C=C | | H H	

		H H H
		| | |
		H—C=C—C—H
		        |
		        H	

butane			
	C_4H_8		
		H OH H | | | H—C—C—C—H | | | H H H	
	HCOOH		
	C_3H_7COOH		

25 Quantitative chemistry 2

Specification points

This chapter covers specification points 2.6.5 to 2.6.8. It covers calculations of solution concentration and atom economy. Some C2 calculations can also be found in Chapter 17.

Concentration of solutions in mol/dm³

Two solutions of copper(II) sulfate are shown. The one on the right is more concentrated.

Chemists can measure the concentration of a solution by considering the number of moles of solute that dissolve in 1 dm³ of solution.

Concentration (mol/dm³) **is the number of moles of a solute dissolved in 1 dm³ of solution.** The unit is mol/dm³. For example, a solution that has a concentration of 0.2 mol/dm³ has 0.2 mol of solute dissolved in 1 dm³ of solution.

To calculate concentration in mol/dm³ the expression below can be used:

$$\text{Concentration (mol/dm}^3) = \frac{\text{moles}}{\text{volume (dm}^3)}$$

If the volume is given in cm³ it must be converted to dm³ by dividing by 1000.

Tip

The formula triangle may help you. Cover up the quantity you want to find to show the equation you need to use.

Tip

If the volume is given in cm³, you can use also use the equation
$$\text{Concentration (mol/dm}^3) = \frac{\text{moles} \times 1000}{\text{volume (cm}^3)}$$

Tip

Remember that 1 dm³ = 1000 cm³

$$\text{concentration (mol/dm}^3) = \frac{\text{moles}}{\text{volume (dm}^3)}$$

▲ **Figure 25.1** A formula triangle

Example

Find the concentration of each of the following solutions in mol/dm³.

1 10 mol of solute dissolved in 4 dm³.
2 0.20 mol of solute dissolved in 100 cm³.
3 49 g of H_2SO_4 dissolved in 2 dm³.

Answer

1 concentration = $\dfrac{\text{moles}}{\text{volume(dm}^3)}$ = $\dfrac{10}{4}$ = 2.5 mol/dm³

2 volume of solution (dm³) = $\dfrac{100}{1000}$ = 0.10 dm³

concentration = $\dfrac{\text{moles}}{\text{volume (dm}^3)}$ = $\dfrac{0.20}{0.10}$ = 2.0 mol/dm³

Alternatively, you can use

concentration (mol/dm³) = $\dfrac{\text{moles} \times 1000}{\text{volume (cm}^3)}$ = $\dfrac{0.20 \times 1000}{100}$ = 2.0 mol/dm³

3 In this example you must first work out the number of moles of sulfuric acid in 49 g.

moles H_2SO_4 = $\dfrac{\text{mass}}{M_r}$ = $\dfrac{49}{98}$ = 0.5 mol

concentration = $\dfrac{\text{moles}}{\text{volume (dm}^3)}$ = $\dfrac{0.5}{2}$ = 0.25 mol/dm³

Example

Calculate the concentration of a solution containing 73 g of HCl dissolved in 200 cm³ of solution.

Answer

moles HCl = $\dfrac{\text{mass}}{M_r}$ = $\dfrac{73}{36.5}$ = 2 mol

concentration (mol/dm³) = $\dfrac{\text{moles} \times 1000}{\text{volume (cm}^3)}$ = $\dfrac{2 \times 1000}{200}$ = 10 mol/dm³

Test yourself

1 Calculate the concentration of the following solutions in mol/dm³.
   a) 4 mol of solute dissolved in 1 dm³ of solution.
   b) 0.2 mol of solute dissolved in 250 cm³ of solution.
   c) 3 mol of solute dissolved in 2000 cm³.
   d) 0.4 mol of solute dissolved in 500 cm³.
   e) 4 mol dissolved in 8 dm³ of solution.
   f) 30 mol dissolved in 10 dm³ of solution.
   g) 0.36 mol dissolved in 250 cm³ of solution.
   h) 0.10 mol dissolved in 50 cm³ of solution.
2 Calculate the concentration of the following solutions in mol/dm³.
   a) 120 g of NaOH dissolved in 2 dm³ of solution.
   b) 1.7 g of NH_3 dissolved in 200 cm³ of solution.
   c) 2.1 g of $NaHCO_3$ in 250 cm³ of solution.

Show you can

What mass of sodium carbonate Na_2CO_3 would be dissolved in:

1 100 cm³ to make a solution with a concentration of 0.1 mol/dm³?
2 500 cm³ to make a solution with a concentration of 0.80 mol/dm³?

Calculating the number of moles in a solution

A useful rearrangement of the expression:

$$\text{concentration (mol/dm}^3) = \frac{\text{moles}}{\text{volume (dm}^3)}$$

is

$$\text{moles} = \text{volume (dm}^3) \times \text{concentration (mol/dm}^3)$$

This expression is used to find the number of moles of a solute in a solution.

> **Tip**
>
> If the volume is given in cm^3 you can use the expression
> $$\text{moles} = \frac{\text{volume (cm}^3) \times \text{concentration (mol/dm}^3)}{1000}$$

> **Test yourself** 🖊
>
> Calculate the number of moles of solute in the following solutions.
>
> 3 $4\,dm^3$ of a $0.30\,mol/dm^3$ solution.
> 4 $1.5\,dm^3$ of a $2.0\,mol/dm^3$ solution.
> 5 $100\,cm^3$ of a $1.50\,mol/dm^3$ solution.
> 6 $25\,cm^3$ of a $0.50\,mol/dm^3$ solution.
> 7 $250\,cm^3$ of a $1.0\,mol/dm^3$ solution.
> 8 $2\,dm^3$ of a $1.25\,mol/dm^3$ solution.
> 9 $1500\,cm^3$ of a $0.750\,mol/dm^3$ solution.

> **Example**
>
> Calculate the number of moles of solute in each of the following solutions.
>
> 1 $20\,dm^3$ of a $0.50\,mol/dm^3$ solution.
> 2 $25\,cm^3$ of a $1.50\,mol/dm^3$ solution.
>
> **Answer**
>
> 1   moles = volume (dm^3) × concentration = 20 × 0.50 = 10 mol
> 2   volume of solution (dm^3) = $\frac{25}{1000}$ = 0.025 dm³
>
>   moles = volume (dm^3) × concentration = 0.025 × 1.5 = 0.0375 mol

Atom economy

Chemical reactions are used to make specific products that have a use or purpose. In some reactions there is only one product, and so all the atoms in the reactants are present in the product. For example, when ethene (C_2H_4) reacts with steam to make ethanol (C_2H_5OH), all the atoms in the reactants end up in the desired product ethanol.

    ethene + steam → ethanol

    $C_2H_4 + H_2O \rightarrow C_2H_5OH$

However, in many reactions, there are one or more other products as well as the desired product. In these reactions not all the atoms in the reactants end up in the desired product. For example, when ethanol is made by fermentation from glucose, carbon dioxide is formed as well as ethanol. This means that some of the atoms from the glucose do not end up in the ethanol.

    glucose → ethanol + carbon dioxide

    $C_6H_{12}O_6 \rightarrow 2C_2H_5OH + 2CO_2$

The atom economy of a reaction is a way of measuring what percentage of products is theoretically useful.

    atom economy = $\frac{\text{mass of desired product}}{\text{total mass of products}} \times 100$

> **Tip**
>
> To find the mass of the desired product use the relative molecular mass and the number of molecules from the balanced symbol equation.

In a reaction with only one product, all the atoms from the reactants end up in the product, so the atom economy must be 100%.

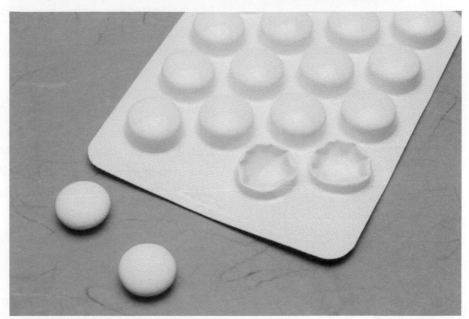

▲ **Figure 25.2** Ibuprofen was originally made in a six-step process with an atom economy of 40%. It is now made in a three-step process with an atom economy of 77%

Example

Calculate the atom economy when ethanol is made from glucose by the following reaction:

$$C_6H_{12}O_6 \rightarrow 2C_2H_5OH + 2CO_2$$

Give your answer to three significant figures.

Answer

Mass (M_r) of desired product = sum of M_r of desired products

The desired product is ethanol which has $M_r(C_2H_5OH) = 46$

The reaction produces $2C_2H_5OH$, so

Mass of desired product = 2 × 46 = 92

Total mass of products = sum of M_r of products

$2M_r(C_2H_5OH) + 2M_r(CO_2) = 92 + (2 × 44) = 180$

Atom economy = $\dfrac{92}{180}$ × 100 = 51.1%

Tip

If the desired product is an atom, then the relative atomic mass is used in the equation.

Example

Iron is extracted from its ore using carbon:

$$2Fe_2O_3 + 3C \rightarrow 4Fe + 3CO_2$$

Calculate the atom economy of this reaction. Give your answer to one decimal place.

Answer

Mass of desired product = sum of A_r of desired product

The reaction produces 4Fe and $A_r(Fe) = 56$

Mass of desired product = $(4 \times 56) = 224$

Total mass of products = sum of M_r of all products

$224 + 3 \times M_r(CO_2) = 224 + (3 \times 44) = 356$

Atom economy = $\dfrac{224}{356} \times 100 = 62.9\%$

Tip

Remember to use the balancing numbers in the equation when working out the total mass (M_r) of the reactants and the desired product.

Test yourself

10 Calculate the atom economy when making calcium oxide by the thermal decomposition of calcium carbonate in the following reaction:

$$CaCO_3 \rightarrow CaO + CO_2$$

11 Calculate the atom economy when making aluminium by electrolysis in the following reaction:

$$2Al_2O_3 \rightarrow 4Al + 3O_2$$

12 Calculate the atom economy when making dichloromethane (CH_2Cl_2) by reaction of methane with chlorine in the following reaction:

$$CH_4 + 2Cl_2 \rightarrow CH_2Cl_2 + 2HCl$$

13 a) Calculate the mass of waste material in a reaction with a 60% atom economy if 1 kg of reactants is used.

b) Calculate the mass of waste material in a reaction with an 80% atom economy if 1 kg of reactants is used.

Show you can (?)

Titanium oxide is found in the naturally occurring ore rutile. It is possible to extract titanium metal from this ore by a displacement reaction with magnesium or by electrolysis of the ore.

Method 1: This method uses a more reactive metal to displace the titanium:

$$TiO_2 + 2Mg \rightarrow Ti + 2MgO$$

Method 2: This is electrolysis of the ore. The overall reaction for this method is:

$$TiO_2 \rightarrow Ti + O_2$$

1 Calculate the atom economy for the production of titanium for each reaction.

2 Oxygen is a useful product and can be sold. What is the atom economy of the electrolysis if the oxygen is collected and sold?

Practice questions

H 1 Calculate the concentration in mol/dm³ of the solution formed when 0.25 mol of Na_2CO_3 is dissolved in 200 cm³ of solution. *(1 mark)*

2 Copper can be extracted by heating copper(II) oxide with carbon.

$2CuO + C \rightarrow 2Cu + CO_2$

Calculate the atom economy of this method to produce copper. *(2 marks)*

3 a) Calculate the atom economy, to one decimal place, for the production of hydrogen by the following reaction:

$CH_4 + 2H_2O \rightarrow CO_2 + 4H_2$ *(2 marks)*

b) How could the atom economy of this reaction be improved? *(1 mark)*

Specification points

This section covers specification points 2.7.1 to 2.7.5 and explains what electrolysis is. The products and half equations for the reactions at each electrode are explored for familiar and unfamiliar examples of molten salts. The industrial process of producing aluminium metal by electrolysis is explained.

Electrochemistry is the study of the links between chemical reactions and electricity. Chemical reactions can be used to provide electrical current. For example, two lemons connected by zinc and copper electrodes can provide enough electrical current to power a digital clock. Chemical changes are also brought about by the passage of an electric current. In this chapter you will study how chemicals can be broken down using electricity.

What is electrolysis?

Ionic compounds are made up of positive ions (cations) and negative ions (anions). As solids, the ions are held together in a lattice by strong ionic bonds and cannot move, so solid ionic compounds do not conduct electricity. When ionic compounds are melted into a liquid or dissolved in water, the ions can move and carry charge, so the compound conducts electricity (see page 149).

When ionic compounds conduct electricity they decompose. Electrolysis **is the decomposition of electrolytes using electricity. An electrolyte is an ionic liquid or solution that conducts electricity and is decomposed during the process.** The conduction of electricity by an electrolyte occurs because the ions can move and carry charge.

If two graphite rods (electrodes) are connected to a supply of electricity and placed in an electrolyte (Figure 26.1), the negative ions (anions) are attracted to the positive electrode (anode) and the positive ions (cations) are attracted to the negative electrode (cathode). This happens because opposite charges attract each other.

> **Tip** ↻
>
> Remember that positive ions are called cations and negative ions are called anions.

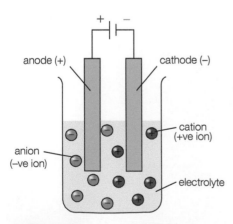

▲ **Figure 26.1** Movement of ions during electrolysis

▲ **Figure 26.2** Remember PANCake – Positive Anode, Negative Cathode.

Tip

Make sure you do not just state that graphite is a good conductor, you must mention 'of electricity'.

An electrode is a solid electrical conductor used to make contact with a non-metallic part of a circuit. **The positive electrode is called the** anode. **The negative electrode is called the** cathode.

Electrodes are often inert, which means they are unreactive. **Inert electrodes are electrodes that will allow the electrolysis to take place but do not themselves react.**

Electrodes can be made of graphite because:

▶ it is a good conductor of electricity

▶ it is unreactive.

Platinum can also be used but it is more expensive.

▲ **Figure 26.3** The graphite in the 'lead' of pencils is used here to form the electrodes in the electrolysis of water.

Test yourself

1 What is electrolysis?
2 What is an electrolyte?
3 Name the electrolytes in the following list:
   • sodium chloride
   • molten lead bromide
   • calcium chloride solution
   • copper
   • graphite
   • magnesium chloride solution
   • potassium iodide.
4 An ionic solid is made up of positive ions and negative ions. State what happens to the ions when the molten solid is electrolysed.
5 Why are electrodes made of graphite?

Show you can

Some molten sodium chloride is electrolysed.

1 What ions are present in molten sodium chloride?

2 Write the formula of the ion that moves to the anode during electrolysis.

Tip

Remember that electrons have a negative charge.

What happens at the electrodes during electrolysis?

When the ions reach the electrodes they are **discharged**. This means that they gain or lose electrons so that they lose their charge and become neutral. Positive ions gain electrons. Negative ions lose electrons.

Negative ions move to the positive electrode (anode) and they discharge by losing electrons and become atoms. Oxidation is loss of electrons, so **oxidation occurs at the anode.** The electrons move around the circuit through the wires to the negative electrode.

Positive ions move to the negative electrode (cathode) and they discharge by gaining electrons and become atoms. Reduction is gain of electrons, so **reduction occurs at the cathode.**

Table 26.1

	Anode (positive electrode)	Cathode (negative electrode
Which ions are attracted?	negative ions (anions)	positive ions (cations)
What happens at the electrode?	negative ions are discharged by losing electrons and becoming atoms, which may combine to form molecules	positive ions are discharged by gaining electrons and becoming atoms
Oxidation or reduction	oxidation (loss of electrons)	reduction (gain of electrons)

What is the difference between an electrolyte and a conductor?

Table 26.2 shows some differences between electrolytes and conductors.

Table 26.2

	Conductor	Electrolyte
Example	metals, e.g. copper, graphite	ionic liquids or solutions, e.g. sodium chloride solution
How do they conduct?	delocalised electrons move and carry charge	ions move and carry charge
What happens when they conduct?	no change	they are decomposed

Predicting the products of the electrolysis of molten salts

Figure 26.4 illustrates what is happening in electrolysis. The diagram can be applied to specific ionic compounds to predict the products at each electrode.

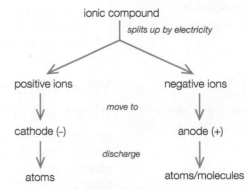

▲ **Figure 26.4** The breakdown of an ionic compound during electrolysis

There are two specific molten compounds that you need to know about in detail for GCSE, but you must also be able to apply your knowledge to other examples.

1 Electrolysis of molten lead(II) bromide

To electrolyse molten lead(II) bromide, or any molten salt, the apparatus is set up as shown in Figure 26.5.

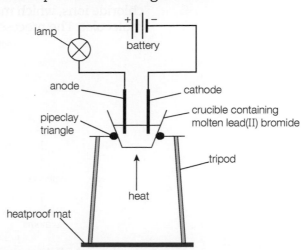

▲ **Figure 26.5** Diagram of laboratory electrolysis of molten salts

The lead(II) bromide is split up into lead ions, which move to the cathode and discharge, forming lead atoms, and bromide ions, which move to the anode and discharge, forming bromine molecules. Figure 26.6 shows how the lead bromide breaks down and Table 26.3 describes what occurs at the cathode and anode.

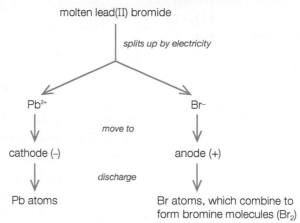

▲ **Figure 26.6** Working out the products from the electrolysis of molten lead (II) bromide

Table 26.3

	Cathode	Anode
Name of product	lead	bromine
Observation	molten grey liquid	red–brown pungent gas, bubbles
Explanation	lead ions (Pb^{2+}) gain electrons and are reduced to form lead (Pb).	bromide ions (Br^-) lose electrons and are oxidised to form bromine (Br_2).
Half equation	$Pb^{2+} + 2e^- \rightarrow Pb$	$2Br^- \rightarrow Br_2 + 2e^-$
Oxidation or reduction	reduction	oxidation

2 Electrolysis of molten lithium chloride

During electrolysis, lithium chloride is split up into lithium ions, which move to the cathode and discharge to form lithium atoms, and chloride ions, which move to the anode and discharge to form chlorine molecules. The process is described in Figure 26.7 and Table 26.4.

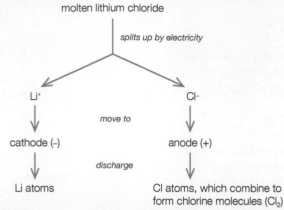

▲ **Figure 26.7** Working out the products from the electrolysis of molten lithium chloride

Table 26.4

	Anode	Cathode
Name of product	chlorine	lithium
Observation	yellow–green pungent gas, bubbles	molten grey liquid
Explanation	chloride ions (Cl–) lose electrons and are oxidised to form chlorine (Cl_2).	lithium ions (Li+) gain electrons and are reduced to form lithium (Li).
Half equation	$2Cl^- \rightarrow Cl_2 + 2e^-$	$Li^+ + e^- \rightarrow Li$
Oxidation or reduction	oxidation	reduction

Tip

For further information on writing half equations, see page 506.

Predicting observations

When a molten halide is electrolysed the halogen will appear as a gas due to the heat present. Bubbles are observed, and you should also note the colour of the gas produced:

▶ chlorine – yellow–green gas

▶ bromine – red–brown gas

▶ iodine – purple gas.

If a molten metal oxide is electrolysed, oxygen is produced at the anode. Bubbles are observed and a colourless, odourless gas is given off.

Most metals appear as grey liquids. Lead is dense and sinks.

Example

Predict the products at each electrode and give half equations and observations for the electrolysis of molten sodium iodide.

Answer

Anode product: iodine
Observation: purple gas
Half equation: $2I^- \rightarrow I_2 + 2e^-$
Cathode: sodium
Observation: grey liquid
Half equation: $Na^+ \rightarrow Na + e^-$

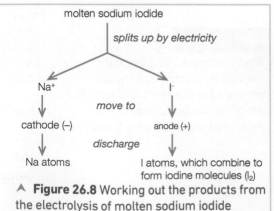

▲ **Figure 26.8** Working out the products from the electrolysis of molten sodium iodide

Test yourself 📝

6 a) Why does lead(II) bromide conduct electricity when molten?
   b) Why does solid sodium chloride not conduct electricity?
   c) Describe what happens in the electrolysis of molten potassium chloride.
7 a) In electrolysis, what happens to negative ions at the positive electrode?
   b) State two differences between conductors and electrolytes.
8 a) Copy and complete Table 26.5 to show the products of electrolysis of some molten ionic compounds.

Table 26.5

Ionic compound (molten)	Product at the negative electrode (cathode)	Product at the positive electrode (anode)	Cathode equation	Anode equation
potassium iodide (KI)				
zinc bromide ($ZnBr_2$)				
magnesium oxide (MgO)				

b) Predict observations at the anode for each of the compounds in a).
c) Predict observations at the cathode for each of the compounds in a).

Show you can ❓

Copy and complete Table 26.6 for the electrolysis of molten lithium iodide.

Table 26.6

	Anode	Cathode
Product		
Observation		
Half equation		
Oxidation or reduction		

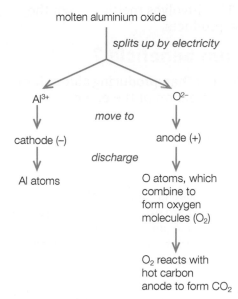

molten aluminium oxide

splits up by electricity

Al^{3+} O^{2-}

move to

cathode (−)   anode (+)

discharge

Al atoms   O atoms, which combine to form oxygen molecules (O_2)

O_2 reacts with hot carbon anode to form CO_2

▲ **Figure 26.9** Working out the products for the electrolysis of molten aluminum oxide

▲ **Figure 26.10** Industrial electrolysis of aluminium ore

The industrial extraction of aluminium

When studying the reactivity series in Chapter 8 you learnt that aluminium cannot be extracted by heating with carbon. As it is a reactive metal it is extracted from its ore by electrolysis.

The main ore of aluminium is **bauxite**. When bauxite is purified aluminium oxide (**alumina**) is formed. For electrolysis to occur the ions must be able to move, and so the alumina must be melted or dissolved – alumina has a very high melting point (2072°C) and the cost of the heat energy to melt it would be very high. The alumina is dissolved in molten **cryolite**. This:

▶ reduces the melting point of alumina to around 900°C, and so saves energy and reduces the cost

▶ increases the conductivity.

The ores are placed in a steel tank called a cell – the cathode is the graphite lining of the tank and the graphite anode dips into the electrolyte. The industrial cell is shown in Figure 26.11.

Aluminium ions (Al^{3+}) are attracted to the negative electrode, where they discharge by gaining electrons to form aluminium metal. Reduction, i.e. gain of electrons, occurs. As it is so hot in the cell, the metal is produced as a liquid and is run off at the bottom.

Cathode half equation: $Al^{3+} + 3e^- \rightarrow Al$

Oxide ions (O^{2-}) are attracted to the positive electrode, where they discharge by losing electrons to form oxygen. Oxidation, i.e. loss of electrons, occurs.

Anode half equation: $2O^{2-} \rightarrow O_2 + 4e^-$

This oxygen reacts with the graphite anode, and the anode burns to produce carbon dioxide. This means that the anode has to be replaced regularly.

Carbon + oxygen → carbon dioxide

$C + O_2 \rightarrow CO_2$

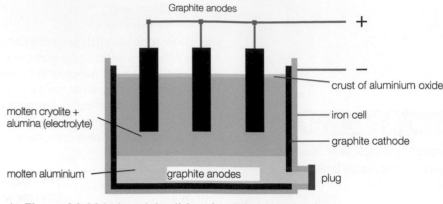

Graphite anodes

+

−

crust of aluminium oxide

molten cryolite + alumina (electrolyte)

iron cell

graphite cathode

molten aluminium

graphite anodes

plug

▲ **Figure 26.11** Industrial cell for aluminium production

A crust of aluminium oxide forms on the top of the electrolyte. This crust:

▶ acts like a lid and keeps heat in
▶ stops impurities entering
▶ prevents the aluminium formed from reacting with the air.

The overall electrolysis process is expensive due to:

▶ the high cost of electricity for the process
▶ the high cost of the heat energy required to melt the metal compounds and keep them molten.

Aluminium metal can be recycled. This involves melting down the product and remoulding it into new products.

Why is recycling aluminium beneficial?

▶ Recycling aluminium is much cheaper than producing aluminium from bauxite because it uses only a fraction of the energy.
▶ Recycling saves waste.
▶ Recycling saves natural resources of bauxite.

RECYCLE

▲ **Figure 26.12** Aluminium should be recycled

▲ **Figure 26.13** Recycling aluminium uses about 5% of the energy required to create aluminium from bauxite.

Test yourself

9 Why can some metals not be extracted by heating with carbon?
10 Name two metals other than aluminium that are extracted by electrolysis.
11 Name the ore of aluminium from which aluminium is extracted.
12 What is the name for the purified ore from which aluminium is extracted?
13 What is the electrolyte in the electrolysis of aluminium?
14 State the operating temperature of the cell.
15 What happens at the negative electrode?
16 Why does the positive electrode have to be replaced regularly?
17 What is electrolysis?

Show you can

Copy and complete Table 26.7 for the electrolysis of molten aluminium oxide.

Table 26.7

	Anode	Cathode
Product		
Half equation		
Oxidation or reduction		

H

Practice questions

1 Ionic compounds like lead(II) bromide conduct electricity when molten.
   a) Explain as fully as possible why ionic compounds conduct electricity when molten. *(2 marks)*
   b) i) Draw a labelled diagram of the assembled apparatus used to carry out the electrolysis of molten lead(II) bromide. *(3 marks)*
      ii) Copy and complete Table 26.8.

   Table 26.8

	Anode	Cathode
Name of product	*(1 mark)*	*(1 mark)*
Observations	*(1 mark)*	*(1 mark)*
Half equation	*(3 marks)*	*(3 marks)*

2 Solid sodium chloride does not conduct electricity; however, when it is molten it is an electrolyte and can be electrolysed. This electrolysis produces a metal and a gas.
   a) What is meant by the term electrolysis? *(1 mark)*
   b) What is meant by the term anode? *(1 mark)*
   c) Name the substance formed at the cathode. *(1 mark)*
   d) Explain why solid sodium chloride does not conduct electricity. *(2 marks)*
   e) What is an electrolyte? *(2 marks)*
   f) Write a half equation for the reaction that occurs at the anode. *(3 marks)*
   g) Name a substance used to make the electrodes. *(1 mark)*

3 Some substances, for example molten lead(II) bromide and aqueous sulfuric acid, are described as electrolytes. Other substances, for example copper metal, are conductors.
   An experiment to investigate the electrolysis of the electrolyte molten lead bromide was set up as shown in Figure 26.14.

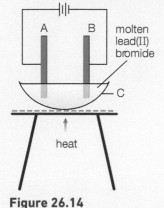

   Figure 26.14

   a) Some pieces of apparatus in the diagram are labelled A–C. State the correct name for each piece of apparatus. *(3 marks)*
   b) Name a piece of apparatus that could be connected in the circuit to show that an electric current is flowing through the molten lead(II) bromide. *(1 mark)*
   c) Copy and complete Table 26.9.

   Table 26.9

Electrode	Name of product	Half equation
A	*(1 mark)*	*(3 marks)*
B	*(1 mark)*	*(3 marks)*

   d) Why does this electrolysis need to be carried out in a fume cupboard? *(1 mark)*
   e) Explain the differences between the conduction of electricity by copper metal and molten lead(II) bromide. *(4 marks)*

4 Aluminium is the most abundant metal in the Earth's crust. Aluminium ore is first purified to give aluminium oxide and the metal is then extracted from the aluminium oxide by electrolysis.
   a) What is meant by the term electrolysis? *(1 mark)*
   b) Name the ore from which aluminium is extracted. *(1 mark)*
   c) The electrolysis of the purified ore is carried out in a Hall–Héroult cell. Figure 26.15 shows the cell used.

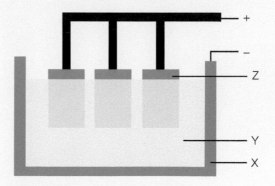

   Figure 26.15

   i) Name X and Z. *(2 marks)*
   ii) Y is the electrolyte. Name the substances in the electrolyte. *(2 marks)*
   iii) What is the temperature at which the electrolysis is carried out? *(1 mark)*

iv) Name the products formed at the positive and negative electrodes. *(2 marks)*

v) Write a half equation for the reaction taking place at the anode. *(3 marks)*

vi) Write a half equation for the reaction taking place at the cathode *(3 marks)*

vii) Which electrode must be replaced regularly? Write a balanced symbol equation to explain your answer. *(3 marks)*

d) Explain in terms of electrons why the extraction of aluminium in this process is a reduction reaction. *(2 marks)*

5 a) In the electrolysis of molten zinc bromide:

i) Name the product formed at the anode. *(1 mark)*

ii) Write a half equation for the process at the cathode and state whether it is an oxidation or a reduction process. *(4 marks)*

iii) Write a half equation for the process at the anode. *(3 marks)*

b) For the following molten salts, name the products at the cathode and the anode:

i) potassium iodide *(2 marks)*

ii) calcium oxide *(2 marks)*

iii) sodium bromide. *(2 marks)*

6 a) Explain what is meant by the following terms:

i) cathode *(1 mark)*

ii) inert electrode. *(1 mark)*

b) Aluminium is produced at the cathode in the electrolysis of alumina.

i) Write a half equation for the production of aluminium at the cathode. *(3 marks)*

ii) Explain, in words, what happens to the oxide ions at the anode during the electrolysis process. *(3 marks)*

iii) Explain why the anodes need to be replaced regularly. *(3 marks)*

iv) State two reasons why recycling aluminium is beneficial. *(2 marks)*

v) Why is the crust of aluminium oxide on the top of the electrolyte useful? *(2 marks)*

vi) Why is the electrolysis of a molten salt expensive? *(1 mark)*

27 Energy changes in chemistry

Some reactions give out heat, for example the combustion of coal. Other reactions take in heat. Sherbet contains sodium hydrogencarbonate, tartaric acid and sugar. When sherbet in sweets comes into contact with water on the tongue, the reaction that occurs takes in energy. The sherbet draws heat energy from the water on the tongue, leading to a cold and tingling sensation. In this chapter you will learn about the energy changes that occur in chemical reactions, some of which take in heat energy and some of which give out heat energy.

Exothermic reactions

An exothermic reaction is one in which heat is given out. This means that the surroundings become hotter and the temperature rises. An exothermic reaction has a **negative** energy change value.

Neutralisation reactions and combustion reactions are exothermic. The burning of fuels, such as methane (CH_4) in natural gas, are highly exothermic combustion reactions. They release a lot of heat energy and the fuel catches fire.

Figure 27.2 shows a self-heating can. Inside these cans the food or drink is separated from a layer containing chemicals used for heating. The chemicals are often calcium oxide and a bag of water. When the user presses a button the bag of water is punctured and the water mixes with the calcium oxide. The calcium oxide reacts with the water in an exothermic reaction. The heat released heats up the food or drink.

> **Tip**
>
> Exothermic heat is given out – remember that heat 'exits'.

▲ **Figure 27.1** An exothermic reaction takes place inside a hand warmer

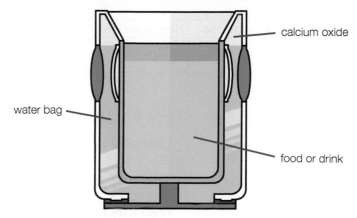

calcium oxide

water bag

food or drink

▲ **Figure 27.2** A self-heating can

Endothermic reactions

An endothermic reaction **is one in which heat is taken in.** The temperature of the surroundings drops. An endothermic reaction has a **positive** energy change value.

Endothermic reactions are less common than exothermic reactions. Thermal decomposition reactions are endothermic, as are the reactions of acids with hydrogencarbonates.

▲ **Figure 27.3** Some sports injury packs contain a packet of ammonium nitrate and a packet of water. When the pack is squeezed the water bag bursts and the ammonium nitrate dissolves in the water in an endothermic process, which cools the injured area and prevents swelling

Test yourself

1 State whether each of the following reactions is endothermic or exothermic.
   a) The temperature started at 21 °C and finished at 46 °C.
   b) The temperature started at 18 °C and finished at 14 °C.
   c) The temperature started at 19 °C and finished at 25 °C.
2 State whether each of the following reactions is endothermic or exothermic.
   a) burning alcohol
   b) thermal decomposition of iron(II) carbonate
   c) reaction of ethanoic acid with sodium hydrogencarbonate
   d) reaction of ethanoic acid with sodium hydroxide.
3 a) Why does the temperature increase when an exothermic reaction takes place in a solution?
   b) Why does the temperature decrease when an endothermic reaction takes place in a solution?
4 Classify the following types of reaction as exothermic or endothermic:
   a) neutralisation
   b) photosynthesis
   c) thermal decomposition
   d) displacement.

Show you can

The reactions P and Q can be classified in different ways.

Reaction P: calcium carbonate → calcium oxide + carbon dioxide

Reaction Q: sodium hydroxide + nitric acid → sodium nitrate + water

Copy and complete Table 27.1 by placing a tick (✓) in each row for reactions P and Q to indicate the terms that apply to each reaction. More than one term can apply to each reaction.

Table 27.1

	Combustion	Decomposition	Neutralisation	Oxidation	Respiration	Exothermic	Endothermic
P							
Q							

Reaction profiles

Chemical reactions can only occur when particles collide with each other with enough energy to react. **The activation energy is the minimum energy needed for a reaction to occur.**

A reaction profile is a graph that shows the relative energy of the reactants and products in a reaction (Figure 27.4). It can also show the activation energy (E_a).

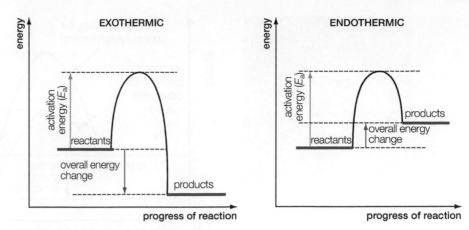

▲ **Figure 27.4** Reaction profiles for exothermic and endothermic reactions

Note that for an exothermic reaction the products have less energy than the reactants, as heat has been given out. This means that the overall energy change is negative. For an endothermic reaction, the products have more energy than the reactants, and are at a higher level on the reaction profile, as heat energy has been taken in. This means the overall energy change is positive.

The reaction pathway is shown as a line from the reactants to the products. Reaction pathways require an input of energy to break bonds in the reactants before new bonds can form in the products. The energy taken in is the minimum amount of energy needed for a reaction to occur (the activation energy, E_a).

Example

Draw a labelled reaction profile diagram for the following reaction:

$CH_4 + 2O_2 \rightarrow CO_2 + 2H_2O$
The energy change for the reaction is −890 kJ.

Answer

The energy change is negative, so it is an exothermic reaction.

Draw the general shape of an exothermic reaction profile.

Then label the reactants (CH_4 and $2O_2$) and the products (CO_2 and $2H_2O$).

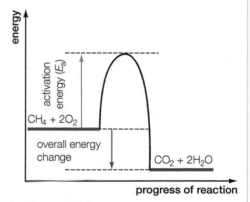

▲ **Figure 27.5**

You learnt in Chapter 10 that a catalyst speeds up a chemical reaction by providing an alternative reaction pathway of lower activation energy. This is represented on the reaction profile shown in Figure 27.7.

Figure 27.6 A successful jump can be compared to a successful reaction occurring when the activation energy (bar) is overcome

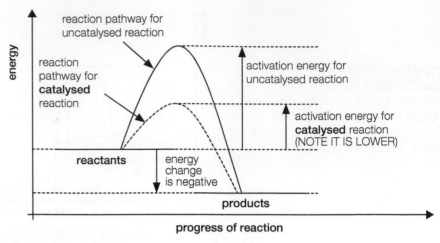

▲ **Figure 27.7** Reaction profile for a catalysed and uncatalysed reaction

Test yourself

5 Look at the following reaction profiles.

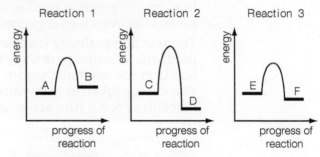

▲ **Figure 27.8**

a) Which reaction(s) is/are exothermic?

b) Which reaction(s) is/are endothermic?

6 a) Sketch an energy profile for the following reaction, which is endothermic:

$$CuCO_3 \rightarrow CuO + CO_2$$

b) Draw an arrow to show the overall energy change for the reaction and label it **E**.

c) Draw an arrow to show the activation energy for the reaction and label it E_a.

7 From the graph in Figure 27.9, using the letters A, B, C or D, identify:

a) the activation energy of the catalysed reaction

b) the energy change of the reaction

c) the activation energy of the uncatalysed reaction.

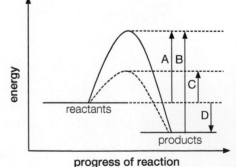

▲ **Figure 27.9**

Show you can

The reaction profile for a reaction is shown below.

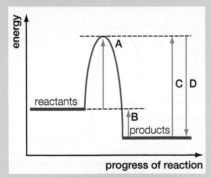

▲ **Figure 27.10**

1 Is it an exothermic reaction or an endothermic reaction?

2 Which letter represents the activation energy for the conversion of reactants to products?

Bond making and bond breaking

Breaking a chemical bond takes in energy (Figure 27.12). **Making a chemical bond releases energy** (Figure 27.13). Bond breaking is endothermic and the energy change has a positive value. Bond making is exothermic and the energy change has a negative value.

For example, 436 kJ of energy is taken in to break one mole of H–H covalent bonds, and this is written as +436 kJ/mol. Forming one mole of H–H covalent bonds releases 436 kJ of energy, and this is written as –436 kJ/mol.

During a chemical reaction:

▶ energy must be supplied to break bonds in the reactants
▶ energy is released when bonds in the products are formed.

▲ **Figure 27.11** Breaking a window takes in energy. Breaking a bond takes in energy

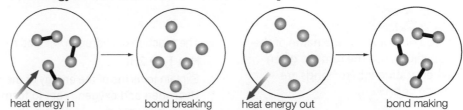

heat energy in   bond breaking   heat energy out   bond making

▲ **Figure 27.12** Endothermic bond breaking

▲ **Figure 27.13** Exothermic bond making

The overall energy change for a reaction equals the difference between the energy taken in to break the bonds in the reactants and the energy released when bonds are formed to make the products.

Energy change = energy taken in breaking bonds in reactants – energy released making bonds in products

Table 27.2 Comparison of exothermic and endothermic reactions

	Exothermic reaction	**Endothermic reaction**
Comparison of bond energies	more energy is released in making new bonds than is taken in to break bonds	more energy is taken in to break bonds than is released in making new bonds
Sign of energy change	–	+

Tip

A useful model is to state: 'The energy taken in to break the bonds in (name reactants) is greater/less than the energy given out when bonds are made in (name products).'

To explain if a reaction is exothermic or endothermic:

▶ state that bond breaking takes in energy and name the reactants in which the bonds are broken
▶ state that bond making gives out energy and name the reactants in which the bonds are made
▶ compare the two energies and state which is greater.

Example

Methane burns in oxygen, releasing energy:

$CH_4 + 2O_2 \rightarrow CO_2 + 2H_2O$

Explain in terms of the energy of the bonds why the reaction is exothermic.

Answer

• The energy taken in to break the bonds in methane and oxygen
• is less than
• the energy given out when bonds are made in carbon dioxide and water.

▲ **Figure 27.14** Natural gas contains methane, which burns exothermically

Example

Photosynthesis is an endothermic reaction.

$6CO_2 + 6H_2O \rightarrow C_6H_{12}O_6 + 6O_2$

Explain in terms of the energy of the bonds why this reaction is endothermic.

Answer

- The energy taken in to break bonds in carbon dioxide and water
- is greater than
- the energy given out when bonds are made in $C_6H_{12}O_6$ and O_2.

▲ **Figure 27.15** Photosynthesis takes in energy from the sun. It is an endothermic reaction

Tip

If you do not know the name of the substance, use the formula given when stating where bonds are made or broken.

Test yourself

8 The reaction of hydrogen with iodine to form hydrogen iodide is endothermic.

$H_2 + I_2 \rightarrow 2HI$

Explain in terms of the energy of the bonds why this reaction is endothermic.

9 Hydrogen and oxygen react to form water.

$2H_2 + O_2 \rightarrow 2H_2O$

Explain in terms of the energy of the bonds why this is an exothermic reaction.

10 The equation for the dehydration of ethanol is

$C_2H_5OH \rightarrow C_2H_4 + H_2O$

Explain in terms of the energy of the bonds why this is an endothermic reaction.

Calculating the energy change in a reaction

The **bond energy** is a measure of the energy required to break one mole of a covalent bond. A higher bond energy value means a stronger bond.

To calculate the energy change in a reaction the method is as follows:

▶ calculate the total bond energy for all the bonds broken in the reactants

▶ calculate the total bond energy for all the bonds made in the reactants

▶ use the equation.

Energy change = energy taken in breaking bonds in reactants – energy released making bonds in products

Example

1 Calculate the energy change in the following reaction using the bond energies given.

Bond energies: C–H, 412; O=O, 496; C=O, 743; O–H, 463 kJ

$$
\begin{array}{ccc}
\quad\;\; \text{H} & & \qquad\qquad\qquad \text{H} \\
\quad\;\; | & & \qquad\qquad\qquad \diagup \\
\text{H} - \text{C} - \text{H} \; + \; 2\,\text{O}{=}\text{O} \longrightarrow \; \text{O}{=}\text{C}{=}\text{O} \; + \; 2\,\text{O} \\
\quad\;\; | & & \qquad\qquad\qquad \diagdown \\
\quad\;\; \text{H} & & \qquad\qquad\qquad \text{H}
\end{array}
$$

2 Explain, using bond energies, why the reaction is exothermic or endothermic.

Answer

1 **Table 27.3**

Bonds broken	Bonds made
4 C–H = 4(412)	2 C=O = 2(743)
2 O=O = 2(496)	4 O–H = 4(463)
Total = 2640 kJ	Total = 3338 kJ

Energy change = energy taken in to break bonds – energy released making bonds

$$= 2640 - 3338$$
$$= -698 \text{ kJ}$$

2 This reaction is exothermic because the energy taken in to break bonds in CH_4 and O_2 is less than the energy given out when bonds are made in CO_2 and H_2O.

Example

Calculate the energy change in the following reaction using the bond energies given.

$2H_2 + O_2 \rightarrow 2H_2O$

Bond energies: H–H, 412; O=O, 496; O–H, 463 kJ

Answer

If the equation is not given structurally, draw it out structurally showing all bonds.

$$2\, H—H + O = O \longrightarrow 2\, H—O—H$$

Table 27.4

Bonds broken	Bonds made
2 H–H = 2(436)	4 O–H = 4(463)
1 O=O = 496	Total = 1852 kJ
Total = 1368 kJ	

Energy change = energy needed to break bonds – energy released making bonds

$$= 1368 - 1852 = -484 \text{ kJ}$$

Example

Calculate the energy change in the following reaction using the bond energies given.

Bond energies: C–C, 348; C–H, 412; C–O, 360; O–H, 463; C=C, 612 kJ

Answer

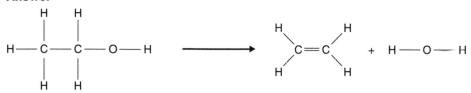

Table 27.5

Bonds broken	Bonds made
5 C–H = 5(412)	4 C–H = 4(412)
1 C–C = 348	1 C=C = 612
1 C–O = 360	2 O–H = 2(463)
1 O–H = 463	Total = 3186 kJ
Total = 3231 kJ	

Energy change = energy needed to break bonds – energy released making bonds

$$= 3231 - 3186 = +45 \text{ kJ}$$

Test yourself

11 Calculate the enthalpy change for each of the following reactions. Use the bond energy values given in the table.

Table 27.6

Bond	C–C	C=C	C–H	C–Br	C=O	O=O	O–H	H–H	H–Br	N≡N	N–H	Br–Br
Bond energy /kJ	348	612	412	276	743	496	463	436	366	944	388	193

a) H——H   +   Br——Br   ⟶   2   H——Br

b) N≡N   +   3 H——H   ⟶   2   H——N——H
                                        |
                                        H

c)
```
   H  H                           H   H
   |  |                           |   |
H——C=C——H   +   Br——Br   ⟶   H——C———C——H
                                  |   |
                                  Br  Br
```

d)
```
   H   H   H
   |   |   |
H——C———C———C——H   +   5  O=O   ⟶   3  O=C=O   +   4  H——O——H
   |   |   |
   H   H   H
```

12 Calculate the enthalpy change for the reaction $H_2 + Cl_2 \rightarrow 2HCl$. Use the following bond energies:
H–H, 436; Cl–Cl, 242; H–Cl, 432 kJ

Show you can

For the reaction shown below the energy change is −30 kJ.

I —— I   +   Cl —— Cl   ⟶   2 I —— Cl

The bond energies are I–I, 150; Cl–Cl, 242 kJ.
Calculate the bond energy of I–Cl.

Practice questions

1 a) For each of the reactions A–E choose the appropriate word from the list below to describe the type of reaction. Each word may be used once, more than once or not at all. *(5 marks)*

combustion   decomposition   neutralisation oxidation   reduction

  A copper(II) carbonate → copper(II) oxide + carbon dioxide

  B ethanoic acid + sodium hydroxide → sodium ethanoate + water

  C magnesium + oxygen → magnesium oxide

  D methane + oxygen → carbon dioxide + water

  E sodium hydroxide + hydrochloric acid → sodium chloride + water

b) Explain the difference between an exothermic reaction and an endothermic reaction. *(2 marks)*

c) For each of the reactions A–E above decide whether it is an exothermic or an endothermic reaction. *(5 marks)*

d) Describe how you would prove experimentally that the reaction between magnesium and hydrochloric acid is an exothermic reaction. *(6 marks)*

2 Photosynthesis is an endothermic process used by plants to produce carbohydrates.

$6CO_2 + 6H_2O \rightarrow C_6H_{12}O_6 + 6O_2$

a) What is meant by the term 'endothermic'? *(1 mark)*

b) Explain what is meant by the term 'activation energy'? *(1 mark)*

c) Draw a labelled reaction profile for this reaction. You must show:
- the position of the reactants and the products
- the activation energy
- the energy change for the reaction. *(3 marks)*

d) Explain in terms of bond making and breaking why this reaction is endothermic. *(3 marks)*

3 Sodium azide (NaN_3) is present in car airbags. During a car crash the airbag rapidly fills with nitrogen gas from the reaction shown below:

$2NaN_3(s) \rightarrow 2Na(s) + 3N_2(g)$

The energy change for the reaction is positive.

a) What is the name given to reactions that have a positive enthalpy change? *(1 mark)*

b) Draw a labelled reaction profile for the reaction. *(3 marks)*

c) Calculate the mass of sodium azide needed to produce 28 g of nitrogen gas. *(3 marks)*

4 Hydrogen reacts with fluorine to form hydrogen fluoride.

$H_2 + F_2 \rightarrow 2HF$

a) Use the bond energies in Table 27.7 to calculate the energy change for this reaction. *(3 marks)*

Table 27.7

Bond	Bond energy/kJ
H–H	436
F–F	158
H–F	568

b) State if the reaction is exothermic or endothermic. *(1 mark)*

5 Figure 27.16 shows the profile for a reaction with a catalyst and without a catalyst.

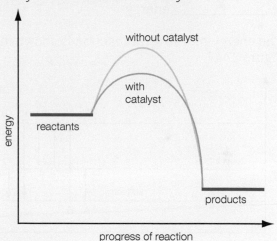

Figure 27.16

a) Is this an exothermic or an endothermic reaction? *(1 mark)*

b) Does a catalyst have an effect on the overall energy change of the reaction? *(1 mark)*

c) On a copy of the diagram, label the activation energy for the catalysed reaction as A and the activation energy for the uncatalysed reaction as B. *(2 marks)*

d) From the information shown in the graph, state the effect of a catalyst on the activation energy of a reaction. *(1 mark)*

6 a) The following reactions all took place in a solution in a beaker. The temperature before and after the chemicals were mixed was recorded in each case. Decide whether each reaction is exothermic or endothermic. *(3 marks)*

Table 27.8

	Start temperature/°C	End temperature/°C
Reaction 1	21	15
Reaction 2	20	27
Reaction 3	22	67

b) Copy the following sentences and fill in the spaces.
In an exothermic reaction, heat energy is transferred from the chemicals to their surroundings and so the temperature _____. In an _____ reaction, heat energy is transferred away from the surroundings to the chemicals and so the temperature _____. *(3 marks)*

c) Copy and complete the spaces in the following sentences.
Chemical reactions can only take place when particles _____ with each other and have enough energy. The minimum energy particles need to react is called the _____ energy. *(2 marks)*

H 7 The energy profile for a reaction is shown in Figure 27.17.

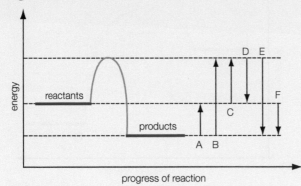

Figure 27.17

a) State the letter of the arrow that shows the activation energy for the reaction. *(1 mark)*

b) State the letter of the arrow that shows the overall energy change for the reaction. *(1 mark)* **H**

c) Is this reaction endothermic or exothermic? *(1 mark)*

8 a) Calculate the energy change for the following reaction using the bond energies given. *(3 marks)*
$H_2 + Cl_2 \rightarrow 2HCl$
Bond energies: H–H = 436, Cl–Cl = 242, H–Cl = 431 kJ

b) Is this reaction endothermic or exothermic? *(1 mark)*

c) Explain your answer in terms of bond energies. *(3 marks)*

9 Calculate the energy change for the following reaction using the bond energies below.

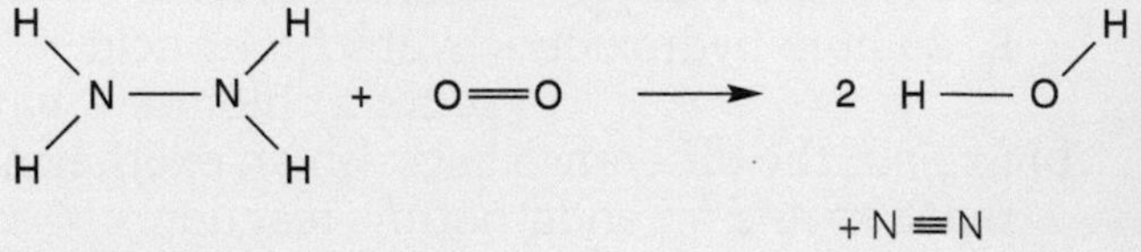

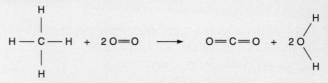

Figure 27.18
Bond energies: N–H, 388; N–N, 158; N≡N, 944; O–H, 463; O=O, 496 kJ *(3 marks)*

10 Methane burns in oxygen, according to this equation:

$$H-\underset{\underset{H}{|}}{\overset{\overset{H}{|}}{C}}-H + 2O=O \rightarrow O=C=O + 2\,O\overset{H}{\underset{H}{}}$$

Figure 27.19

a) Use the bond energies below to calculate the energy change for this reaction. *(3 marks)*
Bond energies: C–H, 412; O=O, 496; C=O, 743; O–H, 463; C–C 348 kJ

b) Write a balanced symbol equation for the combustion of methane. *(1 mark)*

28 Gas chemistry

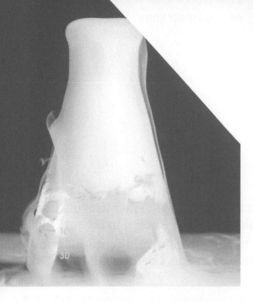

Specification points

This chapter covers specification points 2.9.1 to 2.9.9 and looks at the methods of preparation for carbon dioxide, oxygen and hydrogen, as well as the physical properties and some reactions of these gases. A prescribed practical activity to investigate the preparation, properties, tests and reactions of hydrogen, oxygen and carbon dioxide is also included.

Carbon dioxide gas is unusual, as it sublimes when heated. Dry ice is solid carbon dioxide, and it changes directly to gas (sublimes) when placed in water, producing a white smoke of water droplets, which is used for special effects at concerts. This chapter introduces you to the chemistry of some gases found in the atmosphere, including carbon dioxide and oxygen as well as hydrogen and ammonia.

Gases in the atmosphere

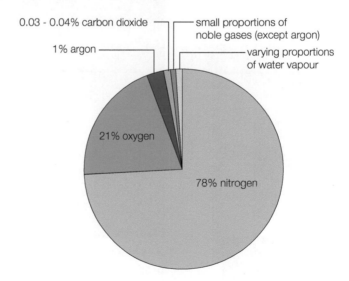

▲ **Figure 28.1** Gas composition of the atmosphere

The atmosphere of the Earth is the layer of gases, commonly known as air, that surrounds the planet. The air is a mixture of gases. Its composition is shown in Table 28.1 and Figure 28.1 (the percentages are approximate).

Table 28.1

Gas	%
nitrogen	78
oxygen	21
argon	1
carbon dioxide	0.03–0.04
noble gases (except argon)	small proportions
water vapour	varying proportions

Nitrogen

Physical properties

▶ colourless, odourless gas
▶ neutral
▶ insoluble in water.

Lack of reactivity

Nitrogen is a diatomic element with the formula N_2. The nitrogen atoms are bonded together by a triple covalent bond, as shown in Figure 28.2.

Nitrogen is an **unreactive** gas due to the **strong triple covalent bond**. Substantial energy is needed to break this bond before the nitrogen atoms can react.

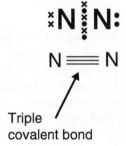

Triple covalent bond

▲ **Figure 28.2** The bonding in a nitrogen molecule

▲ **Figure 28.3** Nitrogen is the gas in packets of crisps

▲ **Figure 28.5** Stem cells are stored in liquid nitrogen

▲ **Figure 28.4** Nitrogen is unreactive at room temperature. The high energy produced by lightning will break the nitrogen–nitrogen triple bond and allow the nitrogen atoms to react with oxygen

Uses of nitrogen

▶ Nitrogen is used in **food packaging** to create an inert atmosphere to keep food fresh. It is safe, unreactive and ensures that delicate foods such as crisps or crackers do not get crushed.

▶ Liquid nitrogen is used to freeze food products and stem cells. It is also used as a **coolant** in refrigerated lorries and in computers.

Ammonia gas

In industry, ammonia is manufactured from nitrogen. Ammonia gas (NH_3) is a very soluble basic gas with a pungent smell.

Test for ammonia

▶ Hold a glass rod dipped in **concentrated hydrochloric acid** near the gas.

▶ If ammonia is present, **white smoke** (ammonium chloride) is formed.

The ammonia reacts with hydrogen chloride gas to form ammonium chloride.

ammonia + hydrogen chloride → ammonium chloride

$$NH_3(g) + HCl(g) \rightarrow NH_4Cl(s)$$

Uses of ammonia

Ammonia is basic, and in industry is reacted with acids to make ammonium salts, which are used as artificial fertilisers. The ammonium salts contain nitrogen, which is good for plant growth.

ammonia + hydrochloric acid → ammonium chloride

$$NH_3 + HCl \rightarrow NH_4Cl$$

ammonia + sulfuric acid → ammonium sulfate

$$2NH_3 + H_2SO_4 \rightarrow (NH_4)_2SO_4$$

ammonia + nitric acid → ammonium nitrate

$$NH_3 + HNO_3 \rightarrow NH_4NO_3$$

▲ **Figure 28.6** A positive test for ammonia. This test should be carried out in a a fume cupboard

Test yourself

1 a) Copy and complete Table 28.2.

Table 28.2

Gas	Symbol/formula	Element or compound	% in atmosphere
nitrogen	N_2		
oxygen			21
argon			1
carbon dioxide			

b) Name one other gas that is a compound and may be present in the atmosphere.

2 Copy and complete Table 28.3.

Table 28.3

Gas	Formula	Dot and cross diagram	Colour	Acidic, basic or neutral
nitrogen				
ammonia				

3 a) Name a gas which is used in food packaging.
   b) Name a gas which is used to make fertilisers.
   c) Name the gas which makes up most of the air.
   d) Describe the test for ammonia gas.
   e) Write an equation for the preparation of ammonium nitrate.
   f) Why is nitrogen an unreactive gas?

▲ **Figure 28.7** A worker using oxyacetylene welding equipment. This equipment burns a mixture of oxygen and acetylene, allowing a flame temperature of about 3300 °C to be attained

Oxygen

Oxygen is a diatomic gas with the formula O_2.

Physical properties

▶ colourless, odourless gas
▶ neutral
▶ slightly soluble in water.

Uses of oxygen

▶ In **welding**, the gas acetylene is mixed with oxygen to increase the temperature at which it burns. Oxyacetylene welding is used to cut and weld steel.
▶ In **medicine**, oxygen is used for patients with breathing difficulties.

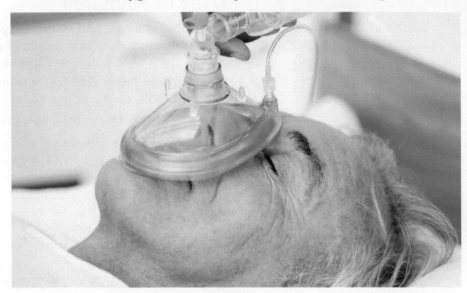

▲ **Figure 28.8** Oxygen is used in hospitals

Test for oxygen

▶ Apply a **glowing** splint.
▶ The glowing splint **relights**.

Laboratory preparation and collection of oxygen

Hydrogen peroxide is a colourless liquid that slowly decomposes (breaks down) to produce water and oxygen. To prepare a sample of oxygen in the laboratory, this decomposition is speeded up using a catalyst of **manganese(IV) oxide** (manganese dioxide), a black powder.

hydrogen peroxide → oxygen + water

$$2H_2O_2 \rightarrow O_2 + 2H_2O$$

Oxygen is produced by the catalytic decomposition of hydrogen peroxide in the apparatus shown in Figure 28.9. It can be collected over water, as it has a low solubility in water. The black solid manganese(IV) oxide catalyst can be filtered off at the end of the reaction, dried and reused.

Tip

Remember that a catalyst is a substance which increases the rate of a reaction without being used up. A catalyst does not feature in a balanced symbol equation.

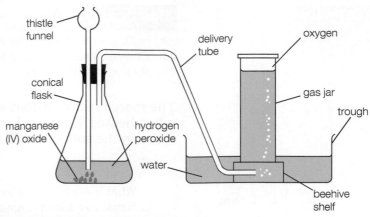

▲ **Figure 28.9** Diagram of the preparation of oxygen

▲ **Figure 28.10** Elephant toothpaste demonstration. When a catalyst such as potassium iodide is added to hydrogen peroxide and some washing up liquid, it rapidly decomposes to water and oxygen. The oxygen is trapped as bubbles in the liquid soap and foams out of the flask. It is often referred to as 'elephant toothpaste'

Burning elements in oxygen

Metals burn in oxygen to produce **basic** oxides. **Non-metals** burn in oxygen to produce **acidic** dioxides. The observations and equations for the burning of some elements are given in Table 28.4.

Table 28.4

Element	Metal or non-metal	Observation before heating	Observation during heating	Observation after heating	Basic oxide or acidic dioxide made?	Symbol equation
magnesium	metal	grey solid	white light	white solid	basic oxide	magnesium + oxygen → magnesium oxide $2Mg(s) + O_2(g) \rightarrow 2MgO(s)$
copper	metal	red–brown solid	glows red blue–green flame	black layer on the copper	basic oxide	copper + oxygen → copper(II) oxide $2Cu(s) + O_2(g) \rightarrow 2CuO(s)$
iron	metal	grey solid	orange sparks	black solid	basic oxide	iron + oxygen → iron oxide $3Fe(s) + 2O_2(g) \rightarrow Fe_3O_4(s)$
sulfur	non-metal	yellow solid	blue flame bad smell melts into a red liquid	colourless, pungent gas	acidic dioxide	sulfur + oxygen → sulfur dioxide $S(s) + O_2(g) \rightarrow SO_2(g)$
carbon	non-metal	black solid	orange sparks	colourless gas	acidic dioxide	carbon + oxygen → carbon dioxide $C(s) + O_2(g) \rightarrow CO_2(g)$

▲ **Figure 28.11** Copper being heated in a Bunsen flame. It becomes covered with a black layer

▲ **Figure 28.12** Yellow sulfur burns with a blue flame

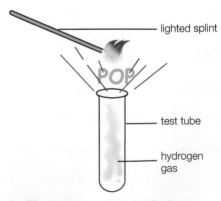

▲ **Figure 28.13** Test for hydrogen

lighted splint

test tube

hydrogen gas

Tips

▶ Fe_3O_4 is a mixed oxide containing FeO and Fe_2O_3.
▶ When giving observations for an element burning it is a good idea to give observations before, during and after the reaction.

The more reactive metals such as magnesium burn completely when heated. However, less reactive metals such as copper only react on the surface and a layer of product forms.

Test yourself

4 a) What is meant by the term 'diatomic'?
  b) Name two diatomic gases found in the air.
5 A colourless liquid is used to prepare oxygen.
  a) Name the liquid.
  b) Write a balanced symbol equation for the preparation of oxygen.
  c) Explain why the reaction in b) is a decomposition.
  d) What is a catalyst?
  e) Name and describe the catalyst used in this preparation.
  f) What is the test for oxygen?
6 a) Write word and symbol equations for the following reactions:
     i) magnesium + oxygen →
     ii) copper + oxygen →
     iii) iron + oxygen →
     iv) sulfur + oxygen →
     v) carbon + oxygen →
  b) Identify the products of the reactions in a) as acidic or basic.
  c) Write observations for reactions i), ii) and iii) in a) above.

Hydrogen

Hydrogen is a diatomic gas with the formula H_2.

Physical properties

▶ colourless, odourless gas
▶ insoluble in water
▶ less dense than air
▶ neutral.

Test for hydrogen

▶ Apply a **lighted** splint to a sample of gas.
▶ A squeaky **pop** indicates the presence of hydrogen (see Figure 28.13).

The reaction that occurs is

$$2H_2 + O_2 → 2H_2O$$

Laboratory preparation and collection of hydrogen

The reagents used to prepare hydrogen in the laboratory are:

▶ dilute hydrochloric acid
▶ zinc metal.

zinc + hydrochloric acid → zinc chloride + hydrogen

Zn + 2HCl → $ZnCl_2$ + H_2

Hydrogen can be collected over water, as it is insoluble. The same apparatus is used as for preparing oxygen and is shown in Figure 28.14.

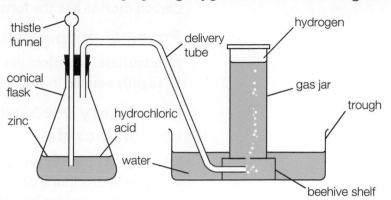

▲ **Figure 28.14** Diagram of the laboratory preparation of hydrogen

Uses of hydrogen

▶ Hydrogen is light, and is used in **weather balloons** to carry weather instruments high into the atmosphere.

▶ It is used in **hardening oils**. Unsaturated vegetable oils are liquid at room temperature, but they can also be 'hardened' by reaction with hydrogen to make them solid at room temperature, and more useful in baking cakes and pastries (see page 290).

Potential as a clean fuel

Hydrogen is a 'clean' fuel as it burns producing only water as a product, which is non-polluting.

hydrogen + oxygen → water

$$2H_2(g) + O_2(g) \rightarrow 2H_2O(g)$$

Fossil fuels are running out, and they contribute to the greenhouse effect. Hydrogen has potential as an alternative 'clean' fuel. It can be used safely in a fuel cell, where the reaction above occurs, producing only water, and fuel cells are being used to power some vehicles. There is not widespread use of hydrogen fuel cells yet due to their significant cost and the other disadvantages given in Table 28.5.

Table 28.5 Advantages and disadvantages of hydrogen as a fuel

Advantages	Disadvantages
Burns to produce water only, so it causes no pollution.	Hydrogen is very flammable, so it is difficult to transport and store.
Hydrogen can be produced from the electrolysis of water. Hence it is renewable, and if the electricity used to electrolyse the water comes from an alternative source, (e.g. solar) the use of hydrogen can help conserve fossil fuels.	The production of hydrogen may cause pollution: • Most hydrogen is made from burning methane, which releases the greenhouse gas carbon dioxide in the process. • It can be made from the electrolysis of water, which may use electricity generated from fossil fuels, which also give off the greenhouse gas carbon dioxide.
	The hydrogen needed for fuel cells must be liquefied or stored under pressure in the vehicles. These methods of storage are expensive.
	Fuel cells are very expensive.

▲ **Figure 28.15** This weather balloon from Cape Canaveral Air Force Station is used to check atmospheric conditions prior to the launch of a satellite. It carries a radiosonde, which measures atmospheric pressure, humidity, temperature and wind speeds, and transmits the data to a receiver

▲ **Figure 28.16** Some vehicles, including buses and cars, are powered by hydrogen fuel cells

Carbon dioxide

Carbon dioxide has the formula CO_2.

Physical properties

▶ colourless, odourless gas
▶ slightly soluble in water

▶ acidic
▶ denser than air.

Laboratory preparation and collection of carbon dioxide

The reagents used to prepare carbon dioxide in the laboratory are:
▶ dilute hydrochloric acid
▶ calcium carbonate (marble chips).

calcium carbonate + hydrochloric acid → calcium chloride + water + carbon dioxide

$$CaCO_3 + 2HCl \rightarrow CaCl_2 + H_2O + CO_2$$

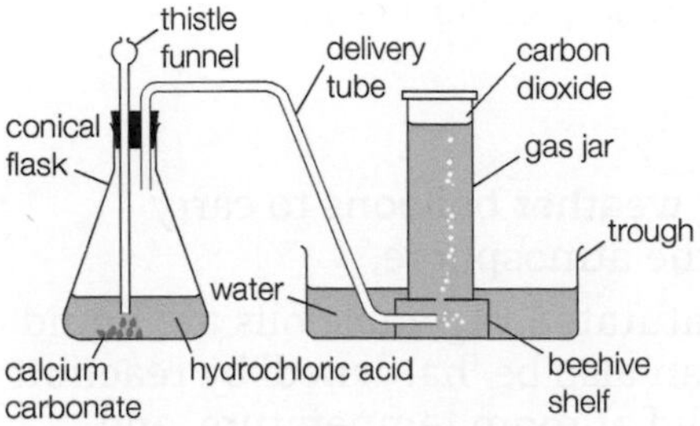

Carbon dioxide can be collected over water, as it has low solubility in water. The same apparatus is used as for preparing oxygen and hydrogen (see Figure 28.17).

▲ **Figure 28.17** Diagram of the laboratory preparation of carbon dioxide

Test for carbon dioxide

▶ Bubble the gas into **colourless limewater**.
▶ If the limewater turns **milky** the gas is carbon dioxide.

If excess carbon dioxide is bubbled into the milky limewater it will eventually go colourless again.

Limewater is calcium hydroxide solution.

carbon dioxide + limewater → calcium carbonate + water

$$CO_2(g) + Ca(OH)_2(aq) \rightarrow CaCO_3(s) + H_2O(l)$$

The calcium carbonate produced is insoluble, so a milky white precipitate forms.

When excess carbon dioxide is added soluble calcium hydrogencarbonate is formed, so the white precipitate disappears and a colourless solution is formed.

calcium carbonate + water + carbon dioxide → calcium hydrogencarbonate

$$\underbrace{CaCO_3(s) + H_2O(l)}_{\text{milky limewater}} + CO_2(g) \rightarrow \underset{\text{colourless solution}}{Ca(HCO_3)_2(aq)}$$

▲ **Figure 28.18** Bubbling carbon dioxide through limewater first produces a white precipitate, which with excess carbon dioxide dissolves to produce a colourless solution

Reaction of carbon dioxide with water

Carbon dioxide reacts with water to form the weak acid, carbonic acid.

carbon dioxide + water → carbonic acid

$$CO_2(g) + H_2O(l) \rightarrow H_2CO_3(aq)$$

Uses of carbon dioxide

▶ To make **drinks fizzy**.
▶ Figure 28.19 shows firefighters using carbon dioxide **fire extinguishers** to put out a fire. Carbon dioxide is denser than air so it sinks over the flames, pushing out the surrounding oxygen and starving the fire of oxygen.

▲ **Figure 28.19** Firefighters

Test yourself

7 Copy and complete Table 28.6.

Table 28.6

	Hydrogen	Carbon dioxide
Formula		
Element or compound?		
Acidic, basic or neutral?		
Lighter or denser than air?		
Test for the gas		

8 a) Name two substances used to prepare hydrogen.
  b) Write a balanced symbol equation for the preparation of hydrogen gas.

9 a) Why is hydrogen described as a 'clean' fuel?
  b) Why are hydrogen fuel cells not used extensively in place of fossil fuels?
  c) Why is hydrogen used when making margarine from olive oil?

10 a) Name two substances used to prepare carbon dioxide.
  b) Write a balanced symbol equation for the preparation of carbon dioxide.
  c) Name the product formed when carbon dioxide reacts with water.
  d) State two uses of carbon dioxide.

Show you can

Copy and complete Table 28.7.

Table 28.7

Gas	Acidic, basic or neutral?	Formula	Solubility in water
nitrogen			
ammonia			
carbon dioxide			
oxygen			
hydrogen			

Prescribed practical

Prescribed practical C6: Investigating the preparation, properties, tests and reactions of the gases hydrogen, oxygen and carbon dioxide

1 Table 28.8 shows some reactions which were carried out in an experiment. Copy and complete the table to give observations for each reaction and the name of the gas produced.

Table 28.8

Reaction	Observations	Name of gas produced
zinc is added to some hydrochloric acid		
calcium carbonate is added to some hydrochloric acid		
hydrogen peroxide is added to some manganese(IV) oxide		

2 In a second part of the experiment some elements were burnt in oxygen. The products were shaken with water and tested with red and blue litmus paper. Copy and complete Table 28.9 to give the symbol equation for each reaction and the colour of red and blue litmus paper after testing the products.

Table 28.9

	Symbol equation	Colour of red litmus paper	Colour of blue litmus paper
A spatula of sulfur was burnt in a gas jar of oxygen.			
A piece of carbon was burnt in a gas jar of oxygen.			
A piece of calcium was burnt in a gas jar of oxygen.			

3 Some carbon dioxide was bubbled into water, and the solution tested with red and blue litmus.
  a) Explain what would be observed.
  b) Describe how a sample of gas would be tested to prove that it is carbon dioxide.

1 Copy and complete Table 28.10 to give information about the three gases.

Table 28.10

	Hydrogen	Carbon dioxide	Oxygen
Formula	*(1 mark)*	*(1 mark)*	*(1 mark)*
Names of substances used to prepare the gas	*(2 marks)*	*(2 marks)*	*(2 marks)*
Test for the gas	*(2 marks)*	*(3 marks)*	*(2 marks)*
Equation for preparation of the gas	*(3 marks)*	*(3 marks)*	*(3 marks)*

2 Hydrogen is a colourless, odourless, tasteless gas.
a) State two other physical properties of hydrogen gas. *(2 marks)*
b) Describe the test for hydrogen, and include the result for a positive test. *(2 marks)*
c) Hydrogen is prepared in the laboratory using an acid and a metal. Complete the following general word equation:

acid + metal → _____ + _____ *(1 mark)*
d) Name the two specific substances used to prepare hydrogen in the laboratory. *(2 marks)*
e) State two observations when the substances named in d) react together. *(2 marks)*
f) Draw a labelled diagram to show the assembled apparatus used to prepare a gas jar of hydrogen. *(5 marks)*
g) Hydrogen is used to power some cars. Why is hydrogen described as a clean fuel? *(2 marks)*
h) Why is hydrogen used in weather balloons? *(1 mark)*

3 Figure 28.20 shows the apparatus used to prepare a gas.

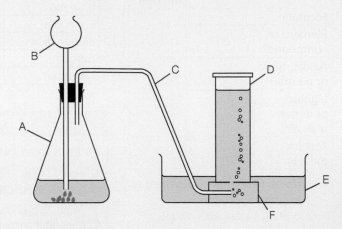

Figure 28.20
a) Name each piece of apparatus A–F. *(6 marks)*
b) Explain what you can deduce about the solubility of the gas produced. *(1 mark)*
c) Name two gases which could be prepared in this apparatus using the method shown. *(2 marks)*
d) For each gas named in c) name the solid used as reactant. *(2 marks)*

4 Nitrogen, oxygen and carbon dioxide are found in air.
a) State the approximate percentage of each of these gases in air. *(3 marks)*
b) Explain why nitrogen is an unreactive gas. *(2 marks)*
c) During a lightning storm, nitrogen reacts with oxygen in the air to form nitrogen dioxide. Write a balanced symbol equation for this reaction. *(2 marks)*
d) Nitrogen reacts with hydrogen, under pressure and high temperature, to form ammonia. Write a balanced symbol equation for this reaction. *(3 marks)*
e) Describe the chemical test for ammonia and state the result for a positive test. *(2 marks)*
f) Ammonia is used to make the fertiliser ammonium sulfate. Write a balanced symbol equation for this reaction. *(3 marks)*
g) Carbon dioxide reacts with water in the atmosphere. Write a balanced symbol equation for this reaction and name the product. *(3 marks)*

h) In the test for carbon dioxide, limewater changes to milky and on adding excess carbon dioxide, changes to colourless.

   i) What is the chemical name for limewater? *(1 mark)*

   ii) What is the colour of limewater? *(1 mark)*

   iii) Why does the limewater change to milky? *(2 marks)*

   iv) Why does the limewater eventually change to colourless? *(2 marks)*

i) Write a balanced symbol equation for the preparation of carbon dioxide. *(3 marks)*

j) State two uses of carbon dioxide. *(2 marks)*

5 Oxygen is used in respiration and in breathing.

   a) State one other use of oxygen. *(1 mark)*

   b) Oxygen is prepared in the laboratory using hydrogen peroxide.

     i) State the appearance of hydrogen peroxide. *(1 mark)*

     ii) State the name and describe the appearance of the catalyst used to prepare oxygen. *(2 marks)*

     iii) Write a balanced symbol equation for the preparation of oxygen. *(3 marks)*

   c) Compare and contrast the reaction of copper and magnesium with oxygen. In your answer you should give:

- the names of the products
- observations
- comparison of reactivity.

In your answer your quality of written communication will be assessed. *(6 marks)*

   d) Some carbon and some sulfur are burnt in oxygen.

     i) State three observations for sulfur burning in oxygen. *(3 marks)*

     ii) Write a symbol equation for carbon burning in oxygen. *(2 marks)*

     iii) Is the product in **ii)** acidic or basic? *(1 mark)*

6 Five samples of gases are carbon dioxide, ammonia, hydrogen, oxygen and nitrogen. Describe how you would carry out tests to identify the gases. *(10 marks)*

29 Motion

Specification points

This chapter covers sections 1.1.1 to 1.1.6 of the specification. It also covers Prescribed practical P1. The relationships between distance, average speed, time and rate of change of speed through practical work are covered. It also introduces graphical methods of describing motion. At the higher tier, students are introduced to the concept of vectors and scalars, and they learn about the terms displacement, velocity and acceleration.

Distance and displacement

Distance is the separation between two points. For example, the distance between Belfast and Coleraine is 75km (see Figure 29.1).

We define displacement as distance in a specified direction. The displacement of Belfast from Coleraine is 75 km south-east. Displacement is represented by an arrow – the length of the arrow is proportional to the distance, and the direction of the arrow is in the same direction as the displacement (see Figure 29.2).

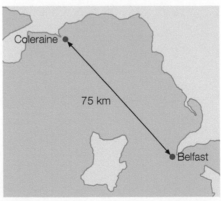

▲ **Figure 29.1** The distance between Coleraine and Belfast

▲ **Figure 29.2** The displacement of Belfast from Coleraine

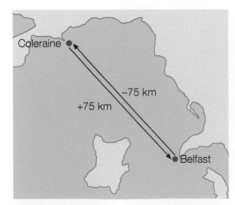

▲ **Figure 29.3** The displacement on a return journey is 0 km

The return journey from Coleraine to Belfast and then back again is a distance of 150 km, but the displacement is 0 km.

The addition of +75 km to –75 km = 0 km (see Figure 29.3).

We say that distance is a scalar quantity – a quantity with size only – whereas displacement is a vector quantity because it has size and direction.

Speed

If a car travels between two points on a road, its speed can be calculated using the formula:

$$\text{speed} = \frac{\text{distance}}{\text{time}}$$

If distance is measured in metres (m) and time in seconds (s), speed is measured in metres per second (m/s).

For example, if a car travels from Coleraine to Belfast in 2 hours, its speed is:

$$\frac{75\,\text{km}}{2\,\text{h}} = 37.7\,\text{km/h}$$

It is unlikely that the car travelled at exactly 37.5 km/h for the whole journey, so this is instead the average speed. The formula for average speed is:

$$\text{average speed} = \frac{\text{total distance}}{\text{time}}$$

To find the actual (or instantaneous) speed at a particular moment in time, we would need to know the distance travelled in a very short interval of time.

time = 0 s   time = 3 s

60 m

▲ **Figure 29.4** Calculating instantaneous speed at a moment in time

Show you can

Explain the difference between:
a) distance and displacement
b) speed and velocity.

Test yourself

1 A car travels 800 m in 40 s.
   a) What is its average speed?
   b) Why is its actual speed usually different from its average speed?
2 A car has a steady speed of 10 m/s.
   a) How far does the car travel in 9 s?
   b) How long does it take the car to travel 220 m?
3 Calculate the average speed of each of the objects shown in Figure 29.5.

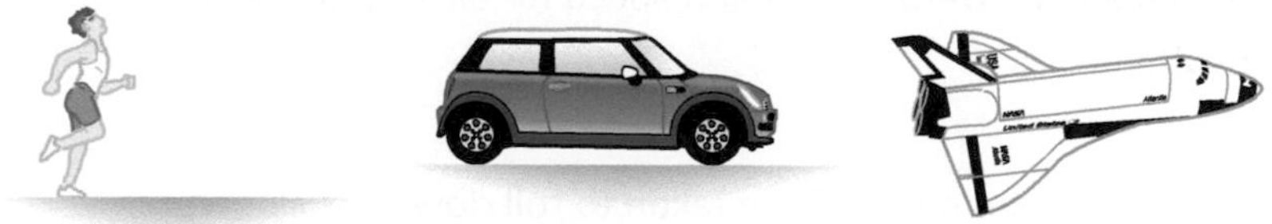

this runner travels 400 m in 44 s   this car travels 210 kilometres in 3 hours   this shuttle travels 43 750 kilometres in 2.5 hours

▲ **Figure 29.5**

   a) a runner who travels 400 m in 44 s
   b) a car that travels 210 km in 3 hours
   c) a shuttle that travels 43 750 km in 2.5 hours.

Prescribed practical

Prescribed practical P1: Investigating how the average speed of an object moving down a runway depends on the slope of the runway

Aims
- to measure average speed
- to investigate how average speed varies with the height of the slope

Apparatus
- steel ball-bearing
- runway
- metre ruler
- stopwatch

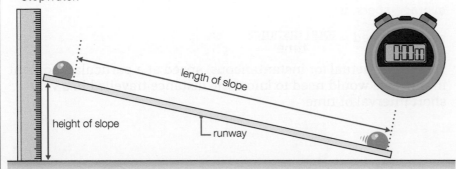

▲ **Figure 29.6** Diagram of apparatus

Method
1 Prepare a table for your results similar to Table 29.1.
2 Measure the length of the slope using the metre ruler.
3 Position the runway so that the ball-bearing is a vertical distance of 20 cm from the desk.
4 Release the ball-bearing and, using a stopwatch, measure the time taken for the ball to travel the length of the slope.
5 Record your result in your table to 2 d.p. in the column headed '1'.
6 Repeat this measurement and record your new result in the column headed '2'.
7 Repeat this process for slope heights of 30 cm, 40 cm and 50 cm.
8 Record all your times in the appropriate column in Table 29.1 to 2 d.p.
9 Calculate an average time and speed for each slope height.

Results
Table 29.1

Height of slope/cm	Time taken to roll down the slope/s			Average speed/cm/s
	1	2	Average	
20				
30				
40				
50				

Conclusion
Comment on the relationship between the height of the slope and the average speed of the ball-bearing.

Rate of change of speed

If the speed changes over a period of time (t) it is possible to calculate the rate of change of speed with respect to time, using the formula:

$$\text{rate of change of speed} = \frac{\text{final speed} - \text{initial speed}}{\text{time taken}}$$

This rate of change of speed is a scalar quantity and its units are m/s².

Velocity

Whereas speed is the distance travelled in unit time, velocity is the distance travelled in unit time in a **specified direction**.

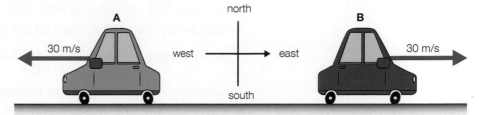

▲ **Figure 29.7** These two cars have the same speed, but different velocities

In Figure 29.7, car A has the same speed as car B, but a different velocity. Car A's velocity is 30 m/s due west, while car B has a velocity of 30 m/s due east.

Speed is a scalar quantity but velocity is a vector quantity.

Since displacement is the distance travelled in a specified direction, we can rewrite the formula for velocity as:

$$\text{average velocity} = \frac{\text{displacement}}{\text{time}}$$

The units for speed and velocity are the same, metres per second (m/s). Occasionally, you will see the units of kilometres per hour (km/h).

The car in Figure 29.8 is moving at a steady, or constant, speed of 20 m/s as it goes around a bend.

The speed of the car at A, B and C is 20 m/s, but the velocity changes as it travels from A to B to C. This is because velocity is a vector quantity, and although the size of the velocity may be constant at 20 m/s, its direction is constantly changing, so its velocity is constantly changing.

Another formula for average velocity is:

$$\text{average velocity} = \frac{\text{initial velocity} + \text{final velocity}}{2}$$

Example

A car moves from 10 m/s to 30 m/s in a time of 5 seconds. Calculate its rate of change of speed.

Answer

initial speed = 10 m/s
final speed = 30 m/s
time = 5 s

$$\text{rate of change of speed} = \frac{\text{final speed} - \text{initial speed}}{\text{time taken}}$$

$$= \frac{30 - 10}{5}$$

$$= 4 \text{ m/s}^2$$

Show you can

Prove that 72 km/h is equivalent to 20 m/s.

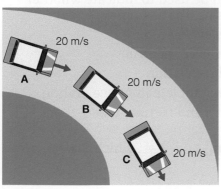

▲ **Figure 29.8** The velocity of the car changes as it goes around the bend

Example

A stone is thrown vertically downwards so that its velocity increases from 2 m/s to 24 m/s. Calculate the average velocity of the stone.

Answer

final velocity = 24 m/s
initial velocity = 2 m/s

$$\text{average velocity} = \frac{\text{initial velocity} + \text{final velocity}}{2}$$

$$= \frac{2 + 24}{2}$$

$$= 13 \text{ m/s}$$

Acceleration

When the velocity of a body increases or decreases, we say it **accelerates**. Consider the example in Figure 29.9.

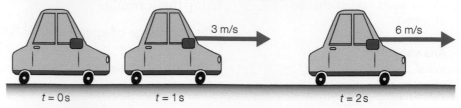

3 m/s

6 m/s

$t = 0\,s$ $t = 1\,s$ $t = 2\,s$

▲ **Figure 29.9** This car is accelerating

The car starts from rest (initial velocity = 0 m/s), but after 1 second its velocity has increased to 3 m/s. After 2 seconds its velocity has increased by 3 m/s to 6 m/s. We say that the car's velocity increases by 3 m/s in 1 second due east – i.e. its acceleration is 3 m/s² due east.

We can define acceleration as the change in velocity in unit time:

$$\text{acceleration} = \frac{\text{final velocity} - \text{initial velocity}}{\text{time}}$$

Acceleration is measured in metres per second squared, written as m/s². Since acceleration is a vector quantity, it can be shown using an arrow of appropriate length and direction.

Most importantly, a '+' or '–' sign can be used to indicate whether the velocity is increasing or decreasing. For example:

+3 m/s² (velocity increasing by 3 m/s every second)

–3 m/s² (velocity decreasing by 3 m/s every second)

A negative acceleration is called a **deceleration**.

A uniform acceleration means a constant (steady) acceleration.

If v = final velocity, u = initial velocity, and t = time taken, then $v - u$ = change in velocity.

So acceleration, a, can be found using $a = \frac{v - u}{t}$, which rearranges to give $v = u + at$.

Show you can

a) A train has an acceleration of 3 m/s². What does this tell you about the velocity of the train?

b) A bus has a deceleration of 2 m/s². What does this tell you about the velocity of the bus?

Example

1 A car starts from rest. After 10 seconds, it is moving with a velocity of 15 m/s. Calculate its acceleration.

Answer

$v = 15\,m/s$

$u = 0\,m/s$

$t = 10\,s$

$a = \frac{15 - 0}{10}$

$= 1.5\,m/s^2$

2 A ball is dropped and accelerates downwards at a rate of 10 m/s² for 5 seconds. By how much will the ball's velocity increase?

Answer

$t = 5\,s$

$a = \frac{v - u}{t}$

$10 = \frac{v - u}{5}$

$v - u = 50\,m/s$

Test yourself

4 A car takes 8 s to increase its velocity from 3 m/s to 30 m/s. What is its acceleration?

5 A motorbike, travelling at 25 m/s, takes 5 s to come to a halt. What is its deceleration?

6 An aircraft on take-off has a uniform acceleration of 4 m/s².
   a) What velocity does the aircraft gain in 5 s?
   b) If the aircraft passes a point on the runway at a velocity of 28 m/s, what will its velocity be 8 s later?

7 A ball is thrown vertically upwards in the air, leaving the hand at 30 m/s. The acceleration due to gravity is 10 m/s². Figure 29.10 shows the positions of the ball over time and a data table.

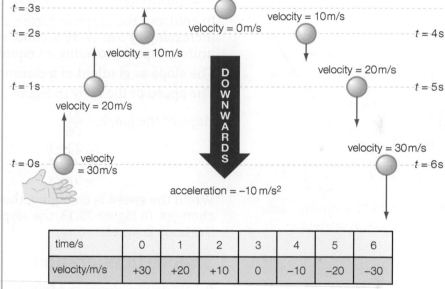

time/s	0	1	2	3	4	5	6
velocity/m/s	+30	+20	+10	0	−10	−20	−30

▲ **Figure 29.10** The changing velocity of a ball thrown into the air

Draw a graph to show the motion of the ball. Plot velocity on the *y*-axis and time on the *x*-axis.

Graphs and motion

Graphs are a very useful way of displaying the motion of objects. There are two main types of graphs used:

▶ distance–time graph
▶ speed–time graph.

Distance–time graphs

A **distance–time graph** is a plot of distance on the y-axis versus time on the x-axis.

The simplest type of distance–time graph is shown in Figure 29.11.

This graph illustrates that although the time increases steadily, the distance travelled does not change. The body must be **stationary**.

A horizontal line on a distance–time graph means that the body is stationary.

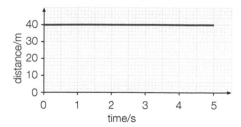

▲ **Figure 29.11** A distance–time graph

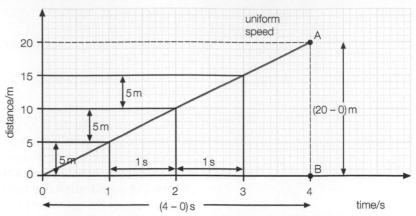

▲ **Figure 29.12** A distance–time graph showing uniform speed

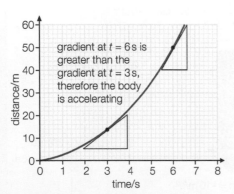

▲ **Figure 29.13** A distance–time graph showing acceleration

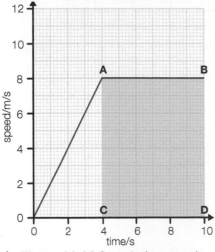

▲ **Figure 29.14** Speed–time graph for a bicycle

In contrast, the graph in Figure 29.12 shows that the distance is increasing by 5 m every second, i.e. the body is travelling with uniform speed, covering an equal distance in equal units of time.

The slope or gradient of a distance–time graph represents the speed. The speed of the body in Figure 29.12 is 5 m/s.

$$\text{slope of the graph} = \frac{AB}{OB}$$
$$= \frac{20 - 0}{4 - 0}$$
$$= 5\,\text{m/s}$$

When the speed is changing, the slope of the distance–time graph changes. In Figure 29.13, the slope is increasing, which means that the body is accelerating.

Speed–time graphs

A speed–time graph is a plot of speed on the y-axis versus time on the x-axis.

An example of a speed–time graph for a bicycle is shown in Figure 29.14.

In the first 4 seconds of its motion, the speed of the bicycle increases steadily. The gradient of the line OA is the rate of change of speed.

$$\text{gradient} = \frac{\text{change in } y}{\text{change in } x}$$
$$= \frac{8 - 0}{4 - 0}$$
$$= 2\,\text{m/s}^2$$

From time $t = 4$ seconds to time $t = 10$ seconds, the speed of the bicycle remains constant (steady) at 8 m/s. There is no rate of change of speed for the bicycle. The area under a speed–time graph represents the distance travelled.

For example, in Figure 29.14 from time $t = 4$ s to time $t = 10$ s:

distance travelled at constant speed = area of rectangle CABD
$$= CA \times CD$$
$$= 8 \times 6$$
$$= 48\,\text{m}$$

The distance travelled by the bicycle during this time is 48 m.

Displacement–time graphs

A **displacement–time graph** is a plot of displacement on the y-axis versus time on the x-axis.

This type of graph is very similar to a distance–time graph. The similarity is that a horizontal line on a displacement–time graph means that the body is stationary.

In contrast, the slope or gradient of a displacement–time graph represents velocity and not speed.

Consider Figure 29.15, which represents the movement of a toy car.

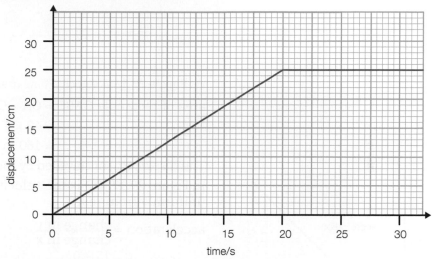

▲ **Figure 29.15** Displacement–time graph for a toy car

From time t = 0 s to time t = 20 s, the displacement of the toy car increases steadily at a rate of 25 cm in 20 s, which is 1.25 cm/s. The velocity is 1.25 cm/s.

From time t = 20 s onwards, the displacement is 25 cm. The toy car is stationary during this time.

Velocity–time graphs

A **velocity–time graph** is a plot of velocity on the y-axis versus time on the x-axis.

The simplest type of velocity–time graph is shown in Figure 29.16.

This graph illustrates that while time is increasing, the velocity remains at a constant (steady) 30 m/s. The car is not accelerating.

In addition, the area under a velocity–time graph represents the displacement travelled.

For example, in Figure 29.16:

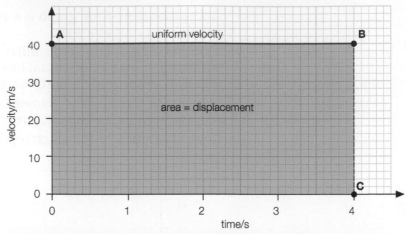

▲ **Figure 29.16** A velocity–time graph showing uniform velocity

area of rectangle OABC = OA × OC

$$= 40 \times 4$$

$$= 160 \text{ m}$$

The displacement travelled by the car is 160 m in a specified direction.

In Figure 29.17, OD is the velocity–time graph for a body accelerating uniformly from rest.

$$\text{acceleration} = \frac{\text{change in } y}{\text{change in } x}$$

$$= \frac{15 - 0}{3 - 0}$$

$$= 5 \text{ m/s}^2$$

The slope or gradient represents the acceleration of the body.

Furthermore, the area of the triangle OCD gives the displacement travelled.

displacement = area of triangle OCD

$$= \tfrac{1}{2} \times \text{OC} \times \text{CD}$$

$$= \tfrac{1}{2} \times 3 \times 15$$

$$= 22.5 \text{ m}$$

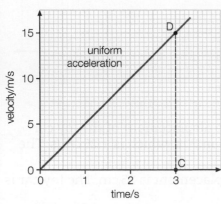

▲ **Figure 29.17** A velocity–time graph showing uniform acceleration

Test yourself

8 Copy and complete Table 29.2.

Table 29.2

Name of quantity	Definition	Scalar or vector
Distance		scalar
Displacement		
Speed	Rate of change of distance with respect to time	scalar
Velocity		vector
Acceleration	Rate of change of velocity with respect to time	

Practice questions

1 Paul and Jim set off at the same time from their separate houses to travel to a nearby shop. Table 29.3 shows the distances travelled by Paul to the shop.

Table 29.3

Distance travelled by Paul/m	0	3	6	9	12	15	18	21
Time elapsed/s	0	1	2	3	4	5	6	7

a) Draw a graph of distance against time for Paul's journey. *(3 marks)*

Table 29.4 shows the distances travelled by Jim to the same shop.

Table 29.4

Distance travelled by Jim/m	0	2	4	6	8	10	12	14
Time elapsed/s	0	1	2	3	4	5	6	7

b) Draw a graph of distance against time for Jim's journey on the same axes. *(3 marks)*

c) Use the graphs to answer the following questions.
 i) Which person is going faster? *(1 mark)*
 ii) How long does it take Paul and Jim to travel 11 m? *(1 mark)*
 iii) How far apart are Paul and Jim after 2.5 s? *(2 marks)*
 iv) Is Paul's speed steady? *(1 mark)*
 v) What is Jim's average speed? *(3 marks)*

2 Study the velocity–time graph (Figure 29.18) and describe in words the motion of the object. *(6 marks)*

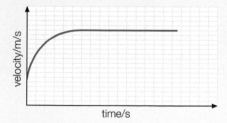

Figure 29.18

3 Figure 29.19 shows a velocity–time graph for a car accelerating away from a junction. Calculate:
 a) the acceleration during the first 5 s *(3 marks)*
 b) the total displacement. *(4 marks)*

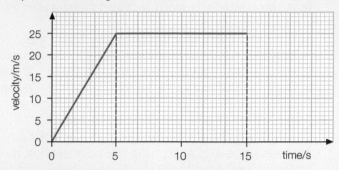

Figure 29.19

4 The graph in Figure 29.20 represents a journey in a lift in a hospital.

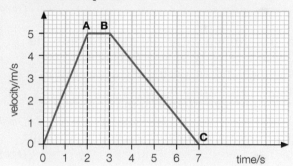

Figure 29.20

a) Briefly describe the motion represented by
 i) OA
 ii) AB
 iii) BC. *(3 marks)*

b) Use the graph to calculate:
 i) the initial acceleration of the lift *(3 marks)*
 ii) the total distance travelled by the lift *(4 marks)*
 iii) the average speed of the lift for the whole journey. *(3 marks)*

Use a graphical method or formula to answer questions 5–11.

5 A car accelerates at $3\,m/s^2$ for 10 seconds. It started with a velocity of $20\,m/s$. Calculate its final velocity. *(3 marks)*

6 An ice-skater moves off from rest (this means that $u = 0\,m/s$) with a uniform acceleration of $3.0\,m/s^2$. What is her speed and distance travelled after 10 s? *(5 marks)*

7 A stone is thrown vertically upwards with a velocity of 20 m/s. Find how high it will go and the time taken to reach this height (assume $a = -10$ m/s²). *(6 marks)*

8 A stone is dropped down an empty mine shaft. It takes 3 seconds to reach the bottom. Assuming that the stone falls from rest and accelerates at 10 m/s², calculate:
  a) the maximum speed reached by the stone before hitting the bottom *(3 marks)*
  b) the average speed of the stone in flight *(3 marks)*
  c) the depth of the mine shaft. *(3 marks)*

9 A helicopter at a height of 500 m drops a package which falls to the ground. Neglecting air resistance and assuming the acceleration is constant and equal to 10 m/s², calculate:
  a) the time taken for the package to reach the ground *(3 marks)*
  b) the average velocity of the package. *(3 marks)*

10 A ball is thrown vertically upwards into the air with a velocity of 50 m/s. Neglecting air resistance and assuming the acceleration is constant and equal to –10 m/s², calculate:
  a) the time taken to reach maximum height *(3 marks)*
  b) the maximum height reached by the ball. *(3 marks)*

11 A cyclist accelerates from rest at 3 m/s².
  a) What is his speed after 5 s? *(3 marks)*
  He then decelerates at 0.5 m/s².
  b) How long will it take for his speed to reach zero? *(3 marks)*
  c) Draw a velocity–time graph for this motion. *(3 marks)*

12 John makes a return journey to his local shop to buy a newspaper. In Figure 29.21, Graph A is the distance–time graph and Graph B is the displacement–time graph for the journey.
  a) Copy and complete Graph B. *(1 mark)*
  b) What is the average velocity for John's journey? *(1 mark)*

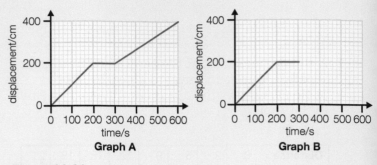

Figure 29.21

13 A sponge ball is allowed to drop from rest and hits the ground after 3.0 seconds. It rebounds to a new height after a further 1.5 seconds. The velocity–time graph of the motion is shown in Figure 29.22.

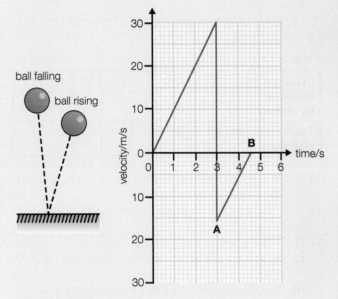

Figure 29.22

The part of the graph AB represents the ball rebounding.
  a) Why is this part of the graph drawn below the time axis? *(1 mark)*
  b) Use the graph to find the height of the rebound. *(3 marks)*
  c) When a ball falls freely under the force of gravity, its acceleration is known as the acceleration of free fall. Use the graph to calculate the acceleration of free fall. *(3 marks)*

Balanced and unbalanced forces

A force has both size and direction. It is another example of a vector quantity. The size of the force is measured in **newtons** (N). When drawing force diagrams, we represent the direction of the force with an arrow and the size of the force by drawing the length of the arrow to scale. By convention, forces acting to the right are positive, and forces acting to the left are negative.

If the forces are equal in size and opposite in direction, then the forces are balanced. Balanced forces do not change the velocity of an object.

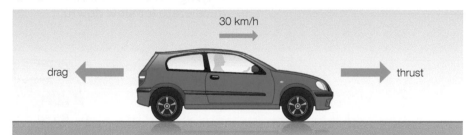

▲ **Figure 30.1** Forces acting upon a car travelling at a steady speed

Figure 30.1 shows a car travelling at a steady speed of 30 km/h in a straight line under the action of two equal and opposite forces: the thrust exerted by the engine, and air resistance (drag).

If an object is stationary (not moving), it will remain stationary.

In a tug of war like that in Figure 30.2, two teams pull against each other. When both teams pull equally hard, the forces are balanced and the rope does not move. But when one team starts to pull with a larger force, the rope moves. This is how we can tell that the two forces are no longer balanced.

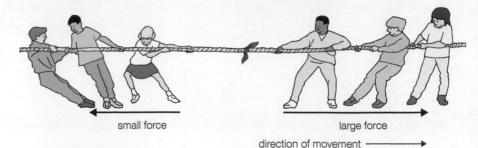

small force

large force

direction of movement ⟶

▲ **Figure 30.2** The forces in this tug of war are unbalanced as the team on the right is pulling with a larger force

Unbalanced forces will change the velocity of an object. Since velocity involves both speed and direction, unbalanced forces can make an object speed up, slow down or change direction.

Unbalanced forces applied to the handlebars will make the cyclist in Figure 30.3 change direction. This means the velocity of the cyclist will change, even though the speed may stay the same.

An object will only accelerate when an unbalanced force acts on it. It then accelerates in the direction of the unbalanced force. In Figure 30.4, if the driving force on a car is greater than the frictional force, the car will accelerate or speed up.

pushing force

pulling force

▲ **Figure 30.3** Unbalanced forces acting upon the handlebars affect velocity

drag

driving force

▲ **Figure 30.4** This car is accelerating

If the driver then decides to apply the brakes (Figure 30.5), the driving force will be smaller than the braking force, and the car will decelerate (or slow down).

braking force

driving force

▲ **Figure 30.5** This car is decelerating

A car is travelling in a straight line along a motorway.

Table 30.1 shows the situations in which there is an unbalanced force on the car.

Table 30.1

Situation	Unbalanced force acting
The car's speed is increasing.	yes
The car's speed is decreasing.	yes
The car's speed is constant.	no
The car starts going round a bend.	yes

Newton's laws

All that we have said about forces so far is summarised by **Newton's first law**:

In the absence of unbalanced forces, an object will continue to move in a straight line at constant speed (its velocity is constant).

Practical activity

An experiment to investigate Newton's first law

Apparatus
▶ linear air track and blower
▶ glider and interrupt card
▶ two light gates and a data logger

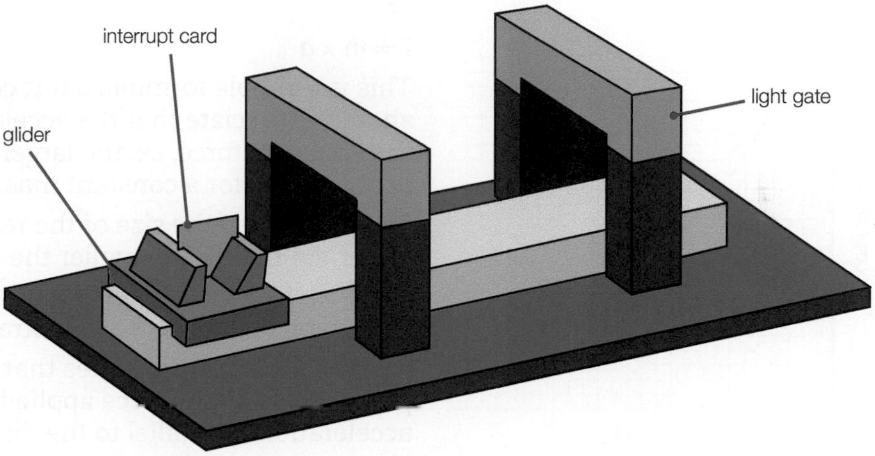

interrupt card

light gate

glider

▲ **Figure 30.6** The apparatus for the experiment

Method

1 Set the linear air track on a flat surface and adjust the feet on the air track to make sure that it is level.

2 Measure the length of the interrupt card and enter this in the data logger.

3 Connect up the light gates so that they measure the velocity of the glider at two points.

4 Give the glider a gentle push so that it passes through both light gates.

5 Confirm by looking at the results that the velocity does not change between the two positions of the light gate and so the glider is obeying Newton's first law of motion.

6 Repeat for other velocities and positions of the light gates.

Linking unbalanced forces, mass and acceleration

Look at Figure 30.7. It is possible for one person to push a car, but the acceleration of the car would be small. The more people pushing the car, the larger the acceleration. So, the larger the force, the larger the acceleration.

Now look at Figure 30.8. Even four people would find it difficult to push a van, because the mass of a van is usually far larger than the mass of a car. The larger the mass, the smaller the acceleration.

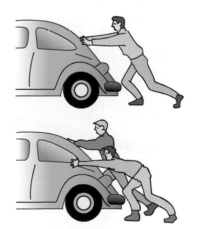

▲ **Figure 30.7** People exerting force upon a car

A

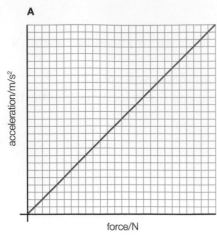

B

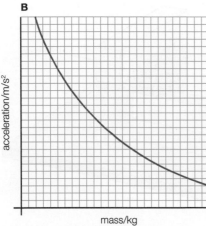

▲ **Figure 30.9** Graphs illustrating Newton's second law

▲ **Figure 30.8** People trying to push a van, a much larger mass

The size of the resultant or unbalanced force needed to accelerate a mass can be worked out using **Newton's second law**.

resultant force (N) = mass (kg) × acceleration (m/s²)

or

$$F = m \times a$$

This is a simple formula, but it contains a lot of information. You should appreciate that the acceleration is directly proportional to the resultant force, i.e. the larger the resultant force, the larger is the acceleration (for a constant mass).

Furthermore, if the size of the resultant force is constant, then the larger the mass, the smaller the acceleration. We say that the mass and the acceleration are inversely proportional to each other.

The graphs in Figure 30.9 illustrate these two concepts.

Newton's second law states that the acceleration of a body is directly proportional to the force applied to it, and the direction of the acceleration is parallel to the direction of the force.

This law also explains why some very large articulated lorries have long braking distances. When the stopping force is constant, the deceleration is inversely proportional to the mass of the lorry.

Practical activity

Investigating Newton's second law

Experiment 1: Investigating the link between acceleration versus force (at constant mass)

Apparatus

▶ runway
▶ trolley
▶ string
▶ double interrupt mask with both sections the same width
▶ light gate and data logger
▶ pulley masses
▶ balance

Method

1 Prepare a table for results as shown in Table 30.2.

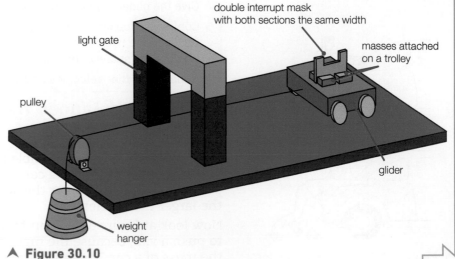

▲ **Figure 30.10**

2 Friction is always present when a trolley rolls along a runway. To compensate for friction, tilt the runway until the trolley moves with a constant speed after it is given a gentle push.

3 Fix the clamped pulley to the end of the bench (see Figure 30.10).

4 Attach a length of string to a dowel rod on the end of the trolley and make a loop in the other end to hang the masses from (the masses falling will provide a constant accelerating force). Pass this over the clamped pulley.

5 Position the light gate in such a way that the mask on top of the trolley passes through it, without hitting anything, before the masses on the end of the string hit the ground

6 Use the light gate to measure the acceleration of the trolley for various driving forces (masses on the mass hanger) from 100 g–500 g, repeating each measurement two times and taking an average. Remember, each 100g mass is equivalent to 1 N.

7 Find the weight of the masses and record this as the value for the resultant force F.

8 Plot a graph of acceleration (y-axis) versus force (x-axis).

Results

Table 30.2

Resultant force F/N	First acceleration/m/s^2	Second acceleration/m/s^2	Average acceleration/m/s^2
1.0			
2.0			
3.0			
4.0			
5.0			
6.0			

Conclusion

▶ The graph is a ....showing that the acceleration is directly proportional to the resultant force.

Is it possible to calculate the mass of the trolley?

Experiment 2: Investigating the link between acceleration and mass (at constant force)

Apparatus

▶ Use the same apparatus as shown in Figure 30.10 but arrange the slotted masses as directed below.

Method

1 Prepare a table for results as shown in Table 30.3.

2 Friction is always present when a trolley rolls along a runway. To compensate for friction adjust the angle of the slope of the runway until the trolley moves with a constant speed after it is given a gentle push.

3 Fix the clamped pulley to the end of the bench.

4 Attach a length of string to the end of the trolley and make a loop in the other end to hang the masses from (the masses falling will provide a constant accelerating force). Pass this over the clamped pulley.

5 Position the light gate in such a way that the mask on top of the trolley passes through it, without hitting the light gate, before the masses on the end of the string hit the ground.

6 Choose a suitable value for the driving force provided by the falling weights, for example 500 g (5 N).

7 Use the light gate to measure the acceleration of the trolley for various masses of trolley by either adding slotted masses to the trolley or stacking trolleys on top of each other.

8 Repeat each measurement three times and take an average.

9 Plot a graph of acceleration (y-axis) versus mass of trolley (x-axis).

Results

Table 30.3

Resultant force F/N	First acceleration/m/s^2	Second acceleration/m/s^2	Average acceleration/m/s^2
1.0			
2.0			
3.0			
4.0			
5.0			
6.0			
7.0			

Conclusion

The graph is non-linear and implies that as the mass increases, the acceleration ....

Example

1 Calculate the force needed to give a train of mass 250 000 kg an acceleration of 0.5 m/s^2.

Answer

$F = m \times a$

$\quad = 250\,000 \times 0.5$

$\quad = 125\,000\,\text{N}$

2 A forward thrust of 400 N exerted by a speedboat enables it to go through the water at constant velocity. The speedboat has a mass of 500 kg. Calculate the thrust required to accelerate the speedboat at 2.5 m/s^2.

Answer

Note the phrase 'at constant velocity'. This is a clue to using Newton's first law. If the thrust exerted by the engine is 400 N, there must be an equal and opposite force of 400 N due to the drag of the water on the boat. To calculate the force to accelerate the speedboat, we should draw a force diagram (Figure 30.11).

unbalanced force = mass × acceleration

$\quad (F - 400) = 500 \times 2.5$

$\quad\quad F - 400 = 1250$

$\quad\quad\quad\quad F = 1650\,\text{N}$

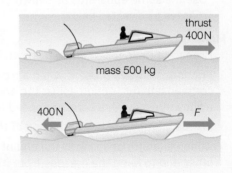

▲ **Figure 30.11** A force diagram

Summary of balanced and unbalanced forces

▶ Balanced forces have no effect on the movement of an object. If it is stationary it will remain stationary; if it is moving it will carry on moving at the same speed and in the same direction.

▶ Unbalanced forces will affect the movement of an object.

▶ An unbalanced force on an object causes its velocity to change (the object accelerates). The greater the force, the greater the acceleration.

▶ The greater the mass of an object, the greater the force needed to make it accelerate.

Test yourself

1 A bicycle and rider have a total mass of 90 kg and travel along a horizontal road at a steady speed. The forward force exerted by the cyclist is 40 N.
   a) Explain why the cyclist does not accelerate.
   b) The rider increases the forward force to 70 N. Calculate the acceleration.

2 A car accelerates at 3.0 m/s² along a road. The mass of the car is 1200 kg and all the resistive forces add up to 400 N. Calculate the forward thrust exerted by the car's engine.

3 Calculate the force of friction on a car of mass 1200 kg if it accelerates at 2 m/s² when the engine force is 3000 N.

4 Figure 30.12 shows the forces on a car of mass 800 kg.
   a) In what direction will the car accelerate?
   b) Calculate the size of the car's acceleration.

5 Look at Figure 30.13. The blades of a helicopter exert an upward force of 25 000 N. The mass of the helicopter is 2000 kg.
   a) Calculate the weight of the helicopter.
   b) Calculate the acceleration of the helicopter.

6 A forward thrust of 300 N exerted by a speedboat engine enables the speedboat to go through the water at a constant speed. The speedboat has a mass of 500 kg. Calculate the thrust required to accelerate the speedboat at 2 m/s².

7 A car and driver are travelling at 24 m/s and the driver decides to brake, bringing the car to rest in 8 seconds. The mass of the car and driver is 1200 kg.
   a) Calculate the deceleration of the car.
   b) Calculate the size of the unbalanced force which brings the car to rest.

8 A cyclist and her bicycle have a combined mass of 60 kg. When she cycles with a forward force of 120 N, she moves at a steady speed. However, when she cycles with a forward force of more than 120 N, she accelerates.
   a) Explain, in terms of forces, why the girl moves at a steady speed when the force is 120 N.
   b) Calculate her acceleration when the forward force is 300 N.

9 A car's brakes are applied and the vehicle's velocity changes from 50 m/s to zero in 5 seconds.
   a) Calculate the acceleration of the car.
   b) The resultant force causing this acceleration is 18 000 N. Calculate the mass of the car.

10 A car of mass 1000 kg is travelling at 20 m/s when it collides with a wall. The front of the car collapses in 0.1 seconds, by which time the car is at rest.
   a) Calculate the deceleration of the car.
   b) Calculate the force exerted by the wall on the car.

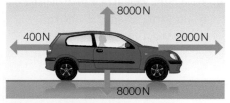

▲ **Figure 30.12** Forces acting upon a car

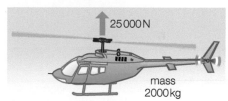

▲ **Figure 30.13** Forces acting on a helicopter

Show you can

a) Copy and complete Table 30.4.

Table 30.4 Comparing mass and weight

	Mass	Weight
1	is an amount of material	
2		varies from place to place
3	is measured in kg	is measured in ....
4	mass = density × volume	

b) What is the weight of a 70 kg man on
   i) the Earth, where $g = 10\,N/kg$?
   ii) the Moon where $g = 1.6\,N/kg$?

Mass and weight

In everyday life, the terms 'mass' and 'weight' are used interchangeably. In physics, however, we must be very careful to distinguish clearly between mass and weight.

What is mass?

Mass is defined as the amount of matter in a body. Mass is measured in kilograms (kg). It is a scalar quantity.

What is weight?

Weight is a force and is a measure of the size of the gravitational pull on an object exerted, in our case, by the Earth. Near the surface of the Earth, there is a force of 10 N on each 1 kg of mass. We say that the Earth's gravitational field strength, g, is 10 N/kg.

The weight, W, of an object is the force that gravity exerts on it. The formula for weight is:

weight/N = mass/kg × acceleration due to gravity/m/s²

or, $W = m \times g$

Weight is measured in newtons (N). It is a vector quantity, so it has direction as well as size.

The value of g is roughly the same everywhere on the Earth's surface. But the further you move away from the Earth, the smaller g becomes (see Figure 30.14).

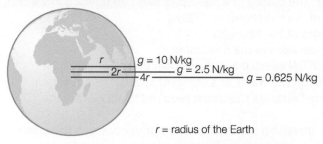

r = radius of the Earth

g = acceleration due to gravity

∧ Figure 30.14 The gravitational field strength, g, decreases with distance from the Earth

The Moon is smaller than the Earth, and pulls objects towards it less strongly. On the Moon's surface, the value of g is 1.6 N/kg.

In deep space, far away from the planets, there are no gravitational pulls, so g is zero, and therefore everything is weightless.

The size of g also gives the gravitational acceleration, because, from Newton's second law:

$$\text{acceleration} = \frac{\text{force}}{\text{mass}}$$

$$g = \frac{W}{m}$$

So an alternative set of units for g is m/s².

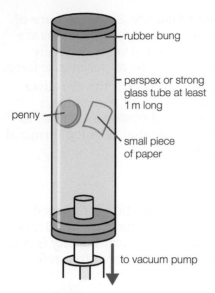

▲ **Figure 30.15** Investigating falling bodies in a vacuum

rubber bung

perspex or strong glass tube at least 1 m long

penny

small piece of paper

to vacuum pump

Free fall

Galileo is said to have shown that two lead balls of different diameter hit the ground at the same instant when dropped from the top of the leaning tower of Pisa.

In the absence of air resistance, all bodies fall at the same rate of 10 m/s² near the surface of the Earth. It is a common misconception to think that a more massive object falls faster than a less massive one. It is true that there is a greater force on the more massive object, but the acceleration, which is the ratio of force to mass, will be the same for both bodies.

$$a = \frac{F}{m} \text{ or } g = \frac{W}{m}$$

This means that, if there is no air resistance, the speed of a falling object will increase by 10 m/s every second, i.e. its acceleration is 10 m/s². This is known as the acceleration of free-fall, the symbol for which is 'g'. In a vacuum, where there is no air resistance, all falling objects accelerate at the same rate.

When the glass tube in Figure 30.15 is evacuated and then turned upside down, the penny and piece of paper fall together. Although the penny has more mass than the piece of paper, gravity will exert a larger force on the penny, giving both objects the same acceleration, i.e. the ratio of weight to mass is the same for both the penny and the piece of paper.

In the Earth's atmosphere, air resistance does act on a falling body. Air resistance can only be ignored if the force it exerts is very small. When a sky-diver jumps from a plane (Figure 30.17), the forces on them are unbalanced, and so the sky-diver accelerates.

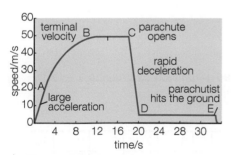

▲ **Figure 30.16** A speed–time graph for a sky-diver

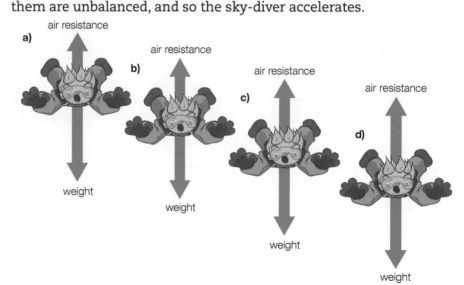

▲ **Figure 30.17** The weight of the sky-diver stays the same, but air resistance increases with speed. Eventually, air resistance = weight, and the sky-diver reaches his terminal velocity

▲ **Figure 30.18** This sky-diver has reached his terminal velocity

Figure 30.16 shows a speed–time graph for the motion of a sky-diver. The faster the sky-diver falls, the larger the air resistance, so the smaller the acceleration. Eventually the downward force due to gravity and the upward force due to air resistance will be balanced. The sky-diver will stop accelerating and start to fall at a constant speed. At this point, the sky-diver has reached his terminal velocity (Figure 30.18).

a) Julie said, 'My weight is 55 kg.' What is wrong with this statement and what do you think her weight really is?

b) A ball-bearing is gently dropped into a tall cylinder of oil which resists its motion. Describe what will happen to the ball-bearing.

c) An astronaut standing on the surface of the Moon releases a hammer and a feather from the same height. What will happen and why?

d) Why does a parachute slow down a falling parachutist?

e) Explain the shape of each section, AB, BC, CD and DE, of the graph in Figure 30.16.

▲ **Figure 30.19**

Opening the parachute increases the air resistance. Since the force of gravity stays the same, it will take less time before the two forces are balanced. This gives the sky-diver a lower terminal velocity (Figure 30.18). Terminal velocity depends on the size of the air resistance force.

Air resistance is affected by the shape of an object. If the sky-diver pulls their arms and legs in line with their body, air resistance is reduced. It will take longer for the upward and downward forces to balance, so the sky-diver reaches a higher terminal velocity. Terminal velocity depends on the shape of an object.

Vertical motion under gravity

When a body is thrown vertically upwards, its motion is opposed by the force of gravity. The velocity of the body will decrease by 10 m/s in each and every subsequent second until its vertical velocity is zero. The body has experienced negative acceleration, more often referred to as **deceleration or retardation**.

Example

1 A ball is thrown vertically upwards with an initial velocity of 50 m/s. How long will the ball take to reach the top of its motion?

Answer

initial velocity $u = 50\,\text{m/s}$ final velocity $v = 0\,\text{m/s}$

time of vertical motion $= t$ $\qquad$ acceleration $a = -10\,\text{m/s}^2$

$$a = \frac{v - u}{t}$$

$$-10 = \frac{0 - 5}{t}$$

$$t = \frac{-50}{-10}$$

$$t = 5\,\text{s}$$

This problem could also have been solved graphically by drawing a velocity–time graph for the motion (Figure 30.19).

Table 30.5

Time/s	0	1	2	3	4	5	6	7	8	9	10
Velocity/m/s	50	40	30	20	10	0	−10	−20	−30	−40	−50

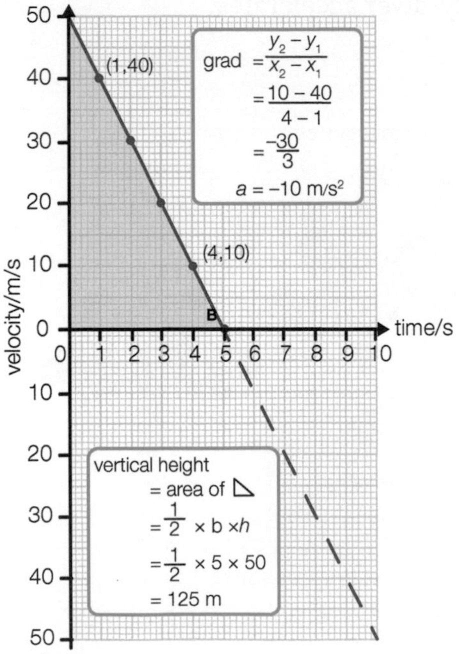

Test yourself ✎

11 Use the graph in Figure 30.19 to determine:
   a) the total time the ball is in the air
   b) the velocity with which it strikes the ground
   c) the total distance the ball travelled in the air
   d) the total displacement of the ball.

Practice questions

1 a) State Newton's second law of motion. *(2 marks)*
A plane accelerates from rest to a velocity of 50 m/s in 25 s just before take-off.
b) Calculate its acceleration. *(3 marks)*
Some of the forces acting on the plane before take-off are shown in Figure 30.20.

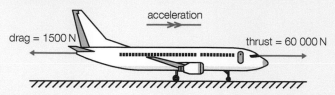

Figure 30.20

c) Use your answer to part (b) to find the mass of the plane. *(3 marks)*

2 a) State Newton's first law of motion. *(2 marks)*
A fighter plane of mass 3000 kg lands on the deck of an aircraft carrier. An arrester cable exerts a force of 90 000 N to oppose the motion of the plane.
b) Calculate the deceleration of the fighter plane. *(3 marks)*
c) The arrester cable brings the fighter plane to rest in 2 seconds. Calculate the velocity with which the fighter plane landed on the deck of the aircraft carrier. *(3 marks)*

H 3 a) Describe an experiment to verify Newton's second law of motion. *(6 marks)*
The graph in Figure 30.21 shows how the acceleration of an off road car depends on the force exerted by the car's engine.

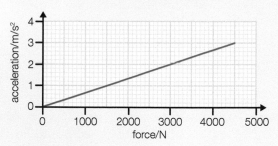

Figure 30.21

b) Use the graph to calculate the mass of the car. *(3 marks)*
On a different journey, the velocity of the car increased from 6 m/s to 18 m/s in a time interval of 4 seconds.
c) Calculate the acceleration of the car. *(3 marks)*

4 An aircraft sits at rest on an aircraft carrier deck. In order to lift off successfully it must reach a speed of 75 m/s. The time interval between the beginning of its motion and lift off speed is 1.5 seconds.
a) Calculate the acceleration of the plane during this time interval. *(3 marks)*
The mass of the aircraft is 22 000 kg.
b) Use your answer to part (a) to calculate the resultant force on the aircraft. *(3 marks)*

5 a) Tim, who has a mass of 65 kg, rides a bicycle of unknown mass. A resultant force of 210 N produces an acceleration of 2 m/s². Calculate the mass of the bicycle. *(3 marks)*
b) A racing car of mass 2500 kg accelerates from rest on the starting grid. The engine exerts a force of 1.5×10^4 N. Calculate the acceleration of the racing car. *(3 marks)*

6 a) Calculate the weight of an object of mass 70 kg on Earth. *(2 marks)*
b) The same object is taken to the Moon, where $g = 1.6$ m/s². Calculate its weight on the Moon. *(3 marks)*
c) On another planet, a mass of 12 kg weighs 105.6 N. Calculate the value of g on this planet. *(2 marks)*
d) Comment on the units for g. *(2 marks)*

7 A bullet is fired vertically upwards from a pistol from the surface of a planet. It rises to a maximum height of 1875 m in a time of 25 seconds.
a) Calculate the average velocity of the bullet during this time. *(3 marks)*
b) Using your answer to part (a), or otherwise, calculate the maximum velocity of the bullet. *(3 marks)*
The bullet then takes another 25 seconds to fall back to the planet's surface.
c) Sketch a graph to show how the velocity of the bullet changes during the entire motion. *(4 marks)*
d) What is the average velocity of the bullet over the entire distance covered? *(1 mark)*
e) Give a reason for your answer. *(1 mark)*
f) Use your sketch to determine the acceleration due to gravity on the planet. *(3 marks)*

8 The boat in Figure 30.22 has a mass of 15 000 kg.

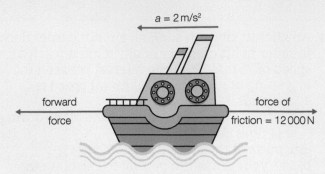

Figure 30.22

The force of friction on the boat is 12 000 N and it is accelerating at 2 m/s². Calculate:
a) the resultant force on the boat *(1 mark)*
b) the forward force from the boat's engines.
(3 marks)

9 An object falls from rest and strikes the ground exactly 1.5 seconds later.
a) At what speed does it hit the ground? *(3 marks)*
b) A ball is thrown vertically upwards with an initial speed of 24 m/s.
  i) How long does it take the ball to reach maximum height? *(3 marks)*
  ii) What is the maximum height? *(3 marks)*

10 The friction force opposing the motion of a locomotive of mass 25 000 kg is 100 000 N.
a) What forward force must the locomotive provide if it is to travel along a straight, horizontal track at a steady speed of 1.5 m/s²? *(1 mark)*
b) What is the acceleration of the locomotive if the forward force increases to 175 000 N and the friction force is unchanged? *(3 marks)*

Specification points

This chapter covers sections 1.2.12 to 1.2.23 of the specification including Prescribed practicals P2 and P3. Students investigate Hooke's law, which introduces them to the idea of proportionality and teaches them that experimental laws are only valid provided certain conditions are met. They are introduced to the idea of pressure along with applications that are dependent on the concept. Students find out how to calculate the moment of a force and how to establish the Principle of Moments through practical investigation. They are introduced to the meaning of centre of gravity and learn how it affects the stability of an object.

Hooke's law

When a helical spring is loaded, it stretches.

The natural length is the normal length of the spring without a load on it.

The extended length is the length of the spring when loaded.

The difference between the extended and natural lengths is known as the extension:

extension = extended length – natural length

Prescribed practical

Prescribed practical P2: Investigating how the extension of a helical spring depends on the applied force

Aims
- to draw a load–extension graph
- to determine the spring constant

Apparatus
- helical spring
- slotted masses
- metre ruler

Method
1 Prepare a table of results as shown in Table 31.1.
2 Measure the natural length of the spring.
3 Add a 100 g (weight = 1.0 N) mass to hang from the spring.
4 Measure the extended length of the spring.

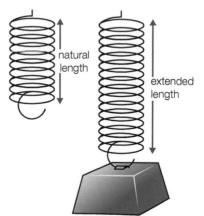

▲ **Figure 31.1** The natural and extended length of a helical spring

5 Calculate and record the extension.

6 Add a second 100 g mass.

7 Repeat measurements and record the results in your table.

Results

Natural length = ... cm

Table 31.1

Load/N	Extended length/cm	Extension/cm
1		
2		
3		
4		
5		
6		

Graphs

Draw a graph of force (N) on the y-axis versus extension (cm) on the x-axis.

The graph produced should show a straight line such as AB in Figure 31.2.

Conclusion

This experiment shows that the extension of a spring is proportional to the load.

A material that behaves in this way is said to obey **Hooke's law**.

The spring constant = ... N/cm

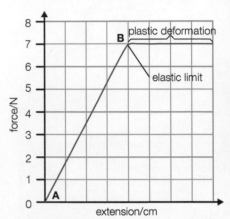

▲ **Figure 31.2** A graph plotting force against extension

Definition of Hooke's law

Extension is proportional to the load, provided the proportional limit is not exceeded.

At point B in Figure 31.2, the spring has reached its **proportional limit**, so no longer obeys Hooke's law. In the region AB of the graph, the spring shows elastic behaviour. This means that when the load is removed, the spring returns to its original length and shape.

If a load of more than 7 N (beyond point B) is applied to this spring, it goes beyond its elastic limit and so changes its shape permanently. When the load is removed, the spring does not return to its original shape. This is called **plastic deformation**.

Formula for Hooke's law

Hooke's law states that the extension of a spring, e, is directly proportional to the load, F.

Mathematically this is expressed as $F \propto e$ or:

$F = k \times e$

where k is the spring constant of the spring.

If k is large, the spring is stiff. If k is small, the spring is easy to extend.

Show you can

a) A student is investigating Hooke's law. She applied different loads to the same helical spring. She obtained the following incomplete set of results.

Table 31.4

Load/N	0	3	6	9	12
Length of spring/cm	6	8	10	12	14
Extension/ cm					

i) Copy the student's table of results and complete the last row.

ii) Explain whether or not Hooke's law was obeyed in this experiment.

Test yourself

1 Table 31.2 shows the total length of a spring obeying Hooke's law when different loads are applied.

Table 31.2

Load/N	Total length/cm
2	12
3	15

a) What extension is produced in the spring by a load of 1 N?

b) Calculate the original length of the spring.

2 The spring in a chest expander has a natural length of 24 cm. A force of 1 N stretches the spring 0.4 cm. Calculate the force needed to stretch the spring to a total length of 60 cm.

3 The following results were obtained from a stretching experiment.

Table 31.3

Force on the spring/N	0	1	2	3	4	5	6
Extension/cm	0	1.5	3.0	4.5	6.0	7.5	9.0
Force on the spring/N	0	1	2	3	4	5	6
Extension/cm	0	3.5	7.5	11.5	15.5	18.5	20

Plot graphs of force against extension for these results and mark any regions that follow Hooke's law.

Pressure

Figure 31.3 illustrates a concrete slab lying on each of its sides on soft ground. The weight of the slab is the same no matter which side it is resting on. However the effect on the soft ground depends on the area of contact.

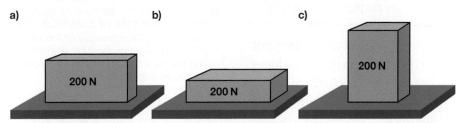

▲ **Figure 31.3** A concrete slab lying on each of its sides

When the slab is lying on side A, the area of contact is very large but the force per metre square is very small, hence the effect on the soft ground will be minimal. In contrast, the force per square metre exerted by side C is very large, so the effect on the soft ground will be large.

We use the term pressure to describe how force is distributed normally (at right angles) over an area. If the force is spread out over a large area, as in the case of side A, we say that the pressure exerted on the ground is small. If the force is concentrated on a small area, as in the case of side C, we say that the pressure is large.

Pressure is defined as the ratio of the normal force to the area of contact.

$$\text{Pressure} = \frac{\text{Force}}{\text{area}} \text{ or } P = \frac{F}{A}$$

Example

A lorry has a side with an area of 40 m². The wind exerts a pressure of 500 Pa on the side of the lorry. Calculate the force exerted by the wind on the lorry.

Answer

$p = \dfrac{F}{A}$

$F = P \times A$

$F = 500 \times 40$

$F = 20\,000\,N$

Show you can

Explain each of the following:
a) You cannot push your thumb through a wooden desk, but with the same force, you can push a drawing pin into the wood.
b) When a firefighter rescues a dog that has fallen through the ice on a frozen lake, they put their ladder on the ice first and then crawl out to the dog on the ladder.
c) A heavy battle tank will not sink into soft ground.
d) A carpenter will sharpen his chisel before he starts work.

When the units of force are newtons (N) and the units of area are square metres (m²), then the units of pressure are N/m², which are commonly called **pascals** (Pa).

The calculations for the pressure exerted by the concrete slab on each of its sides are given in Table 31.5.

Table 31.5 The pressure exerted by the slab on each of its sides A, B and C

Side of block	A	B	C
Weight/N	200	200	200
Area on contact/m²	4	2	1
Pressure/Pa	50	100	200

Often, the normal force will be the weight of an object.

$$\text{pressure} = \frac{\text{weight}}{\text{area}}$$

It is clear from the pressure equation that the pressure that an object exerts is inversely proportional to the area of contact, assuming the force does not change. In other words, as the area of contact increases, the pressure decreases, and vice versa.

This fact has many practical applications, as shown in Figure 31.4.

a) A chef will spend time sharpening a carving knife before cutting a joint of meat, as a small area of contact means enormous cutting pressure.

b) The area of contact between the blade of an ice skate and the ice is very small. This results in very large pressure on the ice, producing a layer of water between the blade and the ice, reducing the frictional force and making skating effortless.

c) The weight of a woman when concentrated on a stiletto heel results in a very large pressure, so large that floors are easily damaged.

d) A digger has very large rear wheels, so its huge weight is spread out. Consequently, the pressure which the digger exerts on the soft ground is small, preventing it from sinking.

e) Snow shoes are used to make walking in snow much easier. The large area of the shoes reduces the pressure on the snow and so prevents sinking.

f) In some places, the ground is so soft that houses are built on rafts of concrete. The large area of concrete spreads the weight of the house, so it doesn't sink into the ground.

▲ **Figure 31.4** Practical applications of the relationship between pressure and area of contact

Moments

Moment of a force

Door handles are usually placed as far from the hinges as possible so that the door opens and closes easily. A much larger force would be needed if the handle was near the hinges. Similarly, it is easier to tighten or loosen a nut with a long spanner than with a short one.

The **turning effect** or moment of a force depends on two factors:

▶ the size of the force
▶ the distance the force is from the turning point or pivot.

The moment of a force is measured by multiplying the force by the perpendicular distance of the line of action of the force from the pivot (Figure 31.6). This can be written as:

moment of a force = force × perpendicular distance from the pivot

The unit of the moment of a force is the newton metre (N m), where the force is measured in newtons (N) and the distance from the pivot to the line of action of the force is measured in metres (m).

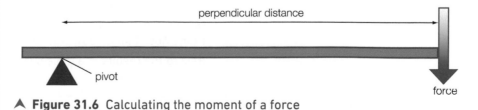

▲ **Figure 31.6** Calculating the moment of a force

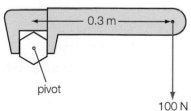

The Principle of Moments

The Principle of Moments is as follows:

When a body is in equilibrium, the sum of the clockwise moments about any point equals the sum of the anticlockwise moments about the same point.

The formula that arises from this definition is:

$$F_1 \times d_1 = F_2 \times d_2$$

Another very important consequence of the fact that the body is in equilibrium is that the forces acting on the body in any direction must balance. The upward forces must balance the downward forces. This idea is very useful when solving problems.

Prescribed practical

Prescribed practical P3: Investigating the Principle of Moments

Aim
- to measure clockwise and anticlockwise moments

Apparatus
- metre ruler
- slotted masses
- thread
- pivot, such as a string loop attached to clamp and retort stand

Diagram of apparatus

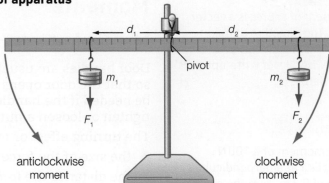

▲ **Figure 31.7** Calculating the moment of a force

Method
1 Suspend and balance a metre ruler at the 50 cm mark using thread.
2 Adjust the position of the thread so that the ruler does not rotate.
3 Hang unequal masses, m_1 and m_2 (100 g slotted masses), from either side of the metre ruler, as illustrated in Figure 31.7.
4 Adjust the position of the masses until the metre rule is balanced (in equilibrium) once again.
5 Gravity exerts forces F_1 and F_2 on the masses m_1 and m_2. Remember that a 100 g slotted mass is equivalent to a weight of 1 N.
6 Record the results in a table such as Table 31.6, and repeat for other loads and distances.

Example

A boy weighing 600 N sits 1.0 m away from the pivot of a see-saw, as shown below.

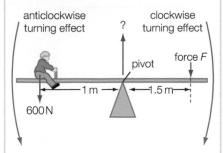

▲ **Figure 31.8** The forces involved in a boy sitting on a see-saw

1 What force 1.5 m from the pivot is needed to balance the see-saw?

Answer

The force F exerts a clockwise turning effect about the pivot, while the boy's weight exerts an anticlockwise turning effect. Since the see-saw is balanced, we can write

$$\frac{\text{clockwise}}{\text{moment}} = \frac{\text{anticlockwise}}{\text{moment}}$$

$$F_1 \times d_1 = F_2 \times d_2$$
$$F_1 \times 1.5\,\text{m} = 600\,\text{N} \times 1.0\,\text{m}$$
$$F_1 = (600 \times 1) \div 1.5$$
$$F_1 = 400\,\text{N}$$

2 Find the size of the upward force exerted by the pivot.

Answer

Since the body is balanced (in equilibrium): the upward force at the pivot = the sum of the downward forces acting on the see-saw

$$= 400\,\text{N} + 600\,\text{N}$$
$$= 1000\,\text{N}$$

7 The force F_1 is trying to turn the metre stick anticlockwise, and $F_1 \times d_1$ is its moment. F_2 is trying to turn the metre stick clockwise, its moment is $F_2 \times d_2$.

8 When the metre stick is balanced (i.e. in equilibrium), the results should show that the anticlockwise moment $F_1 \times d_1$ equals the clockwise moment $F_2 \times d_2$.

Results

Table 31.6

Anticlockwise				Clockwise			
m_1/g	F_1/N	d_1/cm	$F_1 \times d_1$/N cm	m_2/g	F_2/N	d_2/cm	$F_2 \times d_2$/N cm

Conclusion

What do you deduce from the values in columns four and eight of your results?

Test yourself

11 Figure 31.9 shows a car park barrier. The weight of the barrier is 150 N, and its centre of mass is 0.9 m from the pivot.

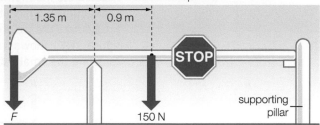

▲ **Figure 31.9** The forces acting upon a car park barrier

a) Calculate the size of the clockwise moment produced by the barrier's weight about the pivot.

b) Calculate the size of the force, F, on the left of the pivot which will just lift the barrier off the supporting pillar.

12 Figure 31.10 shows a uniform metre ruler pivoted at its midpoint. A load of 4 N acts on the right-hand side at a distance of 36 cm from the pivot.

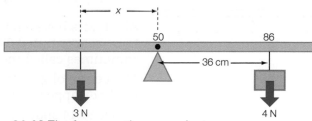

▲ **Figure 31.10** The forces acting on a pivot

Calculate the distance from the pivot where you would place a 3 N weight to balance the metre ruler.

13 Figure 31.11 shows a plan view of a gate pivoted at C. The boy at A is pushing on the gate with a force of 100 N and a man at B is pushing in the opposite direction so that the gate does not move.

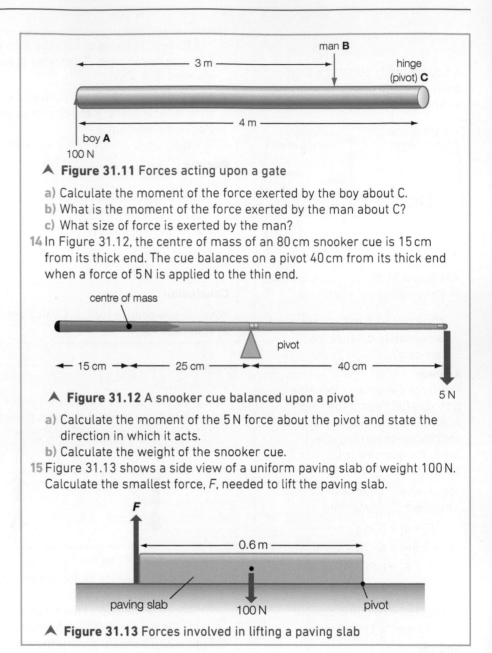

Figure 31.11 Forces acting upon a gate

a) Calculate the moment of the force exerted by the boy about C.
b) What is the moment of the force exerted by the man about C?
c) What size of force is exerted by the man?

14 In Figure 31.12, the centre of mass of an 80 cm snooker cue is 15 cm from its thick end. The cue balances on a pivot 40 cm from its thick end when a force of 5 N is applied to the thin end.

Figure 31.12 A snooker cue balanced upon a pivot

a) Calculate the moment of the 5 N force about the pivot and state the direction in which it acts.
b) Calculate the weight of the snooker cue.

15 Figure 31.13 shows a side view of a uniform paving slab of weight 100 N. Calculate the smallest force, F, needed to lift the paving slab.

Figure 31.13 Forces involved in lifting a paving slab

Centre of gravity

All objects have a point at which we can consider all their weight to be concentrated. This point is referred to as the centre of gravity, sometimes called the **centre of mass** of the object.

The centre of gravity is a point through which the whole weight of the body appears to act.

Figure 31.14 shows a metre ruler that is balanced about its midpoint. You could imagine the metre ruler as consisting of a series of 10 cm sections. The mass of each section is pulled towards the centre of the Earth by the force of gravity, so there are several small forces acting on the metre rule. But it is possible to replace all of these forces by a single resultant force acting through the centre of gravity, G. This force may be balanced by the reaction exerted by the pivot, as illustrated in Figure 31.14b.

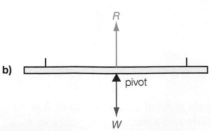

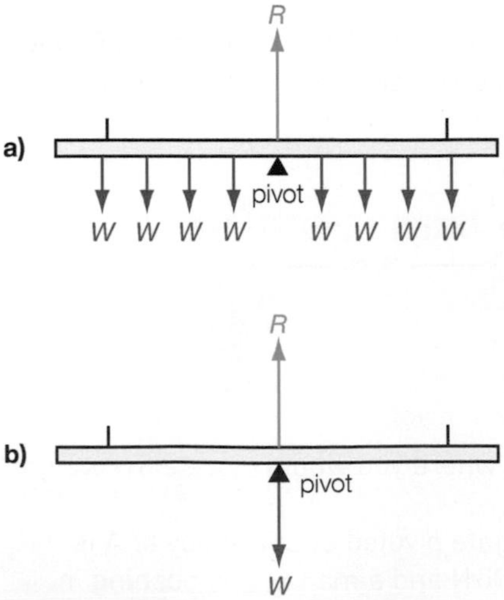

Figure 31.14 Forces acting on a metre ruler

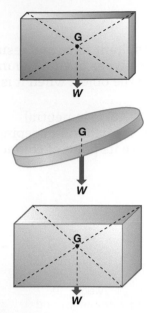

The centre of gravity of a body may be regarded as the point of balance. If a body has a regular shape, such as a flat disc or a rectangular sheet of metal, then the centre of mass is at its geometrical centre (Figure 31.15).

Flat triangular shapes are a little more difficult. In such cases, lines called **medians** are drawn from the corners of the triangle to the midpoints of the opposite sides. Where the medians intersect is the centre of mass (Figure 31.16).

Finding the centre of gravity of an irregularly shaped lamina

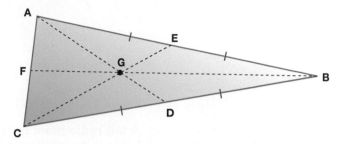

▲ **Figure 31.15** Working out the centre of gravity for regular objects

▲ **Figure 31.16** The point where the medians intersect is the triangle's centre of gravity

A lamina is a body in the form of a flat thin sheet.

Figure 31.17 shows a method for finding the centre of mass of a lamina. It is important to realise that when a body is suspended so that it can swing freely, it will come to rest with its centre of gravity vertically below the point of suspension.

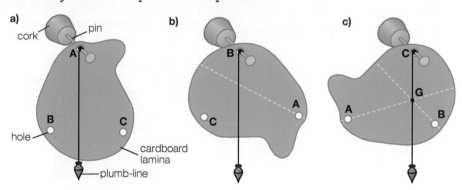

▲ **Figure 31.17** Finding the centre of mass of an irregular lamina

The stages involved in this method are as follows:

1 Hang an irregularly shaped sheet of cardboard from a pin, embedded into a cork.

2 Hang a plumb-line from the same pin.

3 When the cardboard settles, mark the vertical line with a pencil.

4 Repeat from two further points.

The intersection of the vertical lines from the three points of suspension will be the centre of gravity.

Equilibrium and stability

A body is in equilibrium when both the resultant force and resultant turning effect on it are zero. There are three types of equilibrium, which are determined by what happens to the object when it is given a small push.

1 A ball on a flat piece of ground (Figure 31.18) is in neutral equilibrium. When given a gentle push, the ball rolls, keeping its centre of gravity at the same height above its point of contact with the ground.

weight   ground   weight

▲ **Figure 31.18** This ball is in neutral equilibrium with the ground

2 A tall radio mast is in unstable equilibrium (Figure 31.19). It is balanced with its centre of gravity above its base, but a small push from the wind will move its centre of gravity downwards. To prevent the mast toppling, it is stabilised with cables.

3 A car on the road is in stable equilibrium (Figure 31.20a). If the car is tilted (b) the centre of gravity is lifted. In this position, the action of the weight keeps the car on the road. In (c), the centre of gravity lies above the wheels, so the car is in a position of unstable equilibrium. If the car tips further (d) the weight provides the turning effect to turn the car over. Cars with a low centre of gravity and a wide wheelbase are the most stable on the road (Figure 31.21).

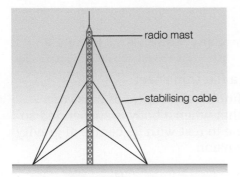

— radio mast

— stabilising cable

▲ **Figure 31.19** This radio mast is in unstable equilibrium

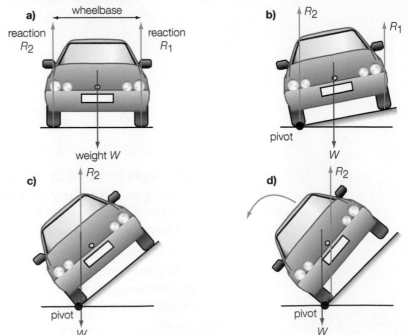

a) wheelbase

reaction R_2 reaction R_1

weight W

b) R_2 R_1

pivot W

c) R_2

pivot W

d) R_2

pivot W

▲ **Figure 31.21** This racing car is extremely stable because of its low centre of gravity and wide wheelbase

▲ **Figure 31.20** As the car tilts further, it becomes more and more unstable until at position (d), it topples over

16 Figure 31.22 shows shapes made from thin sheets of plastic.

▲ **Figure 31.22** 'Shapes made of thin plastic

a) Copy out the first two shapes and draw construction lines to show where the centre of gravity of each shape is located.

b) In the third shape, the central circular portion (white) has been cut out. If the centre of the circle is at the centre of the square, where will the centre of gravity of this plastic sheet be?

17 a) What is meant by the centre of gravity of an object?

b) Figure 31.23 shows a pencil with a plotting compass attached balancing on its point.

   i) Explain why this happens.

   ii) What would happen if the plotting compass were closed slightly?

c) Figure 31.24 shows a piece of cardboard. Copy the diagram exactly and mark a possible position for the centre of gravity.

18 Figure 31.25 shows a racing car.

a) Copy the diagram and mark with a cross the approximate position of the car's centre of gravity.

b) What two features of the car give it great stability?

19 Figure 31.26 shows cross-sections through two drinking glasses.

a) Copy the diagrams and mark with a cross the approximate position of the centre of gravity of each.

b) Which glass is likely to be more stable?

c) Give two reasons for your answer to part b).

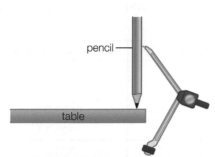

▲ **Figure 31.23** A pencil balancing on its point using a compass

▲ **Figure 31.24** A piece of cardboard

▲ **Figure 31.25** A racing car

▲ **Figure 31.26** A wine glass and a whisky glass

Practice questions

1 a) State Hooke's law. *(3 marks)*

  b) The manufacturers of car seat belts are required by law to test how they behave when different forces are applied to them. A particular seatbelt gave the following results.

Table 31.7

Load/kN	Seatbelt length/cm	Extension/cm
0	120.5	0.5
0.5	122.0	1.5
1.0	123.5	3.0
1.5	125.0	4.5
2.0	126.5	6.0
2.5	129.0	8.5
3.0	132.5	12.0

  i) What is the natural (unstretched) length of this seatbelt? *(2 marks)*

  ii) Up to what load (in kN) does this seatbelt obey Hooke's law? Explain the reason(s) for your answer. *(3 marks)*

  iii) When the load is removed, the seatbelt always gets shorter. In one case, a load of 5 kN is applied and then removed, and in another case, a load of 1 kN is applied and then removed. How (if at all) does the length of the seatbelt change after the load is removed, in each of these two cases? *(2 marks)*

  iv) When a car is involved in a major accident, it is wise to replace the seat belts. Use your knowledge of how materials behave when stretched to suggest why. *(1 mark)*

2 In an experiment with a helical spring, the following results were recorded.

Table 31.8

Load/N	0.0	0.5	1.0	1.5	2.0
Extension/cm	0.0	0.4	0.8	1.2	1.6

  a) Draw a graph of load against extension. *(2 marks)*

  b) Use the graph to find the weight of a metal object that caused an extension of 1.0 cm *(1 mark)*

  c) Calculate the spring constant. *(3 marks)*

  d) Another identical helical spring is connected in series (end to end) with the first spring. Draw a line on the graph to represent the results of the combination of these two springs. *(2 marks)*

3 Table 31.9 shows the extensions of a spiral spring of natural length 50 mm when increasing loads are attached to it.

Table 31.9

Load/N	0	5	10	15	20	25	30	35
Extension/cm	0	2	4	6	8	11	17	26

  a) Draw a graph of extension versus load. *(2 marks)*

  b) Up to what load, according to your graph, does Hooke's law appear to apply? *(1 mark)*

  c) What load should produce an extended length of 25 mm? *(2 marks)*

4) a) Explain what is meant by pressure. *(1 mark)*

  An oil jet is used to cut brittle candy into bars.

  b) The jet has a radius of 0.08 mm at the surface of the candy. Calculate the surface area of the candy in contact with the oil jet, giving your answer in mm² and in m².
  Show clearly how you get your answer. *(4 marks)*

  The pressure of the oil jet on the candy is 180 MPa.

  c) What pressure, in pascals, is exerted by the oil jet on the candy? *(1 mark)*

  d) Use your answers to parts **b)** and **c)** to calculate the force which the oil jet exerts on the candy. *(3 marks)*

5 A lorry trailer is 15 m long by 2 m high, as shown in Figure 31.27 below. The force of the wind on the trailer is 150 000 N.

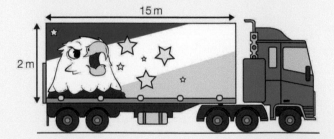

Figure 31.27

  a) Calculate the area of the side of the trailer. *(1 mark)*

  b) Calculate the pressure on the side of the trailer. *(3 marks)*

  c) Explain why similar lorries should avoid high bridges on windy days. *(2 marks)*

H **6** A wheelbarrow and its load together weigh 600N (Figure 31.28). The distance between the pivot and the wheelbarrow's centre of mass is 75 cm.

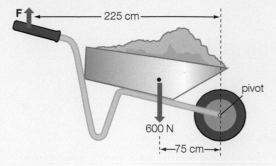

Figure 31.28

The distance between the handles and the pivot is 225 cm. *(3 marks)*
Calculate the size of the smallest force, F, needed to lift the wheelbarrow at the handles.

7 a) Explain what is meant by the centre of gravity of an object. *(1 mark)*
Figure 31.29 below shows a wheelbarrow at rest on level ground. The weight of the wheelbarrow and its contents is 1500 N.
b) Use the values on the diagram to calculate the moment of the 1500N force about the pivot. Show clearly how you get your answer. *(3 marks)*

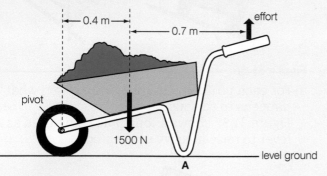

Figure 31.29

c) Use your answer to part **b)** to calculate the effort that must be applied to the handles to lift the wheelbarrow slightly off the ground at A. Show clearly how you get your answer. *(3 marks)*
d) What is the upward vertical reaction (supporting force) from the ground through the pivot when the wheelbarrow is just lifted off the ground by the effort? Show clearly how you get your answer. *(2 marks)*

8 a) State the Principle of Moments. *(2 marks)* **H**
b) A non-uniform plank of wood of length 80 cm is balanced on a pivot as shown in Figure 31.30a.

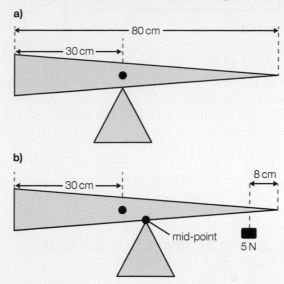

Figure 31.30

This demonstrates that the centre of gravity of the plank is 30 cm from the left-hand end, as indicated by a dot. The plank is now moved and rebalanced at its mid-point using a 5 N weight placed 8 cm from the right-hand end, as shown in Figure 31.30b.
i) Calculate the weight of the plank. *(3 marks)*
ii) Calculate the upward force which is now exerted by the triangular support. *(1 mark)*

9 a) A guillotine is used to cut sheets of paper. A constant downward force of 20 N is exerted on the handle (Figure 31.31).

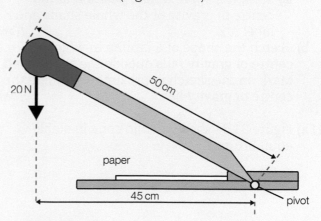

Figure 31.31

Calculate the moment of the 20N force about the pivot. *(3 marks)*
b) A teapot is placed on a tray and the tray is set on a shelf as shown in Figure 31.32. The tray has a weight of 10 N.

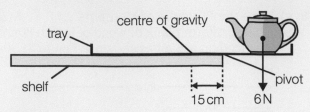

Figure 31.32

The centre of gravity of the tray is 15 cm from the edge of the shelf.

i) Use an arrow to show the direction of the weight of the tray. *(1 mark)*

The teapot weighs 6 N.

ii) Use the principle of moments to find the greatest distance the teapot can be placed from the edge of the shelf without toppling the tray. *(4 marks)*

10 a) Figure 31.33 shows a T-shaped lamina, in which QR is twice as long as AB.

Figure 31.33

Copy the diagram.

i) Mark the centre of gravity of the rectangle ABCD and label it X. *(1 mark)*

ii) Mark centre of gravity of the rectangle PQRS and label it Y. *(1 mark)*

iii) Mark the approximate position of the centre of gravity of the whole shape and label it Z. *(1 mark)*

b) Sketch the shape of a lamina in which the centre of gravity falls outside the shape itself. Mark on the sketch approximately where the centre of gravity lies. *(3 marks)*

11 a) Figure 31.34 shows a solid cone in stable equilibrium.

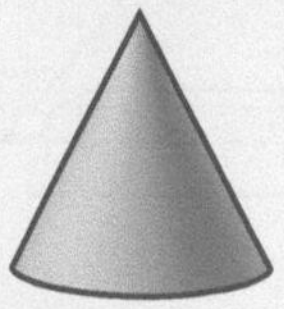

Figure 31.34

Draw two further diagrams to illustrate a solid cone in:

i) unstable equilibrium

ii) neutral equilibrium. *(2 marks)*

b) Figure 31.35 shows the cross-sections of two similarly shaped table lamps, A and B. The bases in each case are solid.

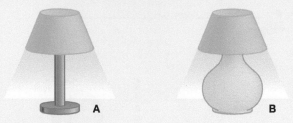

Figure 31.35

i) Copy the diagrams and mark where you might expect the centre of gravity to be. *(2 marks)*

ii) Which lamp is likely to be more stable? *(1 mark)*

iii) Give two reasons for your answer to part **(ii)**. *(2 marks)*

12 Figure 31.36 shows a bus in two positions. The centre of gravity of the bus is marked G.

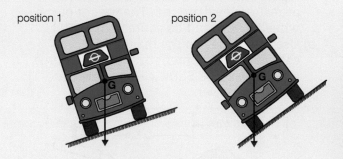

Figure 31.36

a) For each position, describe and explain what happens to the bus. *(4 marks)*

Figure 31.37 shows a long pole being used as a lever to raise a heavy stone block.

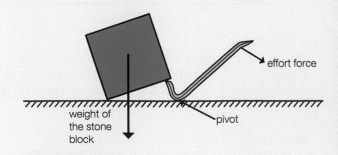

Figure 31.37

The stone block weighs much more than the force the man uses to raise it.

b) Explain carefully how the lever allows him to raise the stone block. *(6 marks)*

13a) Describe, in detail, an experiment to verify the Principle of Moments. In your description you should include:
- the apparatus used
- how the apparatus is used
- the formula you would use to test the Principle of Moments. *(6 marks)*

Wheel-braces are used to remove wheel nuts (Figure 31.38).

b) i) Explain why wheel-braces are designed so that they may be extended. *(1 mark)*

Figure 31.38

ii) Calculate the moment of a force of 20 N when applied by a wheel-brace of length 0.40 m. *(3 marks)*

Density

The spectators at a football match are densely packed on the terraces, whereas the footballers on the pitch are well spread out. In a similar way, different materials have different densities. Some materials, such as lead, have large atoms which are very tightly packed together. We say that lead is a very dense material. In contrast, polystyrene has very small, well-spaced-out atoms.

In physics, we compare materials such as lead and polystyrene using the concept of density.

Density is defined as the mass of unit volume of a substance. It is calculated using the formula:

$$\text{density} = \frac{\text{mass of object made of the substance}}{\text{volume of the object}} \quad \text{or} \quad \rho = \frac{m}{V}$$

The symbol for density is the Greek symbol ρ, which is pronounced 'rho'.

Density is measured in kilograms per cubic metre (kg/m^3) or in grams per cubic centimetre (g/cm^3).

The density of lead is $11\,g/cm^3$, which means that a piece of lead of volume $1\,cm^3$ has a mass of $11\,g$. Therefore, $5\,cm^3$ of lead has a mass of $55\,g$.

If you know the density of a substance, the mass of any volume of that substance can be calculated. This enables engineers to work out the mass (and hence the weight) of a structure if the plans show the volumes of the materials to be used and their densities.

Show you can

Prove that $1\,g/cm^3 = 1000\,kg/m^3$.

Table 32.1 The densities of some common substances

Substance	Density/g/cm³	Density/kg/m³
Aluminium	2.7	2700
Iron	8.9	8900
Gold	19.3	19 300
Pure water	1.0	1000
Ice	0.9	900
Petrol	0.8	800
Mercury	13.6	13 600
Air	0.001 23	1.23

Example

Taking the density of mercury as $14\,\text{g/cm}^3$, find:

a) the mass of $7\,\text{cm}^3$ of mercury and
b) the volume of $42\,\text{g}$ of mercury.

Answer

a) $\rho = \dfrac{m}{V}$

$14 = \dfrac{m}{7}$

$m = 14 \times 7$

$\quad = 98\,\text{g}$

b) $\rho = \dfrac{m}{V}$

$14 = \dfrac{42}{m}$

$v = \dfrac{42}{14}$

$\quad = 3\,\text{cm}^3$

Measuring density

To determine the density of a substance, we need to measure a) its mass and b) its volume. The density, ρ, will then be given by the ratio of its mass (m) to its volume (V), i.e.

$$\rho = \frac{m}{V}$$

Practical activity

Investigating the relationship between the mass and volume of liquids and regular solids

1) Liquids

Aims
- to find the mass of a liquid
- to find the volume of liquid

Apparatus
- $100\,\text{cm}^3$ measuring cylinder
- digital balance
- $250\,\text{cm}^3$ of water

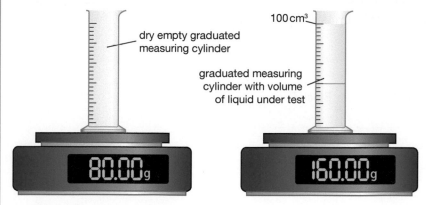

▲ **Figure 32.1** Calculating the density of water

Method
1 Prepare a table for results similar to that shown in Table 32.2.
2 Measure the mass of a $100\,\text{cm}^3$ graduated cylinder, using a digital balance.
3 Zero the balance.
4 Pour $20\,\text{cm}^3$ of water into the measuring cylinder.

5 Record its mass and volume in the table.

6 Record the mass of 40 cm³, 60 cm³, 80 cm³ and 100 cm³ of water in the same way.

7 The density of the water is found by dividing the mass of the water by the volume of the water.

Results

Table 32.2

Volume/cm³	20	40	60	80	100
Mass/g					
Density/g/cm³					

Conclusion

The average value of density of water is ....

2) Regularly shaped objects

Aims

- to find the volumes of regularly shaped cubes
- to find the masses of cubes

Apparatus

- digital balance
- ruler
- cubes of material

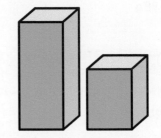

▲ **Figure 32.2** Regularly shaped cubes

Method

1 Prepare a table for results similar to that shown in Table 32.3.

2 Measure and record the dimensions of each cube.

3 Calculate its volume.

4 Measure the mass of each cube using a digital balance.

5 Calculate the density by dividing the mass of the cube by its volume.

Results

Table 32.3

	Block A	Block B	Block C
Length/cm			
Breadth/cm			
Height/cm			
Volume/cm³			
Mass/g			
Density/g/cm³			

Conclusion

The average density of the material is ....

The material is ....

Different volumes of the same material have the same density.

Find the density of irregularly shaped objects

If the shape of the object is too irregular for the volume to be determined using formulae, then a displacement method is used to measure the volume of the irregular solid, as shown in Figure 32.3.

1 The mass of the object is found using a top-pan balance.
2 The volume of the object is equal to the volume of water displaced.
3 The density can be calculated using the formula: $\rho = \dfrac{m}{V}$

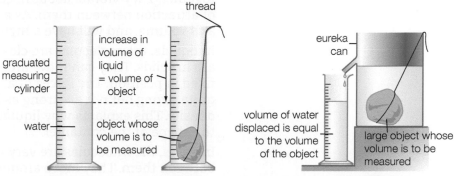

▲ **Figure 32.3a** The volume of a small object can be measured in a measuring cylinder

▲ **Figure 32.3b** Measuring the volume of a large object requires a eureka can

Graphical treatment of density

A graph of mass against volume of a uniform material is always a straight line through the origin. The gradient of the line is equal to the density of a particular substance. The ratio of the co-ordinates of any point on the straight line is the density of that substance (see Figure 32.4).

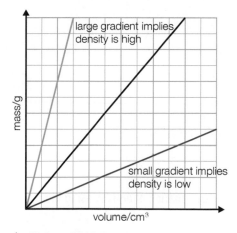

▲ **Figure 32.4** Graph of mass against volume

Test yourself

1 An object with a volume of $3\,cm^3$ weighs $57.9\,g$. Use Table 32.1 on page 385 to find a material that this object could be made from.
2 Aluminium has a density of $2.7\,g/cm^3$.
   a) What is the mass of $20\,cm^3$ of aluminium?
   b) What is the volume of $54\,g$ of aluminium?
3 A piece of steel of mass $120\,g$ has a volume of $15\,cm^3$. Calculate its density.
4 Air has a density of $1.26\,kg/m^3$. Calculate the mass of air in a room of dimensions $10\,m$ by $5\,m$ by $3\,m$.
5 A stone of mass $60\,g$ is lowered into a measuring cylinder, causing the liquid level to rise from $15\,cm^3$ to $35\,cm^3$. Calculate the density of the stone in g/cm^3.
6 The capacity of a petrol tank in a car is $0.08\,m^3$. Calculate the mass of petrol in a full tank if the density of petrol is $800\,kg/m^3$.
7 The mass of an evacuated $1000\,cm^3$ steel container is $350\,g$. The mass of the steel container when full of air is $351.2\,g$. Calculate the density of air.

Explaining the variation in density of solids, liquids and gases using kinetic theory

There are three states of matter – **solids**, **liquids** and **gases**. According to the kinetic theory, matter is made up of very large numbers of atoms and molecules in constant motion.

Look at Figure 32.5. In solids, molecules are packed very closely together. They vibrate about fixed positions and have strong forces of attraction between them. As a result, solids have a fixed shape and volume, and will have a high density.

In liquids, the molecules are close together, but not as close as they are in solids. They can move around in any direction and are not fixed in position. The forces of attraction between them are still quite strong but not as strong as in solids. Liquids have a medium density. This explains why liquids have a fixed volume but take on the shape of the container.

In gases, the molecules are very far apart with large distances between them. They move around very quickly in all directions, and the forces of attraction between them are very weak. Gases have a low density and always completely fill their container.

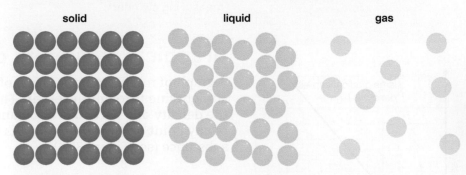

▲ **Figure 32.5** The arrangement of molecules in solids, liquids and gases

Show you can

Copy and complete Table 32.4 to summarise the explanation of density using the kinetic theory.

Table 32.4

Solids	Liquids	Gases
Molecules vibrate about fixed positions. Molecules have strong .... Molecules are packed very close together, so: solids have a .... density.	Molecules .... Forces of attraction between molecules are still quite strong, but not as strong as in solids. Molecules .... together but not as close as in solids, so: liquids have a medium density.	Molecules are very, very far apart. There are .... forces of attraction between molecules, so: gases have a .... density.

Practice questions

1 a) Explain what is meant by density. *(2 marks)*
b) Describe how you could use a measuring cylinder half-filled with water to find the volume of a bracelet. In your description, state what measurements you would make and what calculation you would carry out. *(4 marks)*
c) A necklace has a volume of 2.4 cm³ and a mass of 46 g. Calculate its density. *(3 marks)*
d) The necklace is made from a metal which is almost 100% pure. Use your answer to part c) and Table 32.5 to find out what the metal is.

Table 32.5

Metal	Copper	Gold	Lead	Platinum
Density in g/cm³	8.9	19.3	11.3	21.5

(1 mark)

2 Data relating to a particular concrete slab is shown in Figure 32.6.

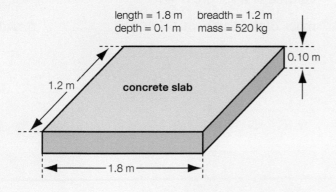

length = 1.8 m   breadth = 1.2 m
depth = 0.1 m   mass = 520 kg
0.10 m
1.2 m   concrete slab
1.8 m

Figure 32.6

a) Use the data to calculate the volume of this concrete slab. *(4 marks)*
b) For bridge construction, the concrete slabs must have a density of at least 2350 kg/m³. Is this particular slab dense enough to be used for bridge construction? *(2 marks)*

3 A brass ingot is 0.6 m wide, 0.5 m tall and 0.2 m long.
a) Find the volume of the brass ingot. *(2 marks)*
   Brass has a density of 8400 kg/m³.
b) Calculate the mass of the ingot. *(3 marks)*

4 a) 1 g of water has a volume of 1 cm³. There are 1 000 000 cm³ in 1 m³ of water.
   i) What is the mass, in g, of 1 m³ of water? *(1 mark)*
   ii) What is the mass, in kg, of 1 m³ of water? *(1 mark)*
   iii) What is the density of water, in kg/m³? *(1 mark)*

b) A hot air balloon is made from a material which has a mass of 150 kg. Its volume when filled with helium is 500 m³. The density of helium is 0.18 kg/m³. Calculate the total mass of the helium-filled balloon. *(4 marks)*

5 a) i) The density of aluminium is 2.7 g/cm³. Explain what this means. *(1 mark)*
   ii) Calculate the number of cm³ in 1 m³. *(1 mark)*
   iii) Calculate the mass in grams of 1 m³ of aluminium. *(1 mark)*
   iv) Calculate the density of aluminium in kg/m³. *(2 marks)*
b) A glass stopper weighs 40 g. It is placed in a measuring cylinder containing a liquid as shown in Figure 32.7. The cylinder gives the volume in cm³.

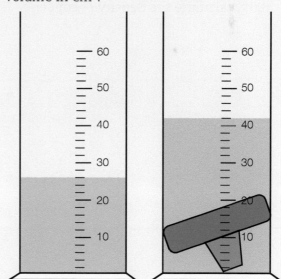

Figure 32.7

By taking readings from the diagrams, find:
i) the volume of liquid. *(1 mark)*
ii) the total volume of liquid and stopper. *(1 mark)*
iii) the volume of the stopper. *(1 mark)*
iv) the density of the stopper. *(3 marks)*

6 100 identical copper rivets are put into an empty measuring cylinder and 50 cm³ of water is added. Figure 32.8 shows the level of the water.

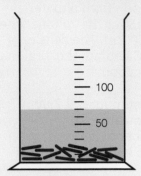

Figure 32.8

a) What is the volume of:
   i) 100 copper rivets   *(1 mark)*
   ii) 1 copper rivet?   *(1 mark)*
b) If all the copper rivets together have a mass of 180 g, calculate the density of copper.   *(1 mark)*

7 A student measured the volumes and masses of six different substances A, B, C, D, E and F. The student plotted their masses on the y-axis and their volumes on the x-axis (Figure 32.9).

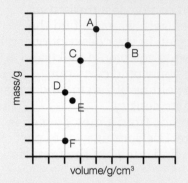

Figure 32.9

Which substances had the same density?   *(1 mark)*

8 Calculate the average density of the Earth, using the following data.

radius of Earth = 6.4×10^6 m

volume of a sphere = $\frac{4}{3} \pi r^3$

mass of the Earth = 6.0×10^{24} kg   *(5 marks)*

33 Energy

Energy forms

It is important to understand the difference between **energy forms** and **energy resources**. Energy forms are the different ways in which energy can appear, such as heat, light, sound, nuclear, kinetic, gravitational potential and chemical energy. Energy resources are the different ways of supplying a particular energy form. Table 33.1 summarises some of the main energy forms.

Table 33.1 Some of the main energy forms

Energy form	Definition	Examples of resources
Chemical	the energy stored within a substance, which is released on burning	coal, oil, natural gas, peat, wood, food
Gravitational potential	the energy a body contains as a result of its height above the ground	stored energy in the dam (reservoir) of a hydroelectric power station
Kinetic	the energy of a moving object	wind, waves, tides
Nuclear	the energy that is stored in the nucleus of an atom	uranium, plutonium

Other common energy forms are electrical energy, magnetic energy and strain potential energy – the energy a body has when it has been stretched or squeezed out of shape and will return to its original shape when the force is removed, such as a wind-up toy.

One of the fundamental laws of physics is the Principle of Conservation of Energy. This states that:

Energy can neither be created nor destroyed, but it can change its form.

We can show energy changes in an **energy flow diagram** (Figure 33.1).

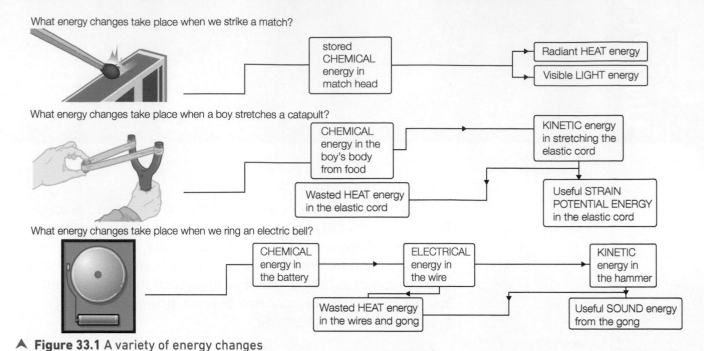

What energy changes take place when we strike a match?

What energy changes take place when a boy stretches a catapult?

What energy changes take place when we ring an electric bell?

▲ **Figure 33.1** A variety of energy changes

Energy resources

Energy resources can be classified in several different ways. One way is to split them into renewable and non-renewable resources. Renewable resources (see Table 33.2) are those that are replaced by nature in less than a human lifetime. Non-renewable resources (see Table 33.3) are those that are used faster than they can be replaced by nature. The UK government has said that 20% of our energy needs, including 30% of the electricity we generate, must come from renewable resources by 2020.

Table 33.2 Renewable sources of energy

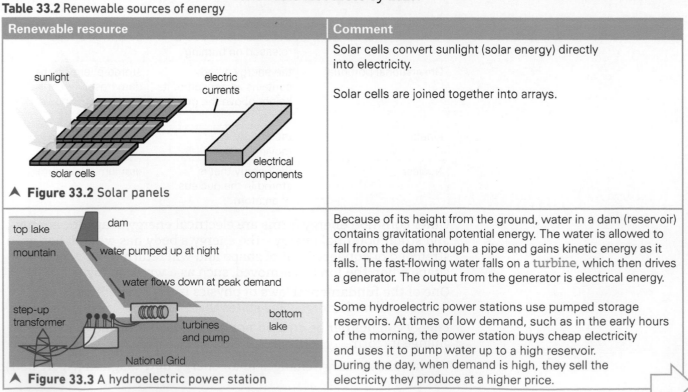

Renewable resource	Comment
▲ **Figure 33.2** Solar panels	Solar cells convert sunlight (solar energy) directly into electricity. Solar cells are joined together into arrays.
▲ **Figure 33.3** A hydroelectric power station	Because of its height from the ground, water in a dam (reservoir) contains gravitational potential energy. The water is allowed to fall from the dam through a pipe and gains kinetic energy as it falls. The fast-flowing water falls on a turbine, which then drives a generator. The output from the generator is electrical energy. Some hydroelectric power stations use pumped storage reservoirs. At times of low demand, such as in the early hours of the morning, the power station buys cheap electricity and uses it to pump water up to a high reservoir. During the day, when demand is high, they sell the electricity they produce at a higher price.

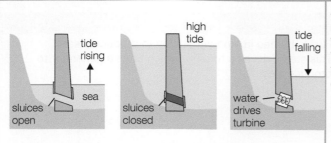

Figure 33.4 Harnessing tidal energy

A tidal barrage is created when a dam is built across a river estuary. The tide rises and falls every 12 hours, and if the water levels on each side of the dam are not equal, water will flow through a gate in the dam. The moving water drives a turbine, which is made to turn a generator. The output from the generator is electrical energy.

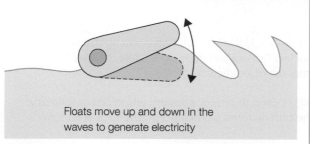

Floats move up and down in the waves to generate electricity

Figure 33.5 Generating electricity using waves

Waves are produced largely by the action of the wind on the surface of water. The wave machine floats on the surface of the water and the up and down motion of the water forces air to drive a turbine and so produces electricity.

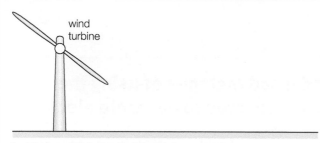

Figure 33.6 A wind turbine

As the wind blows, the large blade turns and this drives a turbine. The turbine drives a generator, which in turn produces electricity.

Large numbers of turbines are often grouped together to form a **wind farm**.

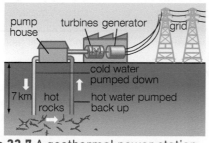

Figure 33.7 A geothermal power station

Geothermal power stations use heat from the hot rocks deep inside the Earth. Cold water is passed down a pipe to the rocks. The water is heated by these rocks and the hot water is then pumped to the surface.

Geothermal energy is often used in power stations or in district heating schemes.

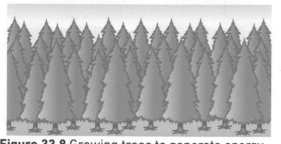

Figure 33.8 Growing trees to generate energy

Fast-growing trees, like willow, are grown on poor-quality land (or land set-aside from food production) and the timber is harvested around every three years. The wood is dried and turned into woodchips, which are then burned in power stations to produce electricity or sold for solid fuel heating.

In Brazil, biomass crops are fermented to produce alcohol. The **alcohol** is added to petrol as a way of extending the life of scarce fossil fuels. The product is called **gasohol**.
Other forms of biomass include rapeseed oil. The oil from the seeds is converted into **biodiesel** for road transport.

Table 33.3 Non-renewable sources of energy

Non-renewable resource	Comment
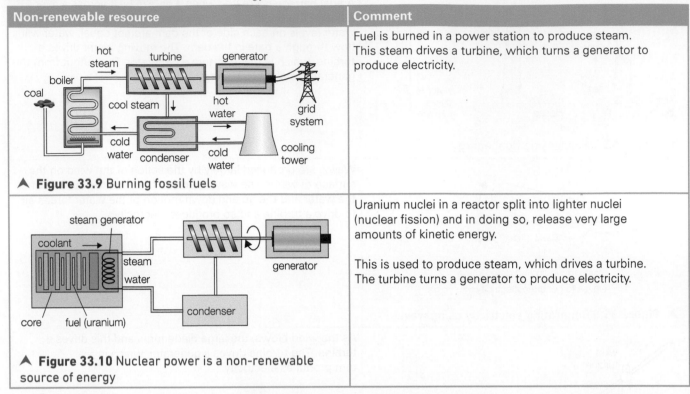 **▲ Figure 33.9** Burning fossil fuels	Fuel is burned in a power station to produce steam. This steam drives a turbine, which turns a generator to produce electricity.
▲ Figure 33.10 Nuclear power is a non-renewable source of energy	Uranium nuclei in a reactor split into lighter nuclei (nuclear fission) and in doing so, release very large amounts of kinetic energy. This is used to produce steam, which drives a turbine. The turbine turns a generator to produce electricity.

Advantages and disadvantages of using the different energy resources to generate electricity

Table 33.4 Advantages and disadvantages of different energy resources

Energy resource	Advantages	Disadvantages	Other comments
Fossil fuels – coal, oil, natural gas, lignite, peat	• relatively cheap to start up • moderately expensive to run • large world reserves of coal (much less for other fossil fuels).	• All fossil fuels are non-renewable. • All fossil fuels release carbon dioxide on burning, and so contribute to global warming. • Burning coal and oil also releases sulfur dioxide gas, which causes acid rain.	• Coal releases the most carbon dioxide per unit of electricity it produces, and natural gas releases the least. • Removing sulfur or sulfur dioxide is very expensive and adds greatly to the cost of electricity production.
Nuclear fuels – mainly uranium	• do not produce carbon dioxide • do not emit gases that cause acid rain.	• The waste products will remain dangerously radioactive for tens of thousands of years. • As yet, no one has found an acceptable method to store these materials cheaply, safely and securely for such a long time. • Nuclear fission fuels are non-renewable. • An accident could release dangerous radioactive material which could contaminate a very wide area, leaving it unusable for decades.	• Nuclear fuel is relatively cheap on world markets. • Nuclear power station construction costs are much higher than fossil fuel stations, because of the need to take expensive safety precautions. • Decommissioning nuclear power stations is a particularly long and expensive process, requiring specialist equipment and personnel.

Energy resource	Advantages	Disadvantages	Other comments
Wind farms	• a renewable energy resource • low running costs • reduces dependency on fossil fuels.	Wind farms are: • unreliable • unsightly • very noisy • hazardous to birds.	• Wind farms take up much more ground per unit of electricity produced than conventional power stations.
Waves	• a renewable energy resource • low running costs • reduces dependency on fossil fuels.	Wave generators at sea are: • unreliable • unsightly • hazardous to shipping.	• Many turbines are needed to produce a substantial amount of electricity.
Tides	• a renewable energy resource • low running costs • reduces dependency on fossil fuels.	Tidal barrages are built across river estuaries and can cause: • navigation problems for shipping • destruction of habitats for wading birds and the mud-living organisms on which they feed.	• Tides (unlike wind and waves) are predictable, but they vary from day to day and month to month. This makes them unsuitable for producing a constant daily amount of electrical energy.

Ireland's natural fuel resources

Ireland has almost no coal or oil resources. The island is rich in peat, but it is important not to over-exploit these resources industrially because of the damage that can be done to habitat. Ireland does, however, have an important fossil fuel resource – natural gas from the Celtic Sea. Northern Ireland has another resource – lignite. This is sometimes called brown coal because it is rocky like coal, but brown like peat. There are millions of tonnes of lignite reserves around Crumlin and under Lough Neagh.

Test yourself

1 Most of Ireland's energy needs are supplied by fossil fuels. Name three fossil fuels.
2 Make a copy of Table 33.5 and tick (✔) those items which are energy forms.

Table 33.5

Quantity	Tick if the quantity is a form of energy
Sound	
Pressure	
Force	
Weight	
Electricity	
Heat	

3 Table 33.6 shows six energy resources. Copy the table and tick (✔) the correct boxes to show whether these resources are renewable or non-renewable.

Table 33.6

Energy source	Renewable?	Non-renewable?
Gas		
Hydroelectricity		
Oil		
Coal		
Wind		
Tides		

4 A model aircraft has its wings covered with solar panels to drive the propellers and charge the battery. Copy and complete the following sentences to show the energy changes which take place in such an aircraft: The solar cells change .... energy into .... energy. The battery stores .... energy. As the propellers turn, they change .... energy into useful .... energy. As the model aircraft gains height, it gains .... energy. If the model aircraft crashes into the ground, it produces wasted heat and .... energy.

5 Explain what is meant by a renewable energy resource.

6 In what ways are the production of electricity in a fossil fuel power station and in a nuclear power station similar? In what ways are these power stations different?

7 Currently, nuclear waste is vitrified (turned into a type of glass), stored in strong metal drums and kept deep underground. Why is this an unsatisfactory solution for the long term?

8 Name the polluting gas that is produced by burning fossil fuels, and which contributes to global warming.

9 Norway has complained that Britain is partly responsible for the destruction of the Norwegian habitat by acid rain How might this have come about?

10 Do you think the UK government's target to have 30% of our electricity production come from renewable resources by 2020 is realistic? What can you and your family do to contribute?

11 Give three reasons for using wind farms to generate electricity.

12 An electricity company might say that electricity is a 'clean' fuel. Why is this statement misleading?

13 For each of the devices or situations shown below, use a flow diagram to show the main energy change that is taking place. The first has been done for you.

Table 33.7

Device/situation	Input energy form		Useful output energy form
Microphone	sound energy	→	electrical energy
Electric smoothing iron	.... energy	→	.... energy
Loudspeaker	.... energy	→	.... energy
Coal burning in an open fire	.... energy	→	.... energy
A weight falling towards the ground	.... energy	→	.... energy
A candle flame	.... energy	→	.... energy and .... energy

Show you can

a) What are the arguments for and against installing a nuclear power station in Ireland?

b) Imagine you are a government scientist. Write about 100 words giving the advantages of building a nuclear power station rather than one which burns fossil fuels.

c) Why do you think Northern Ireland has not yet mined the lignite resources around Crumlin?

The Sun

Almost all energy resources ultimately rely on the energy of the Sun. In the case of fossil fuels, we know that these resources come from the dead remains of plants and animals laid down many millions of years ago. The plants obtained their energy from the Sun by **photosynthesis**. Herbivores ate the plants, and carnivores ate the herbivores. Under the Earth's surface, these remains slowly fossilised into coal, peat, gas and oil.

Other processes also rely on the Sun's energy. Hydroelectric energy depends on the water cycle, and this process begins when ocean water evaporates as a result of absorbing radiant energy from the Sun. Wind and waves rely on the Earth's weather, which is largely controlled by the Sun. Only geothermal and nuclear energy do not depend directly on the energy emitted by the Sun.

1 An electric kettle is rated 2500 W. It produces 2500 J of heat energy every second. The kettle takes 160 seconds to boil some water, and during this time, 360 000 J of heat energy pass into the water. Find the kettle's efficiency.

Answer

useful energy output (passed into water) = 360 000 J

total energy input = 2500 × 160 = 400 000 J

$$efficiency = \frac{useful\ energy\ output}{energy\ input}$$

$$= \frac{360\,000}{400\,000}$$

$$= 0.9$$

Therefore:

* 90% of the electrical energy is used to boil the water
* 10% of the energy supplied is wasted

Most will be passed through the kettle as wasted heat to the surrounding air. A small amount of heat will be lost as some water evaporates.

2 A motor rated 40 W lifts a load of 80 N to a height of 90 cm in 4 s. Find its efficiency.

Answer

useful energy output

= work done by motor

= force × distance

= 80 × 0.9

= 72 J

$$power = \frac{energy\ supplied}{time\ taken}$$

$$40 = \frac{energy\ supplied}{4}$$

$$40 × 4 = energy\ supplied$$

$$160\ J = energy\ supplied$$

$$efficiency = \frac{useful\ energy\ output}{energy\ input}$$

$$= \frac{72}{160}$$

$$= 0.45$$

Efficiency

Efficiency is a way of describing how good a device is at transferring energy from one form to another in an intended way.

If a light bulb is rated 100 W, this means that it normally uses 100 J of electrical energy every second. But it might only produce 5 J of light energy every second. The other 95 J are wasted as heat. This means that only 5% of the energy is transferred from electrical energy into light energy. This light bulb therefore has an efficiency of 0.05, or 5%. If the same light bulb were used as a heater, its efficiency would be 95%, because the intended output energy form would be heat, not light.

Efficiency is defined by the formula:

$$efficiency = \frac{useful\ energy\ output}{energy\ input}$$

As efficiency is a ratio, it has no units. By the Principle of Conservation of Energy, energy cannot be created, so the useful energy output can never be greater than the energy input. However, energy is wasted in every physical process, so the efficiency of a machine is always less than 1.

Test yourself

14 The electrical energy used by a boiler is 1000 kJ. The useful output energy is 750 kJ.
   a) Calculate the efficiency of the boiler.
   b) Suggest what might have become of the energy wasted by the boiler.

15 Explain why the efficiency of a device can never be greater than 1.00 or 100%.

16 A car engine has an efficiency of 0.28. How much input chemical energy must be supplied if the total output of useful energy is 140 000 kJ?

17 Figure 33.11 shows energy transfers in a mobile phone.

▲ **Figure 33.11** Energy transfers in a mobile phone

   a) Use the figures on the diagram to calculate the phone's efficiency.
   b) What principle of physics did you use to calculate the useful sound energy produced?

18 Figure 33.12 shows a rotary engine which has an efficiency of 30%.
   a) Calculate the amount of useful energy it produces when the input chemical energy is 2000 J.
   b) 90% of the wasted energy is heat. What percentage of the input energy is lost as heat?

▲ **Figure 33.12** Energy efficiency in a rotary engine

Work

Work is only done when a force causes movement. Although pushing against a wall might make a person tired, no work is done on the wall because it produces no movement. Similarly, holding a book at arm's length is doing no work on the book. Lifting a book from the floor and placing it on a table is doing work because we are applying a force and producing movement.

We can calculate work using the following formula:

work done = force × distance moved in direction of force or:

$W = F \times d$

The units in this formula are matched. Force must always be measured in **newtons**, but if the distance were in cm, the work would be in N cm. If the distance were in metres, the work would be in N m. The N m occurs so often that physicists have renamed it the **joule** (J).

The joule is defined as the energy transferred when the point of application of a force of 1 newton moves through one metre.

Doing work means 'spending' energy. The more work a person does, the more energy they need. The energy used is equal to the amount of work done.

Example

1 How much work is done when a packing case is dragged 4 m across the floor against a frictional force of 45 N? How much energy is needed?

Answer

Since the case moves at a steady speed, the forward force must be the same size as the friction force. So the forward force is 45 N.

$$\text{So } W = F \times d$$
$$= 45 \times 4$$
$$= 180 \text{ J}$$

Energy needed = work done
$$= 180 \text{ J}$$

2 A crane does 1200 J of useful work when it lifts a load vertically by 60 cm. Find the weight of the load.

Answer

Since the load is being lifted, the minimum upward force is the weight of the load.

$$W = F \times d$$
$$1200 = F \times 0.6 \text{ (convert 60 cm to 0.6 m)}$$
$$F = 1200 \div 0.6$$
$$= 2000 \text{ N}$$

Hence, weight of load = 2000 N

3 How much work is done by an electric motor pulling a 130 N load 6.5 m up the slope shown in Figure 33.13 if the constant tension in the string is 60 N?

Answer

Since the tension and distance moved are both parallel to the slope, they are both used to find the work done. The weight of the load is not used in this question.

work = force × distance moved in direction of the force
$$= 60 \times 6.5$$
$$= 390 \text{ J}$$

---

tension = 60 N, distance = 6.5 m

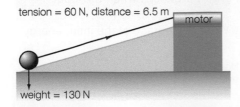

weight = 130 N

▲ **Figure 33.13**

Test yourself

19 Competitors in a strength competition must throw a cement block of mass 100 kg over a wall that is 5.5 m high. How much work is done if the block just clears the top of the wall?

20 A man pushes a lawn mower with a force of 60 N. How much work does he do when he pushes the lawn mower 20 m?

21 Stephen weighs 550 N. How much work does he do in climbing up to a diving board which is 3.0 m high?

Example

1 An electric motor is used to raise a load of 105 N. The load rises vertically 2 m in a time of 6 s. Find the work done and the power of the motor.

Answer

To lift the load, the motor must produce an upward force of at least 105 N.

work done = force × distance

= 105 × 2

= 210 J

$$power = \frac{work\ done}{time\ taken}$$

$$= \frac{210}{6}$$

= 35 W

2 A crane has a power of 2000 W. How much work could it do in an hour?

Answer

In power calculations, the unit of time is the second.

So first convert 1 hour to seconds:

1 hour = 60 minutes

= 60 × 60 seconds

= 3600 seconds

$$2000 = \frac{work\ done}{3600}$$

work done = 2000 × 3600

= 7 200 000 J or 7.2 MJ

Work and energy

Energy is the ability to do work. If a machine has 500 J of stored energy, this means it can do 500 J of work. Similarly, work is sometimes thought of as the amount of energy transferred. Note that both work and energy are measured in joules.

Tip

In all calculations of this type, first write down the appropriate formula, then substitute the values you know. Rearrange the formula if necessary, and carry out the calculations with a calculator. Give your final answer with its unit. Remember to always show your working.

Example

A battery stores 15 kJ of energy. If the battery is used to drive an electric motor, how high could it raise a 750 kg load if it was lifted vertically?

Answer

The battery stores 15 kJ or 15 000 J, so it can do a maximum of 15 000 J of work.

Since a mass of 1 kg weighs 10 N, a mass of 750 kg has a weight of 7500 N. The motor must therefore produce an upward force of at least 7500 N.

$$W = F \times d$$

$$15\,000 = 7500 \times d$$

$$d = 15\,000 \div 7500$$

$$= 2\ m$$

Note that 2 m is the highest that this motor could raise the load. It is likely that it would not raise the load quite this high because some of the energy in the battery is used to produce heat and sound. In our calculation, we have assumed that all the energy in the battery is used to do work against gravity.

Power

Power is the amount of energy transferred in one second, or the amount of work done in one second.

This means that the power of a machine is the work it can do in a second.

The formula for calculating power is:

$$power = \frac{work\ done}{time\ taken} \quad or \quad P = \frac{W}{t}$$

Work is measured in joules and time is measured in seconds, so power must be measured in joules per second or J/s. The J/s is also known as the **watt (W)**, named after James Watt, the Scottish engineer.

1 W = 1 J/s

More generally, power may also be defined as the rate of change of energy transferred.

$$power = \frac{energy\ transferred}{time\ taken}$$

Test yourself

22 A person weighing 550 N runs up the stairs in 3 seconds. The stairs are made of 15 steps each of 14 cm height. Find the person's average power.

23 A nail gun fires a nail with a kinetic energy of 1.8 J into a piece of wood. The average resistive force on the nail is 45 N, and it stops 0.3 s after entering the wood.
Calculate:
a) the distance the nail penetrates into the wood
b) the average power of the resistive forces in stopping the nail.

Prescribed practical

Prescribed practical P4: Investigating personal power

▲ **Figure 33.14** How to measure personal power

Aims
- to measure the weight of a student
- to measure the height of stairs
- to calculate personal power

Apparatus
- bathroom scales
- metre ruler
- stopwatch

Method
1 Prepare a table of results similar to that shown in Table 33.8.
2 Measure mass of a student using bathroom scales.
3 Convert mass to weight using the formula $W = mg$.
4 Find the height of one riser (step) in metres, using a metre rule.
5 Count the number of risers and multiply by the height of one riser. This is the total vertical height of the stairs.
6 Ask another student to measure the time to run up the stairs, using a stopwatch.
7 Repeat and work out the average time for running up the stairs.
8 Use the formulae for work and power to calculate the personal power of the student.

Results

Table 33.8

Mass of student/kg	45
Weight of student/N	450
Height of risers/cm	14.0, 13.8, 13.8, 14.0, 13.9
Average riser height/cm	13.9
Number of risers	30
Vertical height of staircase	$13.9 \times 30 = 417\,cm = 4.17\,m$
Time to run upstairs/s	5.0

Calculations

$$work\ done = force \times distance$$
$$= 450 \times 4.17$$
$$= 1876.5\,J$$
$$power = \frac{work\ done}{time\ taken}$$
$$= \frac{1876.5}{5.0}$$
$$= 375.3\,W$$

Conclusion

The average power was found to be............

> **Tip**
>
> Personal power can also be found by measuring the time, t, for a student to do 50 step-ups onto a single step. The power developed is then $50\,mgh/t$, where m is the student's mass and h is the height of the step.

Kinetic energy

The kinetic energy (KE) of an object is the energy it has because it is moving. It can be shown that an object's kinetic energy is given by the formula:

kinetic energy = ½ × mass × velocity² or $KE = \frac{1}{2}mv^2$

where m is the mass in kg and v is the speed of the object in m/s.

1 A car of mass 800 kg is travelling at 15 m/s. Find its kinetic energy.

Answer

$KE = \frac{1}{2}mv^2$
$= \frac{1}{2} \times 800 \times 15^2$
$= 90\,000\,J$

2 A bullet has a mass of 20 g and is travelling at 300 m/s. Find its kinetic energy.

Answer

$20\,g = 0.02\,kg$
$KE = \frac{1}{2}mv^2$
$= \frac{1}{2} \times 0.02 \times 300^2$
$= 900\,J$

3 Find the speed of a boat if its mass is 1200 kg and it has a kinetic energy of 9600 J.

Answer

$KE = \frac{1}{2}mv^2$
$= \frac{1}{2} \times 1200 \times v^2$

$v^2 = \frac{9600}{600}$
$v = 4\,m/s$

4 The input power of a small hydroelectric power station is 1 MW.
If 18 000 000 kg of water flows past the turbines every hour, find the average speed of the water.

Answer

1 hour = 60 × 60 seconds
= 3600 seconds

Since a 1 MW power station produces 1 000 000 J of electrical energy per second, the minimum KE of the water passing every second is 1 000 000 J.

$KE = \frac{1}{2}mv^2$
$1\,000\,000 = \frac{1}{2} \times 5000 \times v^2$
$v^2 = \frac{1\,000\,000}{2500} = 400$
$v = 20\,m/s$

Test yourself

24 A communications satellite of mass 120 kg orbits the earth at a speed of 3000 m/s. Calculate its kinetic energy.

25 The viewing platform at the Eiffel tower in Paris is about 280 m from the ground. Find the gravitational potential energy of a rubber of mass 50 g on the viewing platform. Compare this to the kinetic energy of a 10 g bullet travelling at 150 m/s. Comment on your answer.

26 An oil tanker has a mass of 100 000 tonnes. Its kinetic energy is 200 MJ. Calculate its speed.
(1 tonne = 1000 kg, 1 MJ = 1 000 000 J)

27 A car of mass 800 kg is travelling at a steady speed. The kinetic energy of the car is 160 000 J. Show carefully that the speed of the car is 72 km/h.

Gravitational potential energy

When any object with mass is lifted, work is done on it against the force of gravity. The greater the mass of the object and the higher it is lifted, the more work has to be done. The work that is done is only possible because some energy has been transferred. This energy is stored in the object as gravitational potential energy (GPE).

When the object is released, it falls back to Earth and the stored energy can be recovered. If the object crashes into the ground, a bang (sound energy) is heard and heat is produced.

Gravitational potential energy is the work done raising a load mass m, against the force of gravity (g) through a height (h), so:

$$GPE = mgh$$

where m is the mass in kg, g is the gravitational field strength in N/kg and h is the vertical height in m.

It is important to remember that 1 kg has a weight of 10 N on the surface of the Earth. This is just another way of saying that the gravitational field strength, g, on Earth is about 10 N/kg.
The value of g is different at different parts of the Universe.
For example, g on the Moon is only about a sixth of its value on Earth, approximately 1.6 N/kg.

Tip

If a process had just GPE at the start and just KE at the end, you know that the GPE must equal the KE.

Example

1 Find the gravitational potential energy of a mass of 500 g when raised to a height of 240 cm.
Take g = 10 N/kg
500 g = 0.5 kg
240 cm = 2.4 m

Answer
$GPE = mgh$
= 0.5 × 10 × 2.4
= 12 J

2 How much heat and sound energy is produced when a mass of 1.2 kg falls to the ground from a height of 5 m? Take g = 10 N/kg.

Answer
Heat and sound energy produced
= original GPE
= mgh
= 1.2 × 10 × 5
= 60 J

3 How much gravitational potential energy is stored in the reservoir of a hydroelectric power station if it holds 5 000 000 kg water at an average height of 80 m above the turbines?

Answer
$GPE = mgh$
= 5 000 000 × 10 × 80
= 4 000 000 000 J

4 A marble of mass 50 g falls to the Earth. At the moment of impact, its kinetic energy is 1 J. From what height did it fall?

Answer
50 g = 0.05 kg
KE at impact = GPE at start
1 = mgh
= 0.05 × 10 × h
= 0.5 × h
h = 2 m

5 A book of mass 500 g has a gravitational potential energy of 3.2 J when at a height of 4 m above the surface of the Moon. Find the gravitational field strength on the Moon.

Answer
$GPE = mgh$
3.2 = 0.5 × g × 4
= 2g
g = 3.2 ÷ 2
= 1.6 N/kg

Test yourself

28 A ball of mass 2 kg at rest then falls from a height of 5 m above the ground. Copy Table 33.9 and complete it to show the gravitational potential energy, the kinetic energy, speed and the total energy of the falling ball at different heights above the ground.

Table 33.9

Height above ground/m	Gravitational potential energy/J	Kinetic energy/J	Total energy/J	Speed/m/s
5.0		0	100	0
4.0				4.47
	64			
1.8		64		
0.0	0			

29 The power of the motor in an electric car is 3600 W. How much electrical energy is converted into other energy forms in 5 minutes?

30 A crane can produce a maximum output power of 3000 W. It raises a load of mass 1500 kg through a vertical height of 12 m at a steady speed.
   a) i) What is the weight of the load?
      ii) How much useful work does the crane do lifting the load 12 m?
   b) How long does it take the crane to raise the load 12 m?
   c) At what speed will the load rise through the air?

31 A barrel of weight 1000 N is pushed up a ramp. The barrel rises vertically 40 cm when it is pushed 1 m along the ramp.
   a) Calculate how much useful work is done when the barrel is pushed 1 m along the ramp.
   b) To push the barrel 1 m along the ramp requires 1200 J of energy. Calculate the efficiency of the ramp.

32 On planet X, an object of mass 2 kg is raised 10 m above the surface. At that height, the object has a gravitational potential energy of 176 J. Details of three planets are given below. Which one of these three planets is most likely to be planet X?

Table 33.10

Planet	Mercury	Venus	Jupiter
Gravitational field strength, g/N/kg	3.7	8.8	26.4

33 A bouncing ball of mass 200 g leaves the ground with a kinetic energy of 10 J.
   a) If the ball rises vertically, calculate the maximum height it is likely to reach.
   b) In practice, the ball rarely reaches the maximum height. Explain why this is so.

Practice questions

1. A satellite orbits the Earth. Name the two main types of energy possessed by the satellite in its orbit. *(2 marks)*

2. Electricity can be generated by wind turbines.
   a) Copy and complete the sentences below to show the useful energy change which takes place in a wind turbine.
   .... energy of the wind is transferred to .... energy. *(2 marks)*
   b) The wind is a renewable energy source. What does this mean? *(1 mark)*
   c) Give two other examples of renewable energy resources. *(2 marks)*

3. In Scotland, hydroelectric power makes a significant contribution as a source of electricity (Figure 33.15).

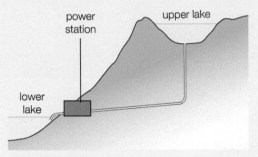

Figure 33.15

Copy and complete the flow diagram in Figure 33.16 to show the energy changes taking place in a hydroelectric power station.

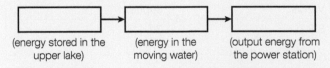

(energy stored in the upper lake) → (energy in the moving water) → (output energy from the power station)

Figure 33.16 *(3 marks)*

4. A tidal barrage in France generates electricity. One environmental effect of using the tides to generate electricity is that it reduces global warming by decreasing the consumption of fossil fuels.
   a) Explain fully how this reduces global warming. *(2 marks)*
   b) Apart from the above environmental issue, state one advantage and one disadvantage of generating electricity from the tides. *(2 marks)*

5. a) The most common energy resources used in Europe today are oil, natural gas, coal, nuclear energy, hydroelectric and wind energy.
   i) Choose one non-renewable energy resource from the list above and say why it is non-renewable. *(2 marks)*
   ii) Choose one renewable energy resource from the list above and say why it is renewable. *(2 marks)*
   iii) Give one advantage that non-renewable energy resources have over renewable energy resources. *(1 mark)*
   b) It has been estimated that 1×10^8 kg (100 000 000 kg) of water flows over Niagara Falls every second. The falls are 50 metres high. **H**
   i) Calculate the gravitational potential energy lost every second by the water flowing over the falls ($g = 10$ m/s^2). *(3 marks)*
   A feasibility study has shown that only 0.008 (0.8%) of the available potential energy could be converted into electrical energy by a hydroelectric power station built on the falls.
   ii) Calculate the maximum power output of such a hydroelectric power station. *(3 marks)*
   iii) Explain why all hydroelectric power stations are dependent on the energy of the Sun. *(2 marks)*
   c) Figure 33.17 shows a vehicle with a winch attached. The winch is connected to a tree by rope. As the winch winds in the rope, the vehicle moves forward towards the tree. The winch uses 500 W of input electrical power. It has an efficiency of 0.6.

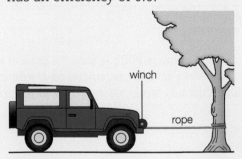

Figure 33.17
   i) Calculate the useful output power of the winch. *(3 marks)*
   ii) Write down the useful work done by the winch in 1 second. *(1 mark)*
   iii) The pulling force in the rope is 1200 N. Calculate the constant speed at which the vehicle moves forward. *(3 marks)*

6 a) How much work is done by a tractor when it lifts a load of 8000N to a height of 1.8m?
(3 marks)

b) The output power of the tractor is 5.2kW. How long does it takes to do 26000J of work?
(3 marks)

c) The efficiency of the tractor is 0.26 (26%). If the output power of the tractor is 5.2kW, calculate the input power. *(3 marks)*

7 Saltburn is a seaside resort in Yorkshire. There is a considerable drop from the cliff top to the beach. In 1884, an inclined tramway was built to carry passengers from the beach to the cliff top. Two identical tramcars were used, each with a water tank underneath. The tramcars were connected by a steel cable which passed around a large pulley at the top (Figure 33.18). The tramcar that happens to be at the top has water added until there is enough to raise the tramcar at the bottom of the tramway.

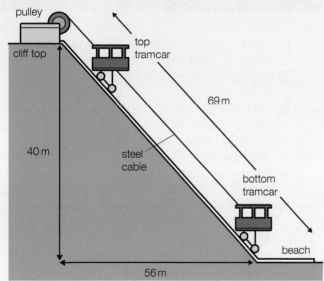

Figure 33.18

a) On one journey, the weight of the lower tramcar and its passengers was 24000N. Ignoring friction, calculate the work done, in kJ, to bring the tramcar from the beach to the cliff top. *(3 marks)*

b) The time for this journey was 20 seconds. Calculate the power needed to raise the tramcar. *(3 marks)*

c) On this journey, the energy provided by the upper car as it descended was 1200kJ. Calculate the efficiency of the tramway on this journey. *(3 marks)*

d) During the journey, certain energy changes take place. Copy and complete the Table 33.11 by stating whether the energy listed in the first column increases, decreases or remains unchanged as the top tramcar descends at a constant speed.

Table 33.11

Energy	Increases/decreases/unchanged
Potential energy of the top tramcar	
Kinetic energy of the top tramcar	
Kinetic energy of the bottom tramcar	
Potential energy of the bottom tramcar	
Heat energy	

(5 marks)

8 a) A basketball player throws a ball up into the air. Copy Table 33.12 and place a tick (✓) in the appropriate column of your table to show what happens to each quantity as the ball rises. Ignore the effects of friction.

Table 33.12

Quantity	Increases	Decreases	Remains constant
Speed of the ball			
Potential energy of the ball			
Total energy of the ball			
Potential energy of the bottom tramcar			
Kinetic energy of the ball			

(4 marks)

b) A heavy ball, of mass 10kg, is dropped from a height of 5 metres.
i) What is the potential energy lost by the ball during this fall? *(3 marks)*
ii) Calculate the velocity of the ball at the bottom of the fall. *(4 marks)*

9 A ball of mass 3kg is dropped from the top of a tall building. The ball loses 60J of energy due to air resistance on its way down. When it strikes the ground, it has a kinetic energy of 600J.
a) Calculate the gravitational potential energy of the ball when it is released from the top of the building. *(1 mark)*
b) Calculate the height of the building. *(3 marks)*
c) Calculate the speed of the ball before it strikes the ground. *(3 marks)*

10 A boulder, of mass 440 kg, rolls down a slope and into the sea. At the edge of the cliff, the boulder has a kinetic energy of 3520 J and a potential energy of 52 800 J.
  a) Calculate the height of the cliff. *(3 marks)*
  b) Calculate the kinetic energy of the boulder as it strikes the water. Assume no energy losses. *(1 mark)*
  c) Calculate the velocity of the boulder when it hits the water. *(4 marks)*

11 A crane uses a wrecking ball to demolish an old building. The diagram shows the motion of the ball. The crane moves the ball from its rest position X up to position Y, where it comes momentarily to rest before falling to collide with the wall of the building.

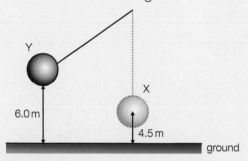

Figure 33.19
  a) Calculate the work done on the ball to move it from its rest position X to position Y. The ball has a mass of 800 kg. *(3 marks)*
  b) What is the loss in potential energy of the ball as it swings from position Y to position X? *(1 mark)*
  c) For the ball to be effective, it must have a minimum kinetic energy of 4900 J. Calculate the velocity of impact for this energy. *(4 marks)*

12 Look at Figure 33.20. A toy car of mass 1.2 kg is released from rest at point A, before it 'loops the loop'.

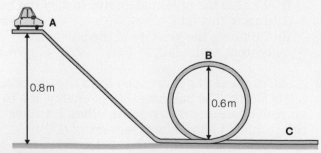

Figure 33.20
  a) Calculate the difference in potential energy of the toy car between points A and B. *(4 marks)*

  b) Calculate the velocity of the toy car at point C, if its kinetic energy at C is 4.2 J. Assume no energy losses due to friction. *(2 marks)*

13 a) Figure 33.21 shows a solar panel. This is made up of a number of photocells. The photocells produce electricity directly from sunlight. Solar panels are placed on the roof of a house. On a cloudless summer day, the solar energy shining on the panel every second is 6000 J. Of this amount, 4800 J are reflected, and the rest is converted to electricity.

Figure 33.21
  i) Calculate the output of electrical energy every second from the solar panel. *(1 mark)*
  ii) Calculate the efficiency of the solar panel. *(2 marks)*
  iii) On a certain summer day, the panel generated electricity for 10 hours. Calculate the number of kilojoules generated on this day by the solar panel. *(2 marks)*
  iv) State one advantage and one disadvantage of using solar panels. *(2 marks)*
  v) A family of four would use on average 54 000 kJ of electrical energy per day. State two things they could do to make up the difference between what the solar panel produces and what they need. *(2 marks)*
  b) John uses a weights machine in a gym.
  i) When using the machine, John wants to do 300 J of work in each lift. He can vary the weight from 100 N to 500 N in steps of 50 N. He can also vary the distance he lifts the weights from 1.0 m to 2.0 m in steps of 0.5 m.
     State three weights and the corresponding distances that John can use to achieve this.
     1. weight =   distance =
     2. weight =   distance =
     3. weight =   distance =
     (3 marks)
  ii) John repeats the exercise. He does 10 complete lifts in a time of 30 seconds. Calculate the power John produces during this time. *(3 marks)*

34 Atomic and nuclear physics

The structure of atoms

We take it for granted today that all matter is made up of atoms, but what are atoms made of? Experiments carried out by J.J. Thomson and Ernest Rutherford led physicists in the early part of the twentieth century to believe that atoms themselves had a structure.

The structure of the nucleus

An atom is made up of smaller particles. There is a central nucleus made up of protons and neutrons. Around this, **electrons** orbit at high speed. The number of particles depends on the type of atom. Protons have a positive (+) charge. Electrons have an equal negative (−) charge. Normally, atoms are neutral. So an atom must have the same number of electrons as protons. Protons and neutrons are collectively called **nucleons**. Each is about 1800 times more massive than an electron, so virtually all of an atom's mass is in its nucleus. Electrons are held in orbit by the force of attraction between opposite charges.

The relative masses and charges of the particles that make up the atom are given in Table 34.1.

Table 34.1

Particle	Location	Relative mass*	Relative charge*
Proton	Within the nucleus	1	+1
Neutron	Within the nucleus	1	0
Electron	Orbiting the nucleus	$\dfrac{1}{1840}$	−1

*Mass and charge are measured relative to the proton

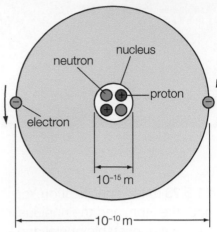

▲ **Figure 34.1** A helium atom

Figure 34.1 shows a simple representation of a helium atom. Since there are two orbiting electrons, there must also be two protons in the nucleus. Note that the diagram is not to scale: the diameter of the atom (about 1×10^{-10} m) is about 100 000 times greater than that of the nucleus (about 1×10^{-15} m).

Atomic number and mass number

The number of protons in the nucleus of an atom determines what type the atom is. All hydrogen nuclei have one proton, all helium nuclei have two protons, and all lithium nuclei have three protons, and so on. The number of protons is called the **atomic number** and is given the symbol **Z**. The atomic number also tells you the number of electrons in the atom.

As the mass of the electrons is negligible, the total number of particles in the nucleus determines the total mass of an atom. The **mass number** (or nucleon number) is the sum of the number of protons and the number of neutrons. The mass number is given the symbol **A**.

atomic number, **Z** = number of protons

mass number, **A** = number of protons + number of neutrons
= number of nucleons

Every nucleus can therefore be written in the form: $^{A}_{Z}X$

where **X** is the chemical symbol, A is the mass number and Z is the atomic number.

For example, the element uranium has the chemical symbol U. All uranium nuclei have 92 protons in the nucleus. One form of uranium, called uranium-235, has a mass number of 235. This means it has 92 protons and 143 neutrons (235 – 92 = 143). A uranium nucleus is given the symbol:

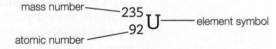

mass number ⎯⎯ $^{235}_{92}$U ⎯⎯ element symbol
atomic number ⎯⎯

It is important to realise that this is the symbol for the nucleus of the atom. Orbiting electrons are completely ignored.

You will notice that the top number gives the mass of the nucleus, and the bottom number gives the charge. This same system can also be used to describe protons, neutrons and electrons.

proton $^{1}_{1}p$ neutron $^{1}_{0}n$ electron $^{0}_{-1}e$

Isotopes

Not all the atoms of the same element have the same mass. For example, one form of helium (helium-3) has three nucleons, and another form (helium-4) has four nucleons (Figure 34.2). But all helium nuclei have two protons. So, helium-3 has two protons and one neutron, helium-4 has two protons and two neutrons.

Atoms with the same number of protons but a different number of neutrons are called isotopes.

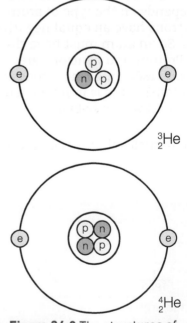

▲ **Figure 34.2** The structures of helium-3 and helium-4

1 An atom contains electrons, protons and neutrons. Which of these particles:
   a) are outside the nucleus?
   b) are uncharged?
   c) have a negative charge?
   d) are nucleons?
   e) are much lighter than the others?
2 How many protons, neutrons and electrons are in the nucleus of carbon-14 if its symbol is $^{14}_{6}C$?
3 The element sodium has the chemical symbol Na. In a particular sodium isotope, there are 12 neutrons. In a neutral sodium atom, there are 11 orbiting electrons. Write down the symbol for the nucleus of this isotope.
4 In what way are the nuclei of isotopes the same? In what way are they different?

Isotopes are atoms of the same element that have the same atomic number but different mass numbers.

The main isotopes of helium are $^{3}_{2}He$ and $^{4}_{2}He$.

Nuclear radiation

In 1896, the French scientist Henri Becquerel discovered by accident that certain rocks containing uranium gave out strange radiation that could penetrate paper and affect photographic film. He called the effect radioactivity. His students, Pierre and Marie Curie, later went on to identify three separate types of radiation. Unsure of a suitable name, the Curies called them alpha (α), beta (β) and gamma (γ) radiation after the first three letters of the Greek alphabet. The Curies and Henri Becquerel were jointly awarded the Nobel Prize for Physics in 1903 for their work on radioactivity.

For very heavy elements such as uranium or plutonium, the large number of protons and neutrons can make the nucleus unstable and cause their nuclei to randomly and spontaneously emit radiation. The atoms that emit such radiation are said to be radioactive. The particles and waves are referred to as nuclear radiation. The materials are called radioactive materials. The disintegration is called radioactive decay.

Ionising radiation

Ions are charged atoms (or molecules). Atoms become ions when they lose (or gain) electrons.

Nuclear radiation can become dangerous by removing electrons from atoms in its path, so it has an ionising effect (see Figure 34.3).

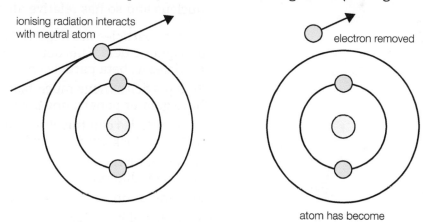

ionising radiation interacts with neutral atom

electron removed

atom has become a positive ion

▲ **Figure 34.3** Ionisation caused when an electron is removed from an atom by radiation

When this happens with molecules of genes in living cells, the genetic material of a cell is damaged and there is a small chance that the cell may become cancerous. Other forms of ionising radiation include ultraviolet and X-rays.

Nature and properties of nuclear radiations

The three main types of nuclear radiation are:

Alpha radiation α or $^4_2$He

▶ Alpha radiation is made up of a stream of alpha particles emitted from large nuclei.

▶ An alpha particle is a helium nucleus with two protons and two neutrons, and so has relative atomic mass of 4.

▶ Alpha particles are positively charged and so will be deflected in a magnetic field.

▶ Alpha particles have poor powers of penetration and can only travel through a few centimetres of air. They can easily be stopped by a sheet of paper.

▶ Alpha radiation has the strongest ionising power.

▶ Alpha radiation is not as dangerous if the radioactive source is outside the body, because it cannot pass through the skin and is unlikely to reach cells inside the body.

▶ Alpha radiation will damage cells if the radioactive source has been breathed in or swallowed.

An example of decay by alpha emission is shown in Figure 34.4.

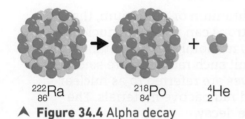

$^{222}_{86}$Ra $\rightarrow$ $^{218}_{84}$Po $+$ $^4_2$He

▲ **Figure 34.4** Alpha decay

Beta radiation β or $^0_{-1}$e

Beta radiation is made up of a stream of beta particles emitted from nuclei where the number of neutrons is much larger than the number of protons.

▶ A beta particle is a fast electron which has been formed in the nucleus and so has relative atomic mass of about $\frac{1}{1840}$.

▶ As beta particles are negatively charged, they will be deflected in a magnetic field. This deflection will be greater than that of alpha particles as beta particles have a much smaller mass.

▶ Beta particles move much faster than alpha particles, and so have greater penetrating power.

▶ Beta particles can travel several metres in air, but will be stopped by 5 mm thick aluminium foil.

▶ Beta radiation has an ionising power between that of alpha and gamma radiation.

▶ Beta radiation can penetrate the skin and cause damage to cells.

An example of beta decay is shown in Figure 34.5.

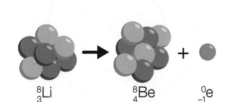

$^8_3$Li $\rightarrow$ $^8_4$Be $+$ $^0_{-1}$e

▲ **Figure 34.5** Beta decay

Gamma radiation γ

▶ Unlike the other types of radiation, gamma radiation does not consist of particles but of high-energy waves.

▶ Like alpha and beta radiation, gamma radiation comes from a disintegrating unstable nucleus. As it is an electromagnetic wave (see Chapter 35), gamma radiation has no mass.

▶ As there are no charged particles, a magnetic field has no effect on gamma radiation.

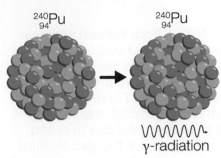

Figure 34.6 Gamma decay

γ-radiation

- Gamma radiation has great penetrating power, travelling several metres through air.
- A thick block of lead or concrete is used to greatly reduce the effects of gamma radiation, but is not able to stop it completely.
- Gamma radiation has the weakest ionising power.
- Gamma radiation can penetrate the skin and cause damage to cells.

An example of gamma emission is shown in Figure 34.6.

The penetrating power of each of these emissions is summarised in Figure 34.7.

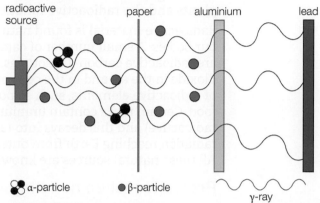

Figure 34.7 Selective absorption of radioactive emissions

Test yourself

5 Name a radioactive isotope that occurs naturally in living things.
6 Copy and complete Table 34.2.

Table 34.2

Property	Alpha particles (α)	Beta particles (β)	Gamma rays (γ)
Nature	each particle is made up of two protons and two neutrons, so it is identical to a nucleus of helium		very high energy electromagnetic waves
Relative charge compared with charge of a proton		−1	0
Mass	high compared to beta particles	low	0
Speed	up to 0.1 × speed of light		speed of light
Ionising effect	strong	weak	very weak
Penetrating effect		penetrating, but stopped by a few millimetres of aluminium or other metal	very penetrating: never completely stopped, though lead and thick concrete will reduce intensity

Show you can

Which of the three types of radiation:

a) is a form of electromagnetic radiation?

b) carries positive charge?

c) is made up of electrons?

d) travels at the speed of light?

e) is the most ionising?

f) can penetrate a thick sheet of lead?

g) is stopped by skin or thick paper?

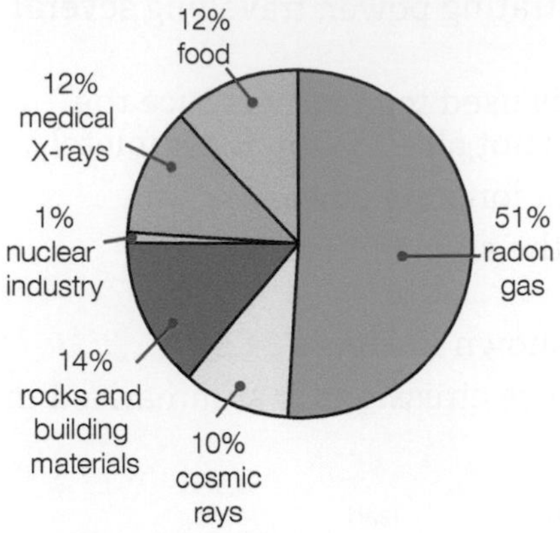

12%
food

12%
medical
X-rays

1%
nuclear
industry

51%
radon
gas

14%
rocks and
building
materials

10%
cosmic
rays

▲ **Figure 34.8** Sources of background radiation

Dangers of radiation

Most radioactive background activity comes from natural sources such as cosmic rays from space or from rocks and soil, some of which contain radioactive elements such as radon gas (see Figure 34.8). Living things and plants absorb radioactive materials from the soil, which are then passed along the food chain. There is little we can do about natural background radiation, although people who live in areas with a high background radiation level due to radon gas require homes to be well ventilated to remove the gas. Human behaviour also adds to the background activity that we are exposed to through medical X-rays, radioactive waste from nuclear power plants and the radioactive fallout from nuclear weapons testing.

Radioactive material is found naturally all around us and inside our bodies. A small number of carbon atoms occurring naturally are radioactive carbon-14 isotopes. They can be found in the carbon dioxide in the air and in the cells of all living organisms. Traces of radioactive elements, such as potassium, can be found in our food. Certain rocks contain uranium (all the isotopes of which are radioactive) and this decays into radon, a radioactive gas. There is also radiation reaching Earth from outer space, referred to as cosmic rays. All these natural sources are known together as background radiation.

Protection when handling

You can minimise the risk to those using radioactive materials by:

▶ wearing protective clothing
▶ keeping the source as far away as possible by using tongs
▶ limiting exposure time to as little as possible
▶ keeping radioactive materials in lead-lined containers.

Nuclear disintegration equations

Symbol equations can be written to represent alpha and beta decay.

The alpha particle can be written as $^4_2$He and the beta particle as $^{\,\,0}_{-1}$e.

Examples

1 Alpha decay of uranium-238

$$^{238}_{92}U \rightarrow\, ^{234}_{90}Th + ^4_2He$$

2 Beta decay of carbon-14

$$^{14}_{6}C \rightarrow\, ^{14}_{7}N + ^{\,\,0}_{-1}e$$

When writing symbol equations, it is important to remember the following:

▶ The sum of the mass numbers on the left-hand side of the equation must equal the sum of the mass numbers on the right-hand side.
▶ The sum of the atomic numbers on the left-hand side of the equation must equal the sum of the atomic numbers on the right-hand side.

If you know the original isotope (referred to as the parent nucleus) and the isotope that was formed by the decay (known as the daughter nucleus), it is possible to determine the type of decay by working out the type of particle emitted.

H

Example

Radium-226 decays to polonium-222. Radium (Ra) has atomic number 86 and polonium (Po) has atomic number 84. What is 'X'? Which type of decay occurs?

$$^{226}_{86}Ra \rightarrow\, ^{222}_{84}Po + ^a_bX$$

Answer

Balancing mass numbers:

$226 = 222 + a$

$a = 4$

Balancing atomic numbers:

$86 = 84 + b$

$b = 2$

So X is a helium nucleus (alpha particle).

This must be alpha decay.

Or, if you know the original isotope and the type of decay, you can work out the isotope that is formed by the decay.

As you can see in the worked example, alpha decay results in the mass number of the parent nucleus decreasing by four and its atomic number decreasing by two.

Alpha decay is exemplified by the equation below.

$$^A_Z X \rightarrow ^{A-4}_{Z-2} Y + ^4_2 He$$

For example, uranium-235 decays by emission of an α-particle into thorium-231:

$$^{235}_{92} U \rightarrow ^{231}_{90} Th + ^4_2 He$$

However, if the mass number of the parent does not change and the atomic number of the daughter nucleus increases by 1, then the reaction must be beta decay.

Beta decay is exemplified by:

$$^A_Z X^* \rightarrow ^A_{Z+1} Y + ^0_{-1} e \ (or \ ^0_{-1} \beta)$$

To take a specific case, radium-228 decays by emitting a β-particle to form actinium-228:

$$^{228}_{88} Ra \rightarrow ^{228}_{89} Ac + ^0_{-1} e$$

In gamma decay, the parent nucleus de-excites, and emits gamma ray(s) in the process. There is no change in the nature of the nucleus, so the mass number and the atomic number stay the same.

The γ-radiation is usually emitted at the same time as the α- and β-particle emissions, and represents the excess energy of the daughter nucleus as it settles down into a more stable condition.

$$^A_Z X \rightarrow ^A_Z X + \gamma$$

Test yourself

7 Copy and complete Table 34.3.

Table 34.3

Radiation	Atomic number (Z)	Mass number (A)
α-emission	decreases by 2	
β-emission		unchanged
γ-emission		

8 Copy and complete the following equations for alpha decay:

a) $^{238}_{92} U \rightarrow \boxed{\ } Th + ^4_2 He$

b) $\boxed{\ }_{94} Pu \rightarrow ^{238}_{\boxed{\ }} U + ^4_2 He$

c) $^{251}_{\boxed{\ }} Cf \rightarrow \boxed{\ }_{96} Cm + ^4_2 He$

d) $\boxed{\ } Hf \rightarrow ^{170}_{70} Yb + ^4_2 He$

e) $\boxed{\ } Bi \rightarrow ^{207}_{81} Tl + ^4_2 He$

f) $^{190}_{\boxed{\ }} Pt \rightarrow \boxed{\ }_{76} Os + ^4_2 He$

9 Copy and complete the following equations for beta decay:

a) $^{14}_{6}C \rightarrow \boxed{}N + ^{0}_{-1}e$

b) $\boxed{}_{1}H \rightarrow \boxed{}^{3}He + ^{0}_{-1}e$

c) $^{137}_{\boxed{}}Cs \rightarrow \boxed{}_{56}Ba + ^{0}_{-1}e$

d) $\boxed{}_{19}K \rightarrow \boxed{}^{40}Ca + ^{0}_{-1}e$

e) $\boxed{}Co \rightarrow ^{60}_{28}Ni + ^{0}_{-1}e$

f) $^{32}_{15}P \rightarrow \boxed{}S + ^{0}_{-1}e$

10 Work out the type of decay in each of the following examples:

a) bismuth-213 to polonium-213

b) radium-226 to radon-222

c) francium-221 to actinium-217

11 a) How does the value of the mass number change in alpha decay?

b) How does the value of the atomic number change in alpha decay?

c) How does the value of the mass number change in beta decay?

d) How does the value of the atomic number change in beta decay?

Radioactive decay

For very heavy elements such as uranium or plutonium, the large number of protons and neutrons can make the nucleus unstable and cause these nuclei to undergo radioactive decay in a **random** and **spontaneous** manner. Random means that we cannot predict when a particular nucleus will disintegrate. Spontaneous means that the rate of decay is unaffected by any physical changes such as temperature, pressure or chemical changes. Some types of nuclei are more unstable than others and decay at a faster rate.

Rate of decay and half-life

As a radioactive isotope decays, the activity of the sample decreases. This is because as the atoms of the original sample disintegrate, there will gradually be fewer and fewer original atoms left to disintegrate. To illustrate this process, consider radioactive iodine-131, which has a half-life of eight days. Imagine there are 40 billion atoms present at an instant in time. Look at Figure 34.9. In this diagram, one dot represents 1 billion atoms.

After eight days, 20 billion atoms will have decayed, leaving only 20 billion radioactive atoms. After another eight days, 10 billion more would have decayed, leaving only 10 billion radioactive atoms. After a further eight days, 5 billion more would have decayed, leaving only 5 billion radioactive atoms, and so on. As a result, the activity (the number of particles decaying in a particular time) decreases.

The half-life of a radioactive substance is the time taken for half the nuclei in any sample of the substance to decay. It follows that half-life of a substance is the time taken for the count rate to fall to half its original value.

The half-life of a radioactive isotope is defined as the time taken for its activity to fall by half.

Each isotope has a specific and constant half-life. Some half-lives are very short – a matter of seconds or even a fraction of a second – and

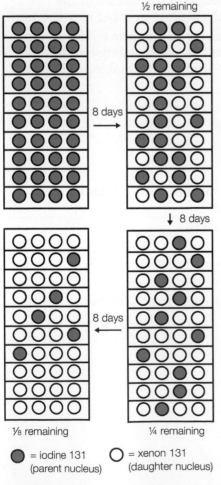

½ remaining

8 days

↓ 8 days

8 days

⅛ remaining   ¼ remaining

⬤ = iodine 131 (parent nucleus)   ◯ = xenon 131 (daughter nucleus)

▲ **Figure 34.9** Illustration of half-life

others can be thousands of years. Figure 34.10 shows a graph of the radioactive activity of a sample of an isotope with a half-life of 2 hours. Table 34.4 gives the half-lives of some common radioactive isotopes.

Table 34.4

Isotope	Half-life
uranium-238	4 500 000 000 years
carbon-14	5730 years
phosphorous-30	2.5 minutes
oxygen-15	2.06 minutes
barium-144	114 seconds
polonium-216	0.145 seconds

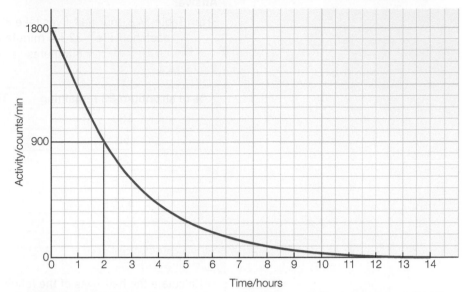

▲ **Figure 34.10** The radioactive decay curve for a substance with a half-life of 2 hours

The unit of radioactivity

The unit for radioactivity is the Becquerel (Bq).
1 Bq = 1 disintegration per second.

So if a radioactive material emits 1800 alpha particles every minute and 30 nuclei decay every second, its activity is 30 Bq.

When doing calculations, it can be very helpful to know the following:

▶ After one half-life, half the radioactive material remains.
▶ After two half-lives, a quarter of the radioactive material remains.
▶ After three half-lives, an eighth of the radioactive material remains.
▶ After four half-lives, a sixteenth of the radioactive material remains, and so on.

Example

1 What mass of nitrogen-13 would remain if 80 g were allowed to decay for 30 minutes? Nitrogen-13 has a half-life of 10 minutes.

Answer

Calculations like this can be easily done using a table. Take note of the headings in Table 34.5.

Table 34.5

Mass of nitrogen-13 remaining	Time/half-lives	Time/minutes
80 g	0	0
40 g	1	10
20 g	2	20
10 g	3	30

So 10 g would remain after 30 minutes.

2 Strontium-93 takes 32 minutes to decay to 6.25% of its original mass. Calculate the value of its half-life.

Answer

From Table 34.6, four half-lives take 32 minutes.

each half-life = 32 ÷ 4 minutes

= 8 minutes

Table 34.6

% of strontium-93 remaining	Time/half-lives	Time/minutes
100	0	0
50	1	8
25	2	16
12.5	3	24
6.25	4	32

Test yourself

12 Calculate the half-lives of the following samples:
a) A sample of iodine-123 whose activity falls from 1000 Bq to 250 Bq in 14.4 hours.
b) A sample of technetium-99 whose activity falls from 200 Bq to 25 Bq in 18 hours.
c) A sample of strontium-90 whose activity falls from 500 Bq to 62.5 Bq in 86.4 years.

13 Calculate how long it would take for the following to decay to an activity of 1 Bq:
a) A sample of cobalt-60 (half-life = 5.27 years) whose original activity is 64 Bq.
b) A sample of iodine-131 (half-life = 8 days) whose original activity is 128 Bq.
c) A sample of polonium-210 (half-life = 138 days) whose original activity is 32 Bq.

14 How long would it take for 20 g of cobalt-60 to decay to 5 g? The half-life of cobalt-60 is 5.26 years.

15 When a radioactive material with a half-life of 24 hours arrives in a hospital, its activity is 1000 Bq. Calculate its activity 24 hours before and 72 hours after its arrival. (Hint: Draw up a table as shown below.)

Table 34.7

Activity in Bq	Time/half-lives	Time/hours
	−1	−24
1000 (start from here)	0	0
500	1	24

16 Plot a graph of activity (y-axis) against time (x-axis) using the data in question 15. Start the graph from time = 0 and activity = 1000 Bq. Use the graph to find the activity 36 hours after the material arrives at the hospital.

Uses of radiation

In medicine

Gamma radiation from the cobalt-60 isotope can be used to treat tumours (Figure 34.13).

Different radioisotopes are used to monitor the function of organs by injecting a small amount into the bloodstream and detecting the emitted radiation. The tracers used in this case must have a short half-life.

Iodine-131 is used in investigations of the thyroid gland (Figure 34.11).

Surgical instruments and hospital dressings can be sterilised by exposure to gamma radiation (Figure 34.12). The source should have a very long half-life so that it does not need to be replaced on a regular basis.

Great care must be taken when using radioactive isotopes because the radiation can damage living cells by altering the structure of the cell's chemicals. Protective clothing must be worn, and the amount of time that the worker is exposed to the radiation must be strictly controlled. Radioactive isotopes that are taken internally are usually not alpha emitters (as they are such powerful ionisers), and they must have a short half-life so that they do not remain for too long in body tissues.

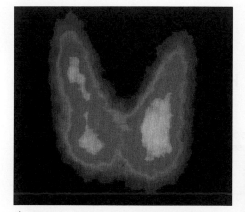

▲ **Figure 34.11** This scan shows radioactive iodine–131 localised in the thyroid gland

▲ **Figure 34.12** Sterilising equipment using gamma rays

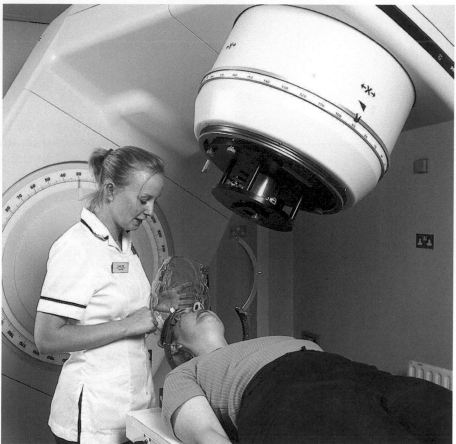

▲ **Figure 34.13** Radiotherapy involves the use of radioactive materials to treat cancers

In agriculture

Gamma radiation can be used to treat fresh food (Figure 34.14a). By killing bacteria on the food, the radiation helps the food to have a longer shelf-life. It is important to remember that this irradiation does not make the food itself radioactive. The use is controversial, however, as many people are worried about the long-term effects on the human body of eating irradiated food. Ideally, the radioisotopes used in food processing plants should have a very long half-life so that it is a long time before they need to be replaced.

The ease with which a plant absorbs a fertiliser can be found by putting a small amount of radioactive isotope in the fertiliser (Figure 34.14b). You can tell how much fertiliser has been taken up by the plant by checking different parts of the plant for radioactivity.

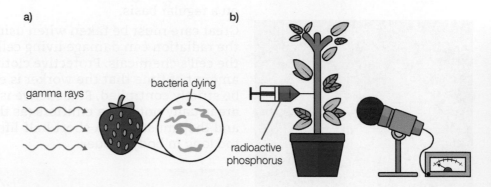

a) b)

gamma rays

bacteria dying

radioactive phosphorus

▲ **Figure 34.14** a) Sterilising food and b) measuring fertiliser take-up by a plant

In industry

Beta radiation can be used to monitor the thickness of a sheet of paper or aluminium (Figure 34.15). An emitter is placed on one side of the sheet and a detector on the other. As the sheet moves past, the activity detected will be the same as long as the thickness remains unchanged.

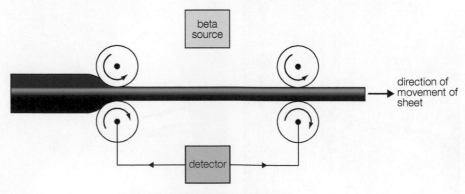

beta source

direction of movement of sheet

detector

feedback control to change thickness

▲ **Figure 34.15** A long half-life beta source is used to control the thickness of an aluminium sheet

Radioactive tracers

A suitable radioactive isotope can be used to provide information about fluid movement and mixing to monitor things like leaks in underground pipes (Figure 34.16). The radiation needs to penetrate many centimetres of soil to reach the detectors, which means that it must be a gamma emitter, because this is the only type of radiation with sufficient penetrating power. To avoid dangerous radioactive materials being in the ground for a long time, the source should have a short half-life.

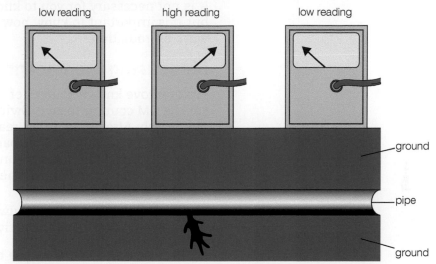

▲ **Figure 34.16** Radioactive tracers can be used to locate a leak in a pipe

In the home

In an ionisation smoke detector (Figure 34.17), a source of alpha radiation (often americium-241) is placed in the detector close to two electrodes. Ions are formed in the air around the radioactive source, and these allow a tiny current to flow.

If there is a fire, smoke will block the path of the ions, causing the current to fall. The fall in current is detected electronically, and a siren is sounded.

Practical work with radioactive materials

Students under the age of 16 are expressly forbidden from handling radioactive sources. However, you should know how practical work can be carried out.

The most common type of radiation detector is the Geiger–Müller tube (GM tube) connected to a counter (Figure 34.18).

When alpha, beta or gamma radiation enters the GM tube, it causes some of the argon gas inside to ionise and give an electrical discharge. This discharge is detected and counted by the counter. If the counter is connected to its internal speaker, you can hear the click when radiation enters the tube.

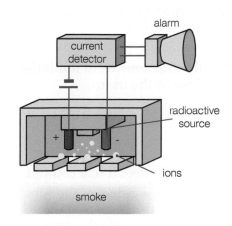

▲ **Figure 34.17** How a domestic smoke detector works

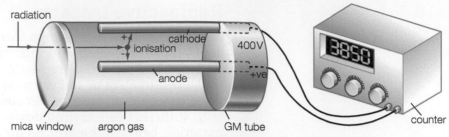

radiation

cathode

ionisation

400V

anode

+ve

mica window   argon gas   GM tube

counter

▲ **Figure 34.18** Section through a GM tube

It is not necessary for you to know how a GM tube works, but it is important to know how it could be used to do practical work on radiation.

Measuring the background radiation

First remove known sources of radiation from the laboratory, then set the GM counter to zero. Switch on the counter and start a stopwatch. After 30 minutes, read the count on the counter. Divide the count by 30 to obtain the background count rate in counts per minute. A typical figure is around 15 counts per minute. Fortunately, the background count in Northern Ireland does not present a serious health risk.

The background count must always be subtracted from any other count when measuring the activity from a specific source.

Safety precautions when using closed radioactive sources in schools

▶ Always store the sources in a lead-lined box, under lock and key, when not required for experimental use.

▶ Always handle sources using tongs, holding the source at arm's length and pointing it away from any bystander.

▶ Wear protective clothing.

▶ Always work quickly and methodically with sources to minimise the time of exposure and hence the dose to the user.

Measuring the approximate range of radiation

Alpha

▶ Place a GM tube on a wooden cradle and connect it to a counter.

▶ Hold an alpha source directly in front of the window of the tube and slowly increase the distance between the source and the tube. At about 3 cm (depending on the source used), the counter reading should fall dramatically to that of background radiation.

▶ Place a thin piece of paper in contact with the window of the GM tube. Bring the alpha source up to the paper so that the casing of the source touches it. The reading on the counter should now be the same as the background count, showing that the alpha particles are unable to penetrate the paper.

Beta

▶ Place a 1 mm thick piece of aluminium in contact with the window of the GM tube.

▶ Bring the beta source up to the aluminium so that the casing of the source touches it.

▶ The reading on the counter should be significantly above the background count, showing that some beta particles have penetrated the aluminium.

▶ Repeat the process with different sheets of aluminium, increasing the width by a millimetre at a time. At about 5 mm, there should be a significant reduction in the count rate, indicating the approximate range of beta particles in aluminium.

Gamma

If the beta particle experiment is repeated with a gamma source, there is practically no reduction in the count rate for a 5 mm thick piece of aluminium. If the aluminium sheets are replaced with lead, it will be found that even school sources will give gamma radiation that can easily penetrate several centimetres of lead (Figure 34.19).

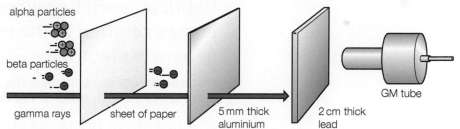

alpha particles

beta particles

gamma rays | sheet of paper | 5 mm thick aluminium | 2 cm thick lead

GM tube

▲ **Figure 34.19** The penetrative range of the three types of radiation

Nuclear fission and nuclear fusion

Nuclear fission

Radioactivity involves the random disintegration of an unstable nucleus. Some heavy nuclei, like those of uranium, can actually be forced to split into two lighter nuclei. The process is called nuclear fission. This usually comes about as a result of the heavy nucleus being struck by a slow neutron (Figure 34.20a). The heavy nucleus splits and the fragments move apart at very high speed, carrying with them a vast amount of energy. At the same time, two or three fast neutrons are also emitted. These neutrons go on to initiate further fission, and so create a chain reaction.

Just how much energy is emitted in fission? The fission of a single uranium nucleus produces about 49 000 000 times more energy than would be produced by a single carbon atom (in coal) reacting with oxygen to produce carbon dioxide. It did not take long for physicists to realise the huge potential of energy production using nuclear fission.

In a nuclear power station, steps are taken to ensure that, on average, just one of the fission neutrons goes on to produce further fission. This is controlled nuclear fission. The heat produced in

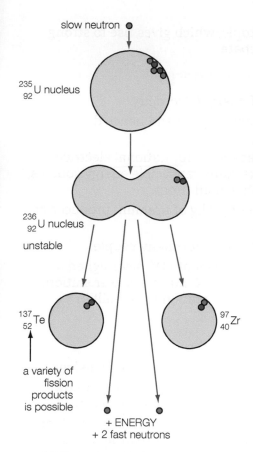

slow neutron

$^{235}_{92}$U nucleus

$^{236}_{92}$U nucleus
unstable

$^{137}_{52}$Te

$^{97}_{40}$Zr

a variety of fission products is possible

+ ENERGY
+ 2 fast neutrons

▲ **Figure 34.20a** The fission of a uranium-235 nucleus

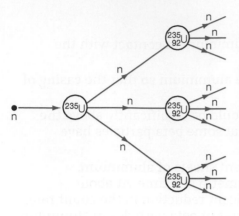

▲ **Figure 34.20b** A chain reaction in uranium-235

the reaction is used to turn water into steam and drive a turbine to generate electricity. In a nuclear bomb, there is no attempt to control the fission process (Figure 34.20b).

A major disadvantage of all fission processes is that the fission fragments are almost always highly radioactive. This type of radioactive waste is extremely dangerous, and expensive measures must be taken to store it until the level of activity is sufficiently small. In some cases, this means the waste must be stored deep underground, in a vitrified (glass-like) state, for tens of thousands of years. The danger is that, over time, the containers may leak and cause underground water pollution.

A further danger comes from earthquakes. Earthquakes may rupture containers of radioactive waste buried underground, causing the radioactive material to leak into the soil and water systems. Even in Britain there are over 200 earthquakes every year, many so weak that they are barely recorded. But as recently as February 2008, there was an earthquake in Lincolnshire of magnitude 5.2. This earthquake lasted for roughly 10 seconds and caused some structural damage. The tremors were felt across a wide area of England and Wales, and as far west as Bangor, Northern Ireland.

Political, social, environmental and ethical issues relating to the use of nuclear energy to generate electricity

Nuclear energy is a controversial topic, which gives rise to strong arguments on both sides of the debate.

Arguments in favour of nuclear energy

▶ It can produce vast amounts of energy/electricity.
▶ It produces very little carbon dioxide (CO_2) and hence does not contribute to global warming.
▶ Nuclear energy provides the 'base-load' for national electricity generation, meaning that unlike some renewable energy sources, such as wind power, it is available at all times.
▶ It is a high density source of energy. 1 kg of uranium produces as much electricity as 20 000 kg of coal.
▶ It provides employment opportunities for many people.

The UK government and French energy giant EDF have signed a contract for the new £18bn Hinkley Point C nuclear power station. This will be the first new nuclear power station to be built in the UK since 1995.

Arguments against nuclear energy

▶ Disposing of nuclear waste can be dangerous and expensive.
▶ Many people are concerned about living close to nuclear power plants and the storage facilities used for radioactive waste.
▶ Past accidents have made people scared of nuclear power, including the disaster at Chernobyl in the Ukraine and the earthquake and tsunami in Fukushima, Japan in 2011, when several reactors were damaged leading to a meltdown and release of radiation.

In both cases, huge economic, health and environmental damage was caused to the area surrounding the power plants.

Although nuclear fission does not release carbon dioxide, the mining, transport and purification of the uranium ore releases significant amounts of greenhouse gases into the atmosphere.

Germany decided in 2011 to phase out all of its nuclear reactors by 2022, prepared instead to invest heavily in renewable energy.

Nuclear fusion

This is the process that goes on in stars like our Sun. At the centre of the Sun, the temperature is about 15 000 000 °C. At this temperature, the nuclei of atoms in a gas are all stripped of their orbiting electrons, and they move at a very high speed. These charged gas nuclei form a state of matter called a plasma. Being positively charged, the nuclei would normally repel each other, but if they are moving fast enough, they can join (or fuse) to form a new nucleus.

In the Sun, hydrogen isotopes known as deuterium (hydrogen-2) and tritium (or hydrogen-3) collide and fuse to create a new nucleus, helium-4 (Figure 34.21). This causes the release of a vast amount of energy, some of which eventually reaches Earth as electromagnetic radiation.

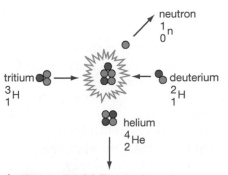

▲ **Figure 34.21** The fusion of deuterium and tritium

The equation representing this process is:

$$^2_1H + {}^3_1H \rightarrow {}^4_2He + {}^1_0n + energy$$

There have been many attempts to obtain controlled nuclear fusion on Earth. The world record for fusion power is held by the European reactor JET. In 1997, JET produced 16 MW of fusion power from a total input power of 24 MW.

Difficulties with fusion

The main problem is how to contain the reacting plasma at a high enough temperature and for a sufficiently long time for the reaction to take place. If we learn how to control nuclear fusion here on Earth, the rewards will be enormous. A large advantage of fusion is that the isotopes of hydrogen, deuterium and tritium (the more favoured isotopes of hydrogen) are widely available as the constituents of sea water, and so are nearly inexhaustible. Furthermore, fusion does not emit carbon dioxide or other greenhouse gases into the atmosphere, since its major by-product is helium, an inert non-toxic gas.

Fusion versus fission

Fusing nuclei together in a controlled way releases four million times more energy per kilogram than a chemical reaction such as the burning of coal, oil or gas, and fusing nuclei together in a controlled way releases four times as much energy as nuclear fission.

Is fusion a solution to the world's energy crisis?

There are many difficulties to overcome before nuclear fusion could provide electricity on a commercial scale, and it may be another 50 years before that happens. Nuclear fusion reactors will be expensive to build, and the system used to contain the plasma will be equally expensive because of the very high temperatures needed for the nuclei to fuse.

1 a) The table below shows the particles that make up a neutral carbon atom. Copy and complete Table 34.8 showing the mass, charge, number and location of the particles. Some information has already been added to the table.

Table 34.8

Particle	Mass	Charge	Number	Location
electron		−1		
neutron	1		6	in the nucleus
proton			6	

(7 marks)

b) Radon is a naturally occurring radioactive gas.
  i) Explain what is meant by radioactive. *(2 marks)*
  ii) Explain the danger of breathing radon gas into the lungs. *(2 marks)*
  iii) Explain the meaning of isotope in terms of the particles that make up the nucleus. *(2 marks)*

c) A student investigates the decay of a radioactive substance. She measures the corrected count rate of the substance every 20 minutes. The half-life of the substance is 20 minutes. At the start, the count rate was 800 counts per minute.
  i) Copy the graph grid shown in Figure 34.22 and plot this point and four more points that she found. *(5 marks)*

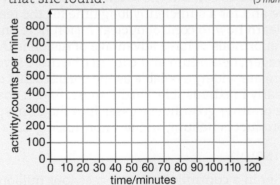

Figure 34.22

  ii) Draw a smooth curve through the plotted points. *(1 mark)*
d) i) The range of beta radiation in aluminium is several millimetres. Explain what this means. *(1 mark)*
  ii) Draw a neat and well-labelled diagram of the assembled apparatus that could be used to measure the range of beta particles in aluminium. *(3 marks)*
  iii) What measurements would be taken during this experiment? *(3 marks)*

  iv) How would you use these measurements to find the range of beta particles in aluminium? *(2 marks)*
  v) Sketch the graph that you would expect to obtain from these measurements and mark on it the range of the beta radiation. Label each axis. *(2 marks)*

2 $^{12}_{6}$C and $^{14}_{6}$C are both isotopes of carbon.
  a) i) Write down one similarity of the nuclei of these isotopes. *(1 mark)*
  ii) Write down one difference between the nuclei of these isotopes. *(1 mark)*
  b) $^{14}_{6}$C is radioactive. It decays to nitrogen by emitting a beta particle. Complete the equation below to describe the decay.
  $$^{14}_{6}C \rightarrow ^{A}_{Z}N + ^{0}_{-1}e$$
  (2 marks)

  c) $^{14}_{6}$C is present in all living materials and in all materials that have been alive. It decays with a half-life of 6000 years.
  i) Explain the meaning of the term 'a half-life of 6000 years'. *(2 marks)*
  ii) The activity of a sample of wood from a freshly cut tree is measured to be 80 disintegrations per second. Estimate the decrease in activity of the sample after 3 half-lives. *(2 marks)*

3 A radioactive isotope of gold emits gamma rays. It is injected into a patient's bloodstream and used to study the working of the patient's heart. The gamma radiation emitted by the gold is detected outside the patient's body by a device called a gamma camera.
  a) Why would a radioactive isotope that emits alpha radiation be unsuitable for this purpose? *(2 marks)*

To check the half-life of this isotope of gold, a radiographer measured the activity of a sample of the isotope every 10 s. He then corrected for the background activity. His measurements are shown in Table 34.9 below.

Table 34.9

Corrected activity/ counts per second	400	320	250	198	160	100	80
Time/seconds	0	10	20	30	40	60	70

b) What causes background activity and how did the radiographer correct his measurements?

(2 marks)

c) Using the measurements above, plot a graph of corrected activity (*y*-axis) against time (*x*-axis). *(5 marks)*

d) Use the graph to find the half-life of this isotope of gold. Show clearly how you use the graph to obtain the best value of this half-life. *(2 marks)*

4 A radioactive decay series can be represented on a graph of mass number, A, against atomic number, Z. Part of a table for such a series is given below:

Table 34.10

Element (symbol)	Atomic number	Mass number	Decays by emitting	Leaving element
U	92	238	α	Th
Th	90	234	β	Pa
Pa	91	234	β	
	92	234	α	
	90	230		Ra
Ra	88	226		Rn
Rn	86			Po
Po		218	α	Pb
Pb				Bi
Bi	83			Po

a) In what ways do mass number and atomic number change in
  i) α decay
  ii) β decay? *(4 marks)*
b) Copy and complete the table. *(7 marks)*
c) Identify two pairs of isotopes using the table.

(2 marks)

5 A sample containing 100 grams of a uranium isotope arrives at a factory. Table 34.11 shows how the mass of the isotope changes over time.

Table 34.11

Mass of isotope in grams	Time in days
100	0
72	10
52	20
37	30
27	40

a) Explain the meaning of the following words:
  i) half-life
  ii) isotope. *(2 marks)*

b) Plot a graph of mass of isotope (*y*-axis) against time (*x*-axis). *(4 marks)*
c) From the graph, calculate as accurately as possible the half-life of this isotope. *(2 marks)*
d) Estimate the mass of the uranium present in the sample three weeks before it arrived in the factory. *(2 marks)*

6 A certain material has a half-life of 12 minutes. What proportion of that material would you expect still to be present an hour later? *(3 marks)*

7 A detector of radiation is placed close to a radioactive source that has a very long half-life. In four consecutive 10 second intervals, the following number of counts were recorded: 100, 107, 99, and 102. Why were the four counts different? *(2 marks)*

8 a) The full symbol for a nucleus of carbon-14 is $^{14}_{6}C$. Copy and complete Table 34.12 by naming the particles in a nucleus of carbon-14 and give the quantity of each in the nucleus of carbon-14.

Table 34.12

Particle	Quantity in the nucleus

(4 marks)

b) Four unknown nuclei are labelled X1, X2, X3 and X4. Their full symbols are given below.

$^{30}_{15}X1$ $^{30}_{16}X2$ $^{32}_{17}X3$ $^{33}_{16}X4$

  i) Which, if any, of these nuclei are isotopes of the same element? *(1 mark)*
  ii) Explain your answer. *(1 mark)*

9 Fission and fusion are nuclear reactions. The table below is intended to show a number of significant differences between the two reactions. Copy and complete Table 34.13 using the following list.
1. building of larger nuclei from small nuclei
2. splitting up of large nuclei
3. nuclear power station
4. requires very high temperatures to start
5. the Sun
6. hydrogen
7. uranium
8. will start at normal temperatures

Table 34.13

Nuclear reaction	Fusion	Fission
Where the process can be found happening		
Fuel used		
Description of the reaction		
Conditions required to starts		

(4 marks)

10 Cobalt-60 is a beta emitter which decays to nickel. The nickel produced is a gamma emitter.
a) Copy and complete the decay equation below for the full beta decay process by writing the correct values in the boxes. *(2 marks)*

$$^{60}_{27}\text{Co} \rightarrow \boxed{}\boxed{}\text{Ni} + ^{0}_{-1}\text{e}$$

After 15 years, the measured activity of a cobalt-60 source is found to have fallen from 120 counts per minute to 15 counts per minute.
b) What is the half-life of cobalt-60? *(3 marks)*
c) State two possible uses of gamma radiation. *(2 marks)*

11 Uranium is an alpha emitter.
a) Explain why wearing heavy gloves to handle uranium rods could be sufficient protection. *(1 mark)*
b) State two alternatives to specialised clothing as ways of protecting people from the radiation from radioactive sources. *(2 marks)*

12 a) Describe the process of nuclear fusion. Your description should include:
- the particles involved
- what happens when nuclear fusion takes place
- where nuclear fusion occurs naturally. *(6 marks)*
b) A great deal of money is being invested on research into nuclear fusion.
  i) Suggest a reason why. *(1 mark)*
  ii) Give two practical difficulties which must be overcome before fusion reactors become viable. *(2 marks)*

13 Figure 34.23 shows the decay curve for a sample of iodine-123.

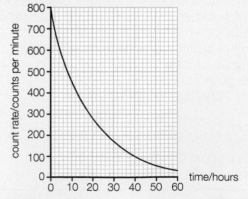

Figure 34.23 Decay curve for iodine-123

a) i) Using the graph, write down the half-life of iodine-123. *(1 mark)*
   ii) What is the count-rate after 24 hours? *(1 mark)*

b) i) If 0.08 g of iodine-123 in solution is injected into a patient, what is the maximum amount still in the body after 39 hours? *(1 mark)*
   ii) Why is it unlikely to actually be this much? *(1 mark)*
c) Iodine-123 is a gamma emitter. Another isotope, iodine-131, which has a half-life of 8 days, is a beta emitter. Give two reasons why iodine 131 is not suitable as a tracer in the body. *(2 marks)*

14 Radioactive sources can be used to treat a person with a cancerous tumour inside their body. The radioactive source is held directly over the tumour, as shown in Figure 34.24.

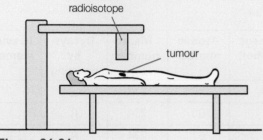

Figure 34.24

a) What type of radiation should the source emit if it is to penetrate the body and reach the tumour? *(1 mark)*
b) Why is this type of radiation effective in the treatment of tumours? *(1 mark)*
c) What disadvantages are there to using radioactivity to treat the tumour? *(1 mark)*
d) What is meant by the half-life of a radioactive source? *(2 marks)*
e) Should the radioactive source used in the treatment of the tumour have a short or long half-life? Explain your answer. *(2 marks)*

35 Waves

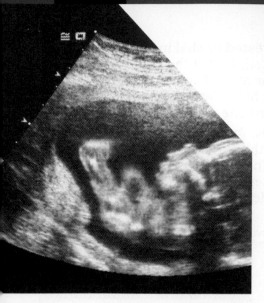

Types of waves

Waves transfer energy from one point to another, but they do not, in general, transfer matter. Radio waves, for example, carry energy from a radio transmitter to your home, but no matter moves permanently in the air as a result.

All waves are produced as a result of vibrations, and they can be classified as longitudinal or transverse. A vibration is a repeated movement, first in one direction and then in the opposite direction.

Longitudinal waves

A longitudinal wave is one in which the particles vibrate **parallel** to the direction in which the wave is travelling. The only types of longitudinal waves relevant to your GCSE course are:

▶ sound waves
▶ ultrasound waves
▶ slinky spring waves
▶ P-type earthquake waves.

▲ **Figure 35.1** A longitudinal wave moving along a slinky spring

It is easy to demonstrate longitudinal waves by holding a slinky spring at one end and moving your hand backwards and forwards parallel to the axis of the stretched spring. Compressions are places where the coils (or particles) bunch together. Rarefactions are places where the coils (or particles) are furthest apart.

All longitudinal waves are made up of compressions and rarefactions. In the case of sound waves, the particles are the molecules of the material through which the sound is travelling. These molecules bunch together and separate just as they do in a longitudinal wave on a slinky spring.

Transverse waves

A transverse wave is one in which the vibrations are **at 90°** to the direction in which the wave is travelling. Most waves in nature are transverse – some examples are:

▶ water waves (Figure 35.2)
▶ slinky spring waves (Figure 35.3)
▶ waves on strings and ropes
▶ electromagnetic waves.

A transverse wave pulse can be created by shaking one end of a slinky. The pulse moves along the slinky, but the final position of the slinky is exactly the same as it was at the beginning (Figure 35.3). None of the material of the slinky has moved permanently. But the wave pulse has carried energy from one point to another.

Figure 35.2 represents water waves. You can see that water waves are transverse. A cork floating on the surface of some water bobs up and down as the waves pass. The vertical vibration of the cork is perpendicular to the horizontal motion of the wave. Energy is transferred in the direction in which the wave is travelling.

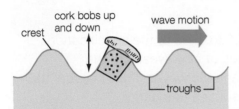

▲ **Figure 35.2** Transverse water waves move the cork up and down

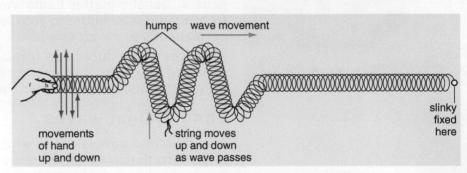

▲ **Figure 35.3** Transverse waves travelling through a slinky

There are many other examples that show that waves carry energy:

▶ Visible light, infrared radiation and microwaves all make things heat up.
▶ X-rays and gamma waves can damage cells by disrupting DNA.
▶ Loud sound waves can cause objects to vibrate (for example, your eardrum).
▶ Water waves can be used to generate electricity.

Describing waves

There are a number of important definitions relating to waves that must be learned.

The frequency of a wave is the number of complete waves passing a fixed point in a second. Frequency is given the symbol f, and is measured in units called hertz (abbreviation Hz).

Sometimes the units kHz and MHz are used. You should remember that $1\,kHz = 1000\,Hz$ and $1\,MHz = 1000\,kHz = 1\,000\,000\,Hz$.

The wavelength of a wave is the distance between two consecutive crests or troughs. Wavelength is given the symbol λ, and is measured in metres. λ is the Greek letter 'l' and is pronounced 'lamda'.

The amplitude of a wave is the greatest displacement of the wave from its undisturbed position. Amplitude is measured in metres. The wavelength and amplitude for a transverse wave are illustrated in Figure 35.4.

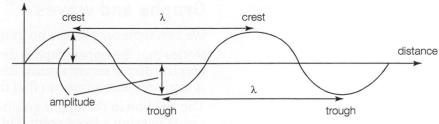

▲ **Figure 35.4** Displacement–distance graph to illustrate wavelength and amplitude

Test yourself

1 Describe the difference between a transverse wave and a longitudinal wave.
2 Give two examples of transverse waves and two examples of longitudinal waves.
3 Define the terms wavelength, frequency and amplitude, and state a unit in which each could be measured.

Show you can

a) Give evidence that microwaves transmit energy.
b) Use a slinky spring to demonstrate:
  i) a longitudinal wave
  ii) a transverse wave.
c) Describe what happens to the compressions in a longitudinal wave on a slinky spring if the wavelength is increased.

Wavelength and amplitude of longitudinal waves

It is much easier to visualise wavelength and amplitude for transverse waves than for longitudinal waves. For a longitudinal wave, the wavelength is the distance between the centre of one compression and the next (Figure 35.5).

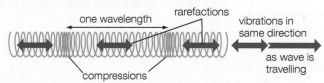

▲ **Figure 35.5** In longitudinal waves, the vibrations are along the same direction as the wave is travelling

But what is the amplitude of a longitudinal wave? Remember that the particles in a longitudinal wave vibrate backwards and forwards parallel to the direction in which the wave is moving. The amplitude of a longitudinal wave is the maximum distance a particle moves from the centre of this motion (Figure 35.6).

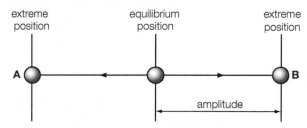

particle vibrates between positions A and B

▲ **Figure 35.6** The amplitude of a longitudinal wave

The wave equation

Imagine a wave with wavelength λ (metres) and frequency λ (hertz). From the definition of frequency, f waves pass a fixed point in 1 second. Each wave has a length λ, so, the total distance travelled every second is $f \times \lambda$.

The distance travelled in a second is the speed. So:

$$\text{wave speed} = \text{frequency} \times \text{wavelength}$$
$$v = f \times \lambda$$

Table 35.1

Frequency	Wavelength	Speed
Hz	cm	cm/s
	m	m/s
	km	km/s

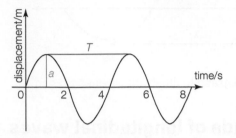

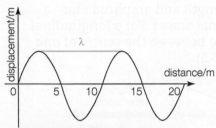

▲ **Figure 35.7** Graphs showing displacement against time and displacement against distance

This is an important equation that you should memorise and learn how to use.

Note that the units used in the wave equation must be consistent, as shown in Table 35.1.

Graphs and waves

We can represent waves on graphs like those shown in Figure 35.7.

Notice that the upper graph is displacement against time. The lower graph is displacement against distance. The vertical axis on both graphs is displacement, so we can find the amplitude from either graph.

The red line in the upper graph shows the time, T, between the crests passing a fixed point. This time is known as the period. Students often wrongly think T is a distance (the wavelength).

The blue line in the lower graph shows the distance between consecutive crests. This is the wavelength, λ.

The period, T, in Figure 35.7 is 4 seconds, and the wavelength, λ, is 10 metres.

How could we find the speed from these data?

The graphs tell us that the wave is travelling 10 metres in 4 seconds.

We know that speed = $\dfrac{\text{distance}}{\text{time}}$.

So, the wave speed = $\dfrac{10\,\text{m}}{4\,\text{s}}$

$= 2.5\,\text{m/s}$

We can also find the frequency from the top graph. The period (the time taken for one wave to pass), T, is 4 seconds, so 0.25 waves must pass every second.

So the frequency, $f = \dfrac{1}{T}$

$= 0.25\,\text{Hz}$

We can use the wave equation to confirm the speed:

$v = f \times \lambda$

$v = 0.25\,\text{Hz} \times 10\,\text{m}$

$= 2.5\,\text{m/s}$

Show you can

a) Recall the wave equation and the units in which speed, wavelength and frequency are measured.

b) Know how to find the frequency of a wave from a graph of displacement against time.

c) Know how to find the wavelength of a wave from a graph of displacement against distance.

Test yourself

4 Copy and complete Table 35.2. Note carefully the unit in which you are to give your answers. The first one has been done for you.

Table 35.2

Wavelength	Frequency	Speed
5 m	200 Hz	1000 m/s
12 m	50 Hz	.... m/s
3 m	60 Hz	.... m/s
...... m	4 Hz	20 cm/s
...... m	5 Hz	2.5 km/s
16 m	...... Hz	80 cm/s
6×10^4 m	...... Hz	3×10^8 m/s

> **5** The vertical distance between a crest and a trough is 24 cm, and the horizontal distance between the first and the fifth wave crests is 40 cm. If 30 such waves pass a fixed point every minute, find the amplitude, frequency, wavelength and speed of the waves.

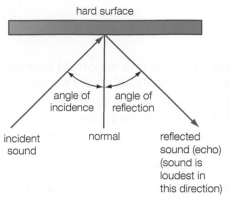

▲ **Figure 35.8** The reflection of sound against a hard surface

Echoes in sound and ultrasound

Like all waves, sound and ultrasound can be made to reflect (Figure 35.8).

This happens in a similar way to the reflection of water waves. As with water waves, the angle of incidence is always equal to the angle of reflection.

Audible sound ranges in frequency from 20 Hz to 20 000 Hz. Sound above 20 000 Hz is called ultrasound and cannot be heard by humans. It can, however, be detected by bats, dogs, dolphins and many other animals.

Reflected sound (and ultrasound) is called an echo. Humans have found clever ways to use ultrasound echoes:

▶ scanning metal castings for faults or cracks (for example in rail tracks)

▶ scanning a pregnant woman's womb to check on the development of her baby

▶ scanning soft tissues to diagnose cancers

▶ locating fish by seagoing trawlers

▶ mapping the surface of the ocean floor in oceanography.

We will look at some of these applications in a little more detail.

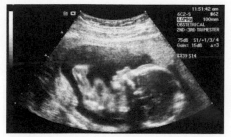

▲ **Figure 35.9** An ultrasound scan of an unborn baby

An application of ultrasound in medicine

In an ultrasound scan of an unborn baby, a probe is moved across the mother's abdomen. The probe sends out ultrasound waves and also detects the reflections. The low wavelength of the ultrasound waves means that, unlike audible sound waves, ultrasound can be sent out in a very narrow beam and focused on the unborn baby. The other end of the probe is connected to a computer.

By examining the reflected waves from the womb, the computer builds up a picture of the foetus (unborn baby) like that in Figure 35.9. Unlike X-rays, ultrasound is now known to be quite safe for this purpose.

Ultrasound can also be used to measure the diameter of the head of the baby as it develops in the womb (Figure 35.10). When the ultrasound reaches the baby's head at A, some ultrasound is reflected back to the detector and produces pulse A on the cathode ray oscilloscope (CRO). Some ultrasound passes through the head to point B, and is then reflected back to the detector. This reflection produces pulse B on the CRO.

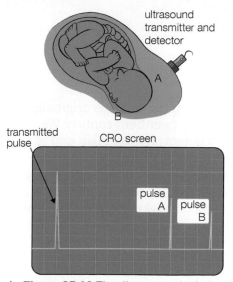

▲ **Figure 35.10** The diameter of a baby's head can be measured using ultrasound

In the diagram of the CRO screen, each horizontal division corresponds to a time of 40 microseconds ($40 \,\mu s = 40 \times 10^{-6} \,s$).

The time interval between the arrival of pulse A and the arrival of pulse B at the detector corresponds to 3 divisions on the CRO. Since each division is 40 μs, this represents a total time of 120 μs. This additional 120 μs is the time taken for ultrasound to travel from A to B and back to A. The time for ultrasound to travel from B to A is therefore half of that or 60 μs.

Now, physicists know that ultrasound travels at a speed of 1500 m/s in a baby's head. So the width of the head can be found as follows:

width of head = speed × time

$$= 1500\,\text{m/s} \times 60 \times 10^{-6}\,\text{s}$$

$$= 0.09\,\text{m}$$

$$= 9\,\text{cm}$$

Scanning metal castings

Railway tracks do not last forever. They wear out. It is important that we find out early if they are developing cracks or flaws below the surface.

Ultrasound scanners attached to specially fitted rail carriages (Figure 35.11) can be used to detect these cracks and flaws. Ultrasound passes through the track. If there is a crack or other flaw it can be imaged using the same science that allows ultrasound technicians to obtain a picture of a baby in the womb.

Alternatively, the depth to the crack can be found using the same science used to measure the diameter of the head of a baby in the womb. To carry out the measurement, we need to know that ultrasound travels at 5000 m/s in steel.

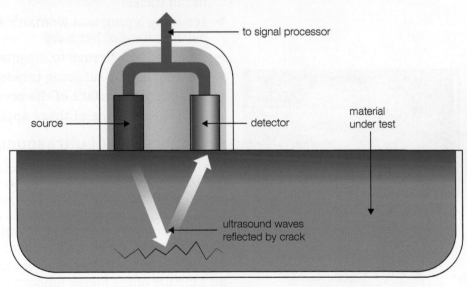

▲ **Figure 35.11** Detecting cracks in metals

Sonar

Sonar stands for **SO**und **N**avigation **A**nd **R**anging and was originally developed to detect submarines in the early twentieth century. We will look at how fishermen use sonar to detect shoals of fish and to measure how far they are below the surface.

The fishing trawler sends out an ultrasound pulse. Ultrasound travels at 1500 m/s in seawater. The ultrasound hits the fish and is reflected as an echo. This echo is detected back on the trawler. By measuring the time between the transmission of the sound and the detection of the echo, we can calculate the depth of the fish below the surface of the sea.

Example

A fishing trawler produces an ultrasound pulse and 0.4 s after it detects an echo reflected from a shoal of fish. Assuming the speed of ultrasound in seawater is 1500 m/s:

i)   Calculate the total distance travelled by the ultrasound.

ii) Calculate the distance from the trawler to the shoal of fish.

iii) Explain why a second echo may be detected after the first.

Answer

i)   total distance = speed × time

     = 1500 m/s × 0.4 s

     = 600 m

ii) distance from the trawler to the fish = ½ × total distance

     = ½ × 600 m

     = 300 m

iii) The first echo is from the fish. The second echo is from the sea bed, which is further away from the trawler than the fish, so the ultrasound from the sea bed takes longer to reach the trawler than the echo from the fish.

Radar

Radar stands for **RA**dio **D**etection **A**nd **R**anging and was originally developed during World War II to detect enemy aircraft and find their distance from the radar station.

Radar waves are in the microwave section of the electromagnetic spectrum. Think of a radar beam as a powerful beam of microwaves. They have wavelengths ranging from a few millimetres to just over a metre. Because radar waves are incredibly fast (3×10^8 m/s), they are used to track very fast objects which may be a large distance away. So, for example, they are used by air traffic controllers to track passenger airliners, by the military to track missiles, and by the coastguard to detect ships.

Radar cannot be used under water. The water absorbs the radar (microwaves) within a metre or so. In addition, radar cannot be used to measure very small distances. This is because radar is so fast that the time taken would be too small to measure easily.

However, the physics of radar is very similar to that of sonar. If you are asked to solve a mathematical problem involving radar, it is likely that numbers will be given in index form as shown in the example.

Electromagnetic waves

Electromagnetic waves are members of a family with common properties called the electromagnetic spectrum. They:

▶ can travel in a vacuum (unique property of electromagnetic waves)
▶ all travel at exactly the same speed in a vacuum
▶ are transverse waves.

Electromagnetic waves also show properties common to all types of wave. They:

▶ carry energy
▶ can be reflected
▶ can be refracted.

We give names to seven broad sections of the electromagnetic spectrum, according to their wavelengths and the effects they produce. The properties of electromagnetic waves depend very much on their wavelength. In Table 35.3 they are arranged in order of increasing wavelength (or decreasing frequency). You need to be able to list these waves in order of increasing (and decreasing) wavelength, but do not need to remember specific wavelengths!

Table 35.3 The electromagnetic spectrum

Electromagnetic wave	Typical wavelength
Gamma (γ) rays	0.01 nm
X-rays	0.1 nm
Ultraviolet light	10 nm
Visible light	500 nm
Infrared light	0.01 mm
Microwaves	3 cm
Radio waves	1000 m

1 nanometre (nm) = 1×10^{-9} m

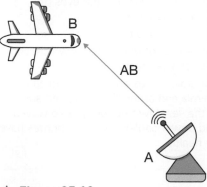

Dangers of electromagnetic waves

Table 35.4

Electromagnetic wave	Dangers
Gamma (γ) rays	damage cells and disrupt DNA, which may lead to cancer
X-rays	damage cells and disrupt DNA, which may lead to cancer
Ultraviolet light	certain wavelengths can damage skin cells, disrupting DNA and potentially leading to skin cancer
Visible light	intense visible light can damage the eyes (for example in snow blindness)
Infrared light	felt as heat and can cause burns
Microwaves	cause internal heating of body tissues which, some say, can lead to eye cataracts
Radio waves	large doses of radio waves are believed to cause cancer, leukaemia and other disorders, and some people claim the very low frequency radio waves from overhead power cables near their homes have affected their health

Test yourself

6 Copy the diagram below and list the sections of the electromagnetic spectrum in order of increasing frequency by writing their names in the boxes. One box has been completed for you.

			visible light			

→ increasing frequency

7 State a property unique to electromagnetic waves.

8 Below are three sections of the electromagnetic spectrum, not arranged in any particular order.

gamma rays, radio waves, ultraviolet light

Copy and complete Table 35.5 writing the name of the missing electromagnetic wave opposite its typical wavelength.

Hint: First identify the missing sections of the spectrum and write them in order of increasing wavelength.

Table 35.5

Wave				
Typical wavelength/m	1×10^{-10}	6×10^{-7}	1×10^{-5}	1×10^{-3}

9 The emitter in Figure 35.13 sends out a pulse of sound. An echo from the object is detected after 2.5 ms. If sound travels at 340 m/s in the air, calculate the distance marked d.

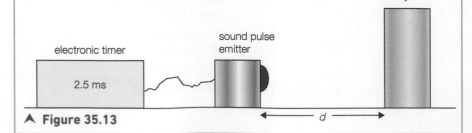

▲ **Figure 35.13**

Practice questions

1 State the difference between a transverse and a longitudinal wave and give an example of each type of wave. *(4 marks)*

2 What is meant by wavelength, frequency and amplitude when applied to longitudinal waves? *(3 marks)*

3 What physical property of a water wave never changes as a result of:
a) reflection and
b) refraction? *(2 marks)*

4 The vertical distance between a crest and a trough of a water wave in the mid-Atlantic is 16 m, and the horizontal distance between the first and the fifth wave crests is 80 m. If 12 such waves pass a fixed point every minute, find the amplitude, frequency, wavelength and speed of the waves. *(8 marks)*

5 A water wave has period $T = 0.25$ s, vertical distance from peak to trough = 25 cm and $\lambda = 1.2$ m. Find the amplitude, frequency and speed of the wave. *(6 marks)*

6 Figure 35.14 shows three wavefronts in shallow water in a ripple tank. The water to the left of the vertical line is deep water.
a) Copy and complete the diagram to show what happens when the waves are refracted. *(3 marks)*
b) In what way, if at all, do the frequency, wavelength and speed of these water waves change when they are refracted? *(3 marks)*

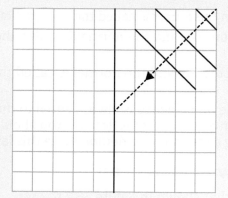

Figure 35.14

7 **a)** What is ultrasound? *(1 mark)*
b) In what way is ultrasound different from the sound of human speech? *(1 mark)*

8 State two differences between ultrasound waves and electromagnetic waves. *(2 marks)*

9 **a)** List the members of the electromagnetic spectrum in order of decreasing wavelength. *(2 marks)*
b) Which three members of the electromagnetic spectrum have been proven to cause cancer? *(3 marks)*

10 State the difference between a transverse and a longitudinal wave and give an example of each type of wave. *(4 marks)*

11 When the frequency of sound is changed, the wavelength also changes. Table 35.6 shows the results of an experiment to measure the wavelength of sound at different frequencies. The unit for $1/\lambda$ is 1/m.

Table 35.6

Wavelength, λ/m	0.7	1.0	1.5	2.5	4.0
Frequency, f/Hz	460	320	210	130	80
$\frac{1}{\lambda}$/1/m				0.40	0.25

a) Complete the table by entering the missing numbers in the third row. *(3 marks)* Two entries have already been done for you.
b) Plot (on graph paper) the graph of f/Hz on the vertical axis against $\frac{1}{\lambda}$/1/m on the horizontal axis and draw the straight line of best fit. *(4 marks)*
c) Find the gradient of your line of best fit and state its unit. *(3 marks)*
d) What is the physical significance of the gradient of the line of best fit? *(1 mark)*
e) Use your graph to find the wavelength of sound of frequency 250 Hz. *(2 marks)*

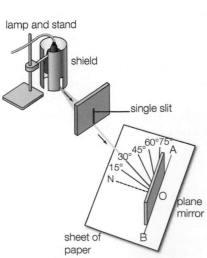

Reflection of light

All of us are familiar with the way light reflects from a straight (plane) mirror.

The essential ideas are shown in Figure 36.1.

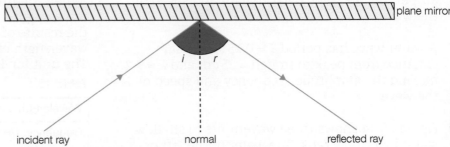

▲ **Figure 36.1** Reflection by a plane mirror

▶ The ray of light travelling towards the mirror is called the **incident ray**.

▶ The perpendicular to the mirror where the incident ray strikes it is called the **normal**.

▶ The ray that travels away from the mirror is called the reflected ray.

▶ The angle between the incident ray and the normal is called the angle of incidence, *i*.

▶ The angle between the reflected ray and the normal is called the angle of reflection, *r*.

Experiments show that the angle of incidence is always equal to the angle of reflection. This is known as the law of reflection. You should be able to describe the following experiment to demonstrate this law.

Demonstrating the law of reflection

We can use apparatus such as that shown in Figure 36.2 to demonstrate the law of reflection.

▶ With a sharp pencil and a ruler, draw a straight line AOB on a sheet of white paper using a ruler.

▶ Use a protractor to draw a normal, N, at point O (90° to the mirror).

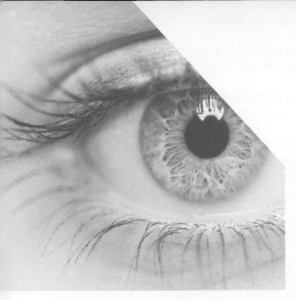

▲ **Figure 36.2** Single slit test illustrating the law of reflection

- With the protractor, draw straight lines at various angles to the normal ranging from 15° to 75°.
- Place a plane mirror on the paper so that its back rests on the line AOB as in Figure 36.2.
- Using a ray box, shine a ray of light along the line marked 15°.
- Mark two crosses on the reflected ray on the paper.
- Remove the mirror and, using a ruler, join the crosses on the paper with a pencil. Extend the line backwards to point O – this line shows the reflected ray.
- Measure the angle of reflection with a protractor.
- Record the angles of incidence and reflection in a table.
- Repeat the experiment for different angles of incidence up to 75°.

Note that when light is reflected off a rough surface, such as paper, the law of reflection still applies. However, because the paper surface is rough (Figure 36.3b), the light is reflected in many different directions and cannot produce a clear image. This is called diffuse reflection.

Where is the image we see in a plane mirror? We know it is behind the mirror, but how far behind?

To answer that question, we need to do another experiment.

Where is the image?

We can use apparatus such as that shown in Figure 36.2 to locate the image produced by a plane mirror.

- Support a plane mirror vertically on a sheet of white paper placed on a horizontal surface, and with a pencil, draw a straight line at the back to mark the position of the reflecting surface.
- Use a ray box to direct two rays of light from point O towards points A and B on the mirror as in Figure 36.4.
- Mark the position of point O with a cross using a pencil.
- Mark two crosses on each of the real reflected rays.
- Remove both the ray box and the mirror.
- Using a ruler, join the crosses with a pencil line to obtain the paths of the real rays from A and B.
- Extend these lines behind the mirror (these are called virtual rays) – they meet at I, the point where the image was formed.
- Measure the distance from the image I to the mirror line (IN), and the distance from the object O to the mirror line (ON). They should be the same.
- Repeat the experiment for different positions of the object O.
- In each case, the object O and its image I should be the same perpendicular distance from the mirror.

The reflected rays get further apart (diverge) and enter the eye. The eye follows the rays back in a straight line. The rays entering the eye appear to come from a point behind the mirror. This point is the image.

Note that the image in a plane mirror is not caused by real rays of light coming to a focus, as happens on a cinema screen. A mirror image is therefore called a virtual image.

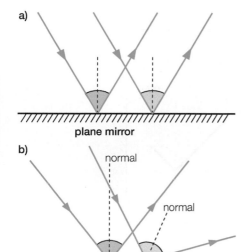

▲ **Figure 36.3** Light reflecting on **a)** a plane mirror and **b)** a rough surface

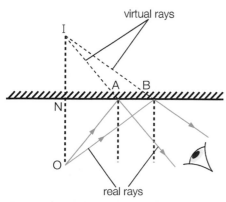

▲ **Figure 36.4** The image in a plane mirror

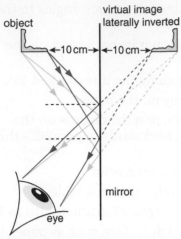

object

virtual image
laterally inverted

mirror

eye

▲ **Figure 36.5** This shows what happens when light from an object strikes a mirror

A mirror image is also reversed. If we hold a left-handed glove in front of a mirror, its image looks like a right-handed glove and vice versa (Figure 36.5). The image is said to be laterally inverted.

The image in a plane mirror is:

▶ virtual
▶ the same size as the object
▶ laterally inverted
▶ the same distance behind the mirror as the object is in front of the mirror.

Tip

Be very precise when asked about the image size and position in a plane mirror. Don't just write: Same size. Remember to add '... as the object'. Don't just write: Same distance behind the mirror. Remember to add '... as the object is in front of the mirror'.

Test yourself

1 State the size of the angle of incidence when the incident ray strikes a plane mirror at 90°.
2 The angle between a plane mirror and the incident ray is 40°. What size is the angle of reflection?
3 The angle between the incident ray and the reflected ray is 130°. What size is the angle of incidence?
4 Write the word 'AMBULANCE' in its laterally inverted form.
   Why is this laterally inverted form sometimes seen painted on real ambulances?
5 Two plane mirrors are inclined at right angles to each other. A ray of light strikes one mirror, M₁, at an angle of incidence of 30°, and the reflected ray from M₁ falls incident on M₂. Find the angle of reflection at M₂. Comment on the direction of the ray incident on M₁ and the reflected ray from M₂.
6 A student stands in front of a mirror and views his image. The student now takes a step backwards so that he is 20 cm further away from the mirror. By how much has the distance between the student and the image increased?
7 A letter L is placed in front of a mirror as shown in Figure 36.6.
   Copy the diagram and use the grid to draw the image of the letter L in the mirror.
8 Two mirrors A and B are arranged at 120° as shown in Figure 36.7. A ray of light is incident on mirror A.
   Calculate the angle of reflection of the ray reflected by mirror B.

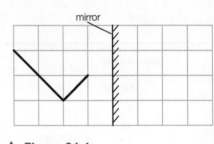

mirror

▲ **Figure 36.6**

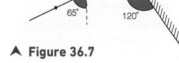

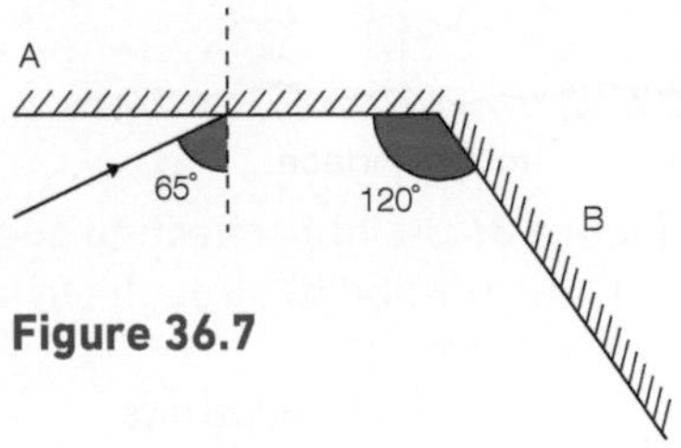

A

65°   120°

B

▲ **Figure 36.7**

Refraction of light

Refraction is the change in direction of a beam of light as it travels from one material to another due to a change in speed in the different materials. Table 36.1 shows the speed of light in various media. It is not necessary to remember the numbers in this table, but you must know that light travels faster in air than in water, and faster in water than in glass. You should also remember that the greater the change in the speed, the greater the angle it bends through (refracts).

Table 36.1

Material	Speed of light/m/s
Air (or vacuum)	3×10^8
Water	2.25×10^8
Glass	2×10^8

▶ The angle between the normal and the incident ray is called the **angle of incidence**.

▶ The angle between the normal and the refracted ray is called the **angle of refraction**.

▶ The angle between the normal and the emergent ray is called the **angle of emergence**.

Figure 36.8 shows the paths of different light rays through a glass block. If the block has parallel sides, the angle of incidence is equal to the angle of emergence.

A ray parallel to the normal does not bend as it enters the block.

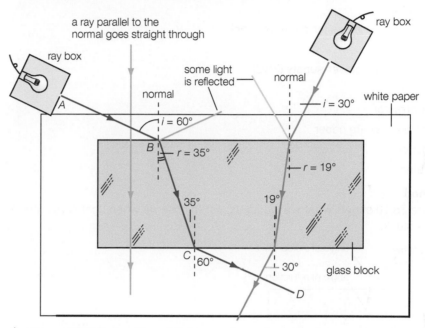

▲ **Figure 36.8** Refraction of light rays by a glass block

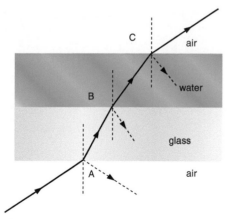

▲ **Figure 36.9** Diffraction through different media

Experiments show that:

▶ when light speeds up, it bends away from the normal

▶ when light slows down, it bends towards the normal.

Remember that this is also what happens to waves travelling from deep water into shallow water.

Figure 36.9 shows what happens when light travels from air through glass, then through water, and finally back into the air. Note the changes of direction in each case.

As the light enters the glass from the air at point A, it slows down, so it bends towards the normal.

As the light passes from glass into water at point B, it speeds up a little, so it bends away from the normal.

As the light passes from water into air at point C, it speeds up even more, so it bends still more away from the normal.

Prescribed practical P5: Using a ray tracing technique to measure angles of incidence and refraction as a ray of light passes from air into glass

Aims

▶ to trace rays through a glass (or perspex) block
▶ to measure angles of incidence and refraction
▶ to plot a graph of angle of incidence (y-axis) against angle of refraction (x-axis)

Apparatus

▶ rectangular glass (or perspex) block
▶ ray box
▶ low voltage power supply (PSU)
▶ leads
▶ protractor
▶ A4 plain white paper
▶ pencil
▶ ruler

Method

Figure 36.10 shows what we would expect to observe when light rays enter a glass block.

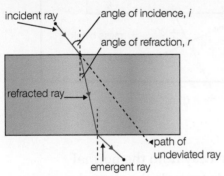

▲ Figure 36.10 Light rays entering a glass block

1 Prepare a table for your results like the one shown on page 441.
2 Place the rectangular glass block in the centre of the sheet of white paper on a drawing board.
3 Draw around the outline of the block with a sharp pencil.
4 Connect the ray box to the PSU with the leads.
5 Switch on the PSU and direct a ray of light to enter the block near the middle of the longest side of the block, so that the angle of incidence is about 10°.
6 Mark the paths of both the incident ray and the emergent ray with two pencil dots, ensuring that the dots on each ray are as far apart as possible.
7 Carefully remove the glass block.
8 With a pencil and ruler, join the dots on the incident ray up to the point of incidence.
9 With a pencil and ruler, join the dots on the emergent ray back to the point of emergence.

10 With a pencil and ruler, draw a straight line between the point of incidence and the point of emergence to show the path of the refracted ray in the glass.

11 With a pencil and protractor, draw the normal at the point of incidence.

12 With the protractor, measure the angle of incidence, i, and the angle of refraction, r. Record the data in a table.

13 Repeat steps 1 to 12 for angles of incidence ranging from 20° up to 80°.

14 Observe that when the angle of incidence is zero (normal incidence), the angle of refraction is also zero.

Results

Table 36.2

Angle of incidence, $i/°$	10	20	29	41	49	58	67	70
Angle of refraction, $r/°$	7	13	19	26	30	34	38	41

The above results are typical of those obtained in this experiment.

Treatment of the results

Plot a graph of angle of incidence (vertical axis) against angle of refraction (horizontal axis). Figure 36.11 shows an example of the type of graph that might be obtained.

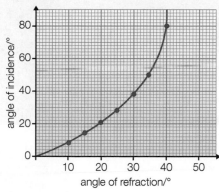

▲ **Figure 36.11** A graph showing angle of incidence against angle of refraction

Conclusion

The graph of angle of incidence against angle of refraction is a curve through the origin with an increasing gradient (the graph gets steeper as the angle of incidence increases). This tells us that i is not directly proportional to r (because the graph is not a straight line), but that i and r have a positive correlation (as i increases, r increases).

Tip

Always remember to put an arrow on real rays of light. Both the normal and the virtual rays are always dotted and never have an arrow.

Show you can

a) Describe what is meant by refraction of light.
b) Explain what happens to the speed of a ray of light if it refracts away from/towards the normal.
c) Describe an experiment to investigate the relationship between the angles of incidence and refraction when light is refracted as it passes from air into glass.
d) State what is meant by dispersion and state the conditions necessary for it to occur.

Dispersion of white light

All colours (frequencies) of light travel at the same speed in air, but different colours of light travel at different speeds in glass. This means that different colours bend by different amounts when they pass from air into glass. When light is passed through a triangular glass block (a prism), the effect is called dispersion and you can see a spectrum showing all the colours of the rainbow (Figure 36.12). Red light is bent (refracted) the least because it travels fastest in glass – it is slowed down the least. Violet light bends the most because it is slowest in glass – it is slowed down the most.

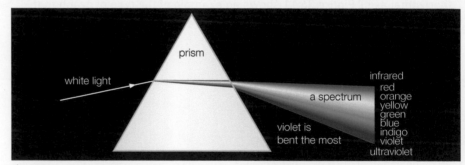

▲ **Figure 36.12** Dispersion of light through a prism

Test yourself

9 a) Figure 36.13 shows a ray of light passing from the air into the cornea of the eye.
  i) State the angle of incidence in air, and the angle of refraction in the cornea.
  ii) In which of the two media does light travel the fastest?
 b) What is the evidence for believing that red light travels faster in glass than blue light?

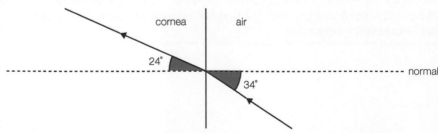

▲ **Figure 36.13** Light passing from the air into the cornea of the eye

10 A beam sound made underwater travels towards the surface. Sound travels faster in water than it does in air. Copy and complete Figure 36.14 to illustrate the refraction that occurs when the sound travels into the air. Mark the normal on your diagram.

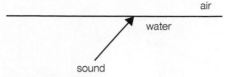

▲ Figure 36.14

11 Different shapes of glass prism are often used to change the direction of light rays.
Copy Figure 36.15 and continue the path of the ray shown until it emerges into the air.

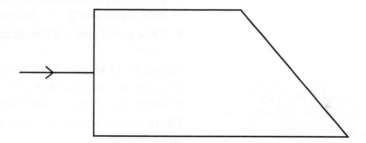

▲ Figure 36.15

12 Table 36.3 shows the speed of light in five different materials.

Table 36.3

Material	Speed/m/s
Air	3.0×10^8
Ice	2.3×10^8
Water	2.2×10^8
Glass	2.0×10^8
Diamond	1.2×10^8

a) Between which pair of materials does light show the greatest change in speed?
b) Between which pair of materials does light refract the most?
c) Does light bend towards or away from the normal when it passes from water into diamond? Give a reason for your answer.

Lenses

Lenses are specially shaped pieces of glass or plastic. There are two main types of lens (Figure 36.16), converging (or convex), and diverging (or concave).

converging (convex) lens

converging lens is thickest at the centre

diverging (concave) lens

diverging lens is thickest at the edges

▲ **Figure 36.16** The shapes of a converging lens and a diverging lens

There are two features of a converging lens that need to be defined:

▶ Rays of light parallel to the principal (central) axis of a converging lens all converge at the same point on the opposite side of the lens.

▶ This point lies on the principal axis, and is called the principal focus.

Figure 36.17 shows how a convex lens refracts light. Note that light refracts at each surface as it enters and leaves the lens, first bending towards the normal, and then away from the normal.

There are two features of a diverging lens (Figure 36.18) that need to be defined:

▶ Rays of light parallel to the principal axis of a diverging lens all appear to diverge from the same point after refraction in the lens.

▶ This point lies on the principal axis, and is called the principal focus.

The distance between the principal focus and the optical centre, C, of any lens is called the focal length.

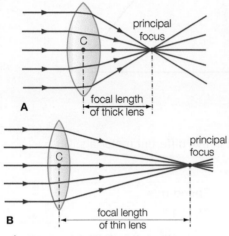

A

principal focus

focal length of thick lens

B

principal focus

focal length of thin lens

▲ **Figure 36.17** How light is refracted through a converging lens

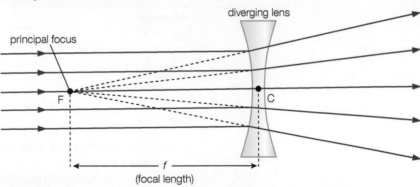

diverging lens

principal focus

F

C

f
(focal length)

▲ **Figure 36.18** How light is refracted through a concave lens

Note that light passing through the optical centre of a convex or concave lens is not bent. It passes straight through without refraction. Of course, light can pass through a lens from left to right or from right to left, so every lens has two principal foci and two focal lengths.

The only lenses you need to learn about at GCSE are equiconvex and equiconcave. This means that the principal foci on each side of the lens are the same distance from the optical centre. This becomes important when drawing ray diagrams to scale to find the position of an image.

Practical activity

This is a non-prescribed activity.

Aims

▶ to measure the focal length, f, of a converging lens using a distant object

Apparatus

▶ convex lens
▶ lens holder
▶ ruler
▶ sticky tape
▶ white screen in a holder
▶ distant object (such as a tree which can be seen through the windows in the laboratory and is at least 20 m away)

Method

1 Tape the ruler to the bench.

2 Place the white screen in its holder at the zero mark (see Figure 36.19).

3 Place the lens in its holder as close as possible to the screen.

4 Slowly move the lens away from the screen until the inverted image of the distant object is as sharp as possible.

5 Using the metre ruler, measure the distance from the centre of the lens to the white screen. This distance is the focal length of the lens, f.

6 Record the measured focal length in a prepared table.

7 For reliability, repeat steps 1 to 6 for four different distant objects, and determine the average value of f.

Diagram of the apparatus

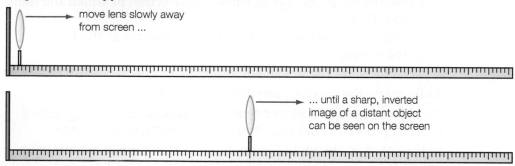

▲ **Figure 36.19** Apparatus used to measure the focal length of a converging lens

Results

Table 36.4

Focal length f/mm	245	240	250	247	243
Average focal length f/mm	245				

(The above results are typical of those obtained using a lens of focal length around 250 mm.)

This is not an experiment where it is appropriate to write a conclusion.

> Note that, regardless of the position of the object, the image in a concave (diverging) lens is always:
> • erect
> • virtual
> • smaller than the object
> • placed between the object and the lens.

Image in a convex (converging) lens

The position and properties of the image in a convex (converging) lens depends on the position of the object. We can find those positions and image properties by drawing a ray diagram.

Rules for drawing ray diagrams

To draw a ray diagram for a convex lens, you must draw at least two of the following rays:

▶ a ray parallel to the principal axis refracted through the principal focus on the other side of the lens

▶ a ray through the optical centre of the lens that does not change its direction (does not refract)

▶ a ray through the principal focus on one side of the lens which emerges so that it is parallel to the principal axis on the other side of the lens.

First steps when drawing a ray diagram for a convex lens:

▶ Using a ruler, draw a horizontal line to represent the principal axis and a vertical line for the lens.

▶ Mark the position of the principal focus with a letter F, the same distance from the optical centre on each side of the lens.

▶ Using a ruler, draw a vertical line touching the principal axis at the correct distance from the lens to represent the object.

▶ Using a ruler, draw at least two of the three construction rays, starting from the top of the object.

▶ Draw arrows on all rays to show the direction in which the light is travelling.

▶ The point where the construction rays meet is at the top of the image.

▶ The bottom of the image lies vertically below on the principal axis.

To illustrate the process, consider the following problem.

An object 5 cm tall is placed 6 cm away from a converging lens of focal length 4 cm. Find the position and height of the image.

In the solution shown in Figure 36.20, circled numbers have been added to show the order in which the lines or rays have been drawn. These numbers are drawn for illustration only and are normally omitted from such ray diagrams.

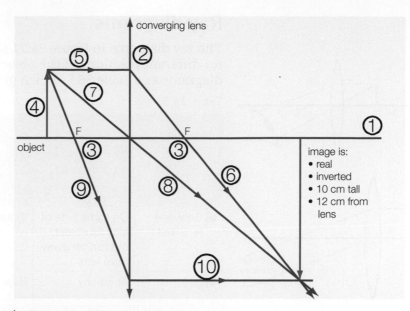

▲ **Figure 36.20** A ray diagram

The image is shown as a continuous line to show that it is real (can be projected on to a screen).

The downward arrow on the image shows it is inverted.

1 Horizontal line representing PA.
2 Vertical line representing lens.
3 Two principal foci marked F, each 4 cm from lens.
4 Object marked 6 cm from lens.
5 Ray from top of object parallel to PA ...
6 ... refracts through F.
7 Ray from top of object through the optical centre ...
8 ... is not refracted.
9 Ray from top of object through F ...
10 ... refracts parallel to PA.

> ### Tip
>
> The Principal Axis (PA) is the imaginary straight line through the centre of the lens, which passes through the principal focus on each side of the lens.

Finally, the image is drawn from the point where the refracted rays meet to the PA.

Now use the ruler to measure the height of the image and its distance from the centre of the lens.

H

a)

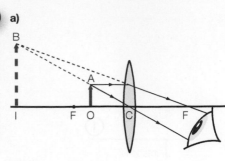

b)

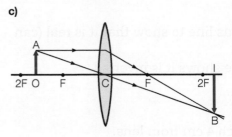

c)

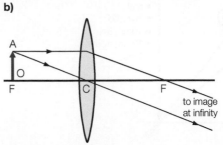

d)

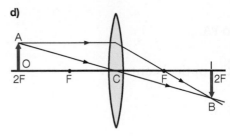

e)

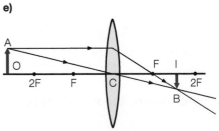

▲ **Figure 36.21** Ray diagrams showing where the image is formed for different positions of the object

Ray diagrams

The ray diagrams in Figure 36.21 show where the image is formed for different positions of the object. You should carefully study the diagrams and Table 36.5, which gives a summary of the information.

Table 36.5

Position of object	Location of image	Properties of image			
		Nature	Erect or inverted	Larger or smaller than object	Application
a) Between lens and F	On same side of lens as object, but further away from lens	Virtual	Erect	Larger	Magnifying glass
b) At F	At infinity	Real	Inverted	Larger	Searchlight
c) Between F and 2F	Beyond 2F	Real	Inverted	Larger	Cinema projector
d) At 2F	At 2F	Real	Inverted	Same size	Telescope – erecting lens
e) Just beyond 2F	Between F and 2F	Real	Inverted	Smaller	Camera
Very far away from lens	At F	Real	Inverted	Smaller	Camera

Tips ⟳

You should pay particular attention to the ray diagrams which illustrate the principles of:
- ▶ the magnifying glass (to give an erect, virtual image)
- ▶ the projector (to give a magnified, real image)
- ▶ the camera (to give a diminished real image)

These diagrams are specifically required by the subject specification.

H

Show you can ?

a) Draw a ray diagram to illustrate how a converging lens can produce:
   i) a real, magnified image
   ii) a diminished image
   iii) a virtual image.
b) State four properties of the image in a converging lens given the location of the object.

H

Practice questions

1 State four properties of the image in a plane mirror. *(4 marks)*

2 Two mirrors, M_1 and M_2, are placed at right angles to one another. Figure 36.22 shows a ray of light incident on mirror M_1 at an angle of 27° to its surface.

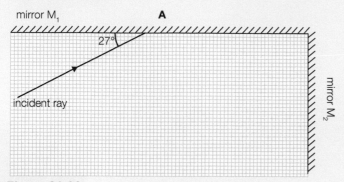

Figure 36.22

a) Copy the diagram onto graph paper and draw the normal to mirror M_1 at point A. *(1 mark)*
b) Calculate the angles of incidence and reflection at point A. *(2 marks)*
c) On your diagram, draw, as accurately as you can, the reflected wave from A and from mirror M_2. *(2 marks)*

3 Figure 36.23 shows a ray of light incident on a glass block. Some of the light is reflected at the top surface, and some of the light passes through the glass and is reflected at the opposite side which has a mirrored surface.

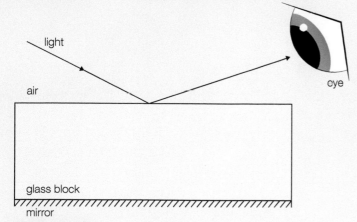

Figure 36.23

Copy the diagram and complete the path of the ray of light through the glass block and back out into the air towards the person viewing it. *(3 marks)*

4 Figure 36.24 shows the refraction of light in glass and water.
   a) Explain why the light bends as it enters both glass and water. *(2 marks)*
   b) How does Figure 36.24 suggest that the speed of light in water is not the same as the speed in glass? *(1 mark)*
   c) Is light faster in glass or in water? Explain your reasoning. *(2 marks)*

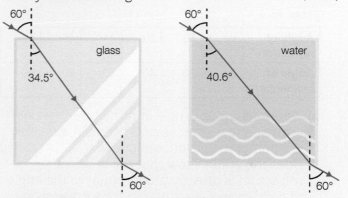

Figure 36.24

5 Figure 36.25 is a ray diagram that shows how a lens can produce an image of an object. Each small square corresponds to a distance of 2 mm.

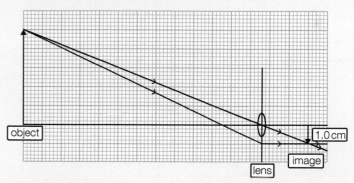

Figure 36.25

a) Which three of the words below best describes the image? *(3 marks)*
   real, diminished, erect, inverted, virtual, enlarged

b) What type of lens is shown in the ray diagram? *(1 mark)*
c) Copy the diagram and mark on it the location of the principal focus. *(1 mark)*
d) Find the focal length of the lens in centimetres. *(1 mark)*

e) The magnification of the image is defined by the equation:

$$\text{magnification} = \frac{\text{height of image}}{\text{height of object}}$$

Use the equation to calculate the magnification of the image. *(1 mark)*

6 A ray of light passes from air into water. Table 36.6 shows how the angle of refraction in water changes for different angles of incidence in air.

Table 36.6

Angle of refraction/°	0	15	29	41	48	49
Angle of incidence/°	0	20	40	60	80	90

a) Plot the graph of angle of refraction/° (*y*-axis) against angle of incidence/° (*x*-axis) and join the points with a smooth curve. *(5 marks)*

b) Does the graph show that the angle of refraction is directly proportional to the angle of incidence? Give a reason for your answer. *(1 mark)*

37 Electricity

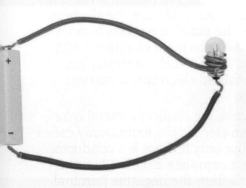

Specification points

This chapter covers sections 2.3.1 to 2.3.16 of the specification, including Prescribed practical P6. It deals with conductors and insulators, simple circuits, symbols, charge, Ohm's law, lamps, series and parallel circuits and resistance.

Electricity

Electricity is an extremely versatile and useful form of energy. Many of our everyday activities depend on the use of electricity. It is hard to imagine life in today's society without it! Simple things such as entertainment, communications, transport and industry would simply grind to a halt if electricity ceased to exist.

When a woollen jumper is taken off over a nylon shirt in the dark, you can hear crackles and see tiny blue electric sparks. The nylon shirt has become charged with **static electricity**.

We now know that there are two types of charge, **positive** charge and **negative** charge. The negative charge is due to the presence of **electrons**, and it is only these particles that we are concerned with in this chapter.

When we connect a battery across a lamp, the lamp lights up. We say that the connecting wire (copper) and the filament of the bulb (tungsten) are both **electrical conductors**. The plastic covering is not an electrical conductor. We say the plastic is an insulator.

How can we tell if a material is a conductor or an insulator? To do that we connect the material in a circuit containing a battery, a bulb and the material being tested (Figure 37.1). If the bulb lights up, the material is a conductor. If the bulb does not light, the material is an insulator.

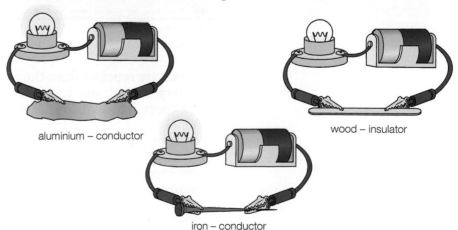

aluminium – conductor

wood – insulator

iron – conductor

▲ **Figure 37.1** Testing if a material is a conductor

In general, all metals are electrical conductors. Almost all non-metals are insulators, but there are a few exceptions. For example, graphite is a non-metal, but it conducts electricity.

Table 37.1 Common conductors and insulators

Good conductors	Gold	Silver	Copper	Aluminium	Mercury	Platinum	Graphite
Insulators	Polythene	Rubber	Wool	Wax	Glass	Paper	Wood

Why are metals good conductors?

An electric current is a flow of electrically charged particles. For the circuits we will consider, the charge involved is always the electron. Most electrons in atoms are bound by the positive nucleus to orbit in the surrounding shells. In metals, the outermost electron is often so weakly held that it can break away. We call such electrons free electrons. Some books call them delocalised electrons. Insulators contain no (or very few) free electrons.

So why does electrical current flow when we connect a metal wire between the terminals of a battery? An electric cell (commonly called a battery) can make electrons move, but only if there is a conductor connecting its two terminals to make a complete circuit. Chemical reactions inside the cell push electrons from the negative terminal round to the positive terminal. Figure 37.2 shows how an electric current would flow in a wire connected across a cell.

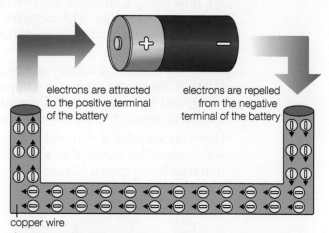

electrons are attracted to the positive terminal of the battery

electrons are repelled from the negative terminal of the battery

copper wire

▲ **Figure 37.2** The flow of electrons in a simple cell

Electrons are repelled from the negative terminal of the cell because like charges repel, and are attracted to the positive terminal because unlike charges attract. This make the electrons flow from the negative terminal to the positive terminal. Scientists in the nineteenth century thought that an electric current consisted of a flow of positive charge from the positive terminal of the cell to the negative terminal. This is now known to be incorrect, but we still refer to the direction of conventional current as flowing from positive to negative.

So, to summarise:

▶ Electrical conductors, like metals, have free electrons.
▶ Electrical insulators, like non-metals, have no free electrons.
▶ Free electrons move from the negative to the positive terminal of the battery.
▶ Conventional current is said to flow from the positive terminal to the negative terminal of the battery.

Standard symbols

An electrical circuit may be represented by a circuit diagram with symbols for components. Some widely used components are listed in Table 37.2. **Circuit diagrams** are easy to draw and are universally understood.

Table 37.2 Components of electrical circuits and their symbols

Component	Symbol	Appearance
switch		
cell	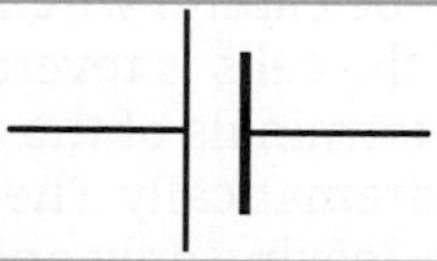	
battery		
resistor		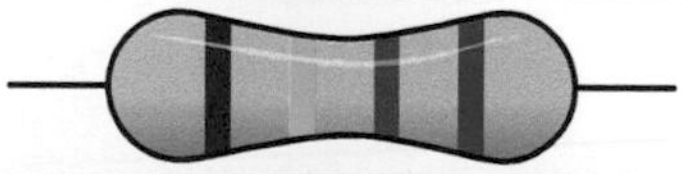
variable resistor (rheostat)	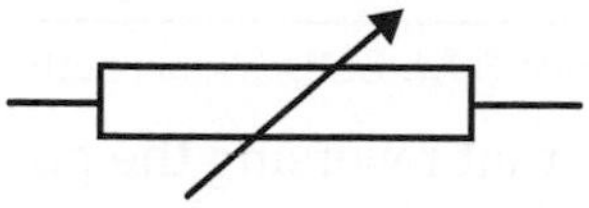	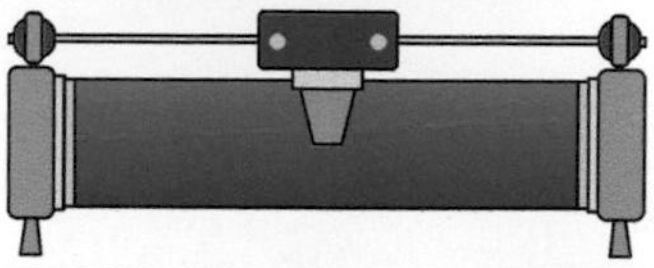
fuse	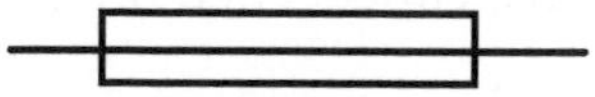	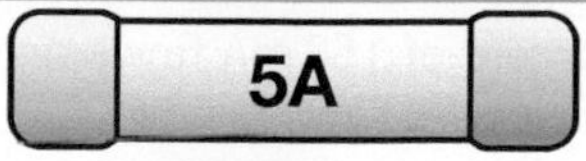 5A
voltmeter	 V	
ammeter	 A	
lamp		

Cell polarity

By convention, the long, thin line in the symbol for a cell is taken as the positive terminal.

The short, bold line is the negative terminal.

Cells can be joined together minus to plus to make a **battery**. This leaves a positive and a negative terminal free to be connected into a circuit. Cells connected in this way are said to be connected in **series**.

Portable stereo systems have a number of cells connected in series. The reason for this is that the system requires a large voltage to operate. Connecting cells in series to make a battery increases the voltage. For example, connecting four 1.5 V cells in this way gives a 6 V battery, as shown in Figure 37.3.

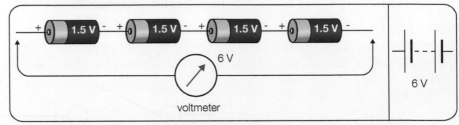

▲ **Figure 37.3** Cells correctly joined in series

You must be careful when connecting cells in series. If the **polarity** of one of the cells is reversed (in other words, if the positive and negative terminals of the cell are swapped) then the voltage is reduced dramatically. The voltages of two of the cells cancel each other out, leaving only one effective cell, as shown in Figure 37.4.

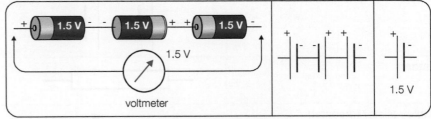

▲ **Figure 37.4** Cells incorrectly joined together

Notice that reversing the polarity of the cell in the middle has the effect of producing a battery of only 1.5 volts. The cells joined positive to positive cancel each other out.

The relationship between charge and current

The unit of charge is the **coulomb** (C). If it were possible to see a coulomb of charge, it would look like a very large assembly of electrons – about six million million million of them.

The unit of current is the **ampere** (A). Currents of around one ampere upwards can be measured by connecting an ammeter in the circuit. For smaller currents, a **milliammeter** is used. The unit in this case is the milliampere (mA). (1000 mA = 1 A). An even smaller unit of current is the **microampere** (µA). (1 000 000 µA = 1 A).

In general, if a steady current of I amperes flows for time t seconds, the charge Q coulombs passing any point is given by:

$$Q = I \times t$$
$$\text{charge} = \text{current} \times \text{time}$$

Example

A current of 150 mA flows around a circuit for 1 minute. How much electrical charge flows past a point in the circuit in this time?

Answer

$I = 150 \, \text{mA}$

$\quad = 0.15 \, \text{A}$

$t = 1 \, \text{minute}$

$\quad = 60 \, \text{s}$

$Q = I \times t$

$\quad = 0.15 \times 60$

$\quad = 9 \, \text{C}$

Tip

Always ensure the units for current, charge and time are in amperes, coulombs and seconds before substituting values into the equation $Q = It$.

Notice that:

▶ charge is measured in coulombs (C) and given the symbol Q
▶ current is measured in amperes (A) and given the symbol I
▶ time is measured in seconds (s) and given the symbol t.

If a charge of 1 C passes a fixed point in 1 s, a current of 1 A is flowing in the circuit.

There are two conditions which must be met before an electric current will flow:

▶ There must be a complete circuit – i.e. there must be no gaps in the circuit.
▶ There must be a source of energy so that the charge may move – this source of energy may be a cell, a battery or the mains power supply.

Show you can

a) Explain the difference between conductors and insulators in terms of free electrons.
b) Determine the direction of electron flow and conventional current given a circuit diagram.
c) Know the standard symbols for electrical components and be able to use them to draw simple circuit diagrams.
d) Know what is meant by cell polarity and calculate the voltage of a battery with regards to cell polarity.

Test yourself

1 The cells in Figure 37.5 are all identical. The total battery voltage is 1.6 V.

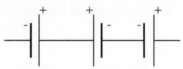

▲ **Figure 37.5** Three battery cells in series

a) Calculate the voltage of each cell.
b) Redraw the battery showing the connections which would give a battery capable of delivering the maximum possible voltage.
c) State the maximum voltage which this battery could deliver.

2 Convert the following currents into milliamperes:
a) 3.0 A
b) 0.2 A
c) 200 µA

3 Convert the following currents into amperes:
a) 400 mA
b) 1500 mA
c) 500 000 µA

4 What charge is delivered if:
a) a current of 6 A flows for 10 seconds?
b) a current of 300 mA flows for 1 minute?
c) a current of 500 µA flows for 1 hour?

5 Calculate the currents that flow when the following charges pass a fixed point in the following times:
a) 100 C, time = 5 s
b) 500 mC, time = 50 s
c) 60 mC, time = 200 s

Resistance

The resistance R of an electrical conductor can be found using the ammeter–voltmeter method. We measure the current I through the conductor when a voltage V is applied across its ends. The resistance R is then calculated using the equation:

$$R = \frac{V}{I}$$

where: R is the resistance in ohms (Ω)

V is the voltage in volts (V)

I is the current in amperes (A)

This is the basis of the Ohm's law experiment which follows.

Prescribed practical

Prescribed practical P6: Obtaining a voltage – current graph for a metal wire and verifying Ohm's law

Aims
▶ to pass an electric current through a wire

▶ to measure the current for different values of the voltage across the wire

▶ to take precautions to ensure the temperature of the wire is kept constant

▶ to plot a graph of voltage across the wire (y-axis) against current in the wire (x-axis)

▶ to use the graph to establish an equation linking voltage and current

▶ to determine the resistance of the wire

Apparatus
▶ low voltage power supply unit (PSU)

▶ rheostat

▶ ammeter

▶ voltmeter

▶ connecting leads

▶ resistance wire

▶ switch

Method
1 Prepare a table for your results like that shown on the next page.

2 Ensure that the PSU is switched off and connect it to the mains socket.

3 Set up the circuit as shown in Figure 37.6. The device marked R represents the wire being tested.

4 Adjust the PSU to supply zero volts.

5 Switch on the PSU.

6 Record the voltage on the voltmeter and the corresponding current on the ammeter.

7 Switch off the PSU immediately after recording in the table values for voltage and current.

8 Wait for about two minutes to ensure the wire cools to room temperature.

9 Switch on the PSU and adjust the voltage (or the rheostat) so that the reading on the voltmeter increases by 1.0 V.

10 Repeat steps 6–9 until readings have been recorded for a voltages ranging from zero to a maximum voltage of 6 V*. This is Trial 1.

11 Repeat the entire experiment to obtain a second set of values for current. This is Trial 2.

Calculate the mean current from the two trials and enter the results in the table.

Plot the graph of voltage against mean current.

* It is necessary to ensure the wire's temperature remains constant (close to room temperature).

We do this by:

▶ keeping the voltage low (so that the current remains small)

▶ switching off the current between readings (to allow the wire to cool).

Table for results
Table 37.3

Voltage/V	0.00	1.00	2.00	3.00	4.00	5.00	6.00
(Trial 1) Current/A							
(Trial 2) Current/A							
Mean current/A							
Ratio of voltage to current/Ω							

Circuit diagram

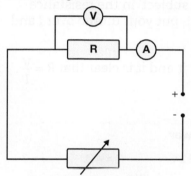

▲ **Figure 37.6** A circuit diagram

Treatment of the results

Plot the graph of voltage/V (vertical axis) against mean current/A (horizontal axis). It should look similar to the graph in Figure 37.7.

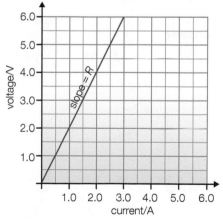

▲ **Figure 37.7** A graph showing voltage against current

Discussion of the results

The graph of V against I is a straight line through the origin. This tells us that the current in a metallic conductor is directly proportional to the voltage across its ends, provided the temperature remains constant.

This result is commonly called **Ohm's law**.

Measuring the resistance

The resistance of the wire does not change when the current and voltage change. The resistance of a wire at constant temperature depends only on three factors:

▶ the material from which the wire is made
▶ the length of the wire
▶ the cross-sectional area of the wire.

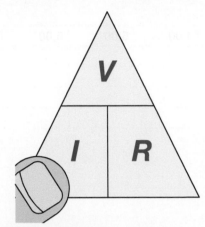

▲ **Figure 37.8** The resistance formula triangle

The graph of V against I is a straight line through the origin. The ratio of V to I is constant throughout the experiment. It also means that, in this case, the slope of the graph of V against I is equal to the resistance of the wire. However, you should understand that measuring the slope of the V–I graph is not, in general, the correct way to measure a resistance.

Earlier, we established the resistance formula:

$$V = I \times R$$
$$\text{voltage} = \text{current} \times \text{resistance}$$

Figure 37.8 may help you to 'change the subject' in the resistance formula. To find the equation for current, put your thumb over I and you can see that $I = \dfrac{V}{R}$.

Similarly to find R, put your thumb over it and it is clear that $R = \dfrac{V}{I}$.

Example

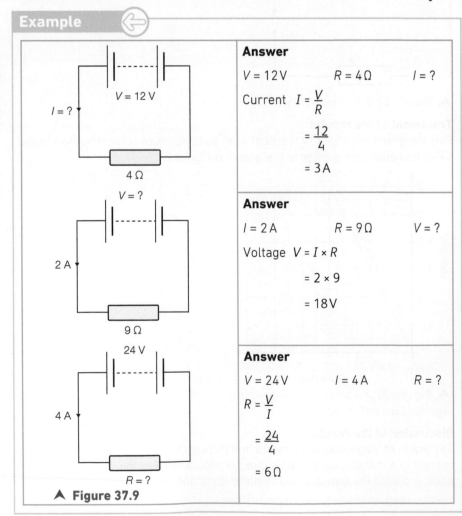

Answer

$V = 12\,V$ $R = 4\,\Omega$ $I = ?$

Current $I = \dfrac{V}{R}$

$= \dfrac{12}{4}$

$= 3\,A$

Answer

$I = 2\,A$ $R = 9\,\Omega$ $V = ?$

Voltage $V = I \times R$

$= 2 \times 9$

$= 18\,V$

Answer

$V = 24\,V$ $I = 4\,A$ $R = ?$

$R = \dfrac{V}{I}$

$= \dfrac{24}{4}$

$= 6\,\Omega$

▲ **Figure 37.9**

Show you can

a) Explain why it is necessary to keep the temperature of the resistance wire constant in the Ohm's law experiment, and describe how this is achieved.

b) Explain why it is necessary to obtain several values of current and voltage in the Ohm's law experiment, and describe how this is achieved.

c) Describe the V–I characteristic graph for a metal wire at constant temperature, and state what conclusion can be drawn from it.

d) State Ohm's law and the condition under which it is valid.

Test yourself

6 Calculate the current flowing through a $10\,\Omega$ resistor which has a voltage of $20\,V$ across it.

7 A resistor has a voltage of $15\,V$ across it when a current of $3\,A$ flows through it.
Calculate the resistance of the resistor.

8 A current of $2\,A$ flows through a $25\,\Omega$ resistor. Find the voltage across the resistor.

9 A voltage of $15\,V$ is needed to make a current of $2.5\,A$ flow through a wire.
a) What is the resistance of the wire?
b) What voltage is needed to make a current of $2.0\,A$ flow through the wire?

10 There is a voltage of $6.0\,V$ across the ends of a wire of resistance $12\,\Omega$.
a) What is the current in the wire?
b) What voltage is needed to make a current of $1.5\,A$ flow through it?

11 A resistor has a voltage of $6\,V$ applied across it and the current flowing through is $100\,mA$. Calculate the resistance of the resistor.

12 A current of $600\,mA$ flows through a metal wire when the voltage across its ends is $3\,V$. What current flows through the same wire when the voltage across its ends is $2.5\,V$?

Practical activity

Resistance of a filament lamp

Aims

▶ to pass an electric current through a filament lamp

▶ to measure the current for different values of the voltage across the lamp

▶ to plot a graph of voltage across the lamp (y-axis) against current in the lamp (x-axis)

▶ to use the graph to determine the resistance of the lamp at two different values of current

Variables

The dependent variable is the current flowing in the lamp.

The independent variable is the voltage across the lamp.

The controlled variables are the length of the filament (inside the lamp) and its cross-sectional area.

Apparatus

▶ low voltage power supply unit (PSU)

▶ rheostat

▶ ammeter

▶ voltmeter

▶ connecting leads

▶ filament lamp in a suitable holder

▶ switch

Method

1 Prepare a table for your results like that shown in Table 37.4 on the next page.

2 Ensure that the PSU is switched off, and connect it to the mains socket.

3   Set up the circuit as shown in Figure 37.10.

4   Adjust the PSU to supply zero volts.

5   Switch on the PSU.

6   Record the voltage on the voltmeter and the corresponding current on the ammeter.

7   Adjust the voltage (or the rheostat) so that the reading on the voltmeter increases by 0.5 V.

8   Repeat steps 6–7 until readings have been recorded for voltages ranging from zero to a maximum voltage of 6 V (for a 6 V lamp). This is trial 1.

9   Repeat the entire experiment to obtain a second set of values for current. This is trial 2.

10 Calculate the mean current from the two trials and enter the results in the table.

11 Plot a graph of voltage against mean current. Your graph should look similar to the graph in Figure 37.11.

Table for results
Table 37.4

Voltage/V	0.00	1.00	2.00	3.00	4.00	5.00	6.00
(Trial 1) Current/A							
(Trial 2) Current/A							
Mean current/A							
Ratio of voltage to current/Ω							

Circuit diagram

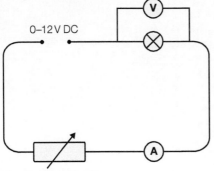

0–12 V DC

▲ **Figure 37.10**

Treatment of the results
Plot the graph of voltage/V (vertical axis) against mean current/A (horizontal axis).

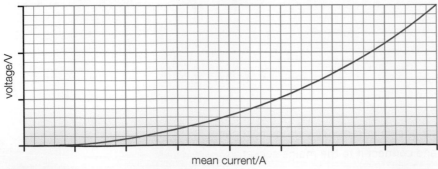

voltage/V

mean current/A

▲ **Figure 37.11**

Discussion of the results

Notice that we tried hard to keep the temperature constant in the previous experiment. This is because the resistance of a metal can rise rapidly when the temperature increases. In this experiment, the temperature of the filament is allowed to rise – at first the lamp only glows orange, but as the current rises, it becomes white hot at its operating temperature. As its temperature rises, so does the resistance.

Measuring the resistance

You should now find the resistance of the filament at two points on your curve. You can do this by calculating the ratio of the voltage to the current for the two different values of the current. You will notice that the resistance rises with increasing current.

Series and parallel circuits

Series circuits

The total resistance of two or more resistors in series is simply the sum of the individual resistances of the resistors.

$$R_{total} = R_1 + R_2 + R_3$$

In Figure 37.12a, the three resistors could be replaced by a single resistor of $4 + 8 + 6 = 18\,\Omega$.

Parallel circuits

Figure 37.12b shows a parallel circuit with two resistors. The total resistance of two equal resistors in **parallel** is half of the resistance of one of them. The total resistance of three equal resistors in parallel is one third of the resistance of one of them, and so on.

When considering two **unequal** resistances, R_1 and R_2, in parallel, we use the 'product over sum' formula:

$$R_{total} = \frac{R_1 \times R_2}{R_1 + R_2}$$

$$= \frac{product}{sum}$$

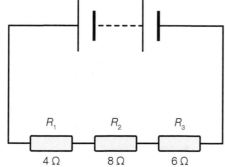

▲ **Figure 37.12a** Calculating the total resistance of three resistors in a series circuit

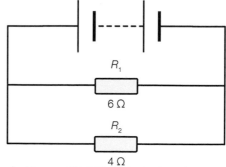

▲ **Figure 37.12b** Calculating the total resistance of two resistors in parallel

Example

1 Find the combined resistance of two 8 Ω resistors
   a) in series and
   b) in parallel.

Answer

a) For resistors in series: $R_{total} = R_1 + R_2$

$$= 8 + 8$$

$$= 16\,\Omega$$

b) The total resistance of two equal resistors in parallel is half of the resistance of one of them, so in this case, the total resistance is $8 \div 2 = 4\,\Omega$.

2 A 6 Ω resistor and a 3 Ω resistor are connected
   a) in series and
   b) in parallel.
   In each case, find the resistance of the combination.

Answer

a) For resistors in series: $R_{total} = R_1 + R_2$

$$= 6 + 3$$

$$= 9\,\Omega$$

b) For two unequal resistors in parallel, we use the product over sum rule. Note that this only works for two resistors.

$$R_{total} = \frac{\text{product}}{\text{sum}}$$

$$= \frac{R_1 \times R_2}{R_1 + R_2}$$

$$= \frac{6 \times 3}{6 + 3}$$

$$= 2\,\Omega$$

3 A 24 Ω resistor, a 12 Ω resistor and a 8 Ω resistor are connected in series. Find the resistance of the combination.

Answer

For resistors in series: $R_{total} = R_1 + R_2 + R_3$

$$= 24 + 12 + 8$$

$$= 44\,\Omega$$

4 Find the total resistance of the combination shown in Figure 37.13.

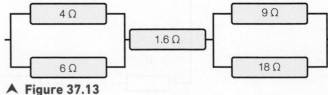

▲ **Figure 37.13**

Answer

Using the product over sum formula, we see that the 4 Ω and 6 Ω parallel combination gives a total resistance of 2.4 Ω. Similarly, the 9 Ω and 18 Ω parallel combination gives a total resistance of 6.0 Ω.

There is, therefore, a series combination of 2.4 Ω + 1.6 Ω + 6.0 Ω, which gives a total resistance of 10 Ω.

Current and voltage in series circuits

When resistors are connected in series:

▶ the current in each resistor is the same
▶ the sum of the voltages across each resistor is equal to the battery or power supply voltage.

1 Resistances of $2\,\Omega$, $4\,\Omega$, and $6\,\Omega$ are connected in series across a $3\,V$ battery. Calculate:
   a) the total resistance of the circuit
   b) the current in each resistor
   c) the voltage across each resistor.

Comment on your answer to part c).

Answer

a) $R_{total} = R_1 + R_2 + R_3$
   $= 2 + 4 + 6$
   $= 12\,\Omega$

This means that the circuit is equivalent to one with $12\,\Omega$ placed across a $3\,V$ battery.

b) $I = \dfrac{V}{R}$

   $= \dfrac{3}{12}$

   $= 0.25\,A$

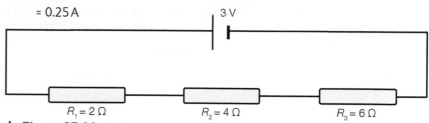

▲ **Figure 37.14**

c) 2Ω $V = I \times R$ 4Ω $V = I \times R$ 2Ω $V = I \times R$
   $= 0.25 \times 2$ $= 0.25 \times 4$ $= 0.25 \times 6$
   $= 0.5\,V$ $= 1.0\,V$ $= 1.5\,V$

Comments

The sum of the voltages ($0.5\,V + 1.0\,V + 1.5\,V$) is $3\,V$, which is exactly the same as the voltage of the battery.

The voltages are in exactly the same proportion as the resistances. So, the $4\,\Omega$ resistor has twice the voltage as the $2\,\Omega$ resistor and the $6\,\Omega$ resistor has three times the voltage as the $2\,\Omega$ resistor.

Voltage in parallel circuits

When resistors are connected in parallel:

▶ the voltage across each resistor is the same as the voltage provided by the battery

▶ the sum of the currents in each resistor is equal to the current coming from the battery.

Example

1 Resistances of 2 Ω and 3 Ω are connected in parallel across a 6 V battery (see Figure 37.15).

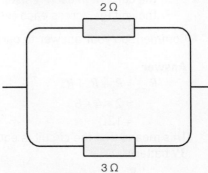

▲ Figure 37.15

a) State the voltage across each resistor.

b) Calculate:
   i)   the total resistance of the circuit
   ii) the current in each resistor
   iii) the total current taken from the battery.

Answer

a) 6 V

b) i) $R_{total} = \dfrac{\text{product}}{\text{sum}}$

$$= \dfrac{R_1 \times R_2}{R_1 + R_2}$$

$$= \dfrac{2 \times 3}{2 + 3}$$

$$= 1.2\,\Omega$$

ii) 2 Ω $I = \dfrac{V}{R}$ 3 Ω $I = \dfrac{V}{R}$

$\qquad = \dfrac{6}{2}$ $\qquad\qquad = \dfrac{6}{3}$

$\qquad = 3\,A$ $\qquad\qquad = 2\,A$

iii) We can do this in two different ways:

$\qquad I_{total} = \dfrac{V_{battery}}{R_{battery}}$ or $I_{battery}$ = sum of currents in resistors

$\qquad\qquad\qquad\qquad\qquad = 3 + 2$

$\qquad\qquad = \dfrac{6}{1.2}$ $\qquad\qquad = 5\,A$

$\qquad\qquad = 5\,A$

Test yourself

13 a) Calculate the value of the current from the cell in each of these circuits.

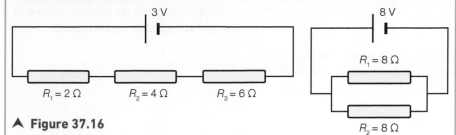

▲ **Figure 37.16**

b) State the voltage across each $8\,\Omega$ resistor in the second circuit.

14 a) In each of the following circuits, calculate the currents I_1, I_2 and I_3 shown by ammeters A_1, A_2 and A_3.
  b) Calculate the voltage across each resistor.

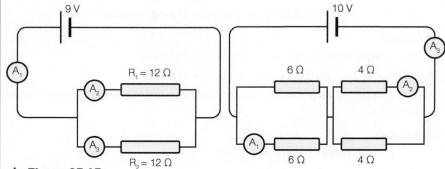

▲ **Figure 37.17**

15 a) In circuits A and B below, all the lamps are identical. Copy and complete Table 37.5.

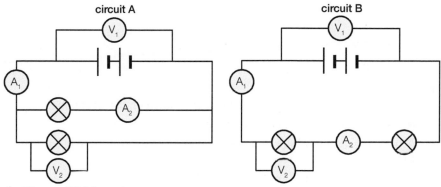

▲ **Figure 37.18**

Table 37.5

circuit	V_1/V	V_2/V	I_1/A	I_2/A
A	3	3	0.4	
B	3		0.1	

b) Calculate the resistance of each lamp.

Resistance and length

Earlier in this chapter, we stated that the resistance of a piece of wire depended on its length. Below is a practical activity that shows the nature of this dependence.

Practical activity

Resistance and length

Aims

▶ to pass an electric current through a wire whose length can be varied
▶ to measure the current and voltage for different lengths of wire
▶ to take precautions to ensure the temperature of the wire is kept constant
▶ to calculate the resistance for each length of wire
▶ to plot a graph of resistance of the wire (*y*-axis) against length of the wire (*x*-axis)
▶ to use the graph to establish an equation linking resistance and length

Variables

▶ The independent variable is the length of the wire.
▶ The dependent variable is the resistance of the wire.
▶ The controlled variables are the temperature and the cross-sectional area of the wire.

Apparatus

▶ low voltage power supply unit (PSU)
▶ rheostat
▶ ammeter
▶ voltmeter
▶ connecting leads
▶ resistance wire
▶ switch
▶ metre stick
▶ sticky tape

Method

1 Prepare a table for your results like that shown on the next page.
2 Measure and cut off one metre of nichrome resistance wire.
3 Attach it with sticky tape to a metre ruler – make sure there are no kinks in the wire.
4 Set up the circuit as shown in Figure 37.19.
5 Ensure that the PSU is switched off and connect it to the mains socket.
6 Adjust the PSU to supply about 1 V.
7 Connect the 'flying lead' so that the length of wire across the voltmeter is 10 cm.
8 Switch on the PSU.
9 Record the voltage on the voltmeter and the corresponding current on the ammeter.
10 Switch off the PSU immediately after recording in the table values for voltage and current.
11 Wait for about 2 minutes to ensure the wire cools to room temperature*.
12 Switch on the PSU again.

13 Repeat steps 7–13 until readings have been recorded for lengths of wire ranging from 10 cm to 90 cm.

14 Calculate the resistance of each length of wire, using $R = \dfrac{V}{I}$.

15 Plot the graph of resistance (y-axis) against length (x-axis).

* It is necessary to ensure the wire's temperature remains constant (close to room temperature).

We do this by:

▶ keeping the voltage low (so that the current remains small)

▶ switching off the current between readings to allow the wire to cool.

Table for results

Table 37.6

Length of wire/cm	10	20	30	40	50	60	70	80	90
Voltage across wire/V									
Current in wire/A									
Resistance of wire/Ω									

Circuit diagram

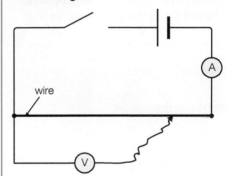

wire

▲ **Figure 37.19**

Treatment of the results

Plot the graph of resistance/Ω (vertical axis) against length/cm (horizontal axis). Your graph should look something like that shown in Figure 37.20.

Discussion of the results

The graph of resistance/Ω against length/cm is a straight line through the origin. This tells us that the resistance of a metal wire is directly proportional to its length, provided its cross-sectional area and the temperature remain constant. This means there is a mathematical relationship between the resistance, R, and the length, L.

The relationship is:

$R = kL$

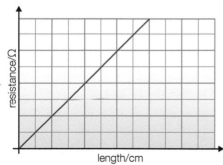

▲ **Figure 37.20**

where k is the gradient of the graph.

Since $k = \dfrac{R}{L}$, the unit for k is Ω/cm (or Ω/m).

Note that the value of k depends on the material of the wire and its cross-sectional area.

Example

A reel of constantan wire of length 250 cm has a total resistance of 15.0 Ω Calculate:

a) the resistance of 1.0 m of wire
b) the length of wire needed to have a resistance of 3 Ω
c) the resistance of a 90 cm length of the wire.

Answer

a) $k = \dfrac{R}{L}$

$= \dfrac{15\,\Omega}{250\,cm}$

$= 0.06\,\Omega/cm$

$= 6\,\Omega/m$, so the resistance of 1.0 m of wire is 6 Ω

b) $L = \dfrac{R}{k}$

$= \dfrac{3\,\Omega}{0.06\,\Omega/cm}$

$= 50\,cm$

c) $R = kL$

$= 6\,\Omega/m \times 0.9\,m$

$= 5.4\,\Omega$

Test yourself

16 A school buys a reel of constantan wire. The supplier's data sheet says that the wire has a resistance of 2.5 Ω/m. Calculate:
   a) the length of wire a technician must cut from the reel to give a resistance of 2 Ω and
   b) the resistance of a 120 cm length of wire cut from the reel.

17 When a current of 0.15 A flows through a 48 cm length of eureka wire the voltage across its ends is 0.90 V. What length of the same type of wire would give a current of 0.36 A when the voltage across its ends is 1.44 V?

18 An 80 cm length of wire A has a resistance of 2.4 Ω. The resistance of a 50 cm length of wire B is 1.2 Ω. A student cuts a 30 cm length, L_1, of wire from a reel of wire A and a 40 cm length L_2 from a reel of wire B.
   a) Which length, L_1 or L_2, has the greater resistance? Explain your reasoning.
   b) Sketch a graph of resistance against length for wire A and wire B, and state which graph has the larger gradient.

19 A technician cuts an 80 cm length of wire from a reel marked 3.0 Ω/m. The technician joins the two free ends of the wire together to form a loop. She then attaches two crocodile clips to the wire at opposite ends of a diameter.
   a) Explain why the total resistance between the crocodile clips is 0.6 Ω.
   b) In what way, if at all, does the total resistance between the crocodile clips change, if one of the clips is moved along the wire towards the other? Explain your reasoning.

Show you can

a) Describe how the resistance of a piece of metal wire depends on:
   i) its length
   ii) the material from which it is made.
b) Describe in detail the experiments which demonstrate how the resistance of a piece of metal wire depends on
   i) its length
   ii) the material from which it is made.

Practice questions

1

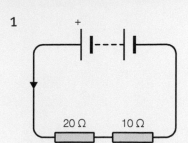

Figure 37.21

a) What flows in the direction indicated by the arrow in Figure 37.21? *(1 mark)*

b) Copy the circuit diagram and mark on it an arrow to show the direction in which charged particles flow through the two resistors. *(1 mark)*

c) What name is given to these charged particles? *(1 mark)*

d) The current in the larger resistor is 0.6 A. State the size of the current in the smaller resistor. *(1 mark)*

e) Show that the electrical charge delivered by the battery in one minute is 36 C. *(3 marks)*

2 Resistors of 20 Ω and 10 Ω are connected in series across a battery. The current in the larger resistor is 0.6 A. Show that the battery voltage is 18 V. *(3 marks)*

3 a) Two identical resistors are connected in parallel across a battery. Their combined resistance is 12 Ω. What is the resistance of each resistor? *(1 mark)*

b) A different pair of identical resistors are connected in series across a battery. Their combined resistance is also 12 Ω. What is the resistance of each resistor? *(1 mark)*

4 In Figure 37.22 a) and b), the lamps are identical and each has an internal resistance of 6 Ω. Each cell in the battery can supply a voltage of 1.5 V. For each circuit, find the reading which would be shown on the four meters. *(4 marks)*

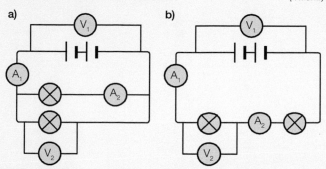

Figure 37.22

5 Three identical resistors are arranged as shown in Figure 37.23.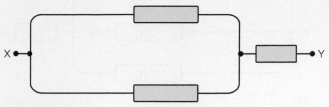

Figure 37.23

The current entering at X is 3 mA, and the voltage between X and Y is 18 mV. Calculate:

a) the total resistance between X and Y *(3 marks)*

b) the current in each resistor *(2 marks)*

c) the voltage across each resistor *(2 marks)*

d) the resistance of each resistor. *(2 marks)*

6 In Figure 37.24, resistors R_1 and R_2 have resistances of 40 Ω and 20 Ω respectively.

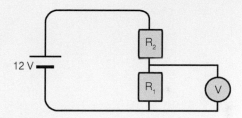

Figure 37.24

a) Calculate the voltage you would expect to observe on the voltmeter. *(3 marks)*

b) What assumption have you made about the resistance of the voltmeter itself? *(2 marks)*

H **7** Look at the circuit in Figure 37.25. Copy and complete Table 37.7 to show the effective resistance between X and Y for the three different switch settings. *(3 marks)*

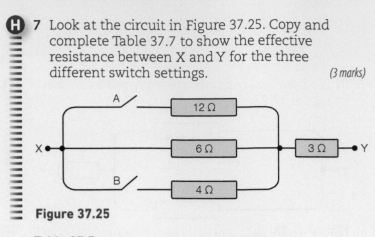

Figure 37.25

Table 37.7

Switch		Effective resistance between X and Y/Ω
A	B	
open	open	
open	closed	
closed	open	

Specification points

This chapter covers sections 2.3.17 to 2.3.25 of the specification. It deals with electrical energy, power and electricity in the home.

In the previous chapter, we saw that electricity generates heat when it passes through a metal wire. Why does that happen?

As electrons pass through the conductor they collide with the atoms. In these collisions the light electrons lose energy and the heavy atoms gain energy. This causes the atoms to vibrate faster. Faster vibrations mean a higher temperature. This is called joule heating.

This heating effect is put to good use in devices such as hairdryers and toasters (Figure 38.1), which contain heating elements usually made from nichrome wire.

Electrical energy

If 1 coulomb of charge gains or loses 1 joule of energy between two points, there is a voltage of 1 volt between those two points.

$$\text{voltage} = \frac{\text{energy transferred}}{\text{charge}}$$

Rearranging this formula gives us:

energy transferred = voltage × charge

It is much easier to measure current than charge.

We know that:

charge = current × time

Substituting for charge gives a very useful formula for **energy transferred** in an electric circuit:

$$\text{energy transferred} = \text{voltage} \times \text{current} \times \text{time}$$
$$E \qquad = \quad V \ \times \quad I \ \times \ t$$

Electrical power

Earlier, you learned that the formula for mechanical power of a machine is defined as the rate at which energy is transferred and is given by:

$$\text{power} = \frac{\text{energy transferred}}{\text{time}}$$

We have seen that in an electrical circuit:

energy transferred = voltage × current × time

▲ **Figure 38.1** Toasters and hairdryers use an electrical current to heat

Substituting for energy transferred:

$$\text{power} = \frac{\text{voltage} \times \text{current} \times \text{time}}{\text{time}}$$

or

electrical power = voltage × current

This is often expressed by the equation $P = I \times V$.

Domestic appliances such as toasters, hairdryers and TVs have a power rating marked on them in watts or in kilowatts (1 kW = 1000 W).

Some widely used electrical devices and their typical power ratings are shown in Figure 38.2.

Example

1 If 0.5 A flows through a bulb connected across a 6 V power supply for 10 seconds, how much energy is transferred?

Answer

energy = $V \times I \times t$

     = 6 × 0.5 × 10

     = 30 J

2 A study lamp is rated at 60 W, 240 V. How much current flows in the bulb?

Answer

power = $V \times I$

  60 = 240 × I

   $I = \dfrac{60}{240}$

     = 0.25 A

Show you can ?

a) Explain, in terms of particle collisions, why an electrical current flowing in a resistance wire generates heat.

b) State the equation relating power, voltage and current.

200 W

20 W

1100 W

2200 W

60 W

▲ **Figure 38.2** A variety of electrical devices found in the home

Example

What power is dissipated in a 10 Ω resistor when the current through it is:

a) 2 A
b) 4 A?

Answer

a) power = I^2R

   $= 2^2 \times 10$

   $= 40\,W$

b) power = I^2R

   $= 4^2 \times 10$

   $= 160\,W$

The three equations for power

We can use the equations for power and voltage to express power in three different ways.

$$P = I \times V \qquad\qquad V = I \times R$$
$$P = IV$$
$$= I^2R$$
$$P = \frac{V^2}{R}$$

This formula (known as Joule's law) can be used to calculate the electrical power **dissipated** (converted) into heat in resistors and heating elements. The heat dissipated is sometimes referred to as the 'ohmic losses'.

The example opposite shows that when the current is doubled, the power dissipated is quadrupled!

This idea has important implications for electricity transmission in the next chapter.

Example

If electricity costs 16 pence per kilowatt hour, find the number of units used by a 2800 W oven when it is switched on for five hours. Calculate the cost of using this oven for that time.

Answer

number of units used
= power rating × time
    (kW)   (hour)

= 2.8 kW × 5 h

= 14 kWh

total cost = number of units used ×
       cost per unit

      = 14 kWh × 16 pence

      = 224 pence

      = £2.24

Paying for electricity

Electricity companies bill customers for electrical energy in units known as kilowatt hours (kWh). These are sometimes called 'units' of electricity.

One kilowatt hour is the amount of energy transferred when 1000 W is delivered for one hour. You should be able to prove for yourself that:

1 kWh = 3 600 000 J = 3.6 MJ

The following two formulae are very useful in calculating the cost of using a particular appliance for a given amount of time:

number of units used = power rating (in kilowatts) × time (in hours)

total cost = number of units used × cost per unit

Figure 38.3 shows an example of an electricity bill.

Northern Electricity Board			Customer account no: 3427 364	
Present meter reading	Previous meter reading	Units used	Cost per unit (incl. VAT)	£
57139	55652	1487	15.0p	£223.05

▲ **Figure 38.3** An electricity bill

The difference between the current reading and the previous reading is the number of units used.

In this particular example, 57 139 − 55 652 = 1487 units (kWh).

1487 units have been used.

If the cost of a unit is known, then the total cost of the electricity used can be determined.

In this particular example: 1487 units at 15.0 pence per unit = 22 305 pence or £223.05.

One-way switches

This kind of switch acts as a make-or-break device to switch a circuit on or off (Figure 38.4). When the switch is open, there is air between the conducting contacts. Since air is an insulator, the circuit is incomplete. No current flows.

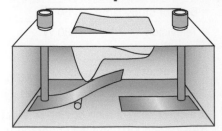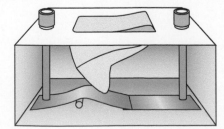

▲ **Figure 38.4** A one way switch

The rocker (the part of the switch that you press) is made of plastic. This is important with high voltages to prevent current flowing through the body of the user. When the rocker is pressed, the conducting contacts are pushed together. There is now a complete circuit, so current can flow.

As we will see later, it is important that the switch is placed in the live side of a circuit.

Two-way switches

In most two-storey houses, you can turn the landing lights on or off from upstairs or downstairs. Two-way switches are used for this.

Figure 38.5a illustrates a two-way switch circuit in one of its two 'on' positions.

When both switches are up (or both are down), the circuit is complete and a current flows through the bulb.

At the top of the stairs, one of the switches is pressed down and the circuit is broken, as shown in Figure 38.5b. The bulb goes dark.

Two-way switches may be used in a number of different situations, such as to control a light from opposite ends of a long corridor.

Figure 38.6 shows the 4 possible states of a two-way switch used in a simple circuit with a battery.

a)

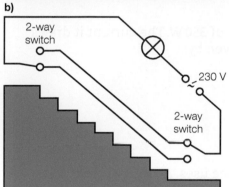

b)

▲ **Figure 38.5** A two-way switch circuit

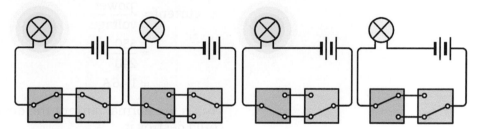

▲ **Figure 38.6** The four possible states of a two-way switch in circuit

How to wire a three-pin plug

▶ The wire with the blue insulation is the neutral wire – connect this to the left-hand pin.

▶ The brown insulated wire is the live wire – connect this to the right-hand pin.

▶ The wire with the yellow and green insulation is the earth wire – connect this to the top pin.

▶ Each of these wires should be wrapped around its securing screw so that they are tightened as the screw is turned.

▶ Insert the correct cartridge fuse into its holder beside the live wire.

▶ Finally, fix the 3-core cable tightly with the cable grip and screw on the plug-top.

Each pin in the plug fits into a corresponding hole in the socket. The earth pin is longer than the others so that it goes into the socket first and pushes aside safety covers, which cover the rear of the neutral and live holes in the socket.

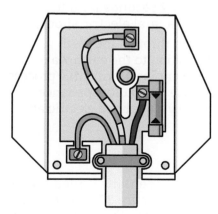

▲ **Figure 38.7** How to wire a three-pin plug

Fuses

A fuse is a device which is meant to prevent damage to an appliance.

The most commonly used fuses are either a 3 A (red) fuse for appliances up to 720 W, or 13 A (brown) fuse for appliances between 720 W and 3 kW.

If a larger-than-usual current flows, the fuse will melt and break the circuit.

Nowadays, residual current circuit breakers (RCCBs) are becoming much more common. They work by detecting any difference between the currents in the live and neutral wires. When a difference is detected due to a fault they break the circuit very rapidly before there is any danger. Unlike fuses, which are designed to protect the appliance, RCCBs protect both the appliance and the user because they are very sensitive and very quick.

Selecting a fuse

Every appliance has a power rating. How much current the appliance will use is found using the power formula:

power = voltage × current

For example, a jig-saw has a power of 350 W. The current it draws when connected to the mains is given by:

$$\text{current} = \frac{\text{power}}{\text{voltage}}$$
$$= \frac{350}{240}$$
$$= 1.46\,A$$

This is the normal current the device uses. A larger current could destroy it.

A 3 A fuse would allow a normal working current to flow and protect the jig-saw from larger currents. A 13 A fuse would allow a dangerously high current to flow without breaking the circuit. It is important to use the correct size of fuse.

Remember that a fuse protects the appliance. It does **not** protect the person using the appliance. It can take 1 to 2 seconds for a fuse to melt – enough time for the user to receive a fatal electric shock.

If an appliance becomes live, a current flows through the earth wire and then from the socket earth connection to the earth via a substantial metal connection such as a water pipe. During this process, the fuse in the plug will blow. Before the fuse in a faulty appliance is replaced, the appliance should be checked by a qualified electrician.

The earth wire

An **earth wire** can prevent harm to the user.

Suppose someone using an electric drill accidentally drilled into a mains electricity wire hidden inside a wall (Figure 38.8). The mains current would flow through the drill bit and into the metal casing of the drill. The casing would become live, and if someone were to touch it they would get a possibly fatal electric shock as the current rushes through their body to earth.

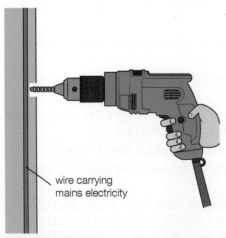

wire carrying mains electricity

▲ **Figure 38.8** If there is no earth wire connected to the casing of the drill, the current will flow through the user

The earth wire can help to prevent this – it offers a low resistance route of escape, enabling the current to go to earth by a wire rather than through a human body. Because of this low resistance, the current through the fuse will be large. After a few seconds, the large current will melt the fuse.

Together, the fuse and the earth wire help to reduce the risk of an electric shock.

Any appliance with a metal casing could become live if a fault developed, so such appliances nearly always have a fuse and an earth wire fitted in the plug.

Double insulation

Appliances such as vacuum cleaners and hairdryers are usually double insulated. The appliance is encased in an insulating plastic case and is connected to the supply by a two-core insulated cable containing only a live and a neutral wire. Any metal attachments that a user might touch are fitted into the plastic case so that they do not make a direct connection with the motor or other internal electrical parts. The symbol for double insulated appliances is shown in Figure 38.9.

▲ **Figure 38.9** The symbol for double insulation

The live wire

Earlier, we said that the switch and fuse must be placed on the live side of the appliance (Figure 38.10). Why is this important? The live wire in a mains supply is at high voltage (effectively around 230 V). The neutral side is at approximately zero volts.

If a fault occurs and the fuse blows, the live, dangerous wire is disconnected. If the fuse were on the neutral side, the appliance would still be live, even when the fuse had blown.

Switches are also placed on the live side for the same reason. If the switch were on the neutral side, the appliance would still be live, even when the switch was in the OFF position.

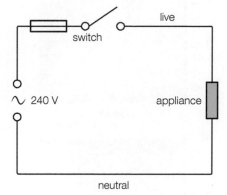

▲ **Figure 38.10** This fuse has been correctly placed on the live side of the power supply

Alternating and direct currents

The mains electrical supply to your home is an alternating current (a.c.). The electrical supply from a battery is a direct current (d.c.). The difference can be seen using a CRO (cathode ray oscilloscope).

A direct current (d.c.) always flows in the same direction: from a fixed positive terminal to the fixed negative terminal of a supply.

A typical d.c. circuit is shown in Figure 38.11. A cell or battery gives a constant, steady, direct current. A graph of voltage versus time for a d.c. supply is shown in Figure 38.12. The current is described as being undirectional.

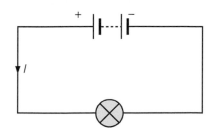

▲ **Figure 38.11** A simple d.c. circuit

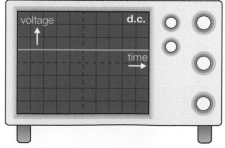

▲ **Figure 38.12** A graph of voltage against time for a d.c. supply

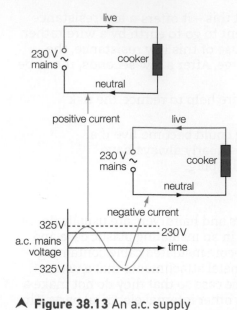

▲ **Figure 38.13** An a.c. supply

The electricity supply in your home is an **alternating current** (a.c.) supply. In an a.c. supply, the voltage changes size and direction in a regular and repetitive way (Figure 38.13).

In fact, the mains voltage changes from +325V to −325V. Considering the sizes of positive and negative voltages together, the effective value of the voltage is 230V. The current changes direction 100 times every second and makes 50 complete cycles per second; hence the frequency of the mains is 50Hz.

It is clear from Figure 38.13 why an a.c. supply is said to be bidirectional.

If the voltage–time graph never crosses the time axis, the supply is d.c. But once the graph crosses the time axis, the supply is a.c.

Test yourself

9 Look at Figure 38.14, which shows five displays on a cathode ray oscilloscope (CRO) screen. Which of the displays illustrate:
   a) d.c. voltages
   b) a.c. voltages?

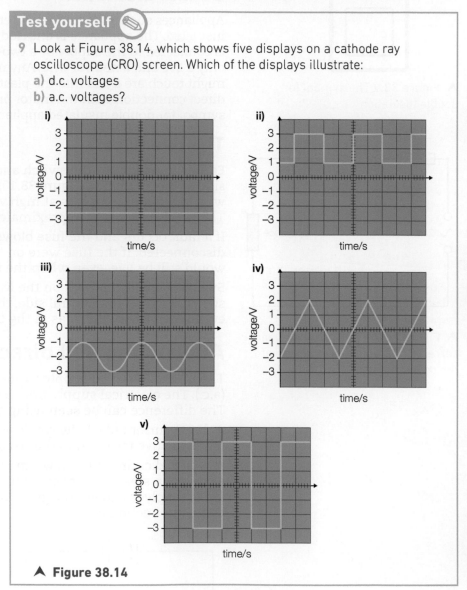

▲ **Figure 38.14**

Practice questions

1 An electric fire is connected to the mains supply by means of a three-pin plug.
   a) The electric fire has a rating of 2000 W when used on a 230 V mains supply. Calculate which fuse would be most suitable for use with this fire: 1 A, 3 A, 5 A or 13 A. *(4 marks)*
   b) Describe the size of the current in the live, neutral and earth wires when the fire is switched on and working properly. *(1 mark)*
   c) The live wire becomes loose and comes into contact with the metal body of the electric fire. Describe the danger that could arise when the electric fire is switched on. *(1 mark)*
   d) How should the earth wire be connected to reduce this danger? *(1 mark)*
   e) How should the fuse be connected so as to reduce this danger? *(1 mark)*
   f) Explain how the action of the earth wire and the fuse reduce this danger. *(1 mark)*

2 a) Mrs Johnston's electricity meter was read at 1 month intervals.
   Reading on 1 April 2017   11 897 kW h
   Reading on 1 May 2017   12 107 kW h
   i) How many units of electricity were used in the Johnston home during April? *(1 mark)*
   ii) If 1 unit of electricity costs 15p, calculate the cost of electricity to Mrs Johnston during April. Show clearly how you get your answer. *(2 marks)*
   b) An electric fire is wired using a three-pin plug as shown in Figure 38.15.

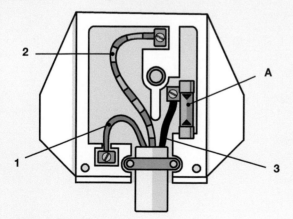

Figure 38.15

   i) Name the part labelled A. *(1 mark)*
   ii) Which of the wires 1, 2 or 3 should be connected to the metal casing of the fire? *(2 marks)*
   iii) State the colours of the insulation on wires 1, 2 and 3. *(3 marks)*

3 a) A television set is marked 240 V, 80 W.
   i) Explain carefully what these numbers mean. *(2 marks)*
   ii) The flex that connects a brand new television to the mains has only two wires inside it. An electrician confirms that there should only be two wires inside the plug. Explain why only two wires are needed. *(2 marks)*
   iii) To which of these wires should the switch on the television be connected? *(1 mark)*
   iv) Apart from allowing the user to switch the television on and off, the switch is connected in this way for another reason. What is this other reason? *(1 mark)*
   v) Explain how the owner of this television is protected from possible electric shock. *(1 mark)*
   b) An electric oven is rated at 8 kW.
   i) Calculate the cost of using the oven to cook for 2 hours. The cost of electricity is 15p per unit. Show clearly how you get your answer. *(3 marks)*
   ii) When the oven is on, the same current passes through the cable as the heating elements. Explain why the cable does not heat up. *(2 marks)*

4 Calculate how much electrical energy, in kilowatt hours, is used for:
   a) a 100 W lamp on for 12 hours *(2 marks)*
   b) a 250 W television on for 4 hours *(2 marks)*
   c) a 2400 W kettle on for 5 minutes. *(2 marks)*

5 An electric shower is rated at 230 V, 15 A.
   a) Calculate the electrical power used by the shower heater. *(2 marks)*
   b) Calculate the cost of a 10-minute shower if 1 kWh costs 12p. *(2 marks)*

6 a) Calculate the amount of electrical energy, in joules, used by a 1000 W electric fire in 1 hour. *(3 marks)*
   b) What common name is given to this quantity of energy? *(1 mark)*

7 Cartridge fuses are normally available in 3 A, 5 A or 13 A.
   a) What could happen if you used a 3 A fuse in the plug for a 3 kW electric fire? *(1 mark)*
   b) Why is it bad practice to use a 13 A fuse in a plug for a 60 W study lamp? *(1 mark)*
   c) What size of fuse would you use for a hairdryer labelled 230 V, 800 W? Explain how you worked out your answer. *(2 marks)*

8 What is the highest number of 60W bulbs that can be run off the 230V mains if you are not going to overload a 5A fuse? *(3 marks)*

9 A 13A socket is designed to allow a current of 13A to be drawn safely from it. Mr White connected the following appliances to a single 13A socket using a 4-way extension lead: 2.4kW electric kettle; a 3kW dishwasher; an 800W television; a 1300W toaster.
  a) Calculate the current through each appliance, assuming that the supply voltage is 230V. *(4 marks)*
  b) Assuming that the plug from the extension lead contained a 13A fuse, what would happen if he attempted to use all the appliances at the same time? *(1 mark)*

10 a) Copy the circuit diagram below and complete the two-way switches A and B so that the bulb lights. *(2 marks)*

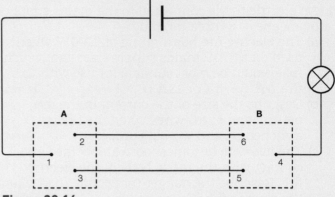

Figure 38.16

  b) Now that you have completed the switches, describe the path of the current from the positive terminal of the battery to the negative terminal by copying and completing the following sentence.
  The current flows from the positive terminal to 1 to ___ to ___ to ___ to the lamp and to the negative terminal. *(2 marks)*

Specification points

This chapter covers sections 2.4.1 to 2.4.3 of the specification. It is about plotting magnetic fields and investigating the factors affecting the strength of a magnetic field.

Magnetic field pattern around a bar magnet

Magnetic fields can be investigated using a small **plotting compass**. The 'needle' in the compass is a tiny magnet which is free to turn on its spindle. When held near a bar magnet, the needle is turned by forces between its poles and the **poles of the magnet**. The needle comes to rest so that the turning effect is zero. Figure 39.1 shows how a plotting compass can be used to plot the field around a bar magnet.

Starting with the compass near one end of the magnet, the needle position is marked using two dots. The compass is then moved so that the needle lines up with the previous dot. Another dot is added and so on. When the dots are joined up, the result is a **magnetic field line**. More lines can be drawn by starting with the compass in different positions. In Figure 39.2, a **magnetic flux** (a selection of field lines) have been used to show the magnetic field around a bar magnet. It should be noted that:

▶ The field lines run from the North pole (N) to the South pole (S) of the magnet.

▶ The field direction, shown by the arrowhead, is defined as the direction in which the force on a North pole would act. The N end of the compass needle would point in this direction.

▶ The magnetic field is strongest where the field lines are closest together at the poles of the magnet. There are no field lines in the middle of the magnet, so there is no magnetic field here.

If two magnets are placed near each other, their magnetic fields combine to produce a single field.

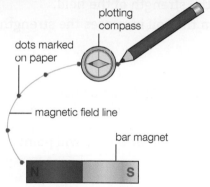

▲ **Figure 39.1** Plotting the field around a bar magnet using a compass

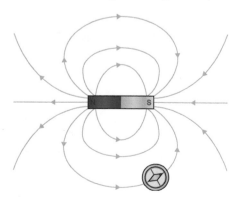

▲ **Figure 39.2** Magnetic field lines around a bar magnet

Magnetic field pattern due to a current-carrying coil

A coil with one turn

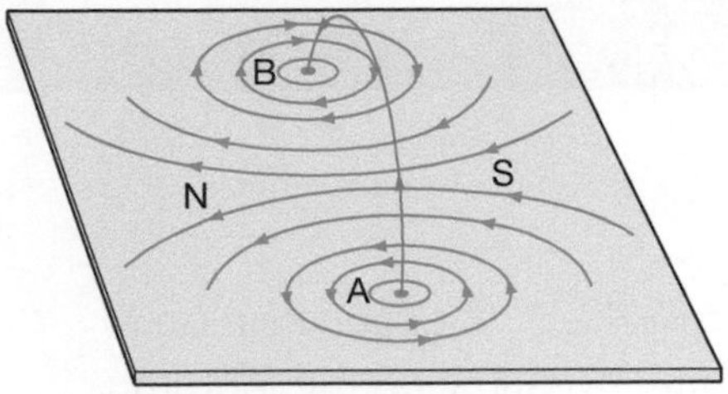

Figure 39.3 shows the magnetic field around a single loop of wire which is carrying a current. Near A, the field lines point anticlockwise and near B, the lines point clockwise. In the middle, the fields from each part of the loop combine to produce a magnetic field running from left to right. This loop of wire is like a very short bar magnet. Magnetic field lines come out of the left-hand side (North pole) and go back into the right-hand side (South pole).

The strength of the magnetic field can be increased by:

▶ increasing the number of turns of wire in the coil
▶ increasing the current through the coil.

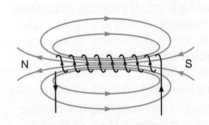

▲ **Figure 39.3** The magnetic field around a single loop of current-carrying wire

A coil with many turns

A stronger magnetic field can be made by wrapping a wire into a long coil. This coil is referred to as a **solenoid**. A current must be passed through it, as illustrated in Figure 39.4.

The magnetic field produced by a solenoid has the following features:

▶ The shape of the field is similar to that around a bar magnet and there are magnetic poles at the ends of the coils.
▶ Increasing the current increases the strength of the field.
▶ Increasing the number of turns on the coil increases the strength of the field.

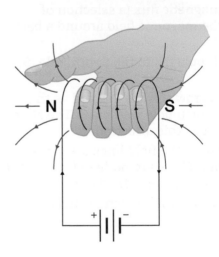

▲ **Figure 39.4** The pattern of magnetic field around a solenoid

Polarity

The **right-hand grip rule**, as shown in Figure 39.5, can be used to help you work out which way round the poles are in a solenoid. Imagine gripping the solenoid with your right hand, so that your fingers point in the direction of the conventional current. Your thumb will point towards the North pole of the solenoid.

▲ **Figure 39.5** The right-hand grip rulet

Test yourself ✏

1 Compare and contrast the magnetic field pattern of a bar magnet and a current-carrying solenoid.
2 Draw a sketch of the magnetic field produced by a bar magnet.
3 Copy the diagrams below (Figure 39.6) and label the poles (A, B, C, D) 'North' or 'South'.

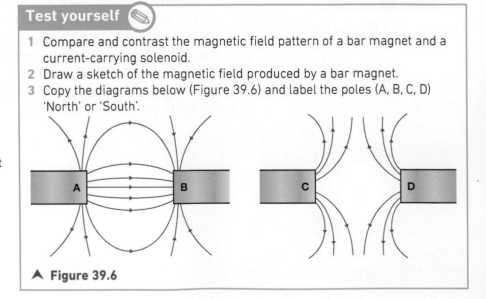

▲ **Figure 39.6**

Show you can ?

a) Plot the field pattern of a single bar magnet using the technique described on page 481. Then, using two bar magnets in the positions shown in Figure 39.7, verify that the field patterns would be as illustrated in the diagram.

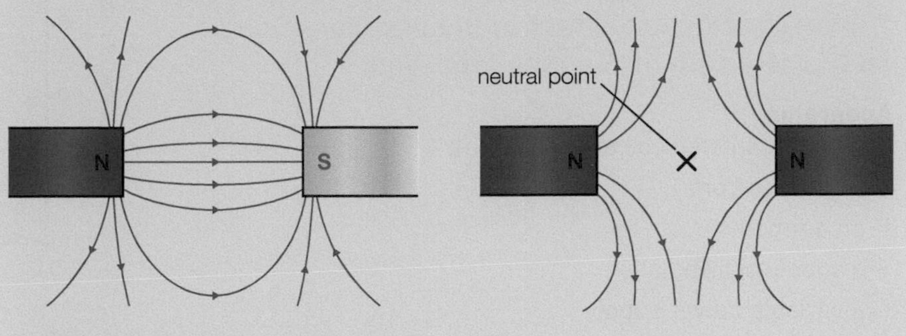

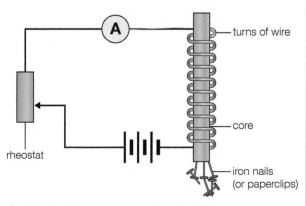

▲ **Figure 39.7**

b) Explain the meaning of the term 'neutral point'.

Practical activity

Investigating the factors affecting the strength of an electromagnet

An electromagnet is a solenoid wrapped around a soft iron core. The strength of an electromagnet, in other words the strength of its magnetic field, is measured by finding the mass of iron the electromagnet will attract. Iron nails or paper clips may be used.

Three factors can vary the strength of the electromagnet and these are discussed in turn below.

1 Investigating the effect of the current on the strength of the magnetic field

Apparatus
- a thick insulated coil of copper wire
- a soft iron core
- an ammeter
- a supply of iron nails
- a variable power supply

Method
1 Construct an electromagnet using 50 turns of insulated wire around a soft iron core.
2 Connect it to the circuit as shown in Figure 39.8.
3 Using the rheostat, increase the current in steps, measuring the number of iron nails attracted to the electromagnet for each current.
4 Record your results in a suitable table, similar to Table 39.1 on the next page.
5 Plot a graph of number of nails lifted on the *y*-axis versus current on the *x*-axis.
6 What does this graph show?

▲ **Figure 39.8** Investigating the strength of an electromagnet

Table 39.1 Results table

Current/A	0.0	0.5	1.0	1.5	2.0	2.5	3.0	3.5
Number of nails lifted								

A graph similar to Figure 39.9 should be obtained.

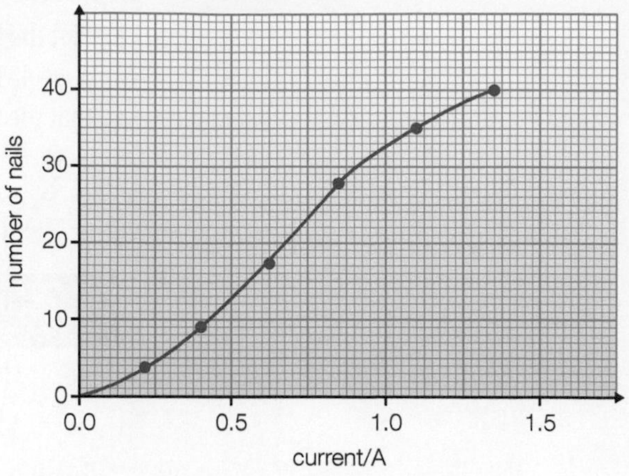

2 Investigating the effect of the number of turns on the strength of the magnetic field

Apparatus
- a thick insulated coil of copper wire
- a soft iron core
- an ammeter
- a supply of iron nails
- a variable power supply

▲ **Figure 39.9** A typical graph of results

Method
1 Keeping the current at 2.5 A and the material of the core constant, increase the number of turns of wire in steps of 10.

2 For each number of turns, measure the number of nails lifted.

3 Record your results in a suitable table, similar to Table 39.2.

4 Plot a graph of number of turns versus number of nails lifted.

Table 39.2 Results table

Number of turns	10	20	30	40	50	60
Number of nails lifted						

3 Investigating the effect of the material of the core on the strength of the magnetic field

Apparatus
- a thick insulated coil of copper wire
- cores made from various materials (at least: soft iron, steel, copper, plastic and wood)
- a supply of iron nails
- a power supply

Method
1 For this third investigation, keep the number of turns of wire and the current constant at 3.0 A but change the material of the core from soft iron to steel, copper, plastic, wood or air (no material).

2 In each case, measure the number of iron nails lifted and record your results in a table.

3 Draw a bar chart of your results and describe your findings.

Table 39.3 Results table

Type of material	Soft iron	Steel	Copper	Plastic	Wood	No material (air)
Number of nails lifted						

Show you can ❓

State three factors which determine the strength of an electromagnet.

Practice questions

1 An electromagnet is a coil of wire through which a current can be passed.
   a) State three ways in which the strength of the electromagnet may be increased. *(3 marks)*
   b) An electromagnet may be switched on and off. Suggest one situation where this would be an advantage over the constant field of a permanent magnet. *(1 mark)*
   c) A coil carrying a current has two magnetic poles.

current

Figure 39.10

   i) Copy the diagram and mark the magnetic poles produced. *(2 marks)*
   ii) On your diagram, draw the magnetic field produced. *(4 marks)*

2 **a)** How could you separate iron filings from a mixture of iron filings and sand using an electromagnet? *(1 mark)*
   b) Why is it better to use an electromagnet rather than a bar magnet for this purpose? *(1 mark)*

3 **a)** Describe in detail how you would use a plotting compass to show the field lines around a bar magnet. *(5 marks)*
   b) How do the field lines show where the field is strongest? *(1 mark)*

Specification points

This chapter covers 2.5.1 to 2.5.15 of the specification. It introduces the variety of objects that make up our Solar System. You will develop an understanding of the forces that keep objects like planets, moons and comets in orbit. It also introduces the life cycle of stars and how the mass of a star determines its final outcome. You will study the Big Bang and its supporting evidence, and discuss the difficulties associated with space travel to other worlds.

The Solar System

We now know that the Earth is one of eight **planets** that travel around the Sun (Figure 40.1). Each planet travels in an elliptical path and, with the exception of Mercury and Venus, they all have at least one moon.

Other objects also orbit the Sun. These are the comets and the asteroids. Table 40.1 gives some data about the planets.

Table 40.1 Some data on the eight planets orbiting the Sun

Planet	Planet diameter compared with Earth	Average distance of planet from Sun compared with Earth	Time to orbit the Sun compared with the Earth	Number of moons
Mercury	0.4	0.4	0.2	0
Venus	0.9	0.7	0.6	0
Earth	1.0	1.0	1.0	1
Mars	0.5	1.5	1.9	2
Jupiter	11.2	5.2	12.0	14
Saturn	9.4	9.5	29.0	24
Uranus	4.1	19.1	84.0	15
Neptune	3.9	30.1	165.0	3

Note that the status of Pluto was changed from a 'planet' to a 'dwarf planet' in 2005.

The orbits of the inner planets are almost circular, with the Sun at the centre. The orbits of Jupiter, Saturn, Uranus and Neptune are much more elliptical (like a rugby ball).

All the planets orbit the Sun in the same plane and travel in the same direction, as a result of the gravitational force between the Sun and the planets. This is evidence that they were formed at around the same time.

The four inner planets have a rocky surface on which it is possible to land a spacecraft. The four outer planets are known as gas giants. They are made up of very dense accumulations of gases, like hydrogen, methane and ammonia.

All of the planets except Mercury and Venus have at least one moon. **Moons** are heavenly bodies which are natural satellites of a planet. **Artificial satellites** are objects put into space by humans. Almost all orbit the Earth, but a few, like Kepler, orbit the Sun, and others, like the Mars Reconnaissance Orbiter, orbit other planets.

Artificial satellites of the Earth have four main purposes:

▶ communications
▶ Earth observation (spying and monitoring rainforests, deserts, crops etc.)
▶ astronomy
▶ weather monitoring.

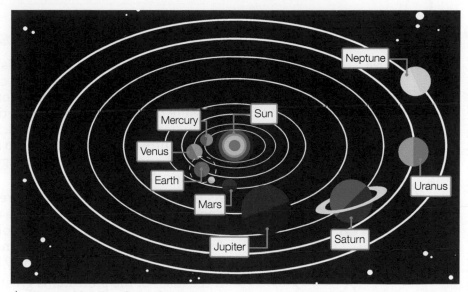

▲ **Figure 40.1** The Solar System

Comets

Comets range from a few hundred metres to tens of kilometres in diameter, and are sometimes called 'dirty snowballs'. At their centres are rock, ice, silicates and some organic compounds. Surrounding this is a 'coma' consisting of gases and dust.

Most comets orbit the Sun in very elliptical paths. As a comet approaches the Sun, solar radiation vaporises some of the frozen gas at the centre, causing the coma greatly to increase in size. Dust and gas stream away from the comet as a long tail, often millions of kilometres long. The radiation from the Sun causes this tail to point away from the Sun. Most astronomers believe that comets in our Solar System originate in the Oort cloud, a region of space between 2000 and 3000 times further from the Sun than Pluto.

Show you can ?

Describe what asteroids are.

Asteroids

Asteroids are large rocks in outer space. Some are very large, while others are as small as a few metres in diameter. Due to their small size, asteroids do not have enough gravity to pull themselves into the shape of a ball. Many are found in the Asteroid Belt, a giant ring between Mars and Jupiter.

Asteroids are left over materials from the formation of the Solar System. These materials were never incorporated into a planet because of the strong gravitational pull of Jupiter.

Test yourself

1 Name all the planets in order of their distance from the Sun, from closest to furthest.
2 Describe the major differences between the inner four planets and the outer four planets.
3 Describe the differences between artificial and natural satellites, and give an example of each.
4 State three uses of artificial satellites of the Earth.
5 State the main features of a comet.

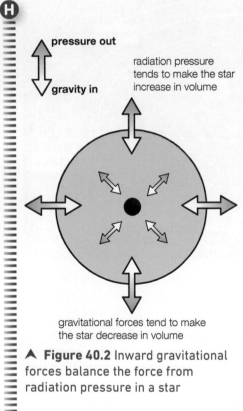

▲ **Figure 40.2** Inward gravitational forces balance the force from radiation pressure in a star

The life cycle of stars

Stars are formed in the cold clouds of hydrogen gas and dust known as stellar nebulae. Gravity causes these gas particles to come together. The gravitational force is greater than the outward pressure due to the particles' kinetic energy, and it brings about the gravitational collapse of the cloud. During this collapse, the material at the centre of the cloud heats up as the gravitational potential energy changes into thermal energy. The hot core at the centre of the cloud is called a protostar.

As the protostar accumulates more and more gas and dust, its density and temperature continue to rise, increasing the outward pressure within the protostar. A point is reached where this outward force is balanced by the gravitational force and the protostar becomes luminous because of its extremely high temperature (Figure 40.2).

If the mass of the protostar is greater than about 8% of our Sun's mass, the temperature will exceed the minimum required for nuclear fusion to begin. There is equilibrium between the inward gravitational force and the outward force from the radiation pressure due to fusion. The star is now in the main phase of its life, so it is called a main sequence star.

The smaller a star is, the longer its life as a main sequence star. Our Sun will have a life of around 10 billion years in the main sequence stage, while more massive stars only live for a few million years as main sequence stars. The life cycle of a star can be seen in Figure 40.3.

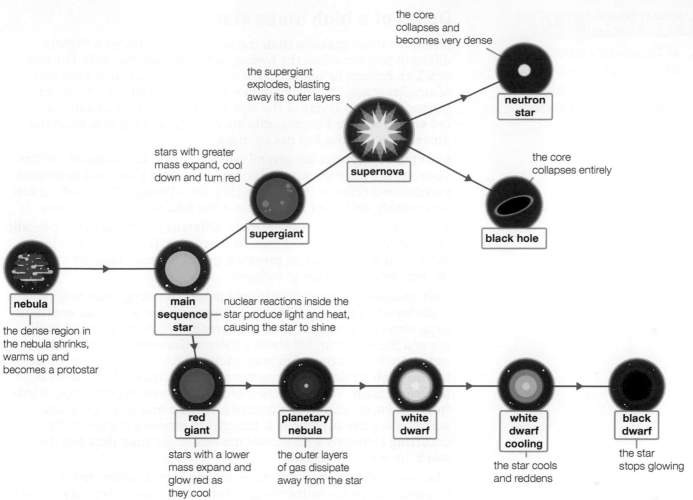

▲ **Figure 40.3** The life cycle of a star

Labels within figure:

the core collapses and becomes very dense

the supergiant explodes, blasting away its outer layers

neutron star

stars with greater mass expand, cool down and turn red

supernova

the core collapses entirely

supergiant

black hole

nebula

the dense region in the nebula shrinks, warms up and becomes a protostar

main sequence star

nuclear reactions inside the star produce light and heat, causing the star to shine

red giant

stars with a lower mass expand and glow red as they cool

planetary nebula

the outer layers of gas dissipate away from the star

white dwarf

white dwarf cooling

the star cools and reddens

black dwarf

the star stops glowing

Death of a star like our Sun

A stage is reached when almost all the hydrogen in the core of the star has been converted into helium by nuclear fusion. Without hydrogen, the energy output from fusion reactions in the core reduces so much that gravity compresses it significantly, but the star doesn't shrink.

Surrounding the core is a layer of hydrogen. Gravitational contraction provides enough energy for nuclear fusion of the hydrogen in this layer. The outward pressure from the nuclear fusion reactions prevents the star from collapsing, instead making it expand to several hundred times its former size. The surface temperature falls, and the starlight is now predominantly orange. We refer to the star as a '**red giant**'.

Within a red giant, other nuclear reactions can take place. Helium, for example, can fuse to become carbon and oxygen. Indeed, all the naturally occurring elements in the periodic table up to iron are formed by nuclear fusion in stars.

Close to the end of the life of a red giant, the gravitational force can no longer hold the outer layers of gas. These outer layers flow out, cool and surround the core to form a nebula. This nebula may eventually contribute to the creation of another star. Over time, the core that remains cools to become a **white dwarf**. Eventually, all fusion stops, and the star cools to become a **black dwarf**.

Show you can **?**

a) Explain what a red supergiant is.
b) Describe what a neutron star is.
c) Describe how black holes get their name, and explain why nothing can escape from them.

Death of a high mass star

If a star is more massive than the Sun, it goes through a slightly different process when the hydrogen fusion process ends. The rate at which helium fusion occurs is much more rapid than for a star of smaller mass. The huge amount of energy from helium fusion pushes the outer layers of the star outwards, and it turns into a **red supergiant**. Red supergiants are among the largest stars in the universe by volume, but not by mass.

Most red supergiants are several hundred times the radius of our Sun. They require huge amounts of energy to sustain them and to prevent gravitational collapse. As a result, they burn through their nuclear fuel very quickly, and most only live for a few tens of millions of years.

A red supergiant successively fuses different elements in the periodic table, up to the creation of iron. At that point, the core can no longer sustain outward radiation pressure, and the force of gravity causes the supergiant to begin to collapse.

This collapse releases gravitational potential energy that heats up the outer layers of the star. These are thrown off the star in an enormous explosion called a **supernova**. It is a really dramatic moment in the life of a massive star. For about a month, the supernova emits more radiation than all the other stars in its galaxy put together! For a relatively short time, the supernova shines with the brightness of ten billion Suns. The interaction of the elements exploding outwards from the supernova with atoms of elements and other particles surrounding the supernova is thought to produce the naturally occurring elements with atomic masses larger than iron. But the star's life is not quite over.

The core of the star is left behind, having been compressed into a **neutron star** by the immense gravitational pressure. For very massive stars, a **black hole** is created.

Neutron stars are composed almost entirely of neutrons, and are the smallest and densest stars known to exist. They typically have a radius of about 10 km, but they can have a mass of about twice that of the Sun.

Black holes are incredibly dense, and so have such enormous gravitational fields that nothing can escape from them, not even light. That is why we call them black holes. We can only infer their presence by the way in which they bend light passing close by.

Experimental evidence

What is the evidence that there is hydrogen, helium and so on in the Sun? The evidence comes from the absorption **spectrum** of sunlight (Figure 40.4). We are all familiar with the continuous spectrum of sunlight. A German physicist called Joseph von Fraunhofer (Figure 40.5) looked carefully at the spectrum and found that it contained many dark lines, which we now call 'Fraunhofer lines'.

▲ **Figure 40.5** Joseph von Fraunhofer

What are the Fraunhofer lines and how are they formed?

When the visible light from below the Sun's surface passes through the layers above, the atoms in the solar 'atmosphere' absorb particular wavelengths, and so these wavelengths are missing in the spectrum we see. When there is no light at a particular wavelength, it appears as a black line in the spectrum (Figure 40.4).

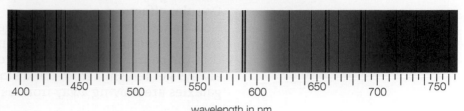

wavelength in nm

▲ **Figure 40.4** Solar absorption spectrum

Each of the black lines corresponds to an element in the Sun's atmosphere. The absorption spectrum tells us that at the moment, the major gases (by mass) in the Sun are hydrogen (71%), helium (27%), and oxygen (1%).

Nuclear fusion in our Sun

Stars are the powerhouses of the Universe. Stars like our Sun get their energy mainly from the fusion of light hydrogen nuclei into heavier helium nuclei. The electrical repulsion between the positive charges on the nuclei means that the nuclei must be moving very fast if fusion is to occur. In fact, fusion requires a temperature of at least 13 million °C and a density of $100\,g/cm^3$. Fortunately, these conditions are met at the core of our Sun.

The Doppler effect

Think about what we hear when a police car passes with its siren sounding. As the car approaches, the sound appears to have a higher pitch (or shorter wavelength) than we would expect. As soon as the car passes, its pitch falls. This is called the Doppler effect.

Look at Figure 40.6a, which shows a source of sound at rest. In Figure 40.6b, this source of sound is moving to the right. Observer A hears sound of high pitch (shorter wavelength) because the waves are being bunched up together. Observer B, to the left of the source, hears a sound of low pitch (longer wavelength) because the waves are being spread out.

A similar effect occurs with light. If the light that we observe from a moving source has a shorter wavelength than expected, it is because the source is moving towards us – we say the light is 'blue shifted'. But if the light we observe has a longer wavelength than expected, it is because the source is moving away from us – we say the light is 'red shifted'. Today, astrophysicists interpret the red shift from distant galaxies as evidence that space itself is expanding.

a) source of sound at rest

b) source of sound moving to the right

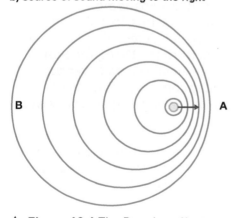

▲ **Figure 40.6** The Doppler effect

The origin of our Universe

Most physicists today believe that our universe began about 14 billion years ago with a 'Big Bang'. Evidence for the Big Bang theory comes from red shift.

Galaxies are huge collections of star systems. Our own galaxy, the Milky Way, contains over 100 billion stars! When we look at the light from distant galaxies, we find that it is shifted to the red (longer wavelength) end of the visible spectrum. This red shift is due to the Doppler effect.

How do we interpret these strange absorption spectrum data? If there is a red shift in the light from another galaxy, this tells us that the source is moving away from us. The fact that we almost always get red shift from the distant galaxies tells us that nearly all galaxies are moving away from us.

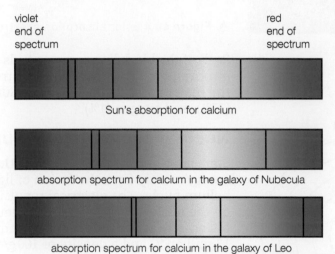

violet
end of
spectrum

red
end of
spectrum

Sun's absorption for calcium

absorption spectrum for calcium in the galaxy of Nubecula

absorption spectrum for calcium in the galaxy of Leo

▲ **Figure 40.7** Red shift in calcium from different galaxies

The red shift in the absorption spectrum for calcium in Figure 40.7, for example, tells us that the galaxies Nubecula and Leo are both moving away from us, and that Leo is moving away faster than Nubecula. If nearly all the galaxies are moving away from each other, we can infer that the universe (space itself) is expanding.

There is a further piece of evidence supporting this. In 1964, two American astrophysicists, Penzias and Wilson, detected microwaves of wavelength 7.35 cm that did not come from the **Earth,** the **Sun** or our closest stars. The microwaves were evenly spread over the sky, and were present day and night. They concluded that these microwaves were coming from outside our own galaxy. These waves come from the cosmos.

Today, we call that radiation cosmic microwave background radiation (CMBR). It represents the signature or 'afterglow' of the Big Bang that occurred 14 billion years ago. Currently, the only model that can give an explanation for CMBR is the Big Bang theory.

Tip

The words cosmic, background, microwave and radiation are all important when talking about CMBR. Be sure to give all four words in your answers!

The Big Bang theory

The argument begins by suggesting that the reason all the galaxies are currently moving away from each other is that they all originated from a common point, called a singularity. About 14 billion years ago, the Universe suddenly came into existence with an enormous explosion, which we call the Big Bang. It immediately went into a short period of very rapid growth, known as **inflation**.

As the Universe expanded, it cooled down and became less dense. Cooling allowed subatomic particles such as neutrons and protons and electrons to form. As the cooling continued, protons and neutrons combined to form simple nuclei. Eventually, around 380 000 years after the Big Bang, and after further expansion and cooling, the first stars came into existence.

Show you can

a) Write a sentence to explain what is meant by 'red shift'.

b) Explain what red shift tells us about neighbouring galaxies.

c) Describe the experimental evidence that there is hydrogen in the Sun.

d) Explain the significance of CMBR.

e) Describe how the first atoms came into existence after the Big Bang.

f) Explain what is meant by nuclear fusion, and where it occurs naturally in the Universe.

g) Write down, according to current estimates, the maximum age of the visible Universe.

Practice questions

1 Figure 40.8 shows a cloud of gas and dust, known as a nebula. The bright spots are stars.

Figure 40.8

a) What force causes the gas to form stars? *(1 mark)*
b) What two gases are the main constituents of stars? *(1 mark)*
c) How do astronomers know this? *(1 mark)*
d) Name the process that supplies the energy in stars. *(1 mark)*
e) Apart from producing energy in stars, what else is produced by this process? *(1 mark)*

2 There are five stages in the lifecycle of a star with the same mass as our Sun.
a) Copy and complete the list below. One stage has been inserted for you.
_____, main sequence star, _____, _____, _____ *(2 marks)*
b) Why do main sequence stars, such as our Sun, remain stable for many billions of years? *(2 marks)*
c) What type of star results in a supernova? *(2 marks)*
d) Some stars end their life as a black hole. Why is this stage of a star's life given this name? *(1 mark)*

3 The speed with which a galaxy is moving away from us is called its recession speed. Table 40.2 shows how the recession speeds of different galaxies change with distance from the Earth.

Table 40.2

Recession speed/m/s	450	700	920	1150	1380	1610
Distance/m	2×10^{20}	3×10^{20}	4×10^{20}	5×10^{20}	6×10^{20}	7×10^{20}

a) On graph paper and starting from a (0,0) origin, plot a graph of recession speed (vertical axis) against distance (horizontal axis), and draw a straight line of best fit through the data points. *(5 marks)*
b) Does your graph show that the recession speed is directly proportional to distance? Give a reason for your answer. *(2 marks)*

c) The gradient of this graph is known as the Hubble constant, H_0. Show that the Hubble constant is approximately 2.3×10^{-18}/s. *(3 marks)*
d) The Canis Major dwarf galaxy is approximately 2.4×10^{20} m away from Earth. Use the Hubble constant (or your graph) to show that this galaxy is moving away from us at approximately 2000 kilometers per hour. *(3 marks)*

4 State the main difficulties that physicists and engineers would have in using manned spacecraft to explore any planet outside our Solar System. *(4 marks)*

5 Describe what happens to a very massive star once it passes out of the main sequence period of its lifecycle. *(5 marks)*

6 Evidence for the expansion of the Universe comes from red shift measurements. Explain what red shift means and how it supports the idea that the Universe is expanding. *(4 marks)*

7 The most widely accepted model for the formation of the Universe is that of the Big Bang. Below is a list of statements or events relating to the formation of the Universe, but they are not in the correct sequence. Copy and complete Table 40.3 and order the events from first to last by writing a number 1–4 in the box beside them.

Table 40.3

Event sequence	Order
Neutrons and protons are formed	
Rapid expansion and cooling occurs	
Further expansion and cooling occurs, allowing hydrogen atoms to form	
More expansion and cooling occurs, allowing hydrogen nuclei to form	

(3 marks)

Symbols, formulae and equations

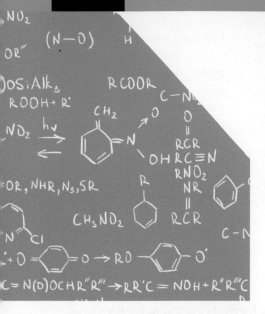

Chemists use formulae and equations as a quick way of identifying substances and showing what happens in chemical reactions. This chapter looks at writing formulae, and understanding what they mean, as well as how to write balanced symbol, ionic and half equations.

Elements

An element consists of one type of atom. Every element is represented by a symbol. A symbol must begin with a capital letter, and if a second letter is part of the symbol it must be lower case.

CA	is incorrect ✗	Ca	is correct ✓
CL	is incorrect ✗	Cl	is correct ✓
h	is incorrect ✗	H	is correct ✓
mg	is incorrect ✗	Mg	is correct ✓

▲ **Figure A.1** Writing symbols for elements

The data leaflet given to you in your examinations will include a Periodic Table, which includes the names and symbols of all elements.

Some elements are made of molecules – where two or more atoms are covalently bonded together. Some elements are **diatomic elements** and have **two** atoms covalently bonded in the molecule. For example, oxygen is diatomic and has the formula O_2. The diatomic elements in the Periodic Table and their formulae are shown in Table A.1.

Table A.1

Diatomic element	Formula of the molecule
hydrogen	H_2
nitrogen	N_2
oxygen	O_2
fluorine	F_2
chlorine	Cl_2
bromine	Br_2
iodine	I_2
astatine	At_2

▲ **Figure A.2** Elements are all represented by symbols between one and three letters long, always beginning with a capital letter

Tip

The diatomic elements in the Periodic Table include the three gases hydrogen, oxygen and nitrogen and all the halogens (Group 7).

Compounds

A compound is two or more elements chemically combined.

It is very useful to know the formula of the common compounds, which are shown in Table A.2.

Table A.2

Compound	Formula	Elements present
water	H_2O	hydrogen, oxygen
carbon monoxide	CO	carbon, oxygen
carbon dioxide	CO_2	carbon, oxygen
sulfur dioxide	SO_2	sulfur, oxygen
sulfur trioxide	SO_3	sulfur, oxygen
nitrogen monoxide	NO	nitrogen, oxygen
nitrogen dioxide	NO_2	nitrogen, oxygen
ammonia gas	NH_3	nitrogen, hydrogen
hydrochloric acid	HCl	hydrogen, chlorine
nitric acid	HNO_3	hydrogen, nitrogen, oxygen
sulfuric acid	H_2SO_4	hydrogen, sulfur, oxygen

> **Tip** ↻
>
> 'Mon' means one, 'di' means two. Carbon *mon*oxide has one atom of oxygen, and carbon *di*oxide has two atoms of oxygen.

Interpreting formulae

A formula gives information about the type and the number of each atom present in a compound.

For example, sulfuric acid has the formula H_2SO_4, which means that it contains two atoms of hydrogen, one atom of sulfur and four atoms of oxygen. The subscript numbers refer to the element that comes in front of them. You will also notice that if there is just one atom in the formula (S in this case) a 1 is not written in the formula.

Sometimes a formula includes brackets. This means the atoms inside the brackets are multiplied by the number outside the brackets. For example, for aluminium sulfate:

$$Al_2(SO_4)_3$$

2 atoms of aluminium $3 \times 4 = 12$ atoms of oxygen   3 atoms of sulfur

Table A.3 shows some formulae and the names and number of each type of atom present.

Table A.3

Formula	Atoms present		
HNO_3	1 atom hydrogen	1 atom nitrogen	3 atoms oxygen
$Ca(OH)_2$	1 atom calcium	2 atoms oxygen	2 atoms hydrogen
$Cu(NO_3)_2$	1 atom copper	2 atoms nitrogen	6 atoms oxygen
H_3PO_4	3 atoms hydrogen	1 atom phosphorus	4 atoms oxygen
$Fe_2(CO_3)_3$	2 atoms iron	3 atoms carbon	9 atoms oxygen

Ion charges

Each column in the Periodic Table is called a **group**. Each group in the Periodic Table forms an ion with a different charge, as shown in Table A.4.

Table A.4

Group	Charge on ion		Group	Charge on ion
1	1+		5	3−
2	2+		6	2−
3	3+		7	1−

Some elements, for example copper and iron, are transition metals. Your data leaflet gives the symbol and charges of some transition metal ions in a table similar to Table A.5.

Table A.5

Name	Symbol
chromium(III)	Cr^{3+}
copper(II)	Cu^{2+}
iron(II)	Fe^{2+}
iron(III)	Fe^{3+}
lead(II)	Pb^{2+}
silver	Ag^+
zinc	Zn^{2+}

A molecular ion **is a charged particle containing more than one element**. Again, your data leaflet will give the charges of some molecular ions in a table similar to Table A.6.

Table A.6

Name	Symbol
ammonium	NH_4^+
carbonate	CO_3^{2-}
dichromate	$Cr_2O_7^{2-}$
ethanoate	CH_3COO^-
hydrogencarbonate	HCO_3^-
hydroxide	OH^-
methanoate	$HCOO^-$
nitrate	NO_3^-
sulfate	SO_4^{2-}

Note that **ammonium is NH_4^+** is a molecular ion and is different from the molecule **ammonia (NH_3)**.

Ammonia (NH_3) is a molecule containing one atom of nitrogen and three atoms of hydrogen that are chemically joined.

Ammonium (NH_4^+) is a molecular ion and is a charged particle, with a 1+ charge, containing one atom of nitrogen and four atoms of hydrogen.

In an ionic compound, a negative ion and a positive ion bond together. Sodium chloride and magnesium nitrate are ionic compounds.

Tip
Remember that the names of negative ions of elements end in '-ide'.

Tip
Sometimes these metals can have several different charges. This is indicated in the name by Roman numerals, for example iron(III) chloride or iron(II) chloride.

Tip
If the name of an ion ends in '-ate' it means that oxygen is present. For example, sulfate SO_4^{2-} contains oxygen, sulfide S^{2-} does not.

Tip
The **charge** is the **top right number**. For example, the charge on the nitrate ion NO_3^- is 1− not 3−. The charge on the ammonium ion NH_4^+ is 1+ not 4+.

Test yourself

1 What is the symbol for each of the following elements?
   a) potassium
   b) phosphorus
   c) silicon
   d) sodium
   e) sulfur
   f) fluorine
   g) zinc
   h) copper
   i) calcium
   j) carbon

2 Name four diatomic gases in the Periodic Table and write their formulae.

3 For each of the following compounds name and give the number of each type of atom present:
   a) $NaOH$
   b) HNO_3
   c) CH_3CH_2COOH
   d) $Mg(OH)_2$
   e) $AlCl_3$
   f) $(NH_4)_2SO_4$
   g) NH_4NO_3
   h) $Al(OH)_3$
   i) $Al_2(CO_3)_3$

4 Write the charge of each of the following ions:
   a) sodium
   b) potassium
   c) chloride
   d) iodide
   e) oxide
   f) aluminium
   g) nitride
   h) sulfide.

5 Write the charge of each of the following ions:
   a) magnesium
   b) nitrate
   c) phosphide
   d) sulfate
   e) ammonium
   f) iron(II)
   g) carbonate
   h) hydroxide
   i) dichromate
   j) ethanoate
   k) lithium.

6 Name the following compounds:
   a) KCl
   b) $NaNO_3$
   c) $CaCl_2$
   d) $CaSO_4$
   e) $Al_2(CO_3)_3$
   f) $Mg(NO_3)_2$
   g) Na_2SO_4
   h) $CaCO_3$
   i) $Fe_2(SO_4)_3$
   j) $Ca(HCO_3)_2$
   k) $K_2Cr_2O_7$
   l) $AgNO_3$
   m) $MgBr_2$
   n) $PbSO_4$
   o) $Zn(OH)_2$
   p) $NaHCO_3$
   q) FeI_2.

Writing formulae for ionic compounds

Ionic compounds contain **positive and negative ions**. The number of positive charges must equal the number of negative charges so that the compound has **no charge overall**. When the positive ion has the same number of charges as the negative ion, it is easy to work out the formula of the compound formed. Sodium chloride contains sodium ions (Na^+) and chloride ions (Cl^-). As both ions have single charges the formula is simply written as $NaCl$, i.e. the positive ion followed by the negative ion with no charges written. Similarly, ammonium chloride is NH_4Cl, magnesium oxide is MgO, and so on. When the number of charges is different, a 'swap and drop' method may help:

1 Use the data leaflet to write down the symbol and charge of each ion in the compound. There should be one positive and one negative charge value.

2 Ignore the sign, and if the number of the charges are the same then write the formula in a 1:1 ratio.

3 If the number of the charges is different, **swap** the charges by crossing sides, **drop** the signs and write the formula.

4 Watch out for **molecular ions** – if there is **more than one** of them then you need **brackets**.

Tip

Always look for the ion in the charges table list first, and if it is not there then use the Periodic Table.

Example

Table A.7

sodium chloride			copper(II) sulfide			calcium carbonate		
Symbol:	Na	Cl	Symbol:	Cu	S	Symbol:	Ca	CO_3
Charge:	+1	−1	Charge:	+2	−2	Charge:	+2	−2
	Same			Same			Same	
	NaCl			CuS			$CaCO_3$	
sodium sulfate			calcium hydroxide			sodium sulfate		
Symbol:	Na	SO_4	Symbol:	Ca	OH	Symbol:	Al	SO_4
Charge:	+1	−2	Charge:	+2	−1	Charge:	+3	−2
	Swap and drop			Swap and drop			Swap and drop	
	2	1		1	2		2	3
	Na_2SO_4			$Ca(OH)_2$			$Al_2(SO_4)_3$	

Show you can

W, X, Y and Z are metals. What is the charge on each of these metals in the following compounds?

1 $W_2(SO_4)_3$
2 XCl
3 YO
4 ZCl_3

Test yourself

Write the formulae for the following ionic compounds:

7 potassium chloride
8 calcium iodide
9 copper(II) oxide
10 aluminium sulfide
11 calcium nitrate
12 copper(II) hydroxide
13 ammonium carbonate
14 iron(II) nitrate
15 iron(II) hydroxide
16 aluminium hydrogencarbonate
17 sodium sulfate
18 calcium sulfide

19 copper(II) sulfate
20 magnesium oxide
21 ammonium chloride
22 sodium nitrate
23 chromium hydroxide
24 iron(II) oxide
25 aluminium dichromate
26 sodium hydrogencarbonate
27 sodium carbonate
28 potassium sulfate
29 sodium sulfide
30 lithium dichromate

31 iron(III) oxide
32 magnesium dichromate
33 copper(II) chloride
34 zinc nitrate
35 ammonium carbonate
36 sodium dichromate
37 iron(II) chloride
38 aluminium hydroxide
39 iron(III) bromide
40 magnesium hydroxide
41 silver nitrate.

Equations

Word equations

In a chemical reaction the substances that react together are called **reactants** and the substances that are made are called **products**. The reactants are used up when products are produced. All reactions can be represented using word equations, which must contain an arrow (→).

reactants → products

For example, when magnesium burns the word equation is:

magnesium + oxygen → magnesium oxide
     reactants   *product*

or when hydrochloric acid reacts with calcium carbonate the word equation is:

hydrochloric acid + calcium carbonate → calcium chloride + water + carbon dioxide
     reactants   *products*

Tip

Check out page 213 for more information on writing word equations for acid reactions and page 237 for reactions of metals with water.

Test yourself

Complete the following word equations:

42 magnesium + hydrochloric acid
43 calcium carbonate + nitric acid
44 sulfuric acid + potassium
45 hydrochloric acid + sodium hydroxide
46 potassium hydroxide + nitric acid
47 magnesium carbonate + sulfuric acid
48 copper(II) oxide + hydrochloric acid
49 ammonia + hydrochloric acid
50 sodium + water
51 magnesium + oxygen
52 potassium + water
53 zinc + oxygen
54 aluminium + oxygen
55 copper + oxygen
56 sodium oxide + sulfuric acid
57 calcium hydroxide + nitric acid
58 potassium carbonate + hydrochloric acid
59 zinc + hydrochloric acid
60 nitric acid + magnesium hydroxide
61 aluminium + hydrochloric acid

It is useful to remember the general word equations shown in the box below.

Example

Acid reactions

acid + base → salt + water
e.g. hydrochloric acid + sodium hydroxide → sodium chloride + water

acid + metal → salt + hydrogen
e.g. sulfuric acid + magnesium → magnesium sulfate + hydrogen

acid + carbonate → salt + water + carbon dioxide
e.g. hydrochloric acid + potassium carbonate → potassium chloride + water + carbon dioxide

acid + ammonia → ammonium salt
e.g. nitric acid + ammonia → ammonium nitrate

Other reactions

element + oxygen → oxide of element
e.g. calcium + oxygen → calcium oxide

metal + water → metal hydroxide + hydrogen (for metals that react with cold water)
e.g. calcium + water → calcium hydroxide + hydrogen

metal + steam → metal oxide + hydrogen
e.g. zinc + steam → zinc oxide + hydrogen

How many atoms are present?

Sometimes in symbol equations you will see numbers **in front of** a formula. For example,

$$2NaOH + H_2SO_4 \rightarrow Na_2SO_4 + 2H_2O$$

The 2 in front of NaOH and in front of H_2O are used to balance the equation, as you will see in the next section, and it indicates that there are 2 molecules of NaOH reacting and 2 molecules of water produced. As a result, to calculate the number of atoms present in the 2 molecules of NaOH, multiply the atoms in NaOH by 2.

For example,

$Mg(OH)_2$ contains 1 atom of Mg, 2 atoms of O and 2 atoms of H

But $2Mg(OH)_2$ contains 2 atoms of Mg, 4 atoms of O and 4 atoms of H

$Ca(NO_3)_2$ contains 1 atom of Ca, 2 atoms of N and 6 atoms of O

But $3Ca(NO_3)_2$ contains 3 atoms of Ca, 6 atoms of N and 18 atoms of O

$Al_2(SO_4)_3$ contains 2 atoms of Al, 3 atoms of S and 12 atoms of O

But $2Al_2(SO_4)_3$ contains 4 atoms of Al, 6 atoms of S and 24 atoms of O

Test yourself

How many atoms of each element are there in the following?

62 $CaCl_2$
63 $Ca(OH)_2$
64 $2H_2SO_4$
65 $2HNO_3$
66 $3H_2$
67 $2CaSO_3$
68 $3Mg(OH)_2$
69 $(NH_4)_2CO_3$
70 $2Al(NO_3)_3$
71 $3K_2SO_4$
72 $4NaOH$
73 $2Al_2(CO_3)_3$
74 $4NaAl(OH)_4$
75 $2Fe_2(CO_3)_3$

Balancing equations

In a chemical reaction no atoms are lost or made but they are rearranged. This means that chemical equations have equal numbers of each atom on each side of the equation.

$$CaCO_3 \rightarrow CaO + CO_2$$

1	Ca	1
1	C	1
3	O	(1+2) = 3

The above equation is said to be **balanced**. It has the same elements on both sides and the same number of atoms of each element.

To balance an equation:

1. Count the number of atoms of each element on each side of the equation.
2. If the numbers of each atom of each element is the same on each side, then the equation is already balanced.
3. If the equation is not already balanced, put numbers in front of the formulae to balance – the numbers multiply the formula and so add in extra molecules to balance the equation.

Example

Balance the following equation:

$$Ca(OH)_2 + HNO_3 \rightarrow Ca(NO_3)_2 + H_2O$$

Answer

- Under the arrow list the symbols of all elements present and write the total number of atoms of each element on each side of the equation.

$$Ca(OH)_2 + HNO_3 \rightarrow Ca(NO_3)_2 + H_2O$$

1	Ca	1
5	O	7
3	H	2
1	N	2

- To balance the equation, put numbers in front of the formulae and readjust the totals.

$$Ca(OH)_2 + 2HNO_3 \rightarrow Ca(NO_3)_2 + H_2O$$

1	Ca	1
8 5	O	7
4 3	H	2
2 1	N	2

- Continue to balance by putting numbers in front and readjust the totals again.

$$Ca(OH)_2 + 2HNO_3 \rightarrow Ca(NO_3)_2 + 2H_2O$$

1	Ca	1
8 5	O	7 8
4 3	H	2 4
2 1	N	2

There is now the same number of each type of atom on each side of the equation, and it is balanced.

Example

Balance the equation:

$$Mg + O_2 \rightarrow MgO$$

Answer

- Under the arrow list the symbols of all the elements present and write the total number of atoms of each element on each side of the equation.

$$Mg + O_2 \rightarrow MgO$$

| 1 | Mg | 1 |
| 2 | O | 1 |

- To balance the equation, put numbers in front of the formulae and readjust the totals.

$$Mg + O_2 \rightarrow 2MgO$$

| 1 | Mg | 1 2 |
| 2 | O | 1 2 |

- Continue to balance by putting numbers in front and readjust the totals again.

$$2Mg + O_2 \rightarrow 2MgO$$

| 2 1 | Mg | 1 2 |
| 2 | O | 1 2 |

There is now the same number of each type of atom on each side of the equation, and it is balanced.

Tip

When balancing an equation, always start with an atom that is unbalanced and only appears in one formula on both sides of the equation. In this example start with N.

Tips

▶ Sometimes it is useful to use half numbers to balance an equation. For example,
$$C_4H_{10} + 6\tfrac{1}{2}O_2 \rightarrow 4CO_2 + 5H_2O$$

This equation can then be multiplied by 2 to get whole numbers:
$$2C_4H_{10} + 13O_2 \rightarrow 8CO_2 + 10H_2O$$

▶ *Never* change a formula when balancing equations.

Balance the following equations:
76 $Mg + N_2 \rightarrow Mg_3N_2$
77 $H_2 + Cl_2 \rightarrow HCl$
78 $MgO + HCl \rightarrow MgCl_2 + H_2O$
79 $N_2 + H_2 \rightarrow NH_3$
80 $Na + O_2 \rightarrow Na_2O$
81 $Ca + O_2 \rightarrow CaO$
82 $K + H_2O \rightarrow KOH + H_2$
83 $Ca + HCl \rightarrow CaCl_2 + H_2$
84 $NaOH + H_2SO_4 \rightarrow Na_2SO_4 + H_2O$
85 $Ca(OH)_2 + NH_4Cl \rightarrow CaCl_2 + NH_3 + H_2O$
86 $C_5H_{12} + O_2 \rightarrow CO_2 + H_2O$
87 $SO_2 + O_2 \rightarrow SO_3$
88 $Ca + HNO_3 \rightarrow Ca(NO_3)_2 + H_2$
89 $C_2H_5OH + O_2 \rightarrow CO_2 + H_2O$
90 $Ca + H_2O \rightarrow Ca(OH)_2 + H_2$
91 $Ca(OH)_2 + HNO_3 \rightarrow Ca(NO_3)_3 + H_2O$
92 $Fe + O_2 \rightarrow Fe_2O_3$
93 $Al + HCl \rightarrow AlCl_3 + H_2$
94 $C_2H_6 + O_2 \rightarrow CO_2 + H_2O$
95 $Mg(OH)_2 + HCl \rightarrow MgCl_2 + H_2O$
96 $Na_2CO_3 + HCl \rightarrow NaCl + H_2O + CO_2$
97 $CaCO_3 + HCl \rightarrow CaCl_2 + H_2O + CO_2$
98 $Na + Cl_2 \rightarrow NaCl$
99 $CH_4 + O_2 \rightarrow CO_2 + H_2O$
100 $Li + HNO_3 \rightarrow LiNO_3 + H_2$
101 $Al + O_2 \rightarrow Al_2O_3$
102 $Pb + O_2 \rightarrow Pb_3O_4$
103 $Na + H_2O \rightarrow NaOH + H_2$
104 $C_2H_4 + O_2 \rightarrow CO_2 + H_2O$
105 $NO + O_2 \rightarrow NO_2$
106 $Zn + HCl \rightarrow ZnCl_2 + H_2$
107 $KHCO_3 + H_2SO_4 \rightarrow K_2SO_4 + CO_2 + H_2O$

Writing balanced symbol equations

To write balanced symbol equations the method is:

Step 1 Write the formula for any compound that you know underneath the word equation (e.g. the formula for hydrogen, carbon dioxide, water or nitric acid).

Step 2 Work out any other formula by the swap and drop method and write it under the equation.

Step 3 Balance the equation.

Example

1 Sodium hydroxide + sulfuric acid → sodium sulfate + water

Step 1: _____ + H_2SO_4 → _____ + H_2O

Working out formula:
Na	OH		Na	SO$_4$
+1	-1		+1	-2
Same			2 ⤬ 1	
NaOH			Na_2SO_4	

Step 2: $NaOH + H_2SO_4 \rightarrow Na_2SO_4 + H_2O$

Step 3: $2NaOH + H_2SO_4 \rightarrow Na_2SO_4 + 2H_2O$

2 Magnesium carbonate + hydrochloric acid → magnesium chloride + water + carbon dioxide

Step 1: _____ + HCl → _____ + $H_2O + CO_2$

Working out formula:
Mg	CO$_3$		Mg	Cl
+2	-2		+2	-1
Same			1 ⤬ 2	
MgCO$_3$			MgCl$_2$	

Step 2: $MgCO_3 + HCl \rightarrow MgCl_2 + H_2O + CO_2$

Step 3: $MgCO_3 + 2HCl \rightarrow MgCl_2 + H_2O + CO_2$

Tips

▶ It is a good idea to have a rough work space to remind you to work out the correct formula.

▶ In general, if a balanced symbol equation is allocated 2 marks it does not need balancing – there is 1 mark for the left-hand side formula and one for the right. A 3 mark equation has an extra mark for the balancing.

Write balanced symbol equations for each of the following:
108 sodium carbonate + sulfuric acid → sodium sulfate + water + carbon dioxide
109 copper(II) carbonate + nitric acid → copper(II) nitrate + water + carbon dioxide
110 aluminium + hydrochloric acid → aluminium chloride + hydrogen
111 sodium + water → sodium hydroxide + hydrogen
112 calcium + water → calcium hydroxide + hydrogen
113 nitric acid + calcium hydroxide → calcium nitrate + water
114 ammonia + nitric acid → ammonium nitrate
115 ammonia + sulfuric acid → ammonium sulfate

116 copper(II) carbonate → copper(II) oxide + carbon dioxide
117 sodium oxide + sulfuric acid → water + sodium sulfate
118 calcium hydroxide + hydrochloric acid → calcium chloride + water
119 potassium carbonate + hydrochloric acid → potassium chloride + water + carbon dioxide
120 magnesium + hydrochloric acid → magnesium chloride + hydrogen
121 nitric acid + magnesium hydroxide → magnesium nitrate + water
122 aluminium + sulfuric acid → aluminium sulfate + hydrogen

State symbols

Chemical equations may include the following state symbols:

▶ (s) = solid
▶ (l) = liquid
▶ (g) = gas
▶ (aq) = aqueous solution.

The equation below includes state symbols:

$$CaCO_3(s) + 2HCl(aq) \rightarrow CaCl_2(aq) + H_2O(l) + CO_2(g)$$

You may be asked to include state symbols in any balanced symbol equation. Your data leaflet may be useful to help determine if a substance is soluble or not, and whether it is (s) or (aq).

Ionic equations

Ionic compounds consist of positive ions and negative ions strongly bonded together in a lattice. When an ionic compound dissolves in water the lattice breaks up and the ions separate out and become surrounded by water molecules (Figure A.3).

When dissolved, ionic compounds react. However, only some of the ions react; others do not react and remain unchanged. These ions are called **spectator ions**. Spectator ions are ions that do not take part in a reaction and are not shown in the ionic equation for the reaction. **Ionic equations** only show what happens to the ions that react.

Ionic equations can be written for different types of reactions:

▶ neutralisation
▶ precipitation
▶ displacement.

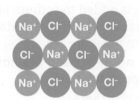
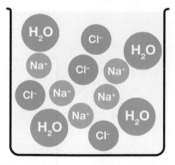

▲ **Figure A.3** Dissolving NaCl(s)

1 Neutralisation

When acids react with alkalis it is the hydrogen ions of the acid that react with the hydroxide ions of the alkali, to form water. The other ions are left unchanged.

For example, in the reaction of hydrochloric acid and sodium hydroxide the hydrogen ions and the hydroxide ions react. The sodium and chloride ions do not react, they are spectator ions and are left unchanged. This is shown in Figure A.4.

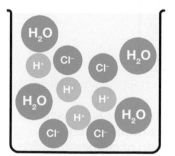
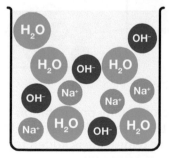

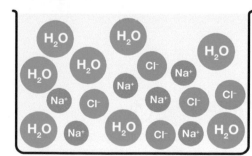

▲ **Figure A.4** Reaction of hydrochloric acid and sodium hydroxide

The ionic equation for neutralisation is:

$$H^+(aq) + OH^-(aq) \rightarrow H_2O(l)$$

From the acid *From the alkali*

When any acid reacts with any alkali the ionic equation is the same. Some examples are shown in Table A.8.

Table A.8

Example	What reacts	Spectator ions	Ionic equation
sulfuric acid (aq) + sodium hydroxide (aq)	H^+ ions from H_2SO_4 OH^- ions from NaOH	SO_4^{2-} ions from H_2SO_4 Na^+ ions from NaOH	$H^+(aq) + OH^-(aq) \rightarrow H_2O(l)$
hydrochloric acid (aq) + potassium hydroxide (aq)	H^+ ions from HCl OH^- ions from KOH	Cl^- ions from HCl K^+ ions from KOH	$H^+(aq) + OH^-(aq) \rightarrow H_2O(l)$
nitric acid (aq) + calcium hydroxide (aq)	H^+ ions from HNO_3 OH^- ions from $Ca(OH)_2$	NO_3^- ions from HNO_3 Ca^{2+} ions from $Ca(OH)_2$	$H^+(aq) + OH^-(aq) \rightarrow H_2O(l)$

2 Precipitation

When two solutions of ionic compounds are mixed, a solid precipitate often forms.

For example, when barium chloride solution is added to sodium sulfate solution the barium ions and the sulfate ions react and a precipitate of insoluble barium sulfate forms. This is shown in Figure A.5.

The chloride ions and the sodium ions do not react and are not included in the ionic equation. They are spectator ions.

The ionic equation for the reaction can be written as:

$$Ba^{2+}(aq) + SO_4^{2-}(aq) \rightarrow BaSO_4(s)$$

From   *From*
barium   *sodium*
chloride   *sulfate*

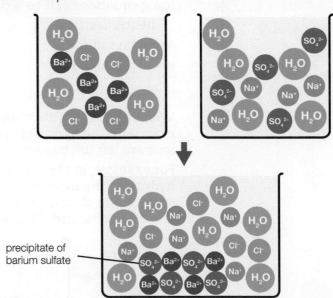

▲ **Figure A.5** Reaction of barium chloride and sodium sulfate to form a precipitate

precipitate of
barium sulfate

You will come across many precipitation reactions, including reactions that are used to test for ions (see Table A.8). For example, when a solution of sodium hydroxide is added to a solution of copper(II) sulfate, a blue precipitate of copper(II) hydroxide forms. The sodium ions and sulfate ions are spectator ions. The ionic equation for this reaction is:

$$Cu^{2+}(aq) + 2OH^-(aq) \rightarrow Cu(OH)_2(s)$$

From   *From*
copper(II)   *sodium*
sulfate   *hydroxide*

Tip

The overall electric charge of the ions in the reactants must equal the overall charge of the ions in the products. Hence in this equation two hydroxide ions are needed to balance the equation.

Table A.9

Example	What reacts	Ions that do not react	Ionic equation
barium chloride (aq) + sodium sulfate (aq)	Ba^{2+} ions from $BaCl_2$ SO_4^{2-} ions from Na_2SO_4	Cl^- ions from $BaCl_2$ Na^+ ions from Na_2SO_4	$Ba^{2+}(aq) + SO_4^{2-}(aq) \rightarrow BaSO_4(s)$
silver nitrate (aq) + potassium iodide (aq)	Ag^+ ions from $AgNO_3$ I^- ions from KI	NO_3^- ions from $AgNO_3$ K^+ ions from KI	$Ag^+(aq) + I^-(aq) \rightarrow AgI(s)$
copper(II) sulfate (aq) + sodium hydroxide (aq)	Cu^{2+} ions from $CuSO_4$ OH^- ions from NaOH	SO_4^{2-} ions from $CuSO_4$ Na^+ ions from NaOH	$Cu^{2+}(aq) + 2OH^-(aq) \rightarrow Cu(OH)_2(s)$

3 Displacement

A displacement reaction occurs when a more reactive metal displaces a less reactive metal from a metal compound. For example, when magnesium reacts with copper(II) sulfate, the magnesium atoms react with the copper ions in the copper(II) sulfate. The sulfate ions are spectator ions and do not appear in the ionic equation (see Table A.9).

The ionic equation for this reaction is:

$$Mg(s) + Cu^{2+}(aq) \rightarrow Cu(s) + Mg^{2+}(aq)$$

From copper(II) sulfate

In these reactions it helps to use the electric charges of the ions to balance the equation. For example, in the second example in Table A.10 two Ag^+ ions are needed giving an overall 2+ charge on the left-hand side of the equation to balance with the 2+ charge on the Fe^{2+} ion on the right-hand side of the equation.

Table A.10

Example	What reacts	Ions that do not react	Ionic equation
zinc (s) + copper(II) sulfate (aq)	Zn atoms in Zn metal Cu^{2+} ions from $CuSO_4$	SO_4^{2-} ions from $CuSO_4$	$Zn(s) + Cu^{2+}(aq) \rightarrow Zn^{2+}(aq) + Cu(s)$
iron (s) + silver nitrate (aq)	Fe atoms in Fe metal Ag^+ ions from $AgNO_3$	NO_3^- ions from $AgNO_3$	$Fe(s) + 2Ag^+(aq) \rightarrow Fe^{2+}(aq) + 2Ag(s)$
aluminium (s) + copper(II) chloride (aq)	Al atoms in Al metal Cu^{2+} ions from $CuCl_2$	Cl^- ions from $CuCl_2$	$2Al(s) + 3Cu^{2+}(aq) \rightarrow 2Al^{3+}(aq) + 3Cu(s)$

Test yourself

Write an ionic equation, with state symbols, for each of the following reactions. Decide if each reaction is a neutralisation, precipitation or displacement reaction.

123 hydrochloric acid (aq) + sodium hydroxide (aq)
124 nitric acid (aq) + potassium hydroxide (aq)
125 The formation of calcium hydroxide (s) from the reaction of calcium nitrate (aq) and sodium hydroxide (aq)
126 The formation of silver bromide (s) from the reaction of silver nitrate (aq) and sodium bromide (aq)
127 copper(II) sulfate (aq) + magnesium

128 silver nitrate (aq) + magnesium
129 The formation of iron(III) hydroxide (s) from the reaction of iron(III) sulfate (aq) and sodium hydroxide (aq)
130 The formation of barium sulfate (s) from the reaction of barium nitrate (aq) and zinc sulfate (aq)
131 The formation of lead(II) iodide (s) from reaction of lead(II) nitrate (aq) and potassium iodide (aq)
132 sulfuric acid (aq) + calcium hydroxide (aq)
133 phosphoric acid (aq) + sodium hydroxide (aq)
134 zinc sulfate (aq) + aluminium

Half equations

Half equations are equations that **include electrons**.

To write half equations the method is:

Step 1   Write down the reactant and product.

Step 2   Balance the atoms.

Step 3   Write the total charge underneath each species in the equation.

Step 4   Balance the charge by adding electrons.

The total charge on the left-hand side of the equation must equal the total charge on the right-hand side of the equation. This can be used to check that a half equation is balanced.

Example

Write a half equation for the conversion of a chlorine molecule into chloride ions.

Answer

Step 1 $Cl_2 \rightarrow Cl^-$

Step 2 $Cl_2 \rightarrow 2Cl^-$

Step 3 $Cl_2 \rightarrow 2Cl^-$

Charge   0   −2

Step 4   To balance the charge, add two electrons to the left-hand side:

$Cl_2 + 2e^- \rightarrow 2Cl^-$

Charge 0   −2   −2

−2 = −2

The charge on both sides is equal.

Example

Write a half equation for the conversion of hydrogen ions into a hydrogen molecule.

Answer

Step 1 $H^+ \rightarrow H_2$

Step 2 $2H^+ \rightarrow H_2$

Step 3 $2H^+ \rightarrow H_2$

Charge +2   0

Step 4   To balance the charge, add two electrons to the left-hand side:

$2H^+ + 2e^- \rightarrow H_2$

Charge +2   −2   0

0 = 0

The charge on both sides is equal.

Example

Write a half equation for the conversion of aluminium ions to aluminium.

Answer

Step 1 $Al^{3+} \rightarrow Al$

Step 2 $Al^{3+} \rightarrow Al$

Step 3 $Al^{3+} \rightarrow Al$

Charge +3   0

Step 4   To balance the charge, add three electrons to the left-hand side:

$Al^{3+} + 3e^- \rightarrow Al$

Charge +3   −3   0

0 = 0

The charge on both sides is equal.

Example

Write a half equation for the conversion of oxide ions into an oxygen molecule.

Answer

Step 1 $O^{2-} \rightarrow O_2$

Step 2 $2O^{2-} \rightarrow O_2$

Step 3 $2O^{2-} \rightarrow O_2$

Charge −4   0

Step 4   To balance the charge, add four electrons to the right-hand side:

$2O^{2-} \rightarrow O_2 + 4e^-$

Charge −4   0   −4

−4 = −4

The charge on both sides is equal.

Test yourself

135 Write a balanced half equation for each of the following conversions:

a) $Mg^{2+} \rightarrow Mg$

b) $S^{2-} \rightarrow S$

c) $K^+ \rightarrow K$

d) $Br^- \rightarrow Br_2$

e) $O^{2-} \rightarrow O_2$

f) $H^+ \rightarrow H_2$

g) $Li \rightarrow Li^+$

h) $Fe^{2-} \rightarrow Fe$

i) $Ca \rightarrow Ca^{2+}$

136 Write a balanced half equation for the conversion of:

a) iodide ions to an iodine molecule

b) an aluminium ion to an aluminium atom

c) nitride ions to a nitrogen molecule

d) a sodium ion to a sodium atom

e) an iron atom to an iron(III) ion

f) a copper(II) ion to a copper atom

g) chloride ions to a chlorine molecule.

Glossary

Abiotic factor A physical (non-living) factor which affects an organism

Absorption The ability of cells to take in substances through their cell membrane

Acceleration The rate at which the speed (or velocity) of a vehicle is changing.

Acid A substance that dissolves in water producing hydrogen ions ($H^+(aq)$)

Activation energy The minimum energy needed for a reaction to occur

Active immunity A type of immunity produced by the body producing antibodies

Active site The part of an enzyme molecule into which the substrate molecule fits because they have complementary shapes

Active uptake (transport) A process which uses energy released by respiration to move substances from a low concentration to a high concentration

Addition polymerisation is the process of joining monomer molecules together to form a long chain molecule.

Aerobic A chemical process which requires oxygen

Algal bloom The excessive growth of green algae on the surface of lakes and rivers

Alkali A soluble base

Allele One of two possible versions of a particular gene

Allotropes Different forms of the same element, in the same state

Alloy A mixture of two or more elements, at least one of which is a metal, and the resulting mixture has metallic properties

Alpha particle A particle emitted in radioactive decay and consisting of two protons and two neutrons.

Alveolus A single bubble-shaped air space in the lungs

Amino acids A group of twenty molecules which join together to form proteins

Amniocentesis A process in which foetal cells are obtained from the amniotic fluid and then examined for the presence of genetic abnormalities

Amnion The lining that contains the amniotic fluid

Amniotic fluid The fluid within the amnion, which cushions the foetus

Amplitude The maximum displacement of a particle in a wave from its undisturbed position.

Anaerobic A chemical process which can take place in the absence of oxygen

Analogy An argument that since there is similarity in some areas, there may be similarity in others.

Angioplasty The process in which dye is injected into the blood to allow examination of (diseased) blood vessels

Anhydrous A substance that does not contain water of crystallisation

Anion A negative ion

Anode The positive electrode

Antibiotic A chemical produced by fungi that kills bacteria

Antibiotic-resistance An antibiotic-resistant bacterium cannot be killed by (at least one type of) antibiotic

Antibody A structure produced by lymphocytes that has a complementary shape (and can attach to) the antigens on a particular microorganism

Anti-bumping granules are added to the mixture in the distillation flask to promote smooth boiling.

Antidiuretic hormone A chemical messenger molecule, produced in the brain which controls the amount of water reabsorbed by the kidneys

Antigen A distinctive marker on a microorganism that leads to the body producing specific antibodies

Aorta The blood vessel that carries oxygenated blood from the heart to the body

Apical Refers to the tip of a plant shoot or root

Aqueous solution A solution in which the solvent is water

Artery A blood vessel that carries blood under high pressure away from the heart

Aseptic technique The procedures used to prevent contamination when culturing microorganisms in the laboratory

Asteroid A very large rock found in space. In our Solar System, many asteroids are found between Mars and Jupiter.

Atomic number The number of protons in an atom

Auxin A plant hormone which controls the growth of plant cells

Average speed The ratio of the total distance travelled to the total time taken.

Axon The extension of the neurone that gives a neurone its long length (and its ability to transmit nerve impulses over a long distance)

Background radiation The radiation all around us coming mainly from outer space and some types of rock below the ground.

Bacteria A group of microorganisms which have a cell wall without cellulose and no nucleus

Balanced forces The forces on an object are balanced if the resultant force is zero.

Base A metal oxide or metal hydroxide that neutralises an acid to produce a salt and water

Base triplet A sequence of three bases in DNA that codes for a particular amino acid

Belt transect A method of sampling used when a habitat changes from one side to the other

Benign tumour A tumour that is surrounded by a capsule and does not spread around the body

Beta particle A fast electron emitted in radioactive decay from an unstable nucleus.

Big Bang Theory This is the theory that the universe began around 14 billion years ago following a huge explosion.

Biodiversity The variety of living organisms in an area

Biotic factor A factor caused by the living organisms in an area

Blood pressure The force exerted by blood in the heart and the circulatory system

Boiling point The temperature at which a liquid changes into a gas

Booster vaccination A second (additional) vaccination that is given in a vaccination programme to combat a particular disease

Breathing rate The number of breaths per unit time

Bronchitis The narrowing of the airways in the lungs, usually caused by smoking tobacco

Cancer Uncontrolled cell division

Capillary A very thin blood vessel through which the exchange of material between blood and cells takes place

Carbohydrate A type of food molecule, including sugars, starch and cellulose, that are made up of only carbon, hydrogen and oxygen and are the main source of energy in the diet

Carbon cycle The cycling of carbon-containing substances in the environment

Cardiac output The volume (amount) of blood the heart pumps per minute

Catalyst Something which speeds up the rate of a reaction but is unchanged by the reaction

Cathode The negative electrode

Cation A positive ion

Cell lysis The bursting of a cell due to too much water being taken in by osmosis

Cell membrane The membrane bounding the outside of living cells controlling the substances entering and leaving the cell

Cell polarity The existence of a positive and negative terminal in a battery.

Cell sap The liquid found in cell vacuoles

Cell The basic building block in the structure and function of living things

Cellulose A complex carbohydrate molecule found in plant cell walls

Cell wall A stiff layer outside the cell membrane of plant, bacterial and fungal cells which provides support

Central Nervous System The term referring to the coordinator in the nervous system, i.e. the brain and the spinal cord

Centre of gravity The centre of gravity is a point through which the whole weight of the body appears to act.

Cervix The opening of the uterus

Chemotherapy A form of cancer treatment in which drugs are used to kill cancer cells

Chlorophyll A green coloured chemical in chloroplasts which contains magnesium and is responsible for absorbing light energy during photosynthesis

Chloroplast A structure in the cytoplasm of plant cells which contains chlorophyll and carries out photosynthesis

Chromosomes Genetic structures usually occurring in functional pairs in the nucleus of cells (except gametes and bacteria)

Circulatory system The body system that includes the heart and blood vessels

Clinical trials The stages involved in testing drugs and medicines that use human volunteers

Cloning A laboratory process which uses a single cell to grow a group of genetically identical cells or organisms

CMBR Cosmic Microwave Background Radiation is the radiation left over from the Big Bang.

Coil of wire A length of wire wrapped in concentric rings or spirals.

Combustion The reaction of fuels with oxygen, forming oxides and releasing heat energy

Comet A satellite of the Sun consisting of rock and ice.

Communicable disease A disease that can be passed from one organism (person) to another

Community All the species normally found in a habitat

Compensation point Occurs in a plant when the environmental conditions [temperature, light intensity and carbon dioxide concentration] cause the rate of photosynthesis to equal the rate of respiration, resulting in no net gas exchange

Competition The interaction between two or more organisms as they try to get sufficient resources to survive

Complementary Shapes which fit into each other as in an enzyme and its substrate

Compound Two or more elements chemically combined

Compressions Places where the particles in a longitudinal wave are packed most tightly together.

Concave (diverging) lens A lens which is thinner at its centre than it is at its edges.

Concentrated acid contains a large number of acid particles dissolved per unit volume.

Concentration (g/dm³) The mass of a solid in grams dissolved in 1 dm³ of solution

Concentration gradient The difference in the concentration of a molecule or ion present in one area compared to another area

Concentration (mol/dm³) The number of moles of a solid dissolved in 1 dm³ of solution

Condensation The change of state from gas to liquid when cooled

Condom A barrier contraceptive method

Conduction Heat travelling through solids.

Conductor A material which allows electricity (or heat) to mass through it easily. Most conductors are metals.

Conservation of energy This is a fundamental principle of Physics. Energy can neither be created nor destroyed, but it can change its form.

Consumer An animal that gains its food from consuming other organisms

Continuity There is continuity where two waves join together in time and space.

Continuous variation The type of variation characterised by gradual change in a characteristic across a population

Contraception A method used to try to avoid pregnancy

Contraceptive pill A contraceptive tablet containing hormones and prevents pregnancy by affecting hormone levels thereby preventing eggs being released

Convection Heat travelling through fluids.

Conventional current The imagined flow of electrical charge from the positive terminal of a battery towards the negative terminal through a circuit.

Convex (converging) lens Commonly called a magnifying glass, this lens has surfaces which bulge outwards like the exterior of a sphere.

Coordinator A part of the nervous system (i.e. brain or spinal cord) that links receptors and effectors

Cornea The transparent front part of the eye that covers the iris and pupil and in which most refraction occurs.

Coronary artery The very narrow arteries that supply the heart tissue with glucose and oxygen

Covalent bond A shared pair of electrons

Cracking is the breakdown of larger saturated hydrocarbons (alkanes) into smaller more useful ones, some of which are unsaturated (alkenes).

Crude oil (petroleum) A mixture containing mainly alkane hydrocarbons

Cuticle A waxy layer on the outer surface of a plant epidermis cell, allowing light through while reducing water loss

Cystic fibrosis A genetic condition caused by having two recessive alleles of a particular gene

Cytoplasm The contents of a cell between the cell membrane and the nucleus where chemical reactions take place

Decomposer An organism which breaks down the tissues of dead organisms and excretory products

Deforestation The cutting down of trees

Delocalised electrons Electrons that are free to move throughout a whole structure

Denaturation An irreversible change in the shape of an enzyme which means it is no longer complementary to the substrate and cannot catalyse the reaction

Denitrification A process carried out by bacteria in anaerobic conditions which changes nitrates into nitrogen gas

Density The density of a material takes into account the size of its atoms/molecules and how tightly packed they are.

Desalination The process of removing dissolved substances from sea water

Destarch A procedure which removes stored starch from plants

Diabetes (type 2) A disease that develops in older people, associated with poor diet and obesity, in which they have difficulty controlling the concentration of their blood glucose because the insulin they produce does not work effectively

Diaphragm A muscular sheet between the thorax and abdomen which helps bring about breathing

Diatomic Two atoms covalently bonded in a molecule

Differentiate The ability of a cell to change into a specialised cell with adaptations to a particular function

Diffuse reflection Reflection from a rough surface.

Diffusion The movement of molecules or ions from where they are in high concentration to where they are in a lower concentration

Digestion Enzymes breaking down large, complex food molecules into small, simple soluble ones which can be absorbed

Dilute acid contains a small number of acid particles dissolved per unit volume.

Diploid The normal chromosome number

Discontinuous variation The type of variation in which all the individuals can be clearly divided into two or more groups and there are no intermediate states

Dispersion The breaking up of light into its component colours (or wavelengths).

Displacement Displacement is distance in a specified direction.

Distance Distance is the separation between two points.

Distillate The liquid that is cooled from the vapour and collected during distillation

Distribution Passing electrical energy from electricity pylons to homes and factories where it is used.

DNA The molecule that forms genes and chromosomes

Dominant In the heterozygous condition, the dominant allele will override the recessive allele

Doppler effect This is the change in the observed wavelength of light (or sound) due to the movement of the source.

Double circulation The type of circulation in humans in which blood travels through the heart twice for each complete circulation of the body

Double helix The structure of DNA

Double insulation This is when electrical equipment is encased in a box of insulating plastic so that users cannot come into contact with live electrical components.

Down's Syndrome A genetic condition in humans caused by having one extra (47) chromosome

Downstreaming The term used to describe the extraction, purification and packaging of insulin during genetic engineering

Ductile The metal can be drawn into a wire

Dynamo A device which converts the kinetic energy of a moving magnet into electricity.

Earth wire A wire of low resistance connected to the metal frame of mains-operated electrical equipment which passes through a plug and ultimately is connected electrically to the ground.

Echoes Reflections of waves, usually sound waves.

Ecosystem An area in which a community of organisms interact with each other and their physical surroundings

Effector An organ (a muscle or gland) which responds to a stimulus by causing a change

Efficiency The fraction of the total input energy into a machine that is actually useful.

Egested The removal of faeces

Egg (ovum) A female gamete

Electric current The flow of charged particles in a circuit.

Electrolysis The decomposition of electrolytes using electricity. An electrolyte is an ionic liquid or solution that conducts electricity and is decomposed during the process.

Electromagnetic induction Producing an electrical voltage by changing the magnetic field near a conductor.

Electromagnetic waves (or spectrum) A family of transverse waves which can all travel trough a vacuum at enormously high speed.

Electron microscope A microscope which uses a beam of electrons instead of light and allows much greater detail of cells to be visible

Electron Negatively charged particle orbiting the nucleus of an atom.

Element A substance that consists of only one type of atom

Elliptical Not circular, but more like the shape of a rugby ball.

Emphysema Damage to the gas exchange surfaces in the lungs, usually caused by cigarette smoke

Empirical formula The simplest whole number ratio of atoms of each element in a compound

Endothermic reaction A reaction in which heat energy is taken in

Energy flow The transfer of energy between the trophic levels in food chains and food webs

Environment The conditions which surround and effect an organism in their habitat

Enzyme specificity The ability of an enzyme to catalyse only one type of substrate

Epidermis An outer layer of cells

Equilibrium A body is in equilibrium when both the resultant force and resultant turning effect on it are zero.

Eutrophication The process occurring in areas of fresh water which have too many nitrates, leading to the death of animal species

Evaporation The change of state from liquid to gas when heated

Evolution The change in a species over time–evolution also leads to the formation of new species

Excretory system An organ system which removes wastes from the body

Exothermic reaction A reaction in which heat energy is given out

Extended length The extended length is the length of the spring when loaded.

Extension Extension = extended length – natural length

Extinction A species is extinct if there are no living members of that species left

Fatty acids Long chain molecules which join to glycerol to form fats and oils (lipids)

Female sterilisation The cutting of the oviducts to prevent pregnancy

Fermentation The breakdown of sugars to produce ethanol and carbon dioxide

Fertilisation The fusion (joining) of a haploid sperm cell and a haploid egg cell to form a diploid zygote

Fertiliser A substance, either a waste product from animals (natural fertiliser) or produced by the chemical industry (artificial fertiliser), which provides plants with the chemicals they need to grow

Field of view The circle of light visible in a light microscope

Filtrate The liquid which passes through the filter paper during filtration

Finite resource A resource that once used, cannot be replaced in a human lifetime

Fission Some heavy nuclei, like those of uranium, can actually be forced to split into two lighter nuclei, emitting energy.

Flagellum The tail of a sperm used for movement

Focal length The distance between the principal focus and the centre of a convex lens.

Food chain A sequence of organisms which feed off each other, passing nutrients and energy

Food web A number of interlinked food chains

Formulation A mixture that has been designed as a useful product. It is formed by mixing together several different substances in carefully measured quantities to ensure the product has the required properties.

Fossil The remains of a living organism that has been preserved (usually in rocks) for millions of years

Fraction The distillate collected at each different temperature in fractional distillation

Free electrons Negatively charged particles which are not attached to atoms and are responsible for the conduction of electricity (and heat) in metals.

Frequency The number of wavelengths which pass a fixed point in a second.

Functional group A reactive group in a molecule

Fuse A safety device consisting of a short length of fine fire which melts if too much current flows through it, thus breaking an electrical circuit.

Fusion Energy is emitted when two light nuclei are fused together.

Galaxy A galaxy is a collection of stars held together by gravity. Our galaxy is called the Milky Way and contains at least 100 thousand million stars. It is thought that there are at least 2 million million galaxies in the Universe.

Gamete Sex cell that contains only one chromosome from each pair

Gamma ray A high energy electromagnetic wave emitted in radioactive decay from an unstable nucleus.

Gas exchange The movements of gases between an organism and its surroundings

Gas giant planets Planets whose surface is a very dense gas like Jupiter, Saturn Uranus and Neptune. These planets are much bigger than the rocky planets.

Gene Short section of DNA (chromosome) that codes for a particular characteristic

Genetic condition A condition caused by problems in genes or chromosomes

Genetic engineering The deliberate modification of the genome (DNA) in an organism to introduce desirable effects

Genetic screening A process used to test people for the presence of particular harmful alleles or other genetic abnormalities

Genome The entire genetic material (all the DNA) in an organism

Genotype The genetic make-up of an organism represented by symbols (letters), e.g. tt

Giant covalent structure A three-dimensional structure of atoms that are joined by covalent bonds

Giant ionic lattice A three-dimensional structure of oppositely charged ions held together by ionic bonds

Global warming The rise in the Earth's temperature

Glycerol A molecule which makes up part of all fats (lipids)

Glycogen A complex carbohydrate molecule used for storage in animals

Gravitational collapse During star formation a nebula gets smaller and smaller in volume as its particles are pulled together by gravity. This is gravitational collapse.

Gravitational field strength This is the size of the force on a mass of 1 kg, placed at a particular point in a gravitational field.

Gravitational potential energy The energy which a body by virtue of its height above the Earth's surface.

Group A vertical column in the Periodic Table

Habitat The place where an organism or population is normally found

Haemoglobin A molecule containing iron that is found in red blood cells

Haemophilia A sex-linked conditions that is almost exclusively found in males

Half-life The half-life of a radioactive material is the time taken for its activity to fall to half of its original value.

Haploid A cell or nucleus with half the normal number of chromosomes

Heart disease Any disease which prevents the heart from functioning normally

Heart valve A structure in the heart that prevents the backflow of blood

Hertz The unit of frequency. 1 Hertz (abbreviation Hz) is equal to 1 vibration per second.

Heterozygous The two alleles of a gene are different (one dominant and one recessive allele)

Homeostasis The ability of the body to maintain an almost constant internal environment

Homologous series A family of organic molecules that have the same general formula, show similar chemical properties, show a gradation in their physical properties and differ by a 'CH$_2$' unit

Homozygous The two alleles of a gene are the same

Huntington's disease A genetic condition caused by the presence of a dominant allele of a particular gene

Hydrated A substance that contains water of crystallisation

Hydrocarbon A molecule /compound consisting of carbon and hydrogen atoms only

Ileum The longest part of the small intestine

Immiscible liquids Liquids which do not mix together but form two distinct layers

Immunity Freedom from disease

Immunotherapy A form of treatment of cancer in which antibodies are injected into the patient. The antibodies attach to cancer cells and help the body's immune system destroy the cancer cells

Implantation The term describing the attachment of the ball of cells (embryo) following fertilisation to the uterus lining

Independent assortment A process that takes place during meiosis, in which chromosomes are reassorted in the formation of gametes

Indicator A substance which is one colour in acid and a different colour in alkali

Indicator species A species which can be used to monitor the level of pollution in a habitat

Infertility The inability of someone to have children

Inhibitor A molecule which fits into the active site of an enzyme and stops the normal substrate entering so reducing the reaction rate of the enzyme

Insoluble substance One which does not dissolve in a solvent

Insulator A material which does not allow electricity (or heat) to mass through it easily. Gases and most liquids are insulators.

Intercellular space A space between cells as in a leaf

Intercostal muscles The muscles in the chest wall that contract and cause the ribs to move out thereby increasing the volume of the thorax during breathing

***In-vitro* fertilisation** Fertilisation outside the body

***In-vitro* testing** Testing of medicines and drugs in the laboratory

Ion A charged particle formed when an atom gains or loses electrons

Ionic bond The attraction between oppositely charged ions

Isotopes Atoms of an element with the same atomic number but a different mass number, indicating a different number of neutrons

Joule heating Heating caused by an electric current flowing in a conductor.

Joule's law An equation for the amount of heat energy per second produced by an electric current, sometimes written $P = IV = I^2R = V^2/R$.

Kilowatt hour (kWh) The amount of electrical energy used by a device of power 1000 W in 1 hour. An amount of energy equivalent to 3 600 000 J.

Kinetic energy The energy of movement.

Lamina A lamina is a body in the form of a flat thin sheet.

Laterally inverted image An image that is inverted from left to right, like an image seen in a mirror.

Law of reflection The angle of incidence is always equal to the angle of reflection.

Left atrium The heart chamber that receives blood from the lungs

Left ventricle The heart chamber that pumps blood around the body

Legume A group of plants including peas, beans and clover

Lethargy Tiredness, a total lack of energy

Leukaemia A type of cancer in which some types of blood cells increase out of control

Light year The distance travelled by a beam of light in one year. This distance is roughly 9.46 million million kilometres or 5.88 million million miles.

Limiting factors Any factor which affects a chemical process such as photosynthesis and is at a level less than optimum will slow or limit the process

Lipid A fat or oil molecule

Live wire A wire in an AC system in which the voltage can be dangerously high.

Lock and key model Model used to explain how an enzyme reacts with its substrate

Lone pair An unbonded pair of electrons

Longitudinal wave A wave in which the particles vibrate parallel to the direction in which the wave is moving.

Lymphocyte A type of white blood cell that produces antibodies

Lysis The process that describes the rupturing of an animal cell when it takes in too much water by osmosis

Magnification The number of times the length of an image is larger than the actual length of the object

Main sequence The main part of a star's life. Our Sun is in the main sequence part of its life-cycle.

Male sterilisation The cutting of the sperm tubes (vasectomy) to prevent pregnancy

Malignant tumour A tumour that is not surrounded by a capsule and capable of spreading around the body

Malleable Can be hammered into shape without breaking

Mass Mass is the amount of matter in a body.

Mass number The total number of protons and neutrons in an atom

Meiosis A type of cell division that produces cells (gametes) that have half the normal chromosome number (haploid cells)

Melting point The temperature at which a solid changes into a liquid

Memory lymphocyte A special type of lymphocyte that can remain in the body for many years and produce antibodies quickly when required

Menstrual cycle The monthly cycle in females of reproductive age that prepares the body for pregnancy

Mesophytic Land plants which grow in temperate climates

Metallic bonding The attraction between the positive ions in a regular lattice and the delocalised electrons

Metallic lattice A three-dimensional structure of positive ions and delocalised electrons bonded by metallic bonds

Milliampere or mA A milliampere is one-thousandth of an ampere. 1 mA = 0.001 A.

Miscible liquids Liquids that mix together, such as alcohol and water

Mitochondria Structures in the cytoplasm where the reactions of respiration occur

Mitosis A type of cell division that produces cells genetically identical to the parent cell and to each other

Mixture Two or more substances mixed together

Molecular formula The actual number of atoms of each element present in a compound

Molecular ion A charged particle containing more than one element

Molecule Two or more atoms covalently bonded together

Moment of a force This is the turning effect of the force about a pivot.

Monochromatic light Light of a single colour (or single wavelength).

Monomer A small molecule that combines with other monomers to make a polymer

MRSA A type of bacterium that is resistant to most antibiotics

Multi-celled An organism made up of many cells

Mutation Random change in the number of chromosomes or type of gene

Nanoparticle A structure that is 1–100 nm in size and contains a few hundred atoms

Natural length The natural length is the normal length of the spring without a load on it.

Natural selection The process in which the better adapted individuals survive (at the expense of the less well adapted individuals) and pass on their genes to their offspring

Nebula A nebula is a huge cloud of gas and dust.

Negative feedback A process involved in homeostasis which, by constant monitoring of an internal factor in the body, causes any change to be reversed, bringing the factor back to normal values

Nerve impulses Small electrical charges that pass along neurones

Neurones The cells (also called nerve cells) of the nervous system

Neutralisation The reaction between the hydrogen ions in an acid and the hydroxide ions in an alkali to produce water

Neutral wire A wire in an AC system in which the voltage is 0 V.

Neutron An uncharged particle (that is, a neutral particle) found in the nucleus of every atom (except one form of hydrogen).

Nicotine The addictive substance in tobacco smoke, which also affects heart rate

Nitrifying bacteria Bacteria that convert ammonium compounds to nitrates in the nitrogen cycle

Nitrogen fixation A process which allows bacteria to change nitrogen gas into nitrates

Non-biodegradeable material A material which is not decomposed by natural bacteria in the environment

Non-renewable energy Energy that we can never replace, so we will eventually run out of it. Fossil fuels are non-renewable.

Nuclear fusion Nuclear fusion is the joining together of two or more light nuclei (such as hydrogen) to form a heavier nucleus (such as helium) with the release of vast quantities of energy. Nuclear fusion is the process by which stars, like our Sun, get their energy.

Nuclear membrane The membrane surrounding the nucleus

Nucleus The large structure in a cell containing chromosomes

Nucleus The tiny central part of every atom where most of the atom's mass is to be found. (Do not confuse this with the nucleus of a cell!)

Nutrient cycle The cycling of substances (e.g. nutrients and elements) in the environment

Obesity Being extremely overweight

Oestrogen The female sex hormone produced by the ovaries, which both causes the repair and build-up of the uterus lining following menstruation and stimulates ovulation

Ohm's law A mathematical relationship linking voltage, current and resistance, sometimes written V = IR.

One mole of a substance has a mass in grams numerically equal to the relative formula mass or relative atomic mass.

Optimum dosage The best amount or concentration of a drug or medicine to use when treating patients

Optimum The value of a factor which allows a reaction to happen at its fastest rate

Ore A rock that contains a metal compound from which the metal can be extracted

Osmoregulation The ability to control the amount of water in the body

Osmosis The diffusion of water molecules from a dilute solution to a more concentrated solution through a selectively permeable membrane

Ovary The female organ that produces eggs (ova)

Oviduct The structure that carries eggs (ova) from the ovary to the uterus

Ovulation The release of an egg by an ovary

Oxidation A reaction in which a substance loses electrons

Palisade mesophyll The upper of the two layers of mesophyll cells in the centre of a leaf, closely packed end on to the upper surface and containing many chloroplasts. They are the main site of photosynthesis

Passive immunity A type of immunity produced by injecting antibodies

Pedigree diagram A diagram that shows how a particular condition is inherited through the different generations in a family

Peer review A process of validation in which other scientists in the same field review research and provide feedback and/or suggest refinements

Penicillin The first antibiotic developed

Penis Organ that introduces sperm into the vagina

Percentage cover A method of estimating the amount of a plant species in a quardat

Period A horizontal row in the Periodic Table

Period The time it takes for one wavelength to pass a fixed point.

Petrochemicals Chemicals made from petroleum (crude oil) and natural gas

Phagocyte A type of white blood cell that destroys microorganisms by engulfing them and digesting them (phagocytosis)

Phenotype The outward appearance of an individual, e.g. tall

Photosynthesis The chemical process in green plants which uses light energy to convert carbon dioxide and water into sugars and oxygen

Phototropism A growth movement in plants in response to light

Phytomining The use of plants to absorb metal compounds from soil as part of metal extraction

Placenta The structure that links the uterus wall to the foetus via the umbilical cord. It is here that exchange of materials takes place between the mother and the foetus

Plane waves A wave whose wavefronts are in infinite parallel planes.

Plasma The liquid part of the blood that functions as a transport medium

Plasmid A small circular ring of DNA in a bacterium

Plasmolysis A plant cell is plasmolysed when it has lost water by osmosis and its membrane separates from the cell wall

Platelets Blood components that help convert fibrinogen to fibrin in the processes of blood clotting and scab formation

Pleural membranes These membranes line the outside of the lungs and the inside of the chest wall

Polymer A long chain molecule made from joining small molecules together

Population A group of organisms of the same species living in an area

Potable water Water that is safe to drink

Potometer A piece of apparatus that can be used to compare rates of transpiration in different conditions and to investigate the factors affecting the rate of water uptake by a plant

Power Rate of doing work. Power is usually expressed in watts or joules per second.

Precipitate A solid formed when two solutions are mixed

Preclinical trials The stages involved in testing drugs and medicines that occur before testing on human volunteers

Pressure Pressure is the ratio of the normal force to area of contact.

Primary response The response of the immune system to the first exposure of a particular type of microorganism / antigen

Principal focus A point on the principal axis (PA) through which rays of light parallel to the PA all pass after refraction in the lens.

Producer An organism that produces food, a plant, which is at the start of a food chain

Product A molecule produced during a chemical reaction

Progesterone The female hormone that maintains the build-up of the uterus lining and prepares the uterus for pregnancy

Prostate gland The male gland that adds fluid to nourish the sperm

Protein A type of food molecule, made up of carbon, hydrogen, oxygen and nitrogen, formed by long chains of amino acids that are important in the structure and functioning of cells

Proton A positively charged particle found in the nucleus of every atom.

Protostar During star formation the collapsed nebula gets very, very hot vat its centre. At this stage it is a protostar.

Pulmonary artery The blood vessel that carries deoxygenated blood from the heart to the lungs

Pulmonary vein The blood vessel that carries oxygenated blood from the lungs to the heart

Punnett square A grid (table) used to work out the offspring in a genetic cross

Pure substance A single element or compound not mixed with any other substance

Putrefying bacteria Decomposing bacteria

Quadrat An apparatus, usually a square frame, used to sample an area

RADAR The use of RAdio waves for Detection And Ranging.

Radiation Heat travelling in self-supporting electromagnetic waves.

Radioactivity The decay of unstable nuclei by the emission of alpha particles, beta particles or gamma rays.

Radiotherapy The use of high energy X-rays in the treatment of cancer

Random sampling A method of sampling in which the position of each sample does not depend on the position of the previous sample and the overall sample of the area being studied is unbiased, representative

Rarefactions Places where the particles in a longitudinal wave are least tightly packed together.

Ray diagram A diagram which shows how rays of light form an image.

Reagent A chemical used as a test for the presence of a particular substance

Real image The apparent reproduction of an object, formed by the intersection of real rays of light. Real images can be projected onto a screen.

Receptor A structure which can detect a change (stimulus) in the environment

Recessive An allele that will only show a characteristic if both alleles are present (and there is no dominant allele present)

Recovery rate The time taken for the breathing or heart rate to return to normal after exercise

Red blood cells Blood cells that carry oxygen around the body

Redox reactions Reactions in which oxidation and reduction occur at the same time

Red shift The light from distant galaxies appears to have a longer wavelength than we would expect. This increase in wavelength is called red shift.

Reduction A reaction in which a substance gains electrons

Reflected waves Waves which are returned from a barrier (such as a mirror) back into the medium from which they came.

Reflex action A very fast response to a stimulus by means of a nervous pathway involving a small number of nerve cells

Refraction Where a wave passes from one material into another and a change in the wave speed results in a change in direction.

Relative atomic mass The mass of an atom compared with that of the carbon-12 isotope, which has a mass of exactly 12

Relative formula mass A weighted mean of the mass numbers of a substance

Reliable results Results which, when repeated, are consistent

Renewable energy Energy that is produced by nature in less than a human lifetime, so we will never run out of it. Wind energy is renewable.

Representative sample A sample which has all the same characteristics as the whole population

Residue The solid that remains on the filter paper after filtration

Resistance The opposition by a material to the flow of electrical current. Conductors have a lower resistance than insulators.

Resolution The ability of a microscope to distinguish detail in an image

Respiration Respiration is the release of energy from food

Response The action of an effector

Restriction enzymes Enzymes used in genetic engineering that cut DNA at particular positions

Resultant force The net force on an object which causes it to accelerate.

Reversible reaction One in which the products, once made, can react to reform the reactants

Right atrium The heart chamber that receives blood from the vena cava

Right ventricle The heart chamber that pumps blood to the lungs

Rocky planets Planets which have a rocky surface such as Mercury, Venus, Earth and Mars.

Salt The compound formed when some or all of the hydrogen ions of an acid are replaced by metal or ammonium ions

Sample A small part of an area or population

Saprophyte A bacterium or fungus which decomposes material by releasing enzymes onto the surface and absorbing the breakdown products

Satellite A satellite is an object which orbits another object. The Earth is a satellite of the Sun. The Moon is a satellite of the Earth.

Saturated (in the context of organic chemistry) All the carbon–carbon bonds are single

Saturated solution One in which no more solid can dissolve at that temperature

Scalar A quantity which has magnitude only. Examples of scalar quantities are mass, volume, length etc.

Scale bar A line drawn on or near a magnified image showing a given length magnified by the same amount

Scrotum Sac that holds and protects the testes

Secondary response The rapid immune response upon a second (or additional) infection by a particular microorganism

Selective breeding The selection and subsequent breeding of organisms chosen by man for their desirable characteristics

Selectively permeable The ability of a membrane to allow some substances to pass through while preventing others

Sex chromosome One of the two chromosomes that determines the sex of an individual

Sex linkage The way in which certain genetic conditions are more likely to affect a particular sex

Side effect An unwanted or unplanned effect of a drug on a person

SI units The International System of units based on the metre, kilogram and second

Solar System The Sun and the planets, comets, asteroids and everything else which orbits it.

Solubility curve A graph of solubility in g/100 g water (y axis) against temperature in °C (x axis)

Solubility is the mass of solid which saturates 100 g of water at a particular temperature

Soluble substance One which will dissolve in a solvent

Solute The substance that dissolves in a solvent

Solution A solute dissolved in a solvent

Solvent front The furthest distance travelled by the solvent

Solvent The liquid in which a solute dissolves

Sonar The use of SOund for Navigation And Ranging.

Specialised A cell that has adaptations to a particular function

Speed Rate of change of distance with respect to time.

Sperm A male gamete (sex cell) formed by meiosis

Sperm tubes The structures that carry sperm from the testes to the penis

Spongy mesophyll The lower of the two layers of mesophyll cells in the centre of a leaf, loosely arranged with airspaces. They are the main site of gas exchange

Stain A chemical which, when added to cells, colours some parts of the cells more than others making the structures more noticeable

Standard solution A solution of known concentration

Stellar Stellar means having to do with stars.

Stem cells Simple cells in animals and plants that can continue to divide to produce more stem cells which in turn can change into one or several types of specialised cells

Stent A small mesh-like structure that is inserted into a blood vessel to keep the lumen open

Sticky end The term used to describe the overlapping (and non-paired) strand that is left when DNA is cut by a restriction enzyme

Stimulus A feature of the environment that stimulates a receptor in the nervous system

Stomata The small pores in the surface of plant leaves

Stroke A type of cardiovascular disease that affects the brain

Strong acids Acids that are completely ionised in water

Strong alkalis Alkalis that are completely ionised in water

Sublimation Change of state from solid directly to gas on heating

Substrate A molecule that is acted upon by an enzyme

Superbug A type of bacterium that is resistant to a number of antibiotics

Surface area A measure of the external boundary of an object, a cell or an organism that is exposed to the environment

Surface epithelium cells A single layer of cells covering the outer surface of multi-celled organisms

Sustainable woodlands Woodlands harvested at a rate which allows them to continue growing without damaging the environment

Synapse The small junction (gap) between adjacent neurones

Terminal velocity When a body falls through a fluid balanced forces act on it and it falls at a constant velocity.

Testes The structure that produces sperm in males

Testosterone The male sex hormone produced by the testes

Thermal decomposition Breakdown of a solid using heat

Thermostable enzymes Enzymes which are able to function over a wide range of temperatures without being broken down

Thorax The area between the lungs and the chest wall

Transformer A device which converts high voltages to low voltages and vice versa.

Transmission Passing electrical energy from the generator in a power station into the wires attached to electricity pylons.

Transmitter substance A chemical which diffuses across a synapse

Transpiration The evaporation of water from mesophyll cells followed by diffusion through air spaces and stomata

Transverse wave A wave in which the particles vibrate perpendicular to the direction in which the wave is moving.

Trophic level The level at which an organism feeds in a food chain or web

Turbine A machine in which the wind energy or steam is used to make a large propeller turn round.

Turgid (turgor) The state of a plant cell when it has gained enough water by osmosis for the cell membrane to push against the cell wall making the cell firm. Turgor provides support in plants

Ultrasound waves Sound waves which humans cannot hear because its frequency is too high (above 20 000 Hz).

Umbilical cord The structure containing blood vessels that links the placenta to the foetus

Urethra The tube through which the sperm leaves the penis

Uterus The female organ in which the foetus will develop if pregnancy occurs

Vaccination The injection of dead or modified pathogens (disease-causing microorganisms) with the purpose of raising antibody and memory lymphocyte levels in the blood

Vacuole A liquid filled space in the cytoplasm of a cell which is large and permanent in plant cells

Vagina The part of the female reproductive system into which sperm is deposited during sexual intercourse

Validity The validity of experimental results depends on whether the methods used are actually testing the question asked

Vasectomy (Male sterilisation) A contraceptive method in which the sperm tubes are cut

Vector A quantity which has magnitude and direction. Examples of vector quantities are velocity, acceleration, force, etc.

Vein A blood vessel that carries blood back to the heart

Velocity Rate of change of displacement with respect to time.

Vena cava The vein that returns deoxygenated blood to the heart

Vibration Regular to-and-fro motion of the particles in a medium.

Villi Small finger-like projections lining the wall of the ileum which increase its surface area for absorption

Virtual image An image from which rays of light only appear to diverge, but no light passes through it. Virtual images cannot be projected on to a screen.

Voltage or potential difference or PD Voltage and potential difference mean the same thing. Voltage causes an electric current to flow.

Volume A measure of the amount of space an object, a cell or organism occupies

Voluntary action A response to a stimulus which involves thinking

Water of crystallisation Water that is chemically bonded into the crystal structure

Wavelength The distance between two successive crests or troughs of a transverse wave or the distances between the centres of two compressions of a longitudinal wave.

Wave speed The distance travelled by a wave in a second.

Weak acids Acids that are partially ionised in water

Weak alkalis Alkalis that are partially ionised in water

Weight Weight is a force and is a measure of the size of the gravitational pull on an object exerted, in our case, by the Earth.

White blood cells Blood cells that help defend against disease

Work Work is the product of the force and the distance moved in the direction of the force. Work is measured in Joules.

Zygote The first cell of the new individual following fertilisation

Index

THE PERIODIC TABLE OF ELEMENTS

Group

1	2												3	4	5	6	7	0

Key:

a
x
b

a = relative atomic mass (approx)
x = atomic symbol
b = atomic number

*58 – 71 Lanthanum series
†90 – 103 Actinium series

1 **H** Hydrogen 1																			4 **He** Helium 2

Period / rows:

| 7 **Li** Lithium 3 | 9 **Be** Beryllium 4 | | | | | | | | | | | | 11 **B** Boron 5 | 12 **C** Carbon 6 | 14 **N** Nitrogen 7 | 16 **O** Oxygen 8 | 19 **F** Fluorine 9 | 20 **Ne** Neon 10 |

| 23 **Na** Sodium 11 | 24 **Mg** Magnesium 12 | | | | | | | | | | | | 27 **Al** Aluminium 13 | 28 **Si** Silicon 14 | 31 **P** Phosphorus 15 | 32 **S** Sulfur 16 | 35.5 **Cl** Chlorine 17 | 40 **Ar** Argon 18 |

| 39 **K** Potassium 19 | 40 **Ca** Calcium 20 | 45 **Sc** Scandium 21 | 48 **Ti** Titanium 22 | 51 **V** Vanadium 23 | 52 **Cr** Chromium 24 | 55 **Mn** Manganese 25 | 56 **Fe** Iron 26 | 59 **Co** Cobalt 27 | 59 **Ni** Nickel 28 | 64 **Cu** Copper 29 | 65 **Zn** Zinc 30 | 70 **Ga** Gallium 31 | 73 **Ge** Germanium 32 | 75 **As** Arsenic 33 | 79 **Se** Selenium 34 | 80 **Br** Bromine 35 | 84 **Kr** Krypton 36 |

| 85 **Rb** Rubidium 37 | 88 **Sr** Strontium 38 | 89 **Y** Yttrium 39 | 91 **Zr** Zirconium 40 | 93 **Nb** Niobium 41 | 96 **Mo** Molybdenum 42 | 98 **Tc** Technetium 43 | 101 **Ru** Ruthenium 44 | 103 **Rh** Rhodium 45 | 106 **Pd** Palladium 46 | 108 **Ag** Silver 47 | 112 **Cd** Cadmium 48 | 115 **In** Indium 49 | 119 **Sn** Tin 50 | 122 **Sb** Antimony 51 | 128 **Te** Tellurium 52 | 127 **I** Iodine 53 | 131 **Xe** Xenon 54 |

| 133 **Cs** Caesium 55 | 137 **Ba** Barium 56 | 139 **La*** Lanthanum 57 | 178 **Hf** Hafnium 72 | 181 **Ta** Tantalum 73 | 184 **W** Tungsten 74 | 186 **Re** Rhenium 75 | 190 **Os** Osmium 76 | 192 **Ir** Iridium 77 | 195 **Pt** Platinum 78 | 197 **Au** Gold 79 | 201 **Hg** Mercury 80 | 204 **Tl** Thallium 81 | 207 **Pb** Lead 82 | 209 **Bi** Bismuth 83 | 210 **Po** Polonium 84 | 210 **At** Astatine 85 | 222 **Rn** Radon 86 |

| 223 **Fr** Francium 87 | 226 **Ra** Radium 88 | 227 **Ac†** Actinium 89 | 261 **Rf** Rutherfordium 104 | 262 **Db** Dubnium 105 | 266 **Sg** Seaborgium 106 | 264 **Bh** Bohrium 107 | 277 **Hs** Hassium 108 | 268 **Mt** Meitnerium 109 | 271 **Ds** Darmstadtium 110 | 272 **Rg** Roentgenium 111 | 285 **Cn** Copernicium 112 | | | | | | |

Lanthanum series:

| 140 **Ce** Cerium 58 | 141 **Pr** Praseodymium 59 | 144 **Nd** Neodymium 60 | 145 **Pm** Promethium 61 | 150 **Sm** Samarium 62 | 152 **Eu** Europium 63 | 157 **Gd** Gadolinium 64 | 159 **Tb** Terbium 65 | 162 **Dy** Dysprosium 66 | 165 **Ho** Holmium 67 | 167 **Er** Erbium 68 | 169 **Tm** Thulium 69 | 173 **Yb** Ytterbium 70 | 175 **Lu** Lutetium 71 |

Actinium series:

| 232 **Th** Thorium 90 | 231 **Pa** Protactinium 91 | 238 **U** Uranium 92 | 237 **Np** Neptunium 93 | 242 **Pu** Plutonium 94 | 243 **Am** Americium 95 | 247 **Cm** Curium 96 | 245 **Bk** Berkelium 97 | 251 **Cf** Californium 98 | 254 **Es** Einsteinium 99 | 253 **Fm** Fermium 100 | 256 **Md** Mendelevium 101 | 154 **No** Nobelium 102 | 257 **Lr** Lawrencium 103 |